Fault Diagnosis of Permanent Magnet Machines

IEEE Press

445 Hoes Lane

Piscataway, NJ 08854

Fault Diagnosis of Permanent Magnet Machines

Jawad Faiz
University of Tehran, Tehran, Iran

Ehsan Mazaheri-Tehrani
University of Tehran, Tehran, Iran

Bashir-Mahdi Ebrahimi
Aura Systems Inc, Lake Forest, California, United States

IEEE Press
WILEY

Published by John Wiley & Sons, Inc., Hoboken, New Jersey.

Library of Congress Cataloging-in-Publication Data has been applied for:

Print ISBN: 9781394424979
ePDF ISBN: 9781394424993
ePub ISBN: 9781394424986
oBook ISBN: 9781394425006

Cover Design: Wiley
Cover Image: tvierimaa/Getty Images

Contents

About the Authors

Jawad Faiz, Ph.D., is a Distinguished Professor and the Director of the Center of Excellence on Applied Electromagnetic Systems in the School of Electrical and Computer Engineering at the University of Tehran, Iran, where his research interests are the design, modeling, and fault diagnosis of electrical machines and transformers. He has authored three books published by Springer, IET, and IEEE-Wiley. He has authored/co-authored top 1% most highly cited papers worldwide in his research field and more than 350 articles in international journals and presented more than 340 papers in international conferences. He was a keynote speaker at different international conferences. He is a fellow of IEEE, a fellow of the Iranian Academy of Sciences, and a fellow of the Euro-Med Academy of Arts and Sciences. He was a recipient of many national awards, including 1st prize for fundamental research from the International Kharazmi Festival, Distinguished Professor of the National Foundation of Iranian Elite, Excellent Badge of Knowledge, and the University of Tehran Excellent Research Award. He was also a recipient of the international award of the Silver Einstein Medal for academic research from UNESCO.

Ehsan Mazaheri-Tehrani received the B.Sc. degree from Isfahan University of Technology, Isfahan, Iran, in 2013, and the M.Sc. degree from the University of Tehran, Tehran, Iran, in 2017, both in electrical engineering. He is currently pursuing a Ph.D. degree in electrical engineering at the University of Tehran. He has over five years of industry experience in data science and machine learning, focusing on real-world applications in energy systems. His research interests include modeling, control, and fault diagnosis of electric machines, with an emphasis on predictive maintenance and data-driven diagnostics, as well as the application of machine learning techniques in power engineering and distributed energy resources.

Bashir-Mahdi Ebrahimi, Ph.D., is a globally recognized leader in electric machines and electrified propulsion, with over 15 years of pioneering contributions across aerospace, automotive, and energy sectors. He has authored more than 100 widely cited peer-reviewed publications in top-tier journals and premier international conferences. He is an inventor or co-inventor of over 10 patents in advanced motor topologies, thermal management, and fault-tolerant propulsion architectures. His work has earned multiple international awards for breakthrough innovation and technical excellence in electrical engineering. He has led the development of state-of-the-art electric propulsion systems, achieving record-breaking power density and efficiency for eVTOL, EV, and defense applications. His expertise spans electromagnetic design, multi-physics modeling, advanced cooling, condition monitoring, and system-level optimization of PM and induction machines.

Preface

Permanent magnet (PM) machines have become increasingly prevalent across industrial, automotive, and renewable energy applications due to their superior power density and efficiency. As these machines assume critical roles in modern systems, the ability to detect, diagnose, and predict faults has become essential for ensuring operational reliability and minimizing costly downtime. This book provides a comprehensive treatment of fault diagnosis for permanent magnet machines, integrating theoretical foundations with practical diagnostic techniques to equip readers with the knowledge needed to implement effective condition-monitoring systems.

The book is organized into nine chapters that progress from fundamental concepts to specific fault types. The first four chapters establish the necessary foundation. Chapter 1 introduces the motivation for PM machine fault diagnosis and provides an overview of the field. Chapter 2 covers the fundamentals of PM machine operation, including various configurations such as surface-mounted and interior PM topologies, establishing the baseline understanding essential for comprehending fault mechanisms. Chapter 3 presents analytical and computational modeling tools, including finite element methods and equivalent circuit approaches, used to predict machine behavior under both healthy and faulty conditions. Chapter 4 provides systematic coverage of diagnostic methods applicable across multiple fault types, including motor current signature analysis, motor flux signature analysis, zero-sequence voltage monitoring, and vibration-based techniques.

The remaining five chapters address specific fault categories in detail. Each of these chapters follows a consistent structure covering failure modes and mechanisms, modeling and analysis, and diagnostic approaches. Chapter 5 focuses on eccentricity faults arising from air gap nonuniformity, examining causes such as bearing wear, manufacturing tolerances, rotor imbalance, and shaft deflection. Chapter 6 addresses stator winding faults, progressing from inter-turn short circuits through phase-to-phase and phase-to-ground faults, with examination of insulation degradation mechanisms including thermal aging, partial discharge, and mechanical stress. Chapter 7 treats bearing faults, the most prevalent mechanical failure mode, establishing relationships between bearing geometry, operating speed, and characteristic defect frequencies. Chapter 8 examines demagnetization faults unique to PM machines, covering thermal demagnetization, electrical demagnetization from excessive armature reaction, and physical magnet degradation. Chapter 9 extends the diagnostic scope to power electronic converters, examining failure modes of IGBTs, MOSFETs, gate drivers, and DC link capacitors, and distinguishing converter fault signatures from machine faults.

This book is intended for graduate students and researchers in electrical engineering, as well as practicing engineers responsible for the maintenance and reliability of PM machine systems.

Readers are expected to have foundational knowledge of electrical machines and basic signal processing concepts. Undergraduate students with adequate preparation in these areas will also find the material accessible. For academic use, Chapters 1 through 4 provide suitable material for a course on condition monitoring fundamentals, while Chapters 5 through 9 can serve as references for specialized study or research. Practicing engineers may focus on the diagnostic sections of the fault-specific chapters most relevant to their applications.

The authors have drawn upon two decades of research experience in electrical machine fault diagnosis, incorporating findings from numerous graduate theses and published studies conducted at the University of Tehran. We have endeavored to present the material in a clear and systematic manner, supported by illustrative diagrams and practical examples throughout.

Tehran, Iran
February, 2026

Jawad Faiz
Ehsan Mazaheri-Tehrani
Bashir-Mahdi Ebrahimi

Acknowledgments

We gratefully acknowledge the support of the University of Tehran and the School of Electrical and Computer Engineering, which provided the academic environment that made this work possible.

This book builds upon more than a decade of research conducted by graduate students in our group. We are particularly indebted to Dr. Mehrage Ghods, formerly a Ph.D. student and now an Assistant Professor at the University of Tehran; Dr. Hossein Nejadi Koti, formerly an M.Sc. student and now a Senior Motor Design Engineer at Rockwell Automation in Milwaukee, USA; Mr. Zabihollah Tabarniarami; and Mr. Mohammad Amin Bazrafshan. Their thesis research contributed substantially to the technical content presented in this book.

We are grateful to the reviewers who provided constructive feedback on the initial manuscript. Their comments and suggestions helped improve the clarity and technical accuracy of the final text.

We also thank Wiley-IEEE Press and Ms. R. Kavipriya, Managing Editor of Academic Professional Learning at Wiley, for their guidance and support throughout the publication process.

Jawad Faiz

Acronyms

AC	Alternating Current
ANN	Artificial Neural Network
CNN	Convolutional Neural Network
DC	Direct Current
DWT	Discrete Wavelet Transform
EMF	Electromotive Force
EPVA	Extended Park's Vector Approach
FEM	Finite Element Method
FFT	Fast Fourier Transform
FRM	Field Reconstruction Method
IGBT	Insulated Gate Bipolar Transistor
IPM	Interior Permanent Magnet
MCSA	Motor Current Signature Analysis
MEC	Magnetic Equivalent Circuit
MFSA	Magnetic Field Signature Analysis
ML	Machine Learning
MMF	Magnetomotive Force
MOSFET	Metal-Oxide-Semiconductor Field-Effect Transistor
PM	Permanent Magnet
PMSM	Permanent Magnet Synchronous Motor
PWM	Pulse Width Modulation
RMS	Root Mean Square
SPM	Surface-mounted Permanent Magnet
SVM	Support Vector Machine
THD	Total Harmonic Distortion
VSI	Voltage Source Inverter
ZSVC	Zero-Sequence Voltage Component

1

Introduction

1.1 Motivation and Problem Statement

The modern industrial landscape has witnessed an unprecedented transformation in electromechanical systems, with permanent magnet (PM) motors emerging as cornerstone technologies across diverse sectors. From electric vehicles revolutionizing transportation to wind turbines harnessing renewable energy, from aerospace propulsion systems enabling new frontiers of flight to precision robotics advancing manufacturing capabilities, PM motors have become indispensable components of critical infrastructure. This widespread adoption, while testament to their superior performance characteristics, has simultaneously elevated the stakes of operational reliability to unprecedented levels.

The consequences of unexpected motor failures extend far beyond simple equipment replacement costs. In aerospace applications, a motor failure can compromise flight safety and passenger security. In automotive systems, particularly electric vehicles and hybrid powertrains, motor malfunctions directly impact vehicle operability and driver safety. Industrial automation systems depend on continuous motor operation to maintain production schedules, where even brief interruptions cascade into substantial economic losses. Wind turbines, especially offshore installations in remote and harsh environments, require exceptional reliability as maintenance interventions involve significant logistical challenges and costs. The medical sector, where PM motors power life-critical equipment, demands near-absolute reliability where failures can have direct implications for patient outcomes.

This criticality of PM motor reliability in modern applications creates a compelling imperative for advanced fault diagnosis systems. Traditional reactive maintenance approaches, where interventions occur only after failures manifest, prove inadequate for applications where downtime carries severe consequences. The ability to detect incipient faults before they progress to catastrophic failures enables proactive maintenance strategies that minimize disruption while maximizing asset utilization. Moreover, as PM motors become increasingly integrated into complex systems with sophisticated control strategies and digital interfaces, the opportunities for implementing intelligent diagnostic capabilities expand correspondingly.

The challenge of ensuring PM motor reliability is compounded by the diverse nature of potential failure modes. Unlike simpler electromagnetic machines, PM motors face unique vulnerabilities stemming from their PM excitation systems, including irreversible demagnetization under thermal or electrical stress. The precision air gaps required for optimal performance make them susceptible to various forms of eccentricity fault. High-performance winding designs, while enabling superior power density, introduce challenges related to insulation degradation and short-circuit faults.

Bearing systems, subjected to magnetic forces and high-speed operation, represent frequent failure points. Furthermore, the power electronic converters essential for PM motor operation introduce additional failure modes that can affect motor performance or, conversely, motor faults can stress converter components.

This book addresses the critical need for comprehensive understanding of fault mechanisms, diagnostic methodologies, and prognostic strategies for PM motors. By integrating theoretical foundations with practical diagnostic techniques, it provides engineers, researchers, and practitioners with the knowledge necessary to implement effective condition monitoring systems. The following sections of this introductory chapter establish the context for this comprehensive treatment by examining the importance of fault diagnosis, quantifying the economic impacts of motor failures, reviewing the current state of PM motor technology, categorizing fault types, and outlining the diagnostic challenges that motivate advanced approaches.

1.2 Importance of Fault Diagnosis in PM Motors

The expansion of PM machines in industrial applications has necessitated intensified research on fault mechanisms and the development of effective diagnostic methodologies [1]. The reliability requirements for PM motor drive systems have escalated dramatically as these machines assume increasingly critical roles in essential infrastructure and safety-critical applications [2]. This section examines the multifaceted importance of fault diagnosis from technical, operational, and strategic perspectives.

From a technical standpoint, PM motors exhibit several characteristics that make fault diagnosis both essential and challenging. The PM excitation system, while providing advantages in efficiency and power density, introduces unique failure modes not present in conventional machines. Demagnetization faults, which can occur due to excessive operating temperatures, PM aging, severe flux weakening operations, or inter-turn short circuits in stator windings, represent irreversible degradation that progressively diminishes motor performance [3]. Unlike field-excited machines where excitation can be adjusted, PM motors cannot compensate for magnet degradation through external means, making early detection critical for preventing performance deterioration.

The high power density characteristic of modern PM motors, achieved through aggressive thermal and electromagnetic designs, operates these machines closer to their physical limits. This optimization leaves limited margins for fault tolerance, meaning that incipient faults can rapidly escalate to catastrophic failures if undetected. Thermal management becomes particularly critical, as many fault conditions manifest initially through localized heating that, if allowed to progress, accelerates additional failure mechanisms. Winding insulation degradation, bearing lubricant breakdown, and magnet demagnetization all exhibit temperature-dependent acceleration, creating positive feedback loops where initial faults exacerbate thermal conditions that promote additional degradation.

The integration of PM motors into complex drive systems with power electronic converters and sophisticated control algorithms introduces interdependencies between components that complicate fault diagnosis. A fault originating in the motor can stress converter components through abnormal current patterns or electromagnetic interference. Conversely, converter faults such as switching device failures or gate driver malfunctions can impose electrical or thermal stresses on the motor. Effective fault diagnosis must therefore consider the motor-drive system holistically, distinguishing between faults originating in different subsystems while recognizing their interactions.

Operational considerations further underscore the importance of diagnostic capabilities. In safety-critical applications such as aerospace or medical systems, fault diagnosis serves not merely as a maintenance tool but as an integral safety function. Regulatory frameworks for such applications increasingly mandate condition monitoring capabilities that can detect potential failures before they compromise safety. The certification processes for these systems require demonstrated reliability levels that can only be achieved through continuous health monitoring and fault detection.

Economic factors provide compelling justification for investment in diagnostic systems. For sensitive industrial applications including aviation, transportation, and medical sectors, PM motor defects can lead to extensive breakdowns and risks to human life. The design and implementation of monitoring systems represent relatively modest investments compared to the potential costs of unplanned failures. In industrial settings, where production continuity directly impacts profitability, the ability to schedule maintenance during planned outages rather than responding to emergency failures reduces both direct repair costs and indirect costs associated with production disruption.

The strategic importance of fault diagnosis extends to asset management and lifecycle optimization. Diagnostic data provides insights into actual operating conditions and degradation rates, enabling data-driven decisions regarding maintenance intervals, operating parameter limits, and replacement timing. This information supports transition from time-based preventive maintenance schedules to condition-based or predictive maintenance strategies that optimize the balance between reliability and maintenance costs. Furthermore, diagnostic information from fielded systems provides valuable feedback to design engineers, revealing actual failure modes and rates that inform next-generation designs.

1.3 Industrial Applications and Economic Impact

PM motors have achieved dominant positions in numerous industrial sectors, driven by their superior efficiency, power density, and controllability. This widespread adoption, while enabling technological advances across industries, has created substantial economic exposure to motor failures. Understanding the scale and distribution of these economic impacts provides essential context for justifying investments in diagnostic capabilities and illustrates the tangible benefits of effective fault detection and prognostic systems.

Recent assessments of industrial downtime costs reveal staggering economic consequences of equipment failures. Fortune Global 500 manufacturing and industrial firms lose approximately 3.3 million hours annually to unplanned downtime, translating to financial costs of $864 billion, equivalent to 8% of their annual revenues. More recent analyses indicate this problem has intensified, with losses reaching $1.4 trillion, representing 11% of annual revenues across these enterprises [4, 5]. Large manufacturing plants experience an average of 323 unplanned production hours annually, with downtime costs averaging $532,000 per hour, resulting in approximately $172 million in annual losses per facility [4].

The economic impact varies significantly across industrial sectors, reflecting differences in production values, process characteristics, and dependency on motor-driven equipment. In automotive manufacturing, unplanned downtime averages 29 production hours monthly at costs ranging from $1.3 million to $2.3 million per hour, equating to approximately $600 per second of production interruption. This sector's heavy reliance on automated production lines, where motors drive robotic systems, conveyors, and assembly equipment, creates cascading effects where single motor failures can halt entire production sequences. The fast-moving consumer goods (FMCG) and consumer packaged goods (CPG) sectors experience 25 hours of monthly unplanned downtime at approximately $23,600 per hour, while mining, metals, and heavy industrial operations average 23 hours monthly at $187,500 per hour, reflecting the high value and capital intensity of these processes. Table 1.1 summarizes the economic impact of unplanned downtime across major industrial sectors.

Within these aggregate statistics, electric motor failures represent a substantial contribution to overall equipment downtime. Studies of preventive maintenance programs have documented that in facilities lacking systematic condition monitoring, motor rewinds constituted 85% of total motor repairs. Implementation of preventive maintenance programs with diagnostic capabilities reduced rewinds to approximately 20% of repairs, demonstrating both the prevalence of motor failures and the effectiveness of proactive approaches. More than two-thirds of industrial businesses experience unplanned outages at least monthly, with costs typically approaching $125,000 per hour for medium to large facilities [6].

The economic case for fault diagnosis extends beyond avoiding downtime costs to encompass maintenance optimization. Reactive maintenance strategies, where repairs occur only after failures, incur not only the direct costs of emergency repairs and expedited parts procurement but also the opportunity costs of unplanned production interruptions. Preventive maintenance approaches, based on fixed time or usage intervals, avoid some emergency situations but perform maintenance earlier than necessary, incurring unnecessary maintenance costs and potentially introducing infant mortality failures through unnecessary interventions. Condition-based maintenance, enabled by diagnostic systems, optimizes this balance by performing maintenance based on actual equipment condition, extending intervals when health indicators permit while enabling early intervention when degradation accelerates.

The application domains where PM motors provide critical functionality underscore both the diversity of industrial dependence and the varied consequences of failures. In EVs, PM motors serve as the primary propulsion system, where reliability directly impacts vehicle availability and customer satisfaction. The automotive industry's transition toward electrification amplifies the importance of motor reliability, as electric drivetrains become defining features rather than ancillary systems. Electric power steering systems, now standard in most vehicles, require continuous motor operation where failures compromise vehicle controllability and safety.

Aerospace applications impose among the most stringent reliability requirements, with PM motors increasingly employed in electric aircraft initiatives, actuation systems, and auxiliary power applications. The consequences of in-flight failures, combined with the extreme operating environments including temperature variations, vibration, and atmospheric pressure changes, demand

Table 1.1 Economic impact of unplanned downtime across industrial sectors [4, 6].

Industry sector	Downtime (hrs/mo)	Cost per hour	Cost per second	Annual impact
Automotive	29	$1.3M–$2.3M	$600	$450M–$800M
FMCG/CPG	25	$23,600	$6.56	$7.1M
Mining/metals/heavy industrial	23	$187,500	$52	$51.8M
General manufacturing (large)	323 (annual)	$532,000	$148	$172M
General industry (medium-large)	20+	$125,000	$35	$30M+

exceptional reliability supported by comprehensive diagnostic capabilities. The regulatory certification processes for aerospace systems require extensive reliability demonstration and often mandate redundant systems or fault-tolerant designs, both of which benefit from integrated diagnostics to detect degradation in primary or backup systems.

Renewable energy systems, particularly wind turbines, represent another domain where PM motor reliability carries substantial economic implications. Offshore wind installations, which increasingly employ direct-drive PM generators to eliminate gearbox reliability issues, face severe logistical challenges for maintenance access. Weather windows for offshore maintenance can be limited, and specialized vessels and personnel are required, making maintenance interventions expensive and time-constrained. Turbine downtime directly impacts energy production revenue, while extended outages during high-demand periods incur opportunity costs. Diagnostic systems that provide advance warning of developing faults enable maintenance scheduling during favorable weather conditions and coordination with planned service visits, substantially reducing intervention costs while maximizing availability.

Industrial automation and robotics applications employ PM motors extensively for precise motion control, where performance degradation can affect product quality before causing complete failures. Manufacturing processes with tight tolerances depend on consistent motor performance, where gradual degradation in torque capability or increases in cogging torque can lead to dimensional variations or surface finish defects in produced parts. Diagnostic systems that detect performance drift enable proactive replacement before quality impacts occur, avoiding costs associated with rework or scrap while maintaining process capability.

The medical sector utilizes PM motors in critical applications, including surgical robots, diagnostic imaging equipment, and life support systems. Reliability requirements approach those of aerospace applications, as failures can directly impact patient outcomes. Regulatory requirements mandate extensive testing and often require redundant safety systems. The economic impacts extend beyond equipment replacement to include procedure delays, patient rescheduling, and potential liability exposure, creating strong incentives for comprehensive diagnostic capabilities that ensure continuous availability of critical medical equipment.

1.4 Current State of PM Motor Technology

The technological evolution of PM motors over recent decades has been characterized by dramatic improvements in performance metrics, expansion into new application domains, and increasing sophistication in design and manufacturing approaches. Understanding the current state of PM motor technology provides essential context for appreciating both the capabilities that make these machines attractive for diverse applications and the challenges that necessitate advanced diagnostic capabilities.

Market growth statistics underscore the expanding role of PM motors across industries. The global market for PM motors was valued at $51.1 billion in 2024 and is projected to reach $86.1 billion by 2030, representing a compound annual growth rate of 9.1% [7]. This growth trajectory is driven primarily by the global emphasis on energy efficiency and carbon emission reduction, with PM motors offering substantial efficiency advantages over alternative technologies. The accelerating electrification of transportation, expansion of renewable energy installations, and increasing automation across industrial sectors all contribute to sustained market expansion.

Performance characteristics of contemporary PM motors reflect continuous advancement in magnetic materials, manufacturing processes, and electromagnetic design optimization.

Modern PM motors achieve efficiency levels exceeding 93%, compared to induction motors that typically operate between 78% and 84%efficiency. This efficiency advantage stems from the elimination of rotor losses associated with field excitation, enabling PM motors to maintain high efficiency across wider load ranges. Power losses can be reduced by up to 25% compared to induction alternatives, translating directly to energy cost savings and reduced thermal management requirements in applications ranging from industrial drives to EVs.

Power density represents another area where PM motors demonstrate significant advantages over alternative technologies. PM motors deliver approximately twice the power density of equivalently rated induction machines, producing the same horsepower from motors that are roughly 50% shorter, or alternatively generating higher power output from the same physical envelope. This characteristic proves particularly valuable in applications where volume and weight constraints are critical, such as aerospace, automotive, and mobile robotics. Recent developments have pushed power density to exceptional levels, with advanced designs targeting 9 kW/kg while maintaining efficiency levels of 98%. Commercial implementations have achieved outputs of 200 kW from compact units through innovations including high-efficiency segment conductor designs and advanced casting and processing technologies. Table 1.2 compares the performance characteristics of PM motors with alternative motor technologies.

The enabling foundation for these performance achievements lies in advanced PM materials, particularly NdFeB compounds. These high-energy-product magnets generate strong magnetic fields that enable both high efficiency and compact designs. The magnetic performance of NdFeB materials, combined with their relatively high Curie temperature, makes them suitable for demanding applications despite concerns regarding thermal demagnetization and material cost. Ongoing research into magnet formulations seeks to reduce or eliminate reliance on critical rare earth elements, including dysprosium and terbium, which are added to enhance thermal stability but contribute to material costs and supply chain vulnerabilities. Alternative approaches under development include electrically excited synchronous machines that eliminate PMs entirely, utilizing rotor coils energized through innovative contactless power transfer technologies.

Manufacturing innovations are enabling increasingly sophisticated motor designs that optimize electromagnetic performance while addressing thermal management and mechanical robustness. Additive manufacturing techniques allow creation of complex geometries that enhance cooling

Table 1.2 Performance comparison of PM motors with alternative motor technologies.

Motor type	Efficiency range	Power density (kW/kg)	Relative size	Key advantages	Primary limitations
PM synchronous	90–98%	High (5–9)	100%	Highest efficiency, compact, wide speed range	Magnet cost, demagnetization risk
Induction	78–91%	Medium (2–4)	~200%	Robust, low cost, mature technology	Lower efficiency, larger size
Switched reluctance	80–90%	Med-High (3–5)	~150%	No magnets, fault tolerant	Torque ripple, acoustic noise
Synchronous reluctance	85–94%	Medium (3–5)	~150%	No magnets, good efficiency	Lower power density than PM
Wound rotor sync.	92–96%	Med-High (4–6)	~120%	Controllable excitation, no rare earths	Slip rings, more complex

Source: Data from [7].

effectiveness and optimize magnetic flux paths in ways that traditional manufacturing approaches cannot achieve. Precision fabrication of stator windings using segment conductor techniques increases slot fill factors, improving current density and efficiency. Advanced casting processes enable production of rotor assemblies with integrated magnet retention features that withstand high-speed operation while minimizing parasitic losses.

Digital technologies are transforming both the design process and operational deployment of PM motors. Finite element analysis (FEA) tools with increasing computational power enable detailed electromagnetic, thermal, and structural optimization during design phases, predicting performance across operating conditions and identifying potential failure modes. Digital twin concepts, where virtual replicas of physical motors are maintained and updated with operational data, facilitate both design refinement and in-service condition monitoring. During operation, Internet of Things connectivity enables motors to report performance data, facilitating predictive maintenance approaches that leverage artificial intelligence algorithms to identify degradation patterns and predict remaining useful life span.

Control sophistication has evolved substantially, with modern PM motor drives implementing advanced algorithms that optimize efficiency, extend operating envelopes, and provide diagnostic capabilities. Field-oriented control strategies maximize torque development while minimizing losses across speed and load ranges. Sensorless control techniques estimate rotor position without mechanical sensors, reducing cost and improving reliability while simultaneously providing signals that can indicate developing faults. Model predictive control approaches optimize multiple objectives simultaneously, balancing performance, efficiency, and thermal management while respecting system constraints.

Application-specific motor designs have proliferated as PM motor technology matures, with topologies optimized for particular performance profiles. Surface-mounted PM (SPM) configurations offer manufacturing simplicity and low cogging torque, suitable for applications requiring smooth operation at low speeds. Interior PM (IPM) designs achieve higher saliency ratios enabling field weakening for extended constant-power operating ranges, advantageous for traction applications. Fractional slot concentrated winding machines provide fault tolerance and simplified manufacturing, finding application in safety-critical aerospace systems. Axial flux topologies achieve exceptional power density for applications with planar form factors. The diversity of available topologies enables selection or customization of designs that optimally match application requirements.

Despite these impressive advances, PM motor technology faces several ongoing challenges that motivate continued research and development. Supply chain considerations for rare earth magnetic materials create economic and geopolitical vulnerabilities, stimulating efforts to develop motors with reduced or eliminated rare earth content [8]. Thermal management remains critical, as the drive toward higher power densities pushes thermal limits, particularly for magnets where elevated temperatures risk irreversible demagnetization. Manufacturing costs, while declining through automation and scale, remain higher than alternative technologies, requiring lifecycle cost justification through efficiency benefits and performance advantages. Recycling and end-of-life management of PMs present environmental and economic challenges that require addressing as the installed base of PM motors grows.

The diagnostic implications of these technological trends are substantial. Higher performance motors operating closer to material limits demand more sophisticated monitoring to detect incipient degradation before catastrophic failures occur. Increasing integration of motors into complex systems with digital interfaces provides opportunities for embedding diagnostic capabilities directly into drive electronics. The expansion into safety-critical applications elevates the consequences of failures and often mandates monitoring systems as safety functions rather than optional features. As PM motors assume increasingly critical roles across diverse industries, the imperative for effective fault diagnosis intensifies correspondingly.

1.5 Overview of Fault Types in PM Motors

PM motors are subject to a diverse array of fault mechanisms that can be systematically categorized based on the affected subsystems and physical phenomena involved. Comprehensively understanding this fault taxonomy provides the foundation for developing effective diagnostic strategies and informs the organization of subsequent chapters in this book. This section presents an overview of the principal fault types, their interrelationships, and their relative prevalence in operational environments.

Fault classification schemes typically organize PM motor faults into three primary categories: electrical faults, mechanical faults, and magnet-specific faults [9]. Statistical analyses of motor failures across industrial applications reveal the distribution shown in Figure 1.1, with mechanical faults representing the largest category, followed by electrical faults and magnet-related failures. This statistical distribution, however, varies substantially across applications depending on operating conditions, duty cycles, and environmental factors. Table 1.3 provides a comprehensive classification of PM motor fault types with their characteristics and prevalence.

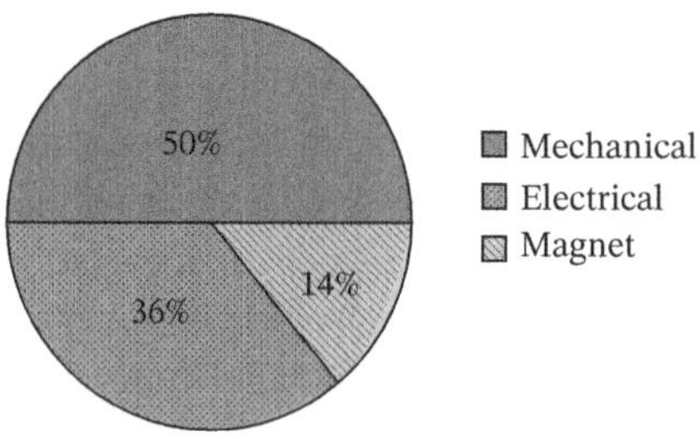

Figure 1.1 PM motor fault distribution. Overall distribution of fault types showing mechanical faults as the dominant category (50%), followed by electrical faults (36%) and magnet-related faults (14%), based on statistical analyses across industrial applications.

Table 1.3 Classification and characteristics of PM motor fault types [1, 9].

Fault category	Specific fault type	Causes and characteristics	Diagnostic signatures
Electrical	Inter-turn short circuit	Insulation degradation from thermal/electrical stress; progressive	Negative sequence current, increased harmonics
	Phase-to-phase fault	Advanced insulation failure; progressive	Severe current imbalance, protection trip
	Phase-to-ground fault	Insulation breakdown, contamination; can be sudden	Ground fault current, protection trip
	Open circuit	Conductor breakage, connection failure; often sudden	Phase current loss, torque reduction

Mechanical	Bearing faults	Wear, lubrication failure, contamination; progressive	Vibration increase, characteristic frequencies
	Static eccentricity	Manufacturing tolerance, misalignment; often present from start	Unbalanced magnetic pull, stationary vibration pattern
	Dynamic eccentricity	Bearing wear, shaft deflection; progressive	Rotating vibration pattern, current harmonics
	Mixed eccentricity	Combination of static and dynamic; progressive	Complex vibration and current signatures
Magnet	Thermal demagnetization	Excessive temperature, cooling failure; can be sudden	Torque reduction, altered harmonic content
	Electrical demagnetization	Excessive armature reaction during faults; progressive or sudden	Back-EMF reduction, performance loss
	Partial demagnetization	Localized heating, uneven magnetic fields; progressive	Torque ripple, rotor asymmetry

Electrical faults primarily involve the stator winding system, where insulation degradation represents the fundamental failure mechanism. The progression of insulation failures typically begins with localized weakening due to thermal, electrical, or mechanical stresses, eventually leading to inter-turn short circuits within a single coil. If undetected, inter-turn faults can propagate as the short-circuit current generates excessive localized heating that degrades adjacent insulation. This progressive failure can extend to phase-to-phase faults or phase-to-ground faults, representing increasingly severe conditions. The short-circuit current flowing in the shorted turns produces substantial Joule heating that not only accelerates local insulation degradation but can also thermally stress nearby PMs, potentially inducing secondary demagnetization faults.

The insulation degradation underlying winding faults results from multiple stress mechanisms acting individually or in combination. Thermal cycling induces differential expansion between conductor and insulation materials, creating mechanical stresses that can produce cracks or delamination. Electrical stress from voltage transients, particularly those associated with power electronic switching, can cause partial discharge activity that gradually erodes insulation material. Mechanical vibration and electromagnetic forces during transient conditions impose cyclic stresses on end windings and slot conductors. Environmental contamination by moisture, chemicals, or conductive particles can accelerate degradation or create leakage paths. The multifaceted nature of insulation aging makes prediction of winding fault occurrence challenging and underscores the value of condition monitoring to detect incipient failures.

Beyond short-circuit faults, winding systems can experience open-circuit faults where conductor breakage interrupts current flow, and high-resistance connection faults where poor terminal connections or internal conductor damage increases resistance. Open-circuit faults typically result from mechanical stress, manufacturing defects, or corrosion at connection points. High-resistance faults, while less immediately catastrophic than short circuits, generate localized heating that can accelerate degradation and progress to more severe fault conditions. Both fault types alter the current distribution in the motor and affect electromagnetic balance, potentially increasing vibration and bearing loads.

Mechanical faults encompass several distinct phenomena, with rotor eccentricity and bearing failures representing the most prevalent modes. Eccentricity fault describes conditions where the rotor rotational axis deviates from the stator bore centerline, creating nonuniform air gap around the rotor circumference. Static eccentricity fault occurs when the rotor rotational axis remains parallel to the stator axis but is laterally displaced, creating a minimum air gap that remains stationary.

Dynamic eccentricity fault involves the rotor rotational axis precessing about the stator axis, causing the minimum air gap position to rotate with the rotor. Mixed eccentricity fault combines both static and dynamic components, representing the most general and common condition in practice.

Eccentricity fault generates unbalanced magnetic pull that loads bearings asymmetrically and can induce vibrations. The magnetic force increases nonlinearly as eccentricity fault worsens, creating positive feedback where bearing wear increases eccentricity fault, which further increases bearing loads. Severe eccentricity fault can cause rotor-stator contact, potentially damaging both surfaces and creating catastrophic failure. Even moderate eccentricity fault affects motor performance by increasing no-load losses, elevating vibration and acoustic noise, and introducing torque ripple. The electromagnetic signature of eccentricity fault appears in stator current spectra, flux density distributions, and vibration patterns, forming the basis for diagnostic approaches.

Bearing failures represent the most frequent mechanical fault mode, resulting from the challenging operating conditions bearings experience in PM motors. High-speed operation imposes centrifugal forces on bearing elements, while magnetic forces from rotor eccentricity or electromagnetic unbalance create additional radial loads. The PMs create attractive forces that bearings must support even during standstill, unlike induction motors where zero-speed magnetic forces are minimal. Inadequate lubrication, contamination, improper installation, and normal fatigue wear all contribute to bearing degradation. Bearing faults typically progress through identifiable stages, beginning with microscopic surface defects that grow into pits or spalls, eventually leading to increased clearances, excessive vibration, and seizure if unaddressed.

The progression of bearing degradation makes this fault type particularly amenable to condition monitoring, as incipient faults generate detectable changes in vibration spectra well before functional failures occur. Characteristic frequencies associated with specific bearing defects, including outer race faults, inner race faults, rolling element defects, and cage faults, appear in vibration signatures and can be identified through spectral analysis. Advanced signal processing techniques enable extraction of bearing fault signatures even when masked by other vibration sources, supporting early detection and prognostic estimation of remaining useful life.

Demagnetization faults represent a unique failure mode specific to PM machines, where the magnetic strength of rotor magnets degrades either partially or completely [1]. Multiple mechanisms can induce demagnetization, with thermal effects representing the most common cause. When magnet temperatures exceed critical thresholds, typically above 150–180 °C for standard NdFeB grades, irreversible loss of magnetization occurs. Thermal demagnetization can result from excessive ambient temperatures, inadequate cooling, or localized heating from nearby electrical faults such as stator short circuits [3]. Electrical demagnetization occurs when armature reaction fields during fault conditions or aggressive field-weakening operation oppose the magnet flux, potentially exceeding the magnet's coercive force and inducing irreversible flux loss. Physical degradation through aging, mechanical shock, or corrosion can also reduce magnetic strength over extended periods.

The consequences of demagnetization include reduced torque capability, decreased efficiency, and altered electromagnetic signatures that affect control performance. Partial demagnetization affecting only certain magnets creates rotor asymmetry that generates torque ripple and vibration. Uniform demagnetization across all magnets primarily reduces output capability while maintaining relative balance. Diagnostic approaches for demagnetization typically analyze signatures in current harmonics, back-EMF waveforms, or torque characteristics that deviate from expected patterns, though distinguishing demagnetization from other fault types can prove challenging.

The interdependencies among fault types complicate both the fault progression and diagnostic processes [10]. Eccentricity-induced forces accelerate bearing wear, while bearing degradation permits increased eccentricity fault. Stator short circuits generate heat that can demagnetize nearby magnets. Demagnetization reduces output torque, potentially causing overcurrent conditions that stress converter components and increase thermal loading. Bearing failures increase vibration that can damage stator windings through mechanical stress. These coupling effects mean that a fault originating in one subsystem can propagate to others, emphasizing the importance of early detection before cascade failures develop.

Power converter faults, while technically external to the motor proper, intimately affect motor operation and must be considered in comprehensive diagnostic systems. Semiconductor device failures, including gate oxide breakdown, bond wire fatigue, and thermal runaway in IGBTs or MOSFETs, can impose abnormal voltages or currents on motor windings. Gate driver circuit failures alter switching patterns, creating excessive harmonic content or DC offset in motor currents. Capacitor degradation in DC links affects voltage regulation and increases ripple. Sensor failures providing feedback to control systems can cause erratic operation or fault conditions. Conversely, motor faults can stress converter components, creating coupled fault scenarios where distinguishing the primary fault source requires sophisticated diagnostics.

1.6 Challenges in Fault Detection and Diagnosis

Developing effective fault diagnosis systems for PM motors presents numerous technical challenges that span signal processing, machine learning, system integration, and practical implementation considerations. These challenges arise from the complex physics governing motor operation, the diverse nature of fault signatures, the variable operating conditions encountered in applications, and the stringent performance requirements for diagnostic systems. Understanding these challenges provides context for the methodologies presented in subsequent chapters and highlights areas where ongoing research continues to advance the state of the art.

The fundamental challenge in fault diagnosis lies in extracting fault signatures from measured signals in the presence of operational variations and measurement noise. Motor currents, voltages, vibrations, and other observable quantities vary substantially with speed, load, and control parameters during normal operation. Fault indicators must be distinguished from these expected variations, requiring diagnostic algorithms that account for operating point dependencies. Traditional approaches based on fixed thresholds or simple pattern matching prove inadequate when operating conditions vary widely, as threshold values appropriate for one condition generate false alarms or miss detections at other conditions. Adaptive approaches that normalize fault indicators relative to expected values for current operating conditions address this challenge but require accurate models of healthy motor behavior across the operating envelope.

Signal-to-noise ratio considerations further complicate fault detection, particularly for incipient faults where signatures are subtle. Electromagnetic interference from power electronic switching, mechanical vibrations from coupled equipment, sensor noise, and quantization effects in data acquisition systems all contribute noise that can mask fault signatures. Advanced signal processing techniques including filtering, time-frequency analysis, and statistical methods help extract fault signatures, but these approaches introduce complexity and computational requirements. The challenge intensifies for bearing faults and eccentricity faults, where characteristic signatures appear at specific frequencies that may coincide with structural resonances or other excitation sources, requiring sophisticated analysis to isolate fault components from background vibration.

The multiplicity of potential fault types and the similarity of some fault signatures create classification challenges [10]. Different faults can produce overlapping signatures in measured quantities, making definitive fault identification difficult. For example, both eccentricity and demagnetization faults can produce increased harmonic content in stator currents, while bearing and eccentricity faults both affect vibration spectra. Multiple concurrent faults compound this difficulty, as the combined signature may not equal the superposition of individual fault signatures due to nonlinear interactions [11]. Conventional diagnostic approaches focusing on single fault types may fail when multiple degradation mechanisms occur simultaneously, a condition increasingly common in aged motors where initial faults stress other components.

The limited availability of fault data for training diagnostic systems represents a practical challenge, particularly for data-driven approaches based on machine learning. Acquiring comprehensive datasets spanning various fault types, severities, and operating conditions requires extensive testing that is expensive and time-consuming. Faults occurring in operational systems are often repaired before progressing through multiple severity levels, limiting data on fault progression. Safety considerations preclude running motors with severe faults to collect data, particularly in applications where failures could cause injury or equipment damage. This data scarcity challenge proves particularly acute for rare fault combinations or unusual operating scenarios not well-represented in training datasets.

Machine learning approaches, while offering powerful pattern recognition capabilities, introduce their own challenges [12, 13]. Conventional supervised learning techniques require large labeled datasets representing all fault classes across operating conditions, a requirement difficult to satisfy given the fault data scarcity previously noted [11]. The computational expense of training complex deep learning models can be substantial, limiting real-time adaptation [14]. Generalization remains problematic, as models trained on specific motor designs or operating conditions may not transfer effectively to different configurations or applications. The "black box" nature of many machine learning models complicates interpretation and trust, particularly in safety-critical applications where regulatory requirements may mandate explainable diagnostic decisions.

Operating condition variability presents challenges beyond simple load and speed variations. Transient conditions during starting, stopping, or load changes create time-varying signatures that can mimic or mask fault indicators. In applications with heavy load cycling characterized by highly variable speed and torque profiles, the diagnostic system must distinguish faults from normal dynamic behavior, a problem receiving limited attention in literatures that predominantly focuses on steady-state conditions. Environmental factors including temperature variations, humidity, and altitude affect motor characteristics and sensor responses, requiring robust diagnostic approaches that maintain performance across environmental conditions.

Sensor selection and placement involve trade-offs between information content, cost, and reliability. Current and voltage measurements, typically already available in motor drives for control purposes, provide convenient fault indicators but may lack sensitivity for certain fault types. Vibration sensors offer excellent bearing fault detection but require mechanical mounting, introduce additional cost, and represent potential failure points. Temperature sensors provide direct thermal monitoring but respond slowly to transient conditions. Flux sensors and search coils can detect electromagnetic anomalies but require integration into motor construction. The optimal sensor suite depends on application requirements, cost constraints, and which fault types pose the greatest risks, requiring application-specific diagnostic system design.

Real-time implementation constraints limit the complexity of diagnostic algorithms that can execute on embedded systems typical of motor drive electronics. Sampling rates, computational throughput, memory capacity, and power consumption all impose limits on algorithm sophistication. Advanced signal processing techniques requiring substantial computation or large data buffers may exceed capabilities of cost-constrained embedded controllers, forcing trade-offs between diagnostic performance and implementation practicality. The development of computationally efficient algorithms that maintain diagnostic effectiveness within embedded system constraints remains an active research area.

The requirement for low false alarm rates and high detection reliability creates stringent performance demands. False alarms that trigger unnecessary maintenance interventions erode user confidence in diagnostic systems and create costs through unnecessary downtime and labor. Missed detections that fail to identify developing faults defeat the purpose of condition monitoring and can lead to catastrophic failures with associated safety and economic consequences. Achieving simultaneously low false alarm and missed detection rates, particularly for incipient faults with subtle signatures, requires carefully designed diagnostic algorithms with appropriate sensitivity tuning. The acceptable trade-off between these error types varies by application, with safety-critical systems prioritizing detection reliability even at the cost of increased false alarms, while other applications may emphasize false alarm reduction.

Integration with existing motor drive systems and control algorithms presents practical challenges. Diagnostic systems must coexist with control functions, sharing computational resources and sensor signals without degrading control performance. Diagnostic algorithms may need to inject test signals or modify control parameters temporarily to enhance fault detectability, requiring coordination with control systems to ensure such interventions do not disrupt operation or violate safety constraints. The diagnostic system may need to interface with higher-level maintenance management systems, requiring standardized communication protocols and data formats.

Prognostic capabilities, estimating remaining useful life and predicting future failures, represent a frontier challenge beyond simple fault detection. Prognosis requires understanding degradation mechanisms, their rates of progression under various operating conditions, and the remaining margin before functional failure occurs. Physics-based prognostic models require detailed knowledge of failure physics and current damage state, information often difficult to obtain. Data-driven prognostic approaches need historical data tracking fault progression over extended periods, data rarely available in sufficient quantity. Uncertainty quantification becomes critical for prognosis, as predictions must convey confidence bounds to support maintenance decision-making. Table 1.4 summarizes the major challenges in PM motor fault diagnosis and current approaches to address them.

Table 1.4 Summary of major challenges in PM motor fault diagnosis with current solutions and remaining research gaps [2, 11, 12].

Challenge category	Key issues	Impact on diagnosis	Current approaches	Remaining gaps
Signal processing	Operating point variations, noise, interference	False alarms, missed detections	Adaptive normalization, advanced filtering, time-frequency analysis	Variable load diagnostics, real-time processing
Fault signature overlap	Multiple faults with similar signatures	Misclassification, ambiguity	Multi-sensor fusion, pattern recognition	Concurrent fault separation
Limited fault data	Expensive testing, safety constraints	Poor ML generalization	Transfer learning, few-shot learning, simulation	Rare fault scenarios, fault progression data

(*Continued*)

Table 1.4 (Continued)

Challenge category	Key issues	Impact on diagnosis	Current approaches	Remaining gaps
Machine learning	Large data requirements, computational cost	Training difficulty, overfitting	Deep learning, CNNs, prototypical networks	Explainability, real-time adaptation
Sensor constraints	Cost, reliability, installation	Limited observability	Current/voltage analysis, sensorless methods	Noninvasive comprehensive monitoring
Computational limits	Embedded system constraints	Algorithm complexity limits	Efficient algorithms, edge computing	Real-time complex analysis
Operating conditions	Transients, variable loads, environment	Masking fault signatures	Condition-based normalization	Heavy load cycling, nonstationary operation
False alarm vs detection	Conflicting requirements	User confidence, safety	Threshold optimization, statistical methods	Optimal trade-off tuning
System integration	Controller interference, resource sharing	Implementation complexity	Coordinated diagnostics, nonintrusive monitoring	Seamless integration
Prognosis	Degradation modeling, RUL estimation	Predictive maintenance difficulty	Physics-informed models, hybrid approaches	Uncertainty quantification, long-term prediction

Despite these substantial challenges, ongoing research continues to advance diagnostic capabilities through novel sensor technologies, sophisticated signal processing algorithms, improved machine learning approaches tailored to limited data scenarios, and increasingly capable embedded computing platforms [2]. The subsequent chapters of this book present the current state of diagnostic methodologies, explaining both fundamental principles and practical implementation considerations while acknowledging areas where challenges persist and further development is needed.

1.7 Book Organization and Chapter Overview

This book provides a comprehensive treatment of fault diagnosis in PM machines, progressing from fundamental principles through specific fault types to practical implementation considerations. The organization follows a logical progression that first establishes foundational knowledge applicable across all fault types, then examines individual fault mechanisms in detail, integrating failure mode analysis, modeling approaches, and diagnostic methodologies for each fault category. Figure 1.2 illustrates this structure, showing how the foundational chapters (Chapters 2–4) provide the theoretical basis that supports the fault-specific chapters (Chapters 5–9), each of which follows a consistent three-part structure covering failure effects, modeling, and diagnosis.

Chapter 2: PM Machines establishes the foundational understanding of PM motor operation essential for comprehending fault mechanisms and diagnostic approaches. This chapter covers the fundamental principles of electromagnetic energy conversion in PM machines, examining various topologies including surface-mounted and IPM configurations, axial and radial flux

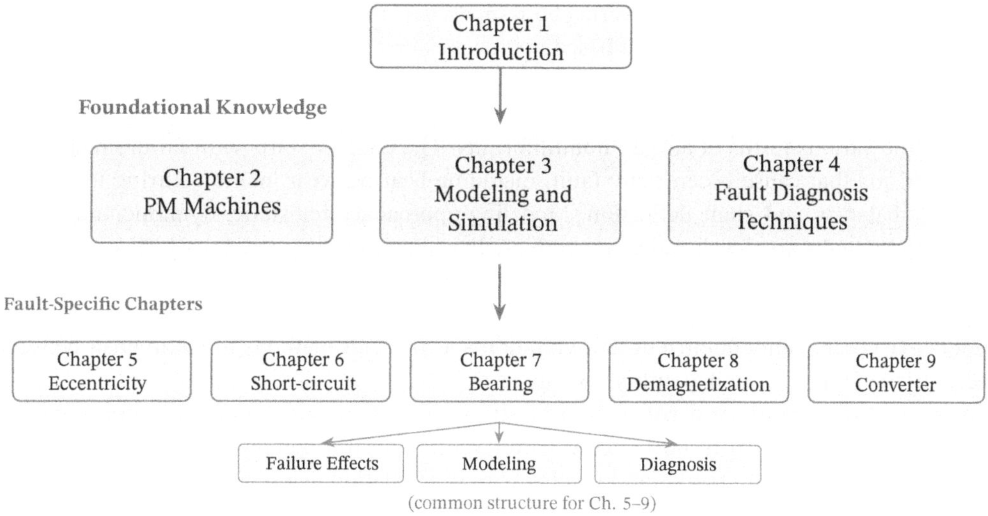

Figure 1.2 Book organization showing the progression from foundational knowledge chapters to fault-specific chapters. Each fault-specific chapter follows a common structure addressing failure effects, modeling approaches, and diagnostic techniques.

arrangements, and fractional slot concentrated winding designs. The treatment includes magnetic circuit analysis, torque development mechanisms, losses and efficiency considerations, and the influence of design parameters on performance characteristics. Understanding these fundamentals enables appreciation of how faults alter motor behavior and what signatures they produce in observable quantities.

Chapter 3: Modeling and Simulation Techniques presents the analytical and computational tools used to predict motor behavior under both healthy and faulty conditions. FEA for detailed electromagnetic field analysis receives comprehensive treatment, including both two-dimensional and three-dimensional formulations appropriate for different motor geometries. Analytical modeling approaches based on magnetic circuit theory and winding function analysis provide computationally efficient alternatives suitable for real-time applications and parametric studies. The chapter examines coupled electromagnetic-thermal modeling for predicting temperature distributions and their effects on motor performance. Machine learning-based modeling approaches that can capture complex behaviors from experimental data without requiring detailed physical models are introduced. These modeling capabilities support both diagnostic algorithm development and fault signature prediction.

Chapter 4: Fault Diagnosis Techniques provides systematic coverage of diagnostic methodologies applicable across multiple fault types. Motor current signature analysis, examining harmonic content and spectral patterns in stator currents, represents one of the most widely applied approaches due to the ready availability of current measurements in motor drives. Magnetic flux signature analysis using external flux sensors or search coils detects electromagnetic anomalies associated with various faults. Zero-sequence voltage component analysis exploits the neutral point voltage that appears during asymmetric fault conditions. Extended Park's vector analysis transforms three-phase quantities into a rotating reference frame where certain faults produce characteristic patterns. Vibration analysis techniques, including time-domain statistical measures and frequency-domain spectral analysis, detect mechanical faults and electromagnetically-induced vibrations. Data-driven approaches employing machine learning for pattern recognition and classification are examined, covering both supervised and unsupervised learning paradigms. The chapter concludes with model-based approaches that compare observed behavior against predictions from motor models to identify discrepancies indicative of faults.

Chapter 5: Eccentricity Faults begins with the detailed examination of specific fault types, focusing on the various forms of air gap nonuniformity. The chapter starts with failure modes and mechanisms that cause eccentricity fault, including bearing wear, manufacturing tolerances, rotor imbalance, and shaft deflection. Modeling approaches for static, dynamic, and mixed eccentricity faults are developed using both analytical permeance models and FEA, predicting the electromagnetic signatures including harmonic patterns in flux density and current spectra. Diagnostic techniques specific to eccentricity faults are examined, including characteristic frequency components in current and vibration spectra, and methods to distinguish between eccentricity fault types and quantify severity levels.

Chapter 6: Short-circuit and Inter-turn Faults addresses electrical faults in stator windings, progressing from inter-turn short circuits within single coils through phase-to-phase and phase-to-ground faults. Failure modes and mechanisms of insulation degradation are examined, covering thermal aging, partial discharge, mechanical stress, and environmental effects. Modeling of short-circuit faults employs coupled circuit-field approaches that capture both the fault current circulating in shorted turns and the resulting electromagnetic field distortions. Diagnostic approaches including negative sequence current monitoring, harmonic analysis, and thermal signature detection are presented. The progressive nature of winding faults and prognosis of fault development receive particular attention.

Chapter 7: Bearing Faults examines the most prevalent mechanical failure mode in PM motors. The chapter begins with bearing kinematics and dynamics, establishing the relationships between bearing geometry, operating speed, and the characteristic frequencies associated with different defect locations. Failure modes and mechanisms including inadequate lubrication, contamination, fatigue, and improper installation are analyzed. Vibration-based diagnostic approaches form the primary focus, including envelope analysis, spectral kurtosis, and wavelet-based techniques for extracting bearing fault signatures. Current-based bearing diagnostics, leveraging the modulation of electromagnetic quantities by bearing-induced vibrations, provide alternative approaches when vibration sensors are unavailable. Bearing fault prognosis and remaining useful life span estimation conclude the chapter.

Chapter 8: Demagnetization Faults treats the unique failure mode specific to PM machines. Failure mechanisms including thermal demagnetization, electrical demagnetization from excessive armature reaction, and physical degradation are examined in detail. Modeling approaches predict both uniform and local demagnetization effects on motor performance using magnetic circuit methods and FEA. Diagnostic techniques exploiting back-EMF signature changes, harmonic pattern variations, and torque characteristic shifts are presented. The challenge of distinguishing demagnetization from other faults producing similar signatures receives particular attention, along with methods for quantifying demagnetization severity.

Chapter 9: Power Converter Faults extends the diagnostic scope beyond the motor proper to the power electronic systems essential for PM motor operation. This chapter examines failure modes of power semiconductor devices including IGBTs and MOSFETs, gate driver circuits, DC link capacitors, and control sensors. The effects of converter faults on motor operation are analyzed, distinguishing signatures of converter faults from motor faults. Diagnostic approaches including switching pattern analysis, thermal monitoring, and device parameter estimation are presented. The integration of motor and converter diagnostics into unified health monitoring systems that recognize the coupled nature of motor-drive faults concludes the chapter.

Throughout the book, emphasis is placed on practical implementation considerations alongside theoretical foundations. Each diagnostic technique is evaluated in terms of sensor requirements, computational demands, sensitivity to operating conditions, and applicability to different motor types and applications. The treatment acknowledges both the successes of current diagnostic approaches and the ongoing challenges that motivate continued research and development.

The comprehensive scope of this book serves multiple audiences. For researchers, it provides thorough coverage of current diagnostic methodologies while identifying open questions and opportunities for advancement. For practicing engineers designing condition monitoring systems, it offers practical guidance on technique selection, implementation, and performance expectations. For students, it presents systematic development from fundamentals through advanced applications, building understanding progressively. The integration of failure mode analysis, modeling, and diagnostics for each fault type provides a complete treatment supporting both understanding and application of PM motor fault diagnosis.

References

1 Jawad Faiz, Mohammad Amin Bazrafshan, and Zabihollah Tabarniarami. Demagnetisation fault analysis and diagnosis based on different methods in permanent magnet machines – An overview. *IET Electric Power Applications*, 18(12):1860–1893, 2024. https://doi.org/10.1049/elp2.12519.

2 Alexandros Sergakis, Marios Salinas, Nikolaos Gkiolekas, and Konstantinos N. Gyftakis. A review of condition monitoring of permanent magnet synchronous machines: Techniques, challenges and future directions. *Energies*, 18(5):1177, 2025. https://doi.org/10.3390/en18051177.

3 Przemyslaw Pietrzak and Marcin Wolkiewicz. Demagnetization fault diagnosis of permanent magnet synchronous motors based on stator current signal processing and machine learning algorithms. *Sensors*, 23(4):1757, 2023. https://doi.org/10.3390/s23041757.

4 Senseye. The true cost of downtime 2022. Technical report, Senseye Predictive Maintenance, Siemens, 2022. https://assets.new.siemens.com/siemens/assets/api/uuid:3d606495-dbe0-43e4-80b1-d04e27ada920/dics-b10153-00-7600truecostofdowntime2022-144.pdf.

5 Stephen Vanderberg. World's largest manufacturers lose $1 trillion/year to machine failure. ISA Blog, 2020. https://blog.isa.org/worlds-largest-manufacturers-lose-1-trillion/year-to-machine-failure.

6 ABB. survey reveals unplanned downtime costs $125,000 per hour. ABB News Center, 2022. https://new.abb.com/news/detail/107660/abb-survey-reveals-unplanned-downtime-costs-125000-per-hour.

7 Research and Markets. Permanent magnet motors market size & forecast to 2030. Market Research Report, 2024. https://www.researchandmarkets.com/report/permanent-magnet-motor.

8 Benjamin Podmiljšak, Boris Saje, Petra Jenuš, Tomaž Tomše, Spomenka Kobe, Kristina Žužek, and Sašo Šturm. The future of permanent-magnet-based electric motors: How will rare earths affect electrification? *Materials*, 17(4):848, 2024. https://doi.org/10.3390/ma17040848.

9 Zia Ullah and Jin Hur. A comprehensive review of winding short circuit fault and irreversible demagnetization fault detection in PM type machines. *Energies*, 11(12):3309, 2018. https://doi.org/10.3390/en11123309.

10 Zhiyuan Li, Qinmu Wu, Shengming Yang, and Xiangping Chen. Diagnosis of rotor demagnetization and eccentricity faults for IPMSM based on deep CNN and image recognition. *Complex & Intelligent Systems*, 8(6):5469–5488, 2022. https://doi.org/10.1007/s40747-022-00764-z.

11 Kai-Jung Shih, Duc-Kien Ngo, Shih-Feng Huang, and Min-Fu Hsieh. Mixed-fault diagnosis for permanent magnet motor with few-shot learning based on the prototypical network. *IET Electric Power Applications*, 19(1):e70081, 2025. https://doi.org/10.1049/elp2.70081.

12 Huizhen Wang, Lei Wang, QiYa Wu, Haoying Pei, and Lijun Diao. A data-driven intelligent fault diagnosis framework for permanent magnet in PMSM. *Alexandria Engineering Journal*, 113:331–346, 2025. https://doi.org/10.1016/j.aej.2024.11.030.

13 Li Li, Shenghui Liao, Beiji Zou, and Jiantao Liu. Mechanism-based fault diagnosis deep learning method for permanent magnet synchronous motor. *Sensors*, 24(19):6349, 2024. https://doi.org/10.3390/s24196349.

14 Chiao-Sheng Wang, I-Hsi Kao, and Jau-Woei Perng. Fault diagnosis and fault frequency determination of permanent magnet synchronous motor based on deep learning. *Sensors*, 21(11):3608, 2021. https://doi.org/10.3390/s21113608.

2

Permanent Magnet Machines

This chapter establishes the foundational knowledge of permanent magnet (PM) machines essential for understanding fault diagnosis techniques. Beginning with historical development and applications, the chapter progresses through machine construction, materials, and performance characteristics. A systematic classification of machine topologies is presented, as different configurations exhibit distinct fault signatures requiring tailored diagnostic approaches. This foundational understanding is crucial for the fault modeling and diagnosis techniques presented in subsequent chapters.

2.1 Historical Development of PM Machines

The advent of advanced PM materials has significantly affected the evolution of PM machines. PM materials with high magnetic properties, such as neodymium–iron–boron (NdFeB) and samarium–cobalt (SmCo), have become available, allowing for the design and development of more efficient and powerful PM machines. The development of PM materials over the past century represents a remarkable progression in magnetic energy density, as illustrated in Figure 2.1. Beginning with AlNiCo in the 1930s, which achieved a maximum energy product $(BH)_{max}$ of approximately 40 kJ/m^3, the field advanced through ferrite magnets in the 1950s and then experienced a significant breakthrough with the introduction of rare-earth magnets. $SmCo_5$ magnets, developed in the mid-1960s, achieved $(BH)_{max}$ values of 130–200 kJ/m^3, followed by Sm_2Co_{17} with approximately 240 kJ/m^3 in 1977. The most significant advancement came with NdFeB magnets in 1983–1984, initially achieving around 290 kJ/m^3 and progressing to modern high-grade variants (N52) exceeding 400 kJ/m^3. This dramatic improvement in magnetic performance has enabled corresponding advances in PM machine applications, from early small DC motors to modern electric vehicles and wind turbines.

One of the key impacts of advanced PM materials on PM machines is the increase in power density. PM machines that utilize high-energy PM materials can generate higher torque and power outputs for a given size, compared to earlier generations of PM machines [1, 2]. This has led to the development of smaller and more compact PM machines with increased power density, making them suitable for various applications where space is limited or weight is a concern.

Moreover, the availability of advanced PM materials has led to the development of high-speed PM machines. PM materials with high coercivity and temperature stability have enabled the design of PM machines that can operate at high speeds without demagnetization or performance degradation. This has expanded the range of applications for PM machines, particularly in high-speed applications such as electric vehicles (EVs) and wind power generation.

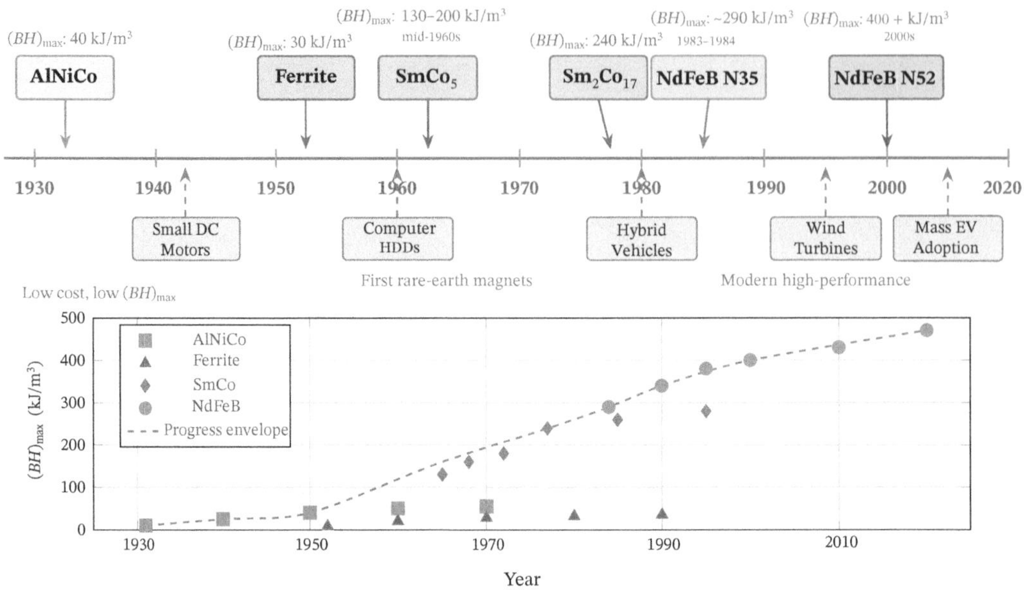

Figure 2.1 Historical development of permanent magnet materials showing key milestones and $(BH)_{max}$ evolution from AlNiCo (1930s) through modern NdFeB grades. The upper timeline displays material introductions with their characteristic energy products, while application milestones indicate the progressive adoption of PM technology in various industries. The lower chart quantifies the $(BH)_{max}$ progression over time for each material family.

Additionally, the use of advanced PM materials has improved the efficiency of PM machines. High magnetic properties of PM materials allow for higher energy density and better utilization of magnetic flux in the machine, resulting in higher efficiency and improved performance. This has made PM machines more attractive in terms of energy savings and reduced operating costs compared to other types of machines, such as induction machines or DC machines.

Furthermore, the availability of advanced PM materials has also facilitated the development of new machine topologies and designs. For example, the use of buried magnet PM motors, where the PMs are embedded within the rotor core, has become more popular due to the improved magnetic properties of PM materials. Buried magnet PM motors offer advantages such as reduced cogging torque, improved thermal performance, and good mechanical robustness. Other designs, such as flux-switching PM machines and magnetically geared machines, have also emerged because of advancements in PM materials.

In addition to the impact on machine performance and design, the availability of advanced PM materials has influenced the development of power electronics devices. The use of high-performance PM materials in PM machines requires advanced power electronics drives that can provide precise control of the PM machine, such as vector control techniques. This has led to advancements in power electronics technology, including high switching frequencies, improved thermal performance, and increased power density.

PM synchronous machines (PMSMs) have emerged as a promising solution for variable speed operations with low maintenance requirements, high efficiency, and power factor compared to conventional DC machines and induction machines [3]. The design characteristics of PMSMs with sinusoidal and square excitation waveforms were initially proposed by Hendershot and Miller in 1994, with further design methods described by Gieras in 1997 [2, 4].

Over the years, numerical methods have been extensively applied to analyze the performance of PMSMs, including PMDC machines [4]. Studies have also focused on optimizing the parameters of SPM motors, showing that PM losses can be significantly reduced, up to 60% lower compared to induction motors of the same size and power [5, 6]. Furthermore, rotor losses and eddy current losses of PM motors operating at their rated frequency and power are often negligible, leading to reduced copper losses and hence, stator-magnetizing current can be neglected.

PMSMs have found wide applications in low power and medium power systems, and extensive research has been conducted on their torque development capability and demagnetization risk for different magnet configurations, including SPM and IPM, as studied by Slemon and Sebastian in 1987 and 1989 [7–9]. It has been concluded that for a PM machine with specified power, it is possible to develop higher stable power and almost twice the peak torque compared to induction motors. Furthermore, it has been shown that the demagnetization risk of PM materials in small and medium-sized motors with SPM and IPM is low, even in the presence of short circuit risk. However, motors with demagnetization protection mechanisms exhibit lower sensitivity to demagnetization in IPM motors compared to IPM motors.

The air gap design of PM machines with V-shape magnets has also been a topic of research, with studies conducted by Nee and Thelia in 1990 and 2000, obtaining numerical calculations of the square air gap magnetic flux density, which takes into account reluctance, axial flux density, and saturation effects. Additionally, inverter-fed PMSMs have been extensively tested, revealing that the ideal number of poles falls between 6 and 12, with speed ranging from 750 to 3000 rpm, and power ratings ranging from 4 to 37 kW. Optimization studies by Thelia in 2002 further emphasized the importance of a large air gap in buried-magnet PM machines to reduce armature reaction and iron losses.

In recent times, PM machines have gained increased attention due to the availability of high-energy PM materials and power electronic devices, particularly buried-magnet configurations [10, 11], which offer improved performance and efficiency in various applications. The historical development of PMSMs has been driven by advancements in design methods, numerical analysis techniques, magnet configurations, and optimization process, making them a prominent choice for modern variable speed motor applications.

2.2 Applications and Advantages of PM Machines

PM machines find a wide range of applications in various sectors due to their advantages such as high efficiency, compact size, high power density, and precise control. Here are some examples of the applications of PM machines in different sectors:

2.2.1 Automotive Sector

EVs extensively use PM machines as traction motors in electric powertrains [12]. The high efficiency translates into better energy utilization and longer driving range, critical for EV performance and usability. The compact size and lightweight characteristics enable easy integration into space-constrained powertrain systems, while high power density enables acceleration and top speeds comparable to conventional internal combustion engine vehicles [13].

PM machines are also used in hybrid powertrains, functioning as electric generators to charge the battery during regenerative braking or as motors to provide additional torque during acceleration. In auxiliary systems, PM machines power steering and air conditioning compressors with efficient and reliable operation.

2.2.2 Industrial Sector

PM machines are used extensively in pumps, fans, compressors, conveyors, and machinery for mechanical power transmission [14, 15]. Variable speed operation enables optimal process control and energy savings in HVAC systems, irrigation, and pneumatic equipment. The precise speed and torque control is particularly valuable in CNC machines, machine tools, and printing presses where production quality depends on accurate motor control. Low maintenance requirements reduce downtime and increase productivity in industrial operations.

2.2.3 Renewable Energy Sector

PM machines are widely used in renewable energy generation systems including wind turbines, hydroelectric generators, and tidal generators [16, 17]. In wind turbines, variable speed operation enables maximum energy capture from varying wind conditions, while robust construction withstands harsh environmental conditions in offshore installations. Hydroelectric and tidal generators benefit from the stable, efficient power conversion characteristics of PM machines operating in predictable flow environments. The growing adoption of PM machines in renewable energy supports the global shift toward cleaner, sustainable energy sources.

2.2.4 Home Appliances and Consumer Electronics Sector

PM machines are widely used in household appliances, including refrigerators, washing machines, air conditioners, and vacuum cleaners, where they drive compressors and drums with efficient, quiet operation [18]. In consumer electronics such as electric fans, blenders, and power tools (drills, saws, and trimmers), PM machines provide the high torque and precise control needed for reliable performance. Their compact size and energy efficiency contribute to the development of environmentally friendly appliances with improved user experience.

2.2.5 Medical Sector

PM machines are used in critical medical devices including MRI machines, CT scanners, and dental drills [19, 20]. In MRI systems, PM machines generate the high-strength, stable magnetic fields necessary for precise imaging of internal body structures. PM machines also power blood pumps, surgical tools, and prosthetic devices, where precise control and reliability are essential for patient safety. The compact size and high precision of PM machines make them integral to advanced medical technology and improved patient care.

2.2.6 Transportation Sector

In electric vehicles, PM machines serve as traction motors for cars, buses, and trains, providing smooth acceleration and regenerative braking capabilities that enhance energy efficiency [21]. The lightweight and compact characteristics are particularly suited to space-constrained vehicle designs. In marine applications, PM machines power electric propulsion systems for ships, ferries, submarines, and underwater drones, offering long operational life and low maintenance requirements. The growing adoption of transportation reflects the demand for sustainable, efficient propulsion technologies.

2.2.7 Other Applications

PM machines are utilized across diverse sectors, including industrial automation (robotics, material handling, machine tools), HVAC systems (air handling, pumps, fans), water treatment (pumps, blowers, mixers), agricultural machinery (irrigation, grain dryers), and energy storage systems (battery and flywheel storage) [22–27]. These applications leverage the rapid response, compact size, and reliable operation characteristics of PM machines to enhance system performance and efficiency.

2.3 PM Machine Construction and Key Components

PM machines consist of several key components that work together to convert electrical energy into mechanical energy (in motors) or mechanical energy into electrical energy (in generators). Understanding the construction and components of PM machines is essential for their design, operation, and fault diagnosis.

2.3.1 Basic Machine Structure

PM machines, like other rotating electrical machines, consist of two main parts: the stator (stationary part) and the rotor (rotating part), separated by an air gap. The general power equation for electrical machines is given by:

$$P_m = \omega_m T_e \tag{2.1}$$

where P_m is the developed or consumed mechanical power of the machine, ω_m is the shaft speed in rpm, and T_e is the developed torque in Nm.

The stator of a PMSM is similar to that of an induction motor or switched reluctance machine. The key difference lies in the rotor structure, where PM machines utilize permanent magnets and iron sheets in their rotor construction, while induction machines have either a cage rotor or a wound rotor.

2.3.2 Stator Construction

The stator is the stationary outer part of the machine and contains the windings that carry the electrical current. The stator construction comprises several important elements:

2.3.2.1 Stator Core

The stator core is constructed using ferromagnetic laminations to reduce core losses. Two sets of materials are commonly employed:

- **Noncrystallized metals** (e.g., meta-glass) with a multijoint chemical structure, which exhibit low hysteresis and eddy current losses, along with nonmetallic elements such as Boron, Silicon, and Carbon.
- **Power materials** (e.g., oriented metal pieces) are suitable for high-frequency applications despite the core's low permeability. They also help to reduce acoustic noise and vibration.

2.3.2.2 Stator Windings

The winding type of the stator determines the arrangement of windings in an electrical machine, as illustrated in Figure 2.2. Windings can be placed in the air (slotless design) or inside slots around

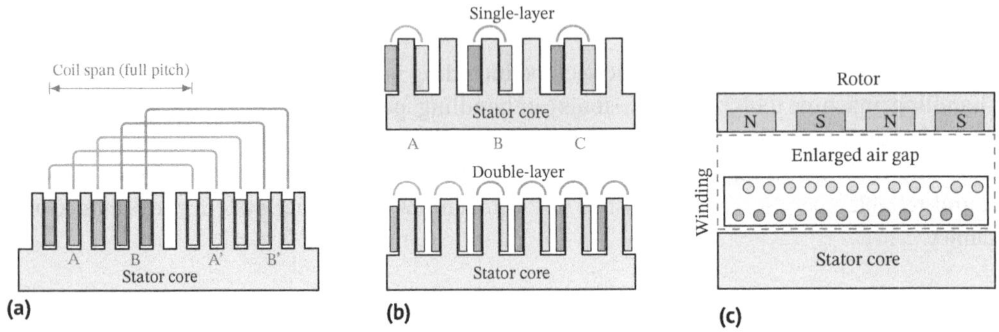

Figure 2.2 Stator winding configurations in PM machines: (a) distributed winding with overlapping coils spanning multiple slots, (b) concentrated winding in single-layer and double-layer arrangements with nonoverlapping coils and short end turns, and (c) slotless winding with conductors placed in the enlarged air gap.

the stator teeth. Winding sets can either be concentrated within a slot or distributed across multiple slots.

Distributed Winding: Commonly employed in rotating machines, distributed windings generate sinusoidal mmf with better electromagnetic performance than concentrated types. However, overlapping conductors increase copper usage, losses, and manufacturing costs [28, 29].

Concentrated Winding: This can be categorized into single-layer and double-layer types. In the single-layer type, one winding is placed in every two teeth, while in the double-layer winding, there is a winding in every tooth. Concentrated winding offers easier and cheaper fixation, requires fewer windings compared to the distributed structure, resulting in a more compact structure (shorter axial length) and significant reduction in copper, thereby lowering costs [28–30].

Slotless Design: In slotless designs, windings are positioned in the air, resulting in a wider air gap between the stator and rotor, which reduces the magnetic flux density from the PMs' side. To compensate for the reduction in flux density, slotless designs require additional PMs and copper windings. Moreover, slotless designs necessitate less iron, leading to reduced hysteresis and eddy current losses. Typically, the power density of a slotless machine is three times lower than that of a machine with slots [29].

2.3.2.3 Stator Slots

Slots in the stator core house the windings and come in different configurations:

Open Slots: The advantage of open slots is their facilitation of automatic insertion of the stator winding. These slots have low leakage reactance. However, they also result in higher cogging torque and increased eddy current losses compared to semi-closed and closed-slot types.

Semi-closed Slots: Semi-closed slots generate fewer slot harmonics, leading to reduced torque ripple. In this structure, the reluctance varies around the stator, resulting in higher leakage inductance compared to open slots, which in turn leads to lower developed power.

Closed Slots: Closed slots are stronger than the aforementioned types. However, inserting windings into these slots is more challenging, and the slots are prone to saturation.

Different slot shapes are employed in the design of electrical machines, such as square, trapezoidal, oval-shaped, and circular [28].

2.3.3 Rotor Construction

The rotor is the rotating inner part of PM machines and contains the permanent magnets that provide the excitation field. Unlike induction machines where the rotor field is induced, PM machines use permanent magnets to create a constant magnetic field. The rotor construction and magnet placement significantly affect the machine's performance characteristics.

PM machines are classified based on their mechanical structures, with the stator and rotor configuration playing a crucial role. The main rotor types include:

- **Internal Rotor Configuration:** The rotor is located inside the stator, which is the most common configuration. This structure features a complex design, high air gap flux density, and low residual flux.
- **External Rotor Configuration:** The wound stator is centrally located inside the rotor, while magnets are fixed around the internal surface of the rotor. This configuration is advantageous for applications such as wind turbines and fans, where blades can be more easily connected to an external rotor machine. The larger rotor cylinder facilitates the insertion of a higher number of poles and enables better cooling of PMs [28, 31, 32].

2.3.4 Air Gap

The air gap is the space between the stator and rotor and plays a critical role in machine performance. The air gap length δ is generally considered to be between 1 and 3 mm. A wider air gap requires more magnetic material to achieve the required air gap magnetic flux density, and the synchronous inductance will be smaller. The air gap magnetic flux density depends on the PM residual flux and the operating point of the machine.

2.4 Permanent Magnet Materials and Properties

The magnetization characteristic of a PM is located in the second quarter of the hysteresis loop, wherein an increase in magnetic field intensity results in a decrease in magnetic flux density. This characteristic is also known as the demagnetization curve. Figure 2.3 illustrates the demagnetization curves for four common PM materials: NdFeB, SmCo, Alnico, and Ferrite. The key parameters including residual flux density (B_r), coercive force (H_c), and the maximum energy product ($(BH)_{\text{max}}$) are indicated on the NdFeB curve. The important parameters of PM are described in the following sections [10, 33, 34].

2.4.1 Residual Magnetic Flux Density B_r and Coercive Force H_c

When a PM is magnetized up to the saturation level and then its magnetic field intensity is reduced to zero, it retains its magnetization property. At this point on the magnetization characteristic, the magnetic flux density is referred to as residual flux density (B_r). If the magnetic field intensity rises in the opposite direction, the magnetic flux density tends to zero. The magnetic field intensity at this point is called the coercive force (H_c), which represents the required magnetomotive force (mmf) for demagnetization.

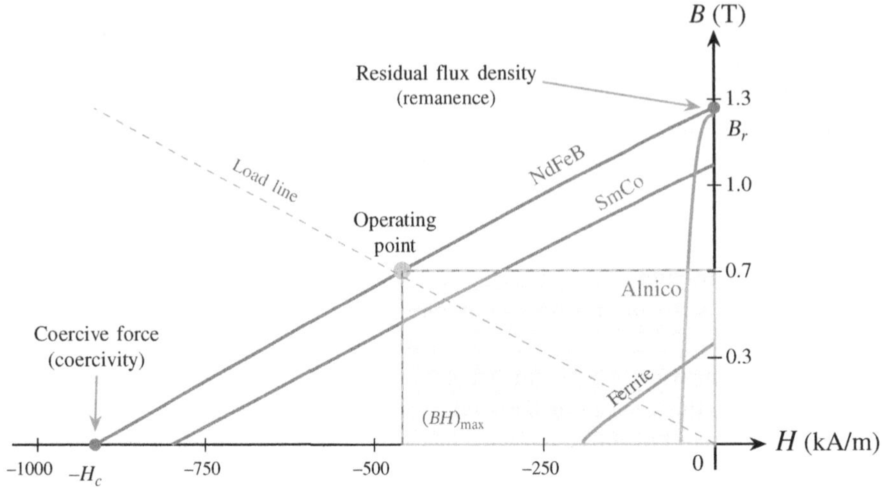

Figure 2.3 Demagnetization curves of common PM materials showing the characteristic B-H relationship in the second quadrant. NdFeB and SmCo exhibit nearly linear curves with high B_r and H_c, while Alnico shows high B_r but very low H_c with a characteristic knee shape. Ferrite has lower B_r but moderate H_c.

2.4.2 Received Energy and its Maximum Value

At any point on the demagnetization curve, the received energy is equal to the product of the magnetic flux density (B) and the magnetic field intensity (H), which indicates the energy density or magnetization power of the PM. The maximum value of this product represents the maximum received energy in joules per cubic meter (J/m^3). Adjusting the operating point of the PM around the maximum received energy allows for the minimization of the required volume of the PM. Hence, selecting a material with the maximum received energy may result in the lowest required PM volume for a specific application.

2.4.3 Curie Temperature

The properties of a PM weaken as the temperature rises. The Curie temperature is the temperature at which the PM loses its magnetization properties and requires re-magnetization to be used again [10, 33, 34].

2.4.4 Classification of PM Materials

Generally, three types of PM materials are commonly used in PM machines [35]:

2.4.4.1 AlNiCo

The first commercially available material was the AlNiCo alloy, consisting of Aluminum, Nickel, Iron, and Cobalt. AlNiCo exhibits high residual magnetic flux density and a low sensitivity factor, which are considered its main features. It possesses appropriate resistance, performance, and good conductivity at high temperatures. However, its weakness lies in its vulnerability to external magnetic fields, which can weaken its magnetic properties. AlNiCo experienced peak demand between 1940 and 1970 [11].

Table 2.1 Comparison of properties of different PM materials.

PM material	Curie temp. (°C)	B_r sensitivity coefficient	H_c sensitivity coefficient	Peak operating temp. (°C)
AlNiCo	830	−0.02	0.01	520
Ferrite	450	−0.20	0.40	400
SmCo	720	0.03–0.045	0.14–0.4	300
NdFeB	310	0.095–0.150	0.14–0.70	150

Source: Data from [10].

2.4.4.2 Ferrites

Also known as ceramics, ferrites are composed of Carbonate or Strontium and iron oxide. These materials are hard and brittle. The most significant advantages of ferrites are their high electrical resistance, thermal resistance, and low cost. They exhibit no eddy current, making their maximum operating temperature 400 °C. Ferrites possess high resistance against external magnetic fields but have a small residual magnetic flux density as their weakness [11, 33, 34].

2.4.4.3 Rare-earth Magnetic Materials

SmCo PM was discovered in 1960. SmCo PMs are strong magnets with a linear magnetization characteristic, resulting in negligible magnetic field losses. They exhibit high residual magnetic flux density, high coercive force, and large received energy. However, their weakness lies in their fragility. The next generation of these materials, such as neodymium, boron, and iron (NdFeB), improved upon this fragility. However, these materials depend strongly on temperature [10, 11]. Table 2.1 compares the properties of different PM materials.

2.5 Losses in PM Machines

Understanding losses is crucial for machine design and efficiency optimization. Losses in PM machines are categorized into stator losses, iron losses, rotor losses, and windage losses.

Stator Losses: These encompass both winding copper losses and core losses. In high-power PM machines, copper losses tend to be significant. The machine's eddy current losses can be attributed to: (a) the skin effect caused by conductors, and (b) the proximity effect resulting from electromagnetic induction between adjacent conductors in the slot.

Iron Losses: These pertain to materials exposed to alternating magnetic fields and consist of hysteresis losses and eddy current losses. Hysteresis losses occur when an alternating magnetic field is applied to the material and are proportional to the hysteresis loop area (BH) of the magnetic materials.

Rotor Losses: Rotor losses in PM machines are primarily attributed to induced eddy currents in the shaft and PM. However, these losses are negligible compared to the total losses of the machine. Dissipating heat from the rotor is more challenging than from the stator to maintain the machine's standard operating temperature [36–38].

Windage Losses: These result from the airflow turbulence between the rotor and stator, which can cause heating. At higher speeds, this turbulence becomes more prominent, leading to higher losses and reduced machine efficiency [36–38].

2.6 Comparison of PMSM and Induction Machines

With an understanding of the loss mechanisms in PM machines, we can now examine how these machines compare to induction machines in terms of efficiency and performance. The superior properties of rare-earth magnets enable performance advantages that distinguish PMSMs from induction machines. PM machines in steady-state operation have no copper losses in the rotor and therefore, most of the losses in these machines occur in the stator, which are lower than those of induction machines of the same size. Additionally, a PMSM operating with a high-power factor can compensate for a portion of the stator losses.

PMSM machines have a smaller pole pitch than induction machines because the magnetic flux density between two magnets restricts it. PMSMs also have a higher number of poles compared to induction motors. As a result, to increase the air gap length, the yoke needs less iron. The torque of the machine increases with a larger air gap diameter, and stator copper losses decrease when winding and copper resistance are reduced. The location, volume, and properties of the PM materials directly affect the torque. PM materials such as NdFeB can increase the air gap flux density up to 1.2 T.

However, in induction machines, the power factor decreases as the number of poles increases due to reduced magnetizing inductance. The magnetizing inductance relationship demonstrates why PMSMs can accommodate more poles than induction machines:

$$L_m = \frac{(N_{\mathrm{ph}}K)^2}{2P}. \tag{2.2}$$

where L_m is directly proportional to the square of the winding coefficient K and the number of turns per phase N_{ph}, and inversely proportional to the number of pole pairs P.

Consequently, in an induction machine, the stator magnetizing current increases quadratically with pole pairs, resulting in reduced efficiency:

$$I_{\mathrm{mag}} \propto (2P)^2. \tag{2.3}$$

The current in an induction motor consists of two components: magnetizing current and torque-developing current. In a PM machine, the PMs produce the air gap magnetic flux, requiring only a small magnetizing current component in the stator. This fundamental difference allows PM machines to achieve higher efficiency at higher pole counts where induction machines would suffer from excessive magnetizing current and reduced power factor. Additionally, the number of poles influences the iron volume, demagnetization risk of PMs, leakage fluxes at the bottom of PMs, and cogging torque. Overall, the number of poles affects the cost and efficiency of the entire system [39].

2.7 Classification of PM Machines and Machine Topologies

Having established the fundamental construction and operational principles, we now systematically classify PM machines based on structural and application criteria. PM machines can be classified based on various criteria including their electrical characteristics, mechanical structure, and application requirements. This section provides a comprehensive overview of the different types and topologies of PM machines.

2.7.1 Classification of Electrical Machines

Electrical machines can be classified into four main categories based on their operating principles and construction: DC machines, AC synchronous machines, AC induction machines, and special machines. Figure 2.4 illustrates this comprehensive classification in a radial layout, showing each category and its subcategories. DC machines are divided by their commutation method (mechanical or electronic), AC synchronous machines by their excitation type (permanent magnet, wound rotor, reluctance, or hybrid), AC induction machines by their rotor construction (squirrel cage or wound rotor), and special machines include switched reluctance motors, stepper motors, and universal motors.

2.7.1.1 AC Machines

AC machines are powered by alternating current (AC) supply. They can be either synchronous or asynchronous machines.

Synchronous Machines: Synchronous machines have a rotating magnetic field that is synchronized with the AC power source. They are used in applications where a constant speed is required, such as electric motors and generators.

- **Permanent Magnet Synchronous Machines (PMSMs):** PMSMs have a stator with windings and a rotor with PMs. The stator windings are energized to create a rotating magnetic field. The PMs on the rotor align themselves with the rotating magnetic field, which creates torque that drives the rotor.
- **Synchronous Reluctance Machines (SRMs):** SRMs have a stator with windings and a rotor with a series of salient poles. The stator windings are energized to create a rotating magnetic field. The rotor poles align themselves with the rotating magnetic field, which develops torque that drives the rotor.
- **Doubly Salient PM Motors (DSPMs):** Switched reluctance motor with PM embedded in the stator or rotor side, where the air-gap magnet flux density is from both the winding and the PM.

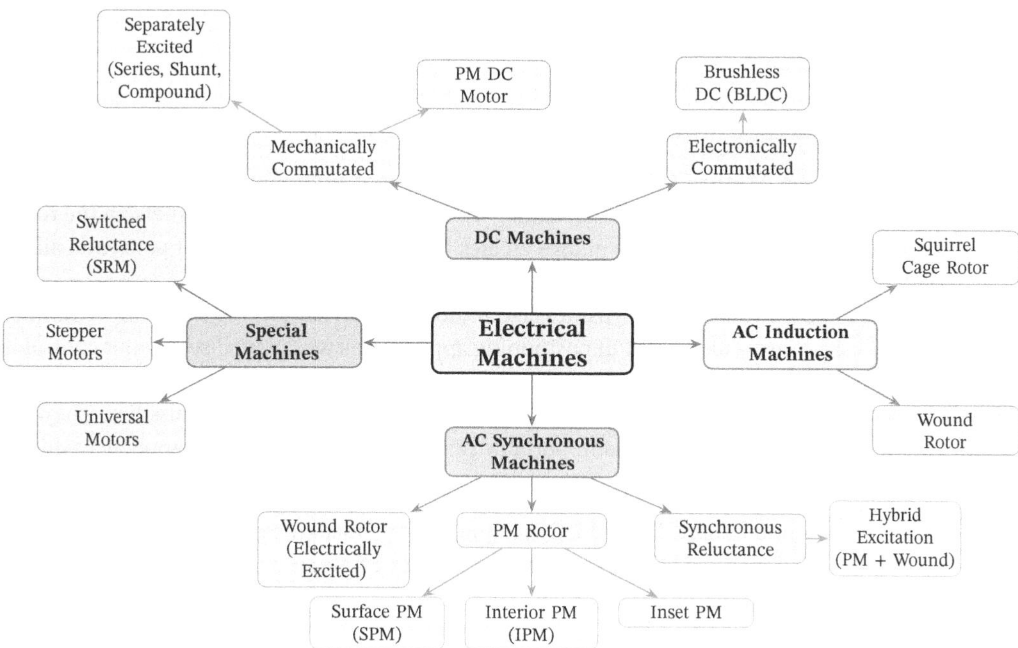

Figure 2.4 Classification of electrical machines into four main categories: DC machines, AC synchronous machines, AC induction machines, and special machines. Each category is further divided based on construction and operating principles.

- **Line-start PM Synchronous Motors (LSPMSM):** These machines combine the features of induction motors and PMSMs.
- **Permanent Magnet Vernier Machines (PMVMs):** These are considered synchronous machines integrated with magnetic gears, such as flux reversal, flux switching, and transverse-flux machines.

Asynchronous Machines: Induction machines do not have a PM on the rotor. The stator windings are energized to create a rotating magnetic field. The rotor is induced to rotate by the magnetic field, which develops torque that drives the rotor.

2.7.1.2 DC Machines

DC machines are powered by direct current (DC) supply. They can be either brushed or brushless.

Brushed Machines: Brushed machines have a commutator and brushes. The commutator is a segmented copper ring that is connected to the rotor windings. The brushes are made of carbon and are pressed against the commutator. The brushes carry the current from the DC power source to the rotor windings.

- **Universal Motors:** Universal motors are brushed DC motors that can be operated on AC power. They have a stator with windings and a rotor with a squirrel cage. The stator windings are energized to create a rotating magnetic field. The rotor squirrel cage aligns itself with the rotating magnetic field, which develops torque that drives the rotor.
- **Brushed PM Motors:** Conventional DC machines with mechanical commutators and brushes where PMs provide the excitation field.

Brushless DC Motors (BLDCMs): BLDCMs are similar to brushed DC motors, but they do not have a commutator or brushes. The stator windings are energized by a controller to create a rotating magnetic field. The PMs on the rotor align themselves with the rotating magnetic field, which develops torque that drives the rotor.

2.7.1.3 Special Machines

Special machines represent a category of electrical machines with unique operating principles that distinguish them from conventional AC and DC machines.

Switched Reluctance Motors (SRMs): SRMs operate on the principle of magnetic reluctance and have a doubly salient structure with neither windings nor permanent magnets on the rotor. The sequential energization of stator phases creates torque by pulling the rotor poles into alignment with energized stator poles.

Stepper Motors: Stepper motors are brushless DC motors that divide a full rotation into a number of equal steps. They are widely used in positioning applications where precise angular control is required without feedback sensors.

Universal Motors: Universal motors can operate on both AC and DC power supplies. They are series-wound motors commonly used in portable power tools and household appliances due to their high speed and compact size.

PM machines are found in both AC and DC machines. In AC machines, the PMs on the rotor are used to create a rotating magnetic field. This rotating magnetic field interacts with the stator windings to develop torque that drives the rotor. The speed of the rotor is determined by the frequency of the AC power supply and the number of poles on the rotor. In DC machines, the PMs on the rotor are used to provide a constant torque. This torque is independent of the speed of the rotor. This makes DC machines ideal for applications where a constant speed is required, such as EVs and power tools. PMSMs are used in a wide range of applications, including EVs, wind turbines, and robotics. BLDCMs are used in a wide range of applications, including electric vehicles, power tools, and consumer electronics.

PM machines offer a number of advantages over other types of electric motors, including:

- **Higher Efficiency:** PM machines are more efficient than other types of electric motors because they do not require a commutator or brushes.
- **Higher Power Density:** PM machines have a higher power density than other types of electric motors because they do not require large windings.
- **Better Dynamic Performance:** PM machines have better dynamic performance than other types of electric motors, because they can be controlled more precisely.
- **Simpler Construction:** PM machines have a simpler construction than other types of electric motors, because they do not require a commutator or brushes.
- **Lower Maintenance:** PM machines require less maintenance than other types of electric motors, because they do not have a commutator or brushes.

Because of these advantages, PM machines are becoming increasingly popular in a wide range of applications. Among the aforementioned PM machines, the most commonly used in the industry are BLDCMs and PMSMs.

2.7.2 Classification Based on Structure and Application

Users of electrical machines try to choose electrical machines with high efficiency based on structure or advanced control system. On the other hand, simple structure, good reliability, and reasonable cost are desirable features. The variable speed drives have been rapidly grown and widely employed in different applications such as robotics and air-conditioning [4].

Generally, the PM machines are classified based on their mechanical structures, applications types or input voltage waveforms. On the other hand, the PMSM itself can be classified based on different criteria consisting of starting method, motoring or generating modes, axial or radial magnetic fields path, rotor type, rotor structures, and machine specifications (maximum power, with sensor or sensor-less). Two main structures of the PMSMs are surface-mounted, in which PM is located on the outer surface of the rotor and buried type in which PM is located inside the rotor magnetic structure. Stator of PMSM is similar to the stator of an induction machine or stator of switched reluctance machine, and their rotors differ largely.

Radial-flux PM (RFPM) synchronous machines are normally used in direct drive applications. From rotor and stator location point of views, these machines are categorized as outer and inner types. RFPMs have the cheapest and simplest structure among PM machines. Although they are larger than the axial-flux PM (AFPM) machines, AFPM machines have unique features. They have low core volume leading to a high torque per volume. The advantages of AFPM machines are their high torque/volume ratio and efficiency compared with conventional structures. Other structures of PM machines include cross-flux, direct-start, segmental and induction PM machine.

One of the significant features of PMSM is that it is direct-drive system. Advantages of PMSMs include high efficiency, high power factor and good construability over low and high speed ranges [5–7].

2.7.2.1 First Classification: Based on Mechanical Structure

The PM machine can be classified into two types: brush and brushless. Additionally, they can also be classified based on their starting method, machine type (motor or generator), magnetic flux path (radial or axial), rotor type, rotor structure, and machine specifications, including maximum power of the machine and whether it has sensors or is sensor-less. The stator of a PMSM is similar to that of an induction motor. When considering SPM machines (PMs on the external surface) and buried-PM machines, it is clear that locating the PMs on the rotor surface is cheaper and simpler. These PMs are typically attached to the surface of the rotor and sometimes protected and secured with a nonconductive material. This design results in a smaller diameter rotor with low inertia, which is suitable for rapid operation. Another type of SPM is a rotor with IPMs, where the pole pitch is shorter than that of the rotor surface case, and the space between PMs is filled with iron pieces. In this salient-pole machine, the high reluctance torque is as good as the developed PM torque. Both SPM and IPM designs are radially magnetized.

Therefore, the PM flux density limits the no-load flux density. If a high air gap flux density is required, there are other types of PMSMs that generate radial and axial flux. Claw-pole machines in car alternators and low inertia disk-rotor machines are examples of axial flux types. Different structures of axial PMSMs include single-stator, single-rotor, single-stator between two rotors (two air gaps), and single-rotor between two stators (two air gaps) [39, 40].

2.7.2.2 Second Classification: Based on Input Current Waveform

These machines are classified based on the input current waveform. PMDC motors are DC-type motors in which the excitation winding is replaced by PMs. The mechanical structure of PMAC is simpler than that of PMDC, because it does not require brushes and a commutation system. The PM placed on the rotor produces a magnetic field. PMACs are categorized based on their back-EMF waveform in each stator phase, which can be trapezoidal or sinusoidal, as illustrated in Figure 2.5. PMACs operating with sinusoidal waveforms are more popular due to their precise speed and torque control. They have precise sensors, higher magnetic flux density, high torque, and efficiency [39–41].

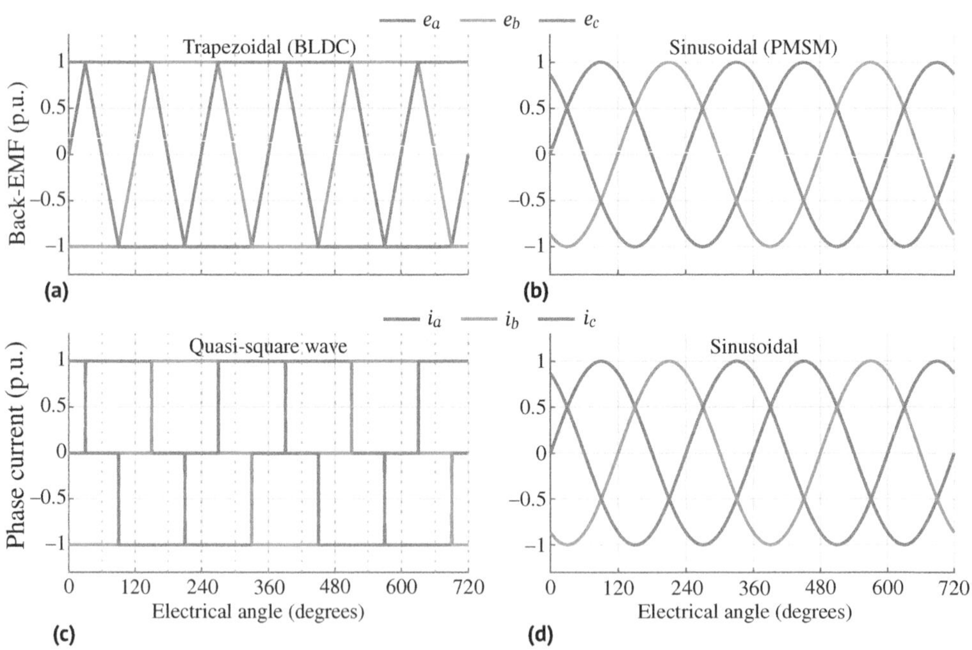

Figure 2.5 Comparison of BLDC and PMSM waveforms: (a) trapezoidal back-EMF of BLDC, (b) sinusoidal back-EMF of PMSM, (c) quasi-square wave phase currents of BLDC with commutation points shown as dashed lines, (d) sinusoidal phase currents of PMSM.

BLDC motors have square or trapezoidal current and back-EMF waveforms, which are generated by a concentrated winding and special-shaped PMs. Since the waveform is trapezoidal, the rotor position sensor is simpler compared to the position sensor of PMAC with a sinusoidal waveform. Additionally, this type of machine requires an input current control sensor. BLDC motors are generally used in small applications where precise control is not necessary, such as hard disks, compressors, and fans.

2.7.3 Rotor Structures and Magnet Placement Topologies

Six main structures of rotors in PM machines are surface-mounted, inset, tangential embedded, inset radial, and buried PMs (U-shaped and V-shaped) [38, 42]. Figure 2.6 illustrates these different rotor configurations.

2.7.3.1 PM Machines with Surface-PM

As depicted in Figure 2.7, PMs are fixed on the surface in this configuration. The main advantage of PMSMs is their simplicity, making them cheaper compared to other PM machines. However, this structure is vulnerable to demagnetization. Additionally, PMs need to withstand centrifugal force, which becomes more problematic at higher speeds [28, 43].

PMSMs can have an internal or external rotor. The internal rotor motor is commonly used in ship propulsion and applications requiring a low diameter and long length. This structure is prevalent in direct-drive wind generators, as the larger rotor diameter allows for the attachment of a higher number of poles. It exhibits increased robustness against centrifugal force and excellent compatibility with wind generators.

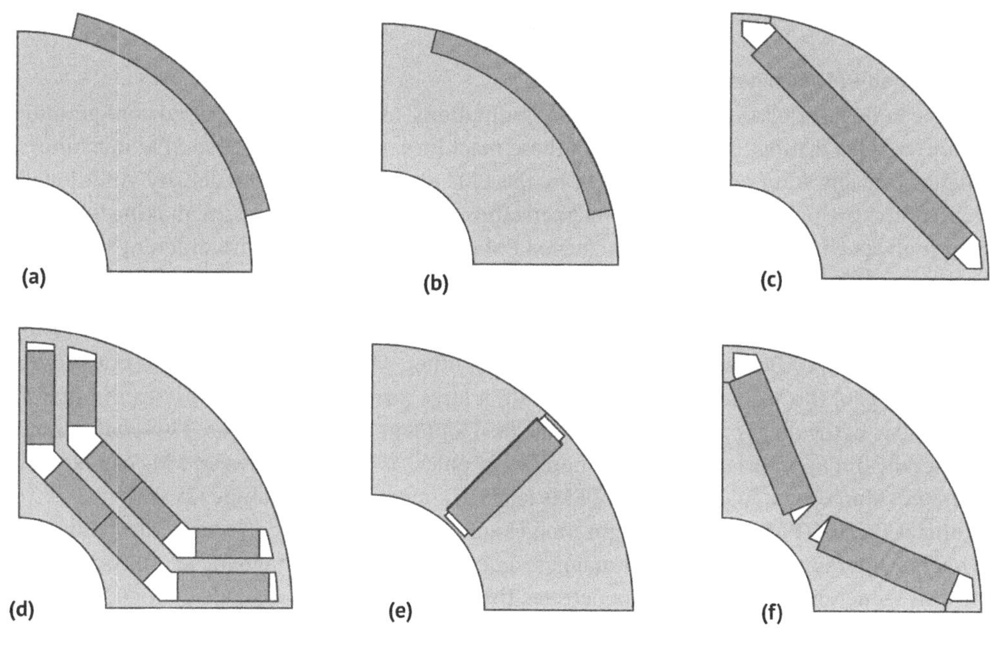

Figure 2.6 Rotor structures and magnet placement topologies: (a) surface-mounted, (b) inset, (c) tangential embedded, (d) buried U-shaped, (e) inset radial, (f) buried V-shaped.

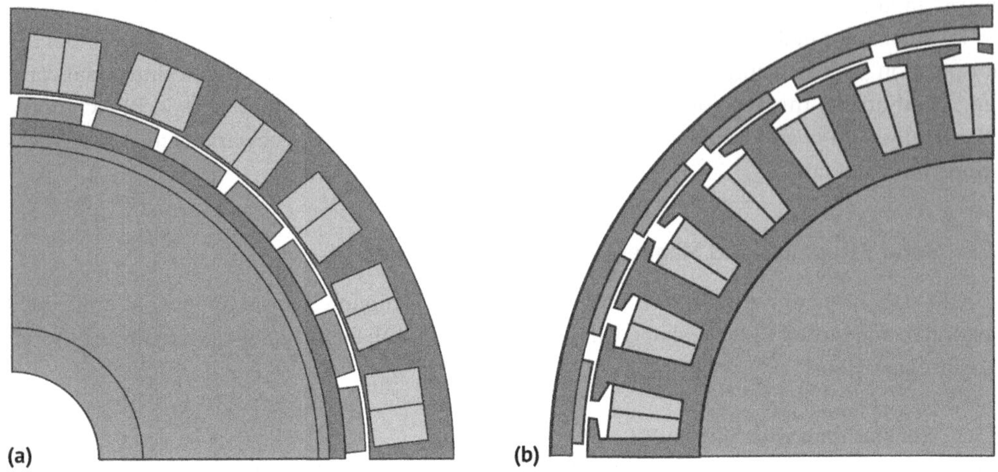

Figure 2.7 Surface-PM machines: (a) internal rotor, (b) outer rotor.

2.7.3.2 Inset-PM Machines

Similar to SPM machines, inset PM machines have PMs fixed on the rotor surface. The difference lies in the presence of iron filling the gaps between the PMs, as illustrated in Figure 2.6b. This iron introduces saliency, leading to the development of reluctance torque in addition to the torque developed by the magnets. These machines find application in direct-drive low-speed scenarios where the additional reluctance torque improves performance characteristics. The impact of reluctance torque on machine performance is an important design consideration when comparing inset configurations with SPM machines [28, 43].

2.7.3.3 Buried-PM Machines

The PMs can be inserted inside the rotor iron laminations, as shown in the buried configurations of Figure 2.6. One distinguishing feature of these machines, when compared to SPM machines, is the higher concentration of flux generated by the PMs and the air gap flux density. Additionally, the buried PMs are protected against demagnetization and mechanical stresses, making them suitable for high-speed applications. Several buried PM configurations exist with different shapes and arrangements [28, 43].

V-shape Buried-PM: As illustrated in Figure 2.6f, two magnets are located in a V-shape configuration with a specific angle with respect to each other. Metallic bridges are present between the two ends of the V-shaped PMs. As a result, a large part of the PM flux passes through the iron instead of the air gap, contributing to the development of higher torque. This configuration is generally not appropriate for a high number of poles, as the iron between PMs can be easily saturated. Moreover, a high number of PMs leads to a more costly machine [28, 43].

Tangential Buried-PM: In this configuration (Figure 2.6c), the rotor consists of alternating iron segments and tangentially-oriented magnets assembled on a ferromagnetic shaft. A portion of the PMs' generated magnetic flux penetrates the shaft. Unlike the V-shape configuration, this structure has no iron bridges and exhibits significantly lower leakage flux. However, the construction requires more iron and magnets compared to surface-mounted designs [28, 43].

U-shape Buried-PM: The U-shape configuration (Figure 2.6d) offers another alternative where magnets are arranged in U-shaped pockets within the rotor laminations. This design provides good flux concentration while offering improved mechanical retention of the magnets at high speeds.

2.7.4 Radial Flux PM Machines

Radial flux PM machines (RFPM) have experienced surprising changes in recent years despite their initial prevalence in low-power applications. ABB initiated the use of large-size PM machines, such as the 5 MW PM machines employed in ships and wind generators. RFPM synchronous machines have found extensive use in direct-driven drives. These machines possess a simple and cost-effective structure, although they are larger compared to axial flux machines [44, 45].

Radial flux PM machines are categorized into two types based on the location of the rotor and stator.

2.7.4.1 Internal Rotor

The internal rotor configuration, illustrated in Figure 2.7a, represents the most common RFPM machine structure. It consists of rotating PM poles positioned on the inner rotor alongside fixed armature windings on the outer stator. The stator comprises laminated layers with distributed windings, while a cylindrical rotor with PMs is supported inside bearings. The air gap flux density depends on the PM residual flux and the operating point of the machine. This configuration features high air gap flux density and is widely used in industrial applications due to its straightforward assembly and maintenance characteristics [28, 31, 46].

2.7.4.2 External Rotor RFPM Machine

The external rotor configuration, shown in Figure 2.7b, positions the wound stator centrally inside the rotor with magnets fixed around the internal rotor surface. While magnetically similar to internal rotor designs, the external rotor configuration facilitates blade connection in wind turbines and fans, improving mechanical integration. The larger rotor cylinder accommodates more poles, reduces the magnetic flux path, and enables better PM cooling, significantly delaying demagnetization [28, 31, 32].

2.7.5 Axial-flux PM Machines

Axial-flux PM machines (AFPM) offer higher torque per volume ratio due to lower core volume, along with reduced vibration and noise levels compared to radial-flux designs [28, 40, 45, 47]. The surface-located air gaps enable easy adjustment and flexible structural configurations.

2.7.5.1 AFPM Machine Structures

AFPM machines are classified based on rotor-stator arrangements and stator construction. The rotor-stator configuration determines magnetic circuit topology and mechanical support structure, while stator design affects electromagnetic performance, power density, and manufacturing complexity. Figure 2.8 presents the classification based on these design aspects.

As shown in Figure 2.8a, AFPM machines are categorized into four main rotor-stator configurations: single-stator single-rotor (SSSR), single-stator double-rotor (SSDR, also known as TORUS), double-stator single-rotor (DSSR, also known as AFIR), and multi-stator multi-rotor (MSMR).

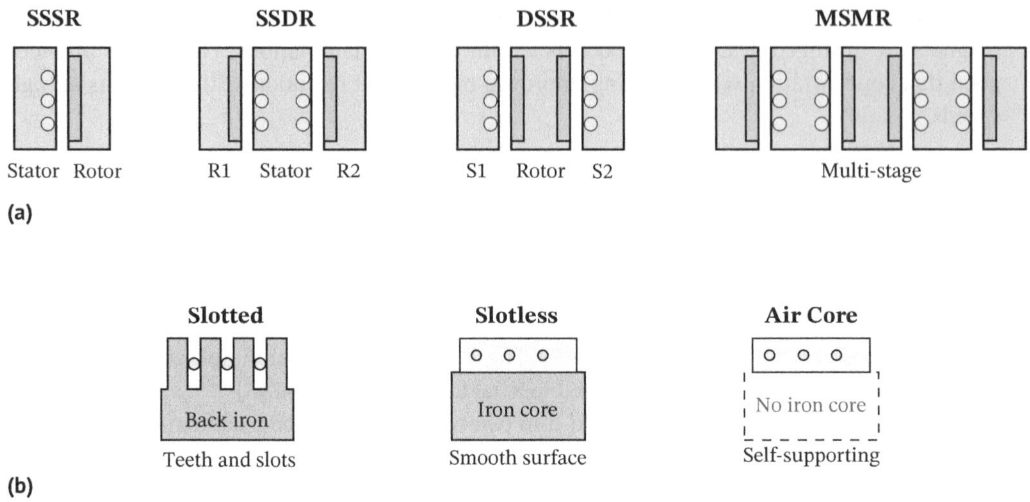

Figure 2.8 Classification of AFPM machines. (a) Rotor-stator configurations showing SSSR, SSDR (TORUS), DSSR (AFIR), and MSMR topologies with permanent magnets positioned at rotor edges facing stator windings. (b) Stator slot types showing slotted construction with teeth and slots, slotless construction with smooth iron core surface, and air-core (coreless) construction with self-supporting windings.

The SSSR configuration is the simplest but suffers from unbalanced axial forces. The SSDR and DSSR configurations provide balanced axial forces and are widely used in practical applications. The MSMR configuration allows for modular construction and higher power density through multiple stages [28, 37, 44].

The stator construction, illustrated in Figure 2.8b, can be slotted, slotless, or air-core (coreless). Slotted stators feature teeth and slots, providing high magnetic flux concentration but introducing cogging torque. Slotless stators have a smooth iron core surface, eliminating cogging torque at the cost of reduced flux density. Air-core stators are self-supporting without iron, offering zero cogging torque and core losses but requiring higher PM volume to compensate for the lack of flux concentration [40, 45].

2.7.5.2 Axial-flux Single-rotor and Single-stator

Figure 2.9 presents a single-rotor and single-stator AFPM machine, which exhibits an unbalanced force between the rotor and stator. In contrast to the balanced axial force structures, this particular structure requires bearings and has a thicker disk rotor. The magnetic force can easily disable the machine. In the slot-less stator, the applied axial force on the iron is lower than that on the copper and windings [28, 37].

2.7.5.3 Axial-flux Single-rotor and Double-stator AFPM Machine

Figure 2.10 illustrates an axial-flux machine known as the internal rotor axial flux machine, which can be either slotless or with slots. The rotor of the TORUS machine can be positioned on the outer or inner part. The primary flux in this machine is axial, either along the rotor plane or around it. Various configurations can be achieved using a single rotor and a single stator [28].

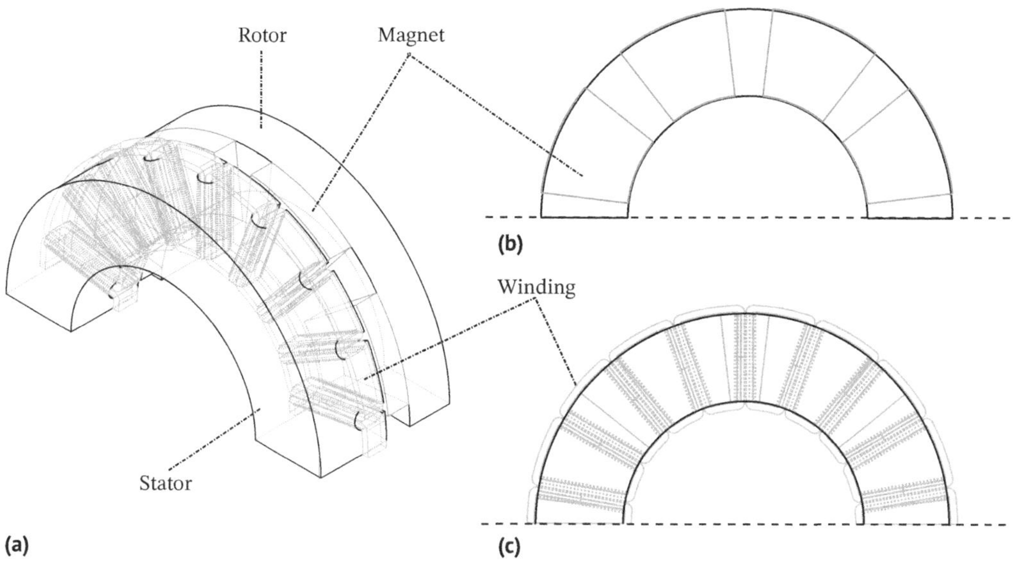

Figure 2.9 Single-stator, single-rotor axial flux permanent magnet machine. (a) Isometric view, (b) top rotor view, (c) top motor view.

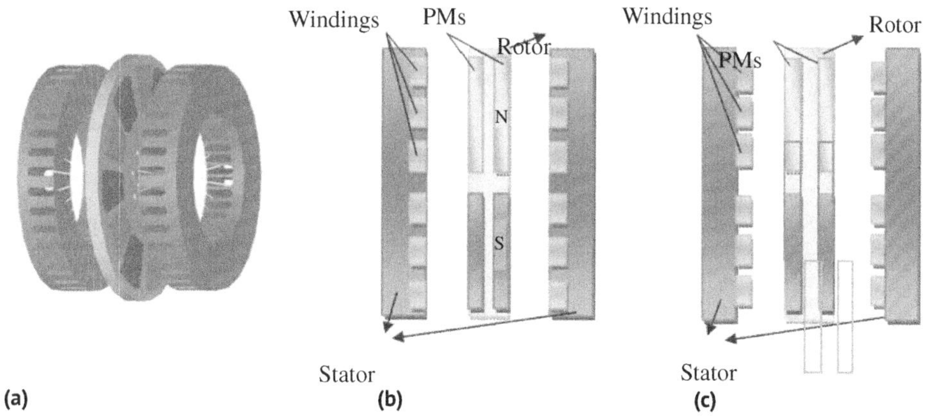

Figure 2.10 (a) A single-rotor double-stator AFPM machine, (b) a single-rotor double-stator AFPM machine with slot, and (c) a single-rotor double-stator slotless AFPM machine.

2.7.5.4 Multi-rotor Multi-stator PM Machine

A multi-rotor multi-stator AFPM machine is created by combining multiple structures assembled on a shaft. The outer diameter of the machine is determined by the desired torque. When space is limited, a multi-stage structure can be employed to attain the desired electromagnetic torque. These machine configurations are commonly used in ship propulsion and high-speed PM generators. Assembling these AFPM machines is comparatively easier than assembling axial-flux multi-rotor multi-stator machines. Typically, they possess one additional rotor plane compared to the stator. The stator windings can be connected in series or parallel, while only the outer rotor is equipped with a core that forms the primary flux path. Figure 2.11 depicts the general view of this machine [43].

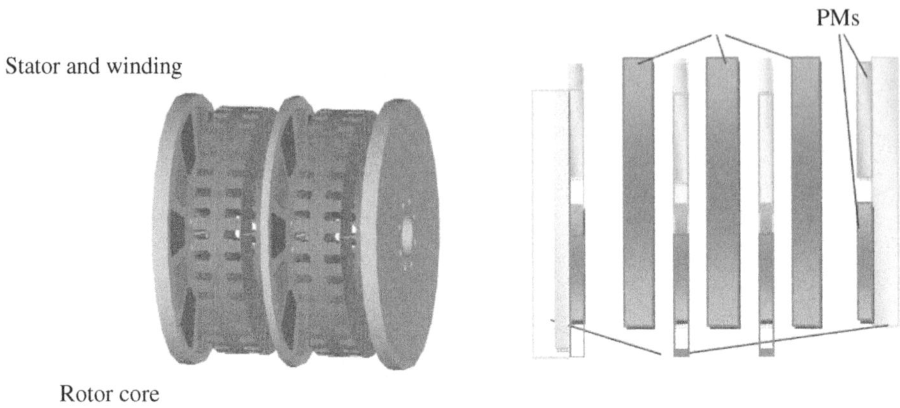

Figure 2.11 Multi-rotor multi-stator AFPM machine.

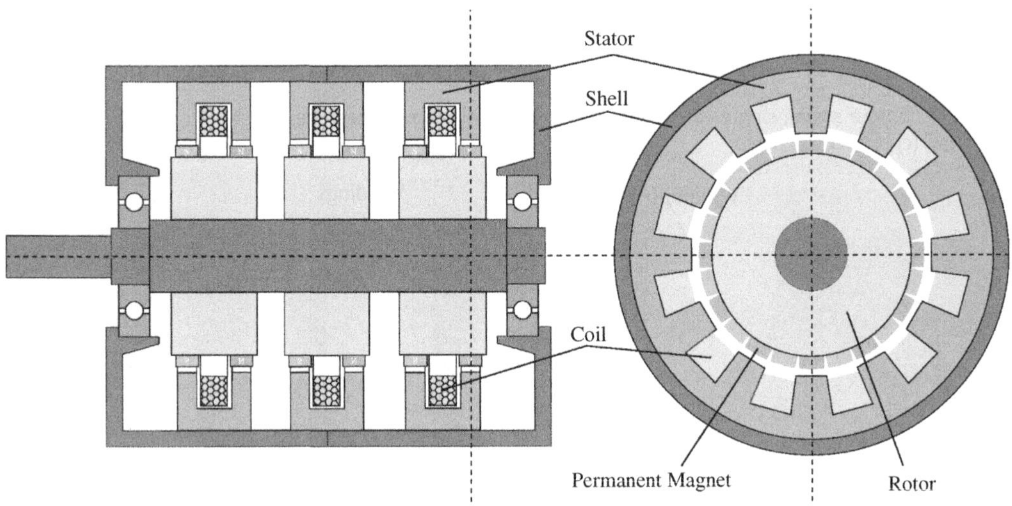

Figure 2.12 Transverse-flux PM machine [48].

2.7.6 Special Topologies of PM Machines

Beyond conventional radial and axial flux configurations, several specialized PM machine topologies have emerged for niche applications. This section focuses on topologies most relevant to industrial applications and fault diagnosis research.

2.7.6.1 Transverse-flux PM Machines

Figure 2.12 illustrates the basic structure of a PM machine with a transverse flux configuration. This type of machine exhibits a small pole pitch and is well-suited for low-speed direct-drive applications. However, it is not very commonly used.

The TFPM machine offers advantages such as a compact structure and high specific torque. Although it excels in torque development, its complex structure results in high fabrication costs. The first TFPM machine prototype was introduced and tested in Germany in 1997, followed by the design of a 20 MW power machine with a rated speed of 180 rpm by the Rose–Royse Company [28, 49] for ship propulsion.

2.7.6.2 Induction PM Machines

Induction PM machines feature two rotating parts: a rotor with a cage bar and a loop holding PMs. A distinctive feature for wind turbine applications is direct grid connection via transformer without power electronics, while the slip characteristic reduces mechanical stress on blades [50, 51].

2.7.6.3 Modular PM Machines

Modularity is desirable in large PMSMs because it facilitates manufacture, assembly, and maintenance. The modular PM machine divides the stator into identical segments that can be manufactured and wound separately before final assembly (Figure 2.13). Fractional-slot concentrated windings (FSCWs) are commonly used in modular designs due to their nonoverlapped coils, which naturally allow stator segmentation. Modular configurations offer advantages including high torque density, smooth output torque, fault tolerance capability, and simplified manufacturing processes. However, conventional modular machines with FSCWs often suffer from high nonworking space harmonic content, which leads to increased rotor losses and reduced efficiency. Advanced winding techniques such as dual three-phase two-slot pitch windings have been developed to reduce space harmonics while retaining modularity [52].

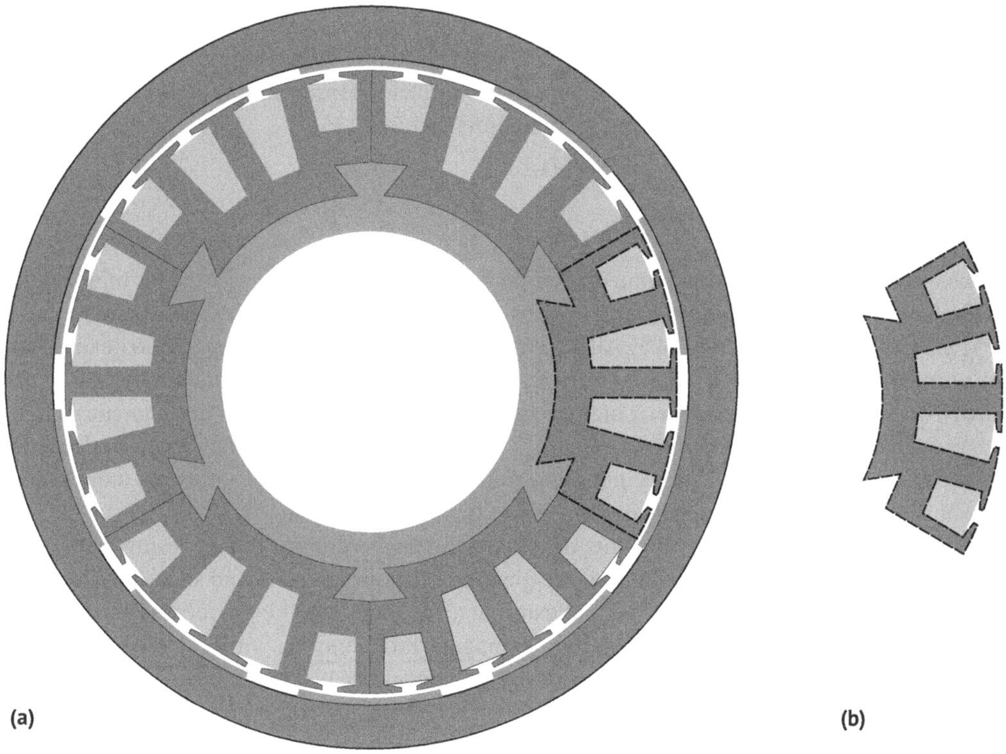

Figure 2.13 Modular PM machine structure: (a) cross-section of the complete machine, (b) single stator module. *Source:* Adapted from [52].

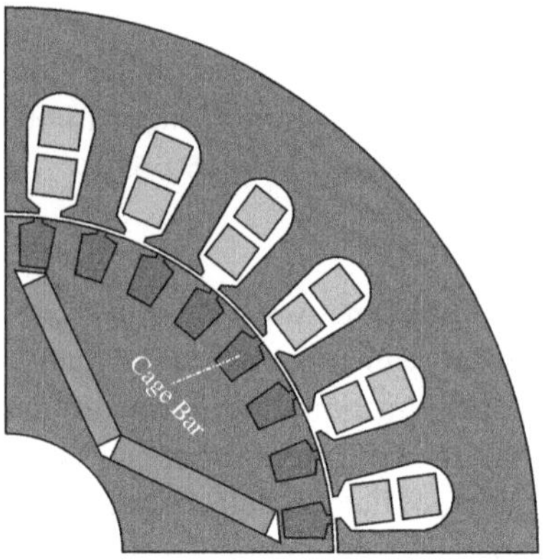

Figure 2.14 Line-start PM machine.

2.7.6.4 Line-start PM Machines

The line-start PM (LSPM) machine is a synchronous PM machine equipped with a bar-cage on the rotor. As shown in Figure 2.14, this machine is directly connected to the mains, and the bar-cage facilitates its start up. Consequently, no converter or inverter is necessary. During steady-state operation, it functions in synchronous mode. The LSPM machine serves as a suitable alternative to induction machines due to its high efficiency and power factor. However, it is more expensive and has a more complex structure compared to induction machines [35].

2.8 Summary

This chapter has provided a foundation on permanent magnet machines essential for understanding the fault diagnosis techniques presented in subsequent chapters. The historical development of PM machines was traced from early AlNiCo-based designs in the 1930s through ferrite magnets, SmCo, and NdFeB, with the maximum energy product progressing from approximately 40 kJ/m^3 for AlNiCo to 400 kJ/m^3 for modern high-grade NdFeB. This advancement in magnetic materials has enabled corresponding improvements in machine power density, efficiency, and operating speed across diverse applications including automotive traction drives, industrial automation, renewable energy generation, medical devices, and transportation systems.

The construction and key components of PM machines were discussed, covering stator core lamination materials, winding configurations (distributed, concentrated, and slotless), slot types, and rotor construction in both internal and external arrangements. The properties of permanent magnet materials were analyzed through their demagnetization curves, with residual flux density, coercive force, maximum energy product, and Curie temperature identified as the key parameters governing material selection. The loss mechanisms in PM machines were categorized into stator copper losses, iron losses, rotor losses, and windage losses. A comparison between PMSMs and induction machines demonstrated the fundamental efficiency advantages of PM machines, arising from the elimination of rotor copper losses, higher power factor, and the ability to accommodate a larger number of poles without the quadratic increase in magnetizing current that limits induction machine pole count.

The classification of PM machines was presented from multiple perspectives, encompassing AC synchronous machines, brushless DC motors, and brushed PM configurations, as well as categorization by back-EMF waveform (sinusoidal versus trapezoidal). Rotor magnet placement topologies were examined including surface-mounted, inset, and various buried configurations, each with distinct trade-offs in flux concentration, demagnetization protection, and mechanical robustness. Radial flux and axial flux machine configurations were discussed along with special topologies including transverse-flux, modular, and line-start PM machines. Understanding these diverse topologies and their constructional features is essential for fault diagnosis, as different configurations exhibit distinct fault signatures and require tailored diagnostic approaches that are the subjects of the remaining chapters.

References

1 Adrian Opritescu. Control of a saturated permanent magnet synchronous motor. Master's thesis, Aalborg University, Denmark, 2010.
2 James R. Hendershot and Timothy John Eastham Miller. *Design of Brushless Permanent-magnet Machines*. Motor Design Books, Venice, FL, USA, 2010.
3 Partha Chattopadhyay Sen. *Principles of Electric Machines and Power Electronics*. John Wiley & Sons, 2021.
4 Jacek F. Gieras. *Permanent Magnet Motor Technology: Design and Applications*. CRC Press, 2002.
5 Gordon R. Slemon. On the design of high-performance surface-mounted PM motors. *IEEE Transactions on Industry Applications*, 30(1):134–140, 1994. https://doi.org/10.1109/28.273631.
6 Gordon R. Slemon and Xian Liu. Modeling and design optimization of permanent magnet motors. *Electric Machines & Power Systems*, 20(2):71–92, 1992. https://doi.org/10.1080/07313569208909572.
7 Tomy Sebastian and Gordon R. Slemon. Transient modeling and performance of variable-speed permanent-magnet motors. *IEEE Transactions on Industry Applications*, 25(1):101–106, 1989. https://doi.org/10.1109/28.18878.
8 Tomy Sebastian and Gordon R. Slemon. Transient torque and short circuit capabilities of variable speed permanent magnet motors. *IEEE Transactions on Magnetics*, 23(5):3619–3621, 1987. https://doi.org/10.1109/TMAG.1987.1065207.
9 Tomy Sebastian, Gordon R. Slemon, and Mohammed A. Rahman. Modelling of permanent magnet synchronous motors. *IEEE Transactions on Magnetics*, 22(5):1069–1071, 1986. https://doi.org/10.1109/TMAG.1986.1064466.
10 Tanja Heikkilä. Permanent magnet synchronous motor for industrial inverter applications-analysis and design. PhD thesis, Lappeenranta University of Technology, Sweden, 2002.
11 Sumio Kozawa. Trends and problems in research of permanent magnets for motors-addressing scarcity problem of rare earth elements. Technical report, NISTEP Science & Technology Foresight Center, 2011.
12 Adrian-Cristian Pop, Johan J. Gyselinck, Diogo E. Pinto, and Ioan Vintiloiu. Optimization of low-power brushless PM-machines for automotive applications with focus on high-volume mass production. *IEEE Transactions on Industrial Electronics*, 64(12):9767–9775, 2017. https://doi.org/10.1109/TIE.2017.2698367.
13 Shuangxia Li, Bulent Sarlioglu, Srdjan Jurkovic, Nirav R. Patel, and Phil Savagian. Comparative analysis of torque compensation control algorithms of interior permanent magnet machines for automotive applications considering the effects of temperature variation. *IEEE Transactions on Transportation Electrification*, 3(3):668–681, 2017. https://doi.org/10.1109/TTE.2017.2684080.

14 Zlatko Kolondzovski, Antero Arkkio, Jorma Larjola, and Pekka Sallinen. Power limits of high-speed permanent-magnet electrical machines for compressor applications. *IEEE Transactions on Energy Conversion*, 26(1):73–82, 2010. https://doi.org/10.1109/TEC.2010.2089459.

15 Xiaodong Lu, Kandadai Lakshmi Vara Prasad Iyer, Kaushik Mukherjee, and Niranjan C. Kar. Development of a novel magnetic circuit model for design of premium efficiency three-phase line start permanent magnet machines with improved starting performance. *IEEE Transactions on Magnetics*, 49(7):3965–3968, 2013. https://doi.org/10.1109/TMAG.2013.2242869.

16 Manuel Pinilla and Santiago Martinez. Selection of main design variables for low-speed permanent magnet machines devoted to renewable energy conversion. *IEEE Transactions on Energy Conversion*, 26(3):940–945, 2011. https://doi.org/10.1109/TEC.2011.2157161.

17 Sandra Eriksson. Permanent magnet synchronous machines. *MDPI*, 12:2830, 2019.

18 Zheng-Qing Zhu and Yongxiang Li. Modularity techniques in high performance permanent magnet machines and applications. *CES Transactions on Electrical Machines and Systems*, 2(1):93–103, 2018. https://doi.org/10.23919/TEMS.2018.8326455.

19 Sanjeev Kumar, Wenping Zhao, Zhaosheng Du, Thomas A. Lipo, and Byung-Il Kwon. Design of ultrahigh speed axial-flux permanent magnet machine with sinusoidal back EMF for energy storage application. *IEEE Transactions on Magnetics*, 51(11):1–4, 2015.

20 Wen-Chih Chi and Ming-Yang Cheng. Implementation of a sliding-mode-based position sensorless drive for high-speed micro permanent-magnet synchronous motors. *ISA Transactions*, 53(2):444–453, 2014. https://doi.org/10.1016/j.isatra.2013.09.017.

21 Ahmed M. Diab, Serhiy Bozhko, Michael Galea, and Chris Gerada. Stable and robust design of active disturbance-rejection current controller for permanent magnet machines in transportation systems. *IEEE Transactions on Transportation Electrification*, 6(4):1421–1433, 2020. https://doi.org/10.1109/TTE.2020.3001042.

22 Dong-Kyun Hong, Won-Ho Hwang, Jin-Yeong Lee, and Byung-Chul Woo. Design, analysis, and experimental validation of a permanent magnet synchronous motor for articulated robot applications. *IEEE Transactions on Magnetics*, 54(3):1–4, 2017.

23 Zhenhua Qiao, Taotao Shi, Yong Wang, Yue Yan, Changliang Xia, and Xiangning He. New sliding-mode observer for position sensorless control of permanent-magnet synchronous motor. *IEEE Transactions on Industrial Electronics*, 60(2):710–719, 2012. https://doi.org/10.1109/TIE.2012.2206359.

24 Ioan Văscan, Florin P. Piglesan, Cornel Marţiş, and Ioan Bârsan. PMSM rotor topologies for automotive HVAC system. In *Electric Vehicles (EV) International Conference*, pages 1–4, 2019. https://doi.org/10.1109/EV.2019.8892852.

25 Franziska Meier. Permanent-magnet synchronous machines with non-overlapping concentrated windings for low-speed direct-drive applications. PhD thesis, KTH, Sweden, 2008.

26 Dong-Hyun Joo, Jong-Ho Cho, Kwangsoo Woo, Byung-Taek Kim, and Dong-Kyun Kim. Electromagnetic field and thermal linked analysis of interior permanent-magnet synchronous motor for agricultural electric vehicle. *IEEE Transactions on Magnetics*, 47(10):4242–4245, 2011. https://doi.org/10.1109/TMAG.2011.2149504.

27 Tien D. Nguyen, King-Jet Tseng, Shu Zhang, and Hieu T. Nguyen. A novel axial flux permanent-magnet machine for flywheel energy storage system: Design and analysis. *IEEE Transactions on Industrial Electronics*, 58(9):3784–3794, 2010. https://doi.org/10.1109/TIE.2010.2089939.

28 Nada Al Fartusi. Permanent magnet synchronous generator (PMSG) to use on board of yachts. Master's thesis, Electrical Power Processing Group (EPP), Mastervolt, 2010

29 Tim D. Strous. Design of a permanent magnet radial flux concentrated coil generator for a range extender application. Master's thesis, Delft University of Technology, Netherland, 2010.

30 Jean Cros and Paul Viarouge. Synthesis of high performance pm motors with concentrated windings. *IEEE Transactions on Energy Conversion*, 17(2):248–253, 2002. https://doi.org/10.1109/TEC.2002.1009476.

31 Maxime Dubois, Henk Polinder, and Jan A. Ferreira. Comparison of generator topologies for direct-drive wind turbines. In *Proceedings of the Nordic Countries Power and Industrial Electronics Conference (NORPIE)*, pages 22–26, 2000.

32 Aldo Cavagnino, Marco Lazzari, Francesco Profumo, and Andrea Tenconi. A comparison between the axial flux and the radial flux structures for PM synchronous motors. *IEEE Transactions on Industry Applications*, 38(6):1517–1524, 2002. https://doi.org/10.1109/TIA.2002.805572.

33 E. Richter. The proper magnet characteristics for industrial PM machines. In *Proceedings of the Motor Convention*, pages 64–581, 1982.

34 Günter Henneberger, Hans Härer, and Wolfgang Schleuter. A new generation of servomotors with rare earth magnets. In *Proceedings of ICEM 86*, page 1083, 1986.

35 Jiabin Wang, Zhen P. Xia, and David Howe. Three-phase modular permanent magnet brushless machine for torque boosting on a downsized ICE vehicle. *IEEE Transactions on Vehicular Technology*, 54(3):809–816, 2005. https://doi.org/10.1109/TVT.2005.847224.

36 Gordon R. Slemon and Xian Liu. Core losses in permanent magnet motors. *IEEE Transactions on Magnetics*, 26(5):1653–1655, 1990. https://doi.org/10.1109/20.104480.

37 Fang Deng. An improved iron loss estimation for permanent magnet brushless machines. *IEEE Transactions on Energy Conversion*, 14(4):1391–1395, 1999. https://doi.org/10.1109/60.815078.

38 Chunting Mi, Gordon R. Slemon, and Richard Bonert. Modeling of iron losses of permanent-magnet synchronous motors. *IEEE Transactions on Industry Applications*, 39(3):734–742, 2003. https://doi.org/10.1109/TIA.2003.810635.

39 Karl J. Strnat. Modern permanent magnets for applications in electro-technology. *Proceedings of the IEEE*, 78(6):923–946, 1990. https://doi.org/10.1109/5.56908.

40 Asko Parviainen. Design of axial-flux permanent-magnet low-speed machines and performance comparison between radial-flux and axial-flux machines. PhD thesis, Lappeenranta University of Technology, Sweden, 2005.

41 Weihua Cai, David Fulton, and Klaus Reichert. Design of permanent magnet motors with low torque ripples: A review. In *International Conference on Electrical Machines (ICEM)*, pages 1384–1388, Espoo, 2000.

42 Juha Kolehmainen and Jari Ikäheimo. Motors with buried magnets for medium-speed applications. *IEEE Transactions on Energy Conversion*, 23(1):86–91, 2008. https://doi.org/10.1109/TEC.2007.914331.

43 Juha Kolehmainen. Dovetail rotor poles in synchronous permanent magnet and reluctance machines. PhD thesis, Aalto University, Finland, 2011.

44 Kiran Sitapati and Ramu Krishnan. Performance comparisons of radial and axial field, permanent-magnet, brushless machines. *IEEE Transactions on Industry Applications*, 37(5):1219–1226, 2001. https://doi.org/10.1109/28.952495.

45 Anyuan Chen, Robert Nilssen, and Arne Nysveen. Performance comparisons among radial-flux, multistage axial-flux, and three-phase transverse-flux PM machines for downhole applications. *IEEE Transactions on Industry Applications*, 46(2):779–789, 2010. https://doi.org/10.1109/TIA.2009.2039914.

46 Jianhua Chen, Chee V. Nayar, and Lin Xu. Design and finite-element analysis of an outer-rotor permanent-magnet generator for directly coupled wind turbines. *IEEE Transactions on Magnetics*, 36(5):3802–3809, 2000. https://doi.org/10.1109/20.908378.

47 Edgar L. C. Arroyo. Modeling and simulation of permanent magnet synchronous motor drive system. PhD thesis, University of Puerto Rico, Mayaguez, Puerto Rico, 2007.

48 James Green. Development of a high performance generator for wind turbines, stage II: The prototype. Technical report, National Wind Technology Center, National Renewable Energy Laboratory, 2010.

49 Guenter Kastinger. Design of a novel transverse flux machine. In *Proceedings of ICEM*, 2002.

50 Eric Spooner and Alan Williamson. Modular, permanent-magnet wind-turbine generators. In *Conference record of the IEEE Industry Applications Conference 21st IAS Annual Meeting*, pages 497–502, 1996. https://doi.org/10.1109/IAS.1996.557073.

51 Eric Spooner, Alan C. Williamson, and Graeme Catto. Large modular PM generators. *IEE Proceedings - Electric Power Applications*, 143(5):388–395, 1996.

52 Kaicheng Wang and Hua Lin. Modular permanent magnet synchronous machine with low space harmonic content. *Energies*, 13(15):3924, 2020. https://doi.org/10.3390/en13153924.

3

Modeling and Simulation Techniques for Permanent Magnet Motors

3.1 Introduction

How can we predict the behavior of a permanent magnet (PM) motor under various fault conditions without physically subjecting it to potentially destructive failure modes? This fundamental question motivates the development and application of modeling and simulation techniques in fault diagnosis systems. Mathematical models serve as the cornerstone of model-based fault diagnosis, enabling engineers to understand, predict, and detect abnormal behavior in PM motors before catastrophic failures occur.

Model-based fault diagnosis relies on creating mathematical representations of the motor that can predict its expected behavior under both healthy and faulty conditions. The core principle is straightforward: by comparing the actual measured outputs of a real motor with the predictions from a mathematical model, discrepancies (called residuals) can be identified that indicate the presence, type, and severity of faults. For instance, if a model predicts that the back-EMF should have a certain waveform under healthy operation, but measurements show significant harmonics or asymmetries, this residual signal can indicate faults such as demagnetization, eccentricity, or winding shorts. Similarly, parameter estimation techniques can track changes in motor parameters (such as winding resistance, inductance, or flux linkage) over time, with deviations from nominal values serving as fault indicators.

The effectiveness of model-based diagnosis depends critically on the accuracy and computational efficiency of the underlying model. This relationship presents engineers with a fundamental trade-off: high-fidelity models that capture detailed electromagnetic phenomena can predict fault signatures with excellent accuracy, but often at the cost of prohibitive computational demands that make them unsuitable for real-time monitoring. Conversely, simplified models may execute quickly enough for online diagnosis but might miss critical fault-related phenomena or produce inaccurate predictions. Understanding this trade-off and selecting the appropriate modeling approach for a given application is essential for successful fault diagnosis system design.

Different modeling techniques serve different purposes across the fault diagnosis workflow. High-fidelity methods such as finite element method (FEM) provide the most accurate predictions by solving Maxwell's equations numerically across detailed geometric representations of the motor. These models are invaluable for offline analysis, research, and understanding fault mechanisms in depth, but their computational cost (often hours for a single operating point) makes them impractical for real-time monitoring. At the other end of the spectrum, lumped parameter models use simplified equivalent circuits that can execute in milliseconds, making them ideal for real-time applications such as model-based observers and online fault detection algorithms, though with reduced accuracy and limited ability to capture spatial phenomena.

Between these extremes lie several semi-analytical and hybrid approaches that attempt to balance accuracy with computational efficiency. The field reconstruction method (FRM) applies superposition principles to achieve 30–150 times faster computation than FEM while maintaining reasonable accuracy for many applications. Magnetic equivalent circuit (MEC) models represent the magnetic structure as reluctance networks, offering good compromise between speed and fidelity while naturally incorporating saturation effects. Winding function theory provides closed-form analytical expressions that are computationally very light while explicitly incorporating machine geometry. More recently, machine learning-based approaches have emerged as powerful tools for capturing complex nonlinear relationships from data, offering a complementary approach that can learn fault signatures directly from measurements.

This chapter presents six major modeling techniques for PM motors, with specific emphasis on their application to fault diagnosis. Each approach is examined through a consistent framework covering theoretical foundations, mathematical formulations, fault modeling applications, advantages and limitations, and the method's role in fault diagnosis workflows. The chapter concludes with essential guidance on simulation validation and verification, which applies across all modeling approaches. The techniques covered include:

1. **Lumped Parameter Models:** Simplified dq-axis equivalent circuit representations ideal for real-time control and observers.
2. **Finite Element Method (FEM):** High-fidelity numerical field simulation serving as reference standard.
3. **Field Reconstruction Method (FRM):** Superposition-based analytical approach balancing speed and accuracy.
4. **Winding Function Models:** Classical machine theory using winding distribution and permeance functions.
5. **Magnetic Equivalent Circuit (MEC):** Reluctance network models handling saturation and geometric details.
6. **Machine Learning-Based Modeling:** Data-driven approaches including neural networks and digital twins.

Following the presentation of these modeling techniques, a dedicated section on **simulation validation and verification** provides systematic approaches to ensure model trustworthiness across all methods.

For each modeling technique, we explore how it can be adapted to represent common fault conditions in PM motors, specifically eccentricity, demagnetization, and inter-turn short circuits. While detailed discussions of fault mechanisms, failure modes, and diagnostic methods appear in Chapters 5–9, this chapter focuses on how each simulation approach models these faults. Inter-turn short circuits can be modeled by modifying winding parameters or splitting winding functions. Demagnetization faults appear as reduced flux linkage in lumped models or altered magnet properties in field-based models. Eccentricity faults require models that can capture air-gap variation, making them better suited to methods that incorporate geometry explicitly. Understanding which fault types can be effectively modeled by each approach guides the selection of appropriate techniques for specific diagnostic applications.

The chapter concludes with comparative guidance on selecting modeling approaches for different scenarios. A comprehensive comparison table summarizes the accuracy, computational cost, real-time capability, and fault applicability of each method. This roadmap helps engineers match modeling techniques to their specific requirements: whether developing real-time embedded diagnostics requiring lightweight lumped models, conducting research investigations leveraging FEM's fidelity, or building hybrid systems that combine multiple modeling approaches for robust fault detection.

By mastering these modeling and simulation techniques, engineers gain powerful tools for understanding PM motor behavior under faults, designing effective diagnostic algorithms, and validating fault detection strategies before deployment in critical applications.

3.2 Lumped Parameter Models

3.2.1 Introduction

Lumped parameter models represent one of the most widely used approaches for analyzing and controlling permanent magnet synchronous machines (PMSMs). Unlike distributed parameter models that account for field variations across space, lumped parameter models simplify the machine's electromagnetic behavior using global, spatially-averaged parameters such as resistance, inductance, and flux linkage. This simplification dramatically reduces computational complexity while maintaining sufficient accuracy for many practical applications, particularly in real-time control and model-based fault diagnosis systems.

The fundamental principle behind lumped parameter modeling is the concept of **lumping**: complex distributed electromagnetic phenomena are approximated by a finite number of equivalent circuit elements. For example, rather than solving Maxwell's equations to determine the magnetic field distribution throughout the machine's air gap, a lumped parameter model represents the entire stator winding of one phase using a single resistance and inductance. Similarly, the permanent magnet flux is represented by an equivalent flux linkage or back-EMF source rather than by detailed field calculations. This approach draws on classical electric circuit theory, making these models intuitive and familiar to engineers while enabling rapid simulation and analysis.

For fault diagnosis applications, lumped parameter models serve two critical but distinct purposes, each requiring different modeling approaches. In **real-time monitoring and control**, lumped models must execute in milliseconds to enable model-based observers, parameter estimators, and fault detection algorithms that run continuously in embedded systems. In **offline fault analysis and signature prediction**, models can be more detailed to capture fault-related phenomena such as harmonic content or parameter asymmetries, even if they cannot run in real-time. This distinction motivates the development of two main categories of lumped parameter models for PMSMs.

The **dq-axis models** transform the three-phase machine equations into a synchronous reference frame using the Park transformation [1]. In this frame, the machine is characterized by direct-axis (d-axis) and quadrature-axis (q-axis) inductances (L_d, L_q), stator resistance (R_s), and permanent magnet flux linkage (ψ_m). Under balanced sinusoidal excitation at synchronous speed, these quantities appear as DC values, greatly simplifying control design and parameter estimation. The dq-axis approach forms the foundation of field-oriented control (FOC) and is universally used in high-performance motor drives. However, this transformation inherently assumes balanced three-phase operation with purely sinusoidal spatial field distributions, making it fundamentally limited when modeling asymmetric faults or machines with significant spatial harmonics [2, 3].

The **abc reference frame models** with voltage-behind-reactance (VBR) structure take a different approach by working directly with natural phase variables (i_a, i_b, i_c) without reference frame transformation. These models represent each phase winding explicitly and characterize the machine using position-dependent inductance matrices $\mathbf{L}(\theta)$ that capture spatial variation in the magnetic coupling between phases. The voltage-behind-reactance formulation treats the permanent magnet contribution as back-EMF voltage sources with potentially nonsinusoidal harmonic content [4, 5]. This natural phase variable representation excels at capturing spatial harmonics, slotting effects, and asymmetric phenomena such as eccentricity-induced unbalanced magnetic pull [6], making it particularly valuable for fault diagnosis where these effects provide diagnostic signatures.

The trade-offs between these two approaches reflect a fundamental tension in fault diagnosis modeling. The dq-axis models offer maximum computational efficiency and seamless integration with existing FOC systems, but their transformation to a rotating frame filters out many fault signatures that appear as harmonics or asymmetries in the stationary frame. The abc models preserve these fault signatures and can represent more complex phenomena, but require solving larger systems of equations with time-varying coefficients. The choice between them depends on the specific fault types of interest, the required computational budget, and whether the model will be used for real-time or offline analysis.

Understanding these two categories, their mathematical foundations, and their respective strengths and limitations is essential for developing effective model-based fault diagnosis systems for permanent magnet machines. The following sections examine each approach in detail, followed by comparative guidance on when to use each modeling strategy.

3.2.2 dq-axis Models

The dq-axis lumped parameter model, also known as the Park transformation model or the synchronous reference frame model, represents the cornerstone of modern permanent magnet synchronous machine analysis and control. This approach transforms the three-phase machine variables into a two-axis rotating coordinate system that rotates synchronously with the rotor, resulting in constant or slowly-varying parameters under steady-state operation [1]. This elegant mathematical transformation converts what would otherwise be time-varying differential equations with periodic coefficients into a much simpler set of equations with constant coefficients, enabling straightforward analysis and controller design.

3.2.2.1 Mathematical Foundation

The transformation from the natural abc phase variables to the dq0 reference frame proceeds in two conceptual steps. First, the Clarke transformation converts the three-phase stationary frame variables into a two-phase stationary orthogonal frame ($\alpha\beta$ frame):

$$\begin{bmatrix} i_\alpha \\ i_\beta \end{bmatrix} = \frac{2}{3} \begin{bmatrix} 1 & -\frac{1}{2} & -\frac{1}{2} \\ 0 & \frac{\sqrt{3}}{2} & -\frac{\sqrt{3}}{2} \end{bmatrix} \begin{bmatrix} i_a \\ i_b \\ i_c \end{bmatrix} \tag{3.1}$$

Second, the Park transformation rotates this stationary frame into the synchronous frame aligned with the rotor position θ_e (electrical angle):

$$\begin{bmatrix} i_d \\ i_q \end{bmatrix} = \begin{bmatrix} \cos\theta_e & \sin\theta_e \\ -\sin\theta_e & \cos\theta_e \end{bmatrix} \begin{bmatrix} i_\alpha \\ i_\beta \end{bmatrix} \tag{3.2}$$

These two transformations are often combined into a single composite Park transformation that directly maps abc to dq0. Under balanced sinusoidal excitation at synchronous speed, this transformation produces DC quantities in the dq frame, eliminating the time-varying nature of the equations.

The voltage equations of a surface-mounted PMSM in the dq reference frame take the form:

$$v_d = R_s i_d + L_d \frac{di_d}{dt} - \omega_e L_q i_q \tag{3.3}$$

$$v_q = R_s i_q + L_q \frac{di_q}{dt} + \omega_e L_d i_d + \omega_e \psi_m \tag{3.4}$$

where R_s is the stator resistance, L_d and L_q are the direct and quadrature axis inductances, ω_e is the electrical angular velocity, and ψ_m is the permanent magnet flux linkage. For surface-mounted PMSMs with nonsalient pole structure, $L_d \approx L_q$, while interior PMSMs exhibit significant saliency with $L_d < L_q$ due to the different magnetic reluctances along the two axes. These voltage equations can be represented as equivalent circuits, as shown in Figure 3.1. The d-axis circuit includes the stator resistance, d-axis inductance, and a cross-coupling voltage source representing the speed-dependent term from the q-axis current. The q-axis circuit similarly includes the stator resistance, q-axis inductance, a cross-coupling term from the d-axis current, and the permanent magnet back-EMF voltage source.

The electromagnetic torque is expressed compactly as:

$$T_e = \frac{3}{2} p[\psi_m i_q + (L_d - L_q) i_d i_q] \tag{3.5}$$

where p is the number of pole pairs. The first term represents the permanent magnet torque component (dominant in surface-mounted machines), while the second term captures reluctance torque (significant in interior PMSMs).

3.2.2.2 Advantages

The dq-axis model provides several crucial advantages that have made it the industry standard for PM machine control:

Computational Efficiency: With only four state variables (i_d, i_q, ω_m, and θ_e), the model can be evaluated in microseconds on modern microcontrollers, enabling high-bandwidth control loops (10–20 kHz) and real-time model-based observers.

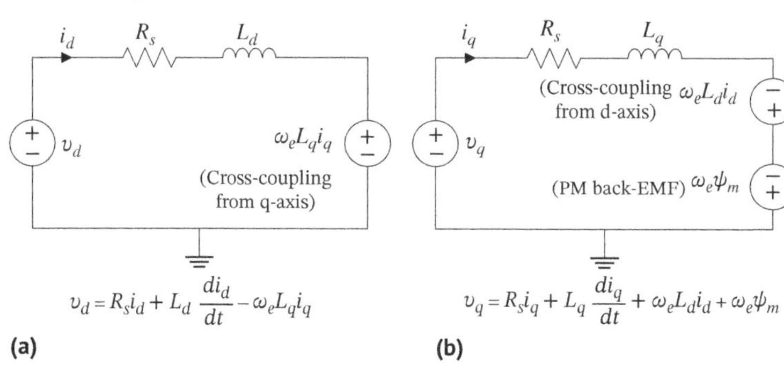

Figure 3.1 Equivalent circuit representation of the dq-axis PMSM model. (a) The d-axis circuit shows the stator resistance R_s, d-axis inductance L_d, and the cross-coupling voltage $\omega_e L_q i_q$ induced by q-axis current. (b) The q-axis circuit includes the stator resistance, q-axis inductance L_q, the cross-coupling voltage $\omega_e L_d i_d$ from d-axis current, and the permanent magnet back-EMF $\omega_e \psi_m$.

Intuitive Physical Interpretation: The d-axis current i_d controls the flux (though typically set to zero for maximum efficiency in surface-mounted PMSMs), while the q-axis current i_q directly controls torque. This decoupling simplifies controller design significantly compared to abc frame control.

DC Quantities at Steady State: Under synchronous operation, dq currents and voltages become DC values, allowing the use of simple PI controllers without the need to track sinusoidal references. This dramatically simplifies control implementation.

Seamless Integration with Field-oriented Control: The dq model naturally aligns with FOC architectures that have become ubiquitous in industrial motor drives, making model-based diagnostic techniques easy to integrate into existing systems [7].

3.2.2.3 Application to Fault Modeling

For fault diagnosis applications, the dq-axis model enables tracking of parameter variations that indicate specific fault conditions. Different faults manifest as changes in different model parameters:

Demagnetization Faults: Partial or uniform demagnetization of the permanent magnets reduces the flux linkage ψ_m. This reduction can be detected by observing a decrease in the back-EMF constant or by using parameter estimation techniques that track ψ_m over time [8]. The relationship is direct: a 10% loss of magnet strength corresponds approximately to a 10% reduction in ψ_m.

Winding Resistance Increase: Inter-turn short circuits or winding degradation typically increase the stator resistance R_s. While difficult to detect at high speeds (where inductive voltage drop dominates), resistance monitoring at low speeds or during startup can indicate winding problems.

Inductance Changes: Some faults alter the machine inductances. For example, stator inter-turn shorts can affect L_d and L_q, while severe demagnetization can also impact inductance due to changes in magnetic saturation levels.

Model-based observers in the dq frame, such as extended Kalman filters or sliding-mode observers, can continuously estimate these parameters during operation. Deviations from nominal values then serve as fault indicators. The key advantage is that such observers execute in real-time on standard motor drive hardware without requiring additional sensors beyond the current and position measurements already present for control.

3.2.2.4 Limitations

Despite its advantages, the dq-axis model suffers from fundamental limitations when applied to fault diagnosis, particularly for faults that introduce spatial asymmetries or harmonic content [2, 3]:

Assumption of Balanced Operation: The Park transformation assumes three-phase symmetry. When faults break this symmetry (e.g., asymmetric demagnetization, single-phase winding shorts, or static eccentricity), the transformation itself becomes problematic, and the dq currents no longer represent simple DC quantities but contain ripple components.

Inability to Capture Spatial Harmonics: The fundamental dq model assumes sinusoidal spatial distribution of flux and windings. Machines with concentrated windings, significant slotting effects, or harmonic magnet magnetization patterns exhibit spatial harmonics that the basic dq model cannot represent. These harmonics are averaged out in the transformation, losing potentially valuable diagnostic information.

Filtering of Fault Signatures: Many faults produce characteristic harmonic signatures in the abc phase currents (e.g., eccentricity produces specific sideband frequencies). The transformation to the rotating dq frame acts as a frequency-shifting filter that can obscure these signatures. For example, a twice-per-revolution eccentricity component appears as a specific harmonic pattern in abc currents but becomes mixed with other components after transformation to dq.

Loss of Phase-specific Information: Because the dq transformation combines all three phases, it cannot distinguish between faults that affect individual phases differently. A short circuit in phase A versus phase B produces different abc current patterns but may appear similar in the dq frame.

These limitations have motivated the development of extended dq models that attempt to incorporate harmonic effects through multiple reference frames rotating at different speeds, or by augmenting the basic dq equations with additional terms [3]. However, such extensions increase complexity and partially sacrifice the computational advantages that made the dq model attractive in the first place. For faults that fundamentally break the balanced sinusoidal assumptions, abc reference frame models often provide a more natural and accurate representation.

3.2.2.5 Role in Fault Diagnosis Workflow

The dq-axis lumped parameter model is most appropriate for fault diagnosis in the following scenarios:

- Real-time embedded diagnostics requiring execution rates of 5–20 kHz on resource-constrained microcontrollers.
- Faults that primarily manifest as parameter changes rather than harmonic signatures (demagnetization, uniform winding degradation).
- Integration with existing field-oriented control systems where dq measurements are already available.
- Applications where computational simplicity and ease of implementation outweigh the need to capture detailed fault signatures.
- Machines with distributed windings and low harmonic content operating under balanced conditions.

For applications requiring detection of spatial harmonics, asymmetric faults, or phase-specific fault signatures, abc reference frame models should be considered as a complement or alternative to dq-axis approaches.

3.2.3 The abc Reference Frame Models with Voltage Behind Reactance

While dq-axis models excel at computational efficiency and control integration, many important fault signatures manifest as spatial harmonics, asymmetries, or position-dependent phenomena that the dq transformation inherently obscures or eliminates. The abc reference frame models with voltage-behind-reactance (VBR) structure address these limitations by working directly with natural phase variables without coordinate transformation, enabling explicit representation of spatial effects and asymmetric operating conditions [4, 5]. This approach has become increasingly important for fault diagnosis applications where capturing harmonic content and asymmetric signatures is essential for reliable detection.

3.2.3.1 Mathematical Foundation

The fundamental voltage equations for a three-phase PMSM in the natural abc reference frame can be written as:

$$\mathbf{v}_{\text{abc}} = R_s \mathbf{i}_{\text{abc}} + \frac{d\boldsymbol{\lambda}_{\text{abc}}}{dt} \tag{3.6}$$

where $\mathbf{v}_{\text{abc}} = [v_a, v_b, v_c]^T$ and $\mathbf{i}_{\text{abc}} = [i_a, i_b, i_c]^T$ are the phase voltage and current vectors, and $\boldsymbol{\lambda}_{\text{abc}}$ is the flux linkage vector. The key difference from dq models emerges in how the flux linkage is expressed.

The total flux linkage combines contributions from the stator currents (through inductances) and the permanent magnets (through position-dependent back-EMF):

$$\boldsymbol{\lambda}_{\text{abc}}(\theta, \mathbf{i}_{\text{abc}}) = \mathbf{L}_{\text{abc}}(\theta)\mathbf{i}_{\text{abc}} + \boldsymbol{\lambda}_{\text{pm}}(\theta) \tag{3.7}$$

The **inductance matrix** $\mathbf{L}_{\text{abc}}(\theta)$ is position-dependent, capturing the spatial variation of magnetic coupling as the rotor rotates:

$$\mathbf{L}_{\text{abc}}(\theta) = \begin{bmatrix} L_{\text{aa}}(\theta) & L_{\text{ab}}(\theta) & L_{\text{ac}}(\theta) \\ L_{\text{ba}}(\theta) & L_{\text{bb}}(\theta) & L_{\text{bc}}(\theta) \\ L_{\text{ca}}(\theta) & L_{\text{cb}}(\theta) & L_{\text{cc}}(\theta) \end{bmatrix} \tag{3.8}$$

For a healthy three-phase machine with symmetric construction, the self-inductances follow periodic patterns:

$$L_{\text{aa}}(\theta) = L_0 + L_2 \cos(2\theta) + L_4 \cos(4\theta) + \cdots \tag{3.9}$$

where L_0 represents the average inductance, and L_2, L_4, etc., capture spatial harmonics due to saliency and slotting effects. The mutual inductances follow similar periodic patterns with 120° phase shifts. Importantly, this formulation naturally includes spatial harmonic content that would be lost in a dq transformation.

The **voltage-behind-reactance** formulation separates the permanent magnet contribution as equivalent voltage sources:

$$\mathbf{v}_{\text{abc}} = R_s \mathbf{i}_{\text{abc}} + \mathbf{L}_{\text{abc}}(\theta)\frac{d\mathbf{i}_{\text{abc}}}{dt} + \frac{d\mathbf{L}_{\text{abc}}(\theta)}{dt}\mathbf{i}_{\text{abc}} + \mathbf{e}_{\text{abc}}(\theta, \omega) \tag{3.10}$$

where $\mathbf{e}_{\text{abc}}(\theta, \omega)$ represents the back-EMF vector. For a healthy machine, this takes the ideal sinusoidal form:

$$\mathbf{e}_{\text{abc}} = \omega \begin{bmatrix} -\psi_m \sin(\theta) \\ -\psi_m \sin(\theta - 2\pi/3) \\ -\psi_m \sin(\theta + 2\pi/3) \end{bmatrix} \tag{3.11}$$

However, unlike the dq model that assumes purely sinusoidal back-EMF, the abc VBR model can incorporate harmonic content:

$$e_a(\theta) = \omega \sum_n k_n \psi_m \sin(n\theta + \phi_n) \tag{3.12}$$

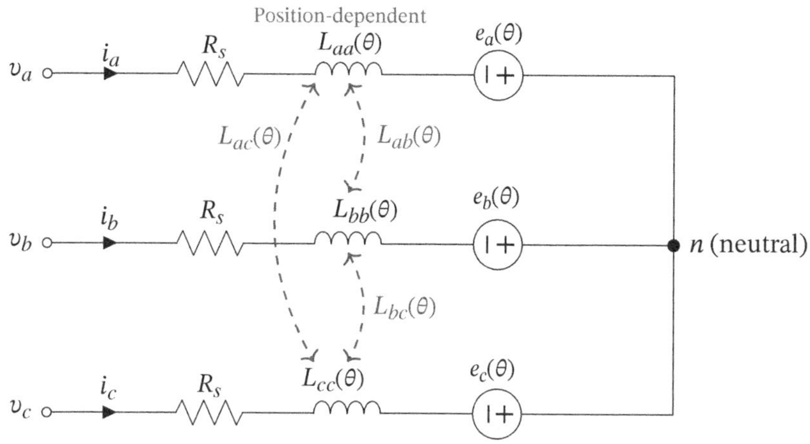

Figure 3.2 Three-phase abc voltage-behind-reactance (VBR) equivalent circuit for PMSM modeling. Each phase includes the stator resistance R_s, position-dependent self-inductance $L_{xx}(\theta)$, and back-EMF source $e_x(\theta)$. The mutual inductances $L_{ab}(\theta)$, $L_{bc}(\theta)$, and $L_{ac}(\theta)$ (shown as dashed coupling lines) capture the magnetic interaction between phases. All inductances and back-EMF sources vary with rotor position θ, enabling representation of spatial harmonics and asymmetric fault conditions.

This capability to represent back-EMF harmonics is crucial for machines with nonsinusoidal magnet magnetization or for modeling harmonic effects introduced by faults. The complete abc VBR equivalent circuit is shown in Figure 3.2, illustrating the Y-connected three-phase structure with position-dependent self-inductances, mutual coupling between phases, and back-EMF voltage sources that can include harmonic content.

3.2.3.2 Advantages

The abc reference frame VBR model provides several critical capabilities for fault diagnosis that are difficult or impossible to achieve with dq models:

Explicit Spatial Harmonic Representation: The position-dependent inductance matrix naturally captures spatial harmonics arising from slotting, saliency, and most importantly, from faults such as eccentricity. These harmonics appear explicitly in the model rather than being averaged out by coordinate transformation [6].

Asymmetric Fault Modeling: Faults that affect phases differently (single-phase winding shorts, partial demagnetization of specific magnet poles, or static eccentricity) can be represented by modifying specific elements of the inductance matrix or specific components of the back-EMF vector. The model preserves phase-specific information that reveals which fault occurred and where.

Direct Interface with Circuit Analysis: The VBR formulation naturally interfaces with external circuit models, making it straightforward to analyze the machine's behavior when connected to power converters, unbalanced loads, or fault conditions in the power electronics [5].

Preservation of Diagnostic Signatures: Fault signatures that appear as characteristic harmonics in measured phase currents remain visible in the model outputs. For example, bearing faults causing rotor eccentricity produce specific sideband frequencies around the fundamental that abc models can predict but dq models typically cannot.

3.2.3.3 Application to Eccentricity and Asymmetric Faults

Eccentricity faults (static, dynamic, or mixed) represent a prime application where abc VBR models significantly outperform dq approaches. When the rotor center does not align with the stator center, the air-gap length varies around the circumference, causing position-dependent changes in the magnetic permeance and hence the inductance matrix.

For **static eccentricity**, where the rotor and stator centers are displaced but the rotor rotates about its own center, the minimum air gap remains at a fixed stator position. This creates inductance variations that repeat once per mechanical revolution:

$$L_{\mathrm{aa}}(\theta) = L_0 + L_2 \cos(2\theta) + L_{\mathrm{ecc}} \cos(\theta - \theta_{\mathrm{ecc}}) \tag{3.13}$$

where L_{ecc} is the inductance perturbation amplitude and θ_{ecc} is the angular position of minimum air gap.

For **dynamic eccentricity**, where the rotor rotates about the stator center, but the rotor's geometric center is displaced from its rotation axis, the air-gap variation rotates with the rotor, producing additional harmonic components in the current spectrum.

A comprehensive abc VBR model for eccentricity can be constructed by parameterizing the inductance matrix as a function of both rotor position and eccentricity magnitude/direction [6]. Such models enable prediction of the characteristic harmonic patterns in phase currents that serve as diagnostic signatures: for static eccentricity, sidebands at $f_s \pm f_r$ (where f_s is supply frequency and f_r is rotational frequency), and for dynamic eccentricity, different sideband patterns that allow discrimination between fault types.

3.2.3.4 Limitations

The primary disadvantage of abc VBR models compared to dq models is increased computational complexity:

Time-varying Coefficients: The inductance matrix varies with rotor position, requiring evaluation of trigonometric functions and matrix operations at each simulation time step. While modern processors handle this efficiently offline, it is more demanding than the constant coefficients of dq models.

Matrix Inversion: Solving for current derivatives requires computing:

$$\frac{d\mathbf{i}_{\mathrm{abc}}}{dt} = \mathbf{L}_{\mathrm{abc}}^{-1}(\theta)\left[\mathbf{v}_{\mathrm{abc}} - R_s\mathbf{i}_{\mathrm{abc}} - \frac{d\mathbf{L}_{\mathrm{abc}}(\theta)}{dt}\mathbf{i}_{\mathrm{abc}} - \mathbf{e}_{\mathrm{abc}}(\theta, \omega)\right] \tag{3.14}$$

The 3×3 matrix inversion $\mathbf{L}_{\mathrm{abc}}^{-1}(\theta)$ must be performed at each time step. While this is computationally cheap for such a small matrix, it can introduce numerical issues if the matrix becomes ill-conditioned.

Larger State Space: With three phase currents as explicit states rather than two dq currents, the state space is 50% larger. This impacts the complexity of state estimators and observers based on these models.

Despite these considerations, abc VBR models remain computationally tractable for offline analysis, fault signature prediction, and moderate-rate online diagnostics (1–5 kHz). Recent work has demonstrated real-time implementations for eccentricity fault detection running at sampling rates sufficient for practical diagnostic systems [6].

3.2.3.5 Parameter Extraction and Validation

A practical challenge in using abc VBR models is obtaining the position-dependent inductance matrix $\mathbf{L}_{abc}(\theta)$ and back-EMF waveforms $\mathbf{e}_{abc}(\theta)$. Several approaches are used:

Finite Element Precomputation: High-fidelity FEA simulations compute flux linkage as a function of rotor position and currents, from which inductance matrices are extracted by numerical differentiation. These lookup tables or harmonic series fits are then used in the abc model.

Experimental Characterization: Standstill tests with rotor locked at various positions measure inductances directly. Back-EMF waveforms are measured during no-load rotation. This approach captures real machine behavior including manufacturing variations but requires physical access to the machine.

Winding Function Theory: Analytical expressions for inductances can be derived from winding distribution and air-gap permeance functions, providing closed-form abc model parameters. This approach enables rapid parametric studies but requires accurate geometric data.

For fault diagnosis applications, it is often sufficient to use approximate abc models where only the dominant harmonics are retained, trading some accuracy for reduced parameter identification effort and faster computation.

3.2.3.6 Role in Fault Diagnosis Workflow

The abc reference frame VBR model is most appropriate when:

- Faults produce spatial harmonics or asymmetric signatures (eccentricity, partial demagnetization, single-phase shorts).
- Detailed prediction of current harmonic spectra is needed for diagnostic algorithm development.
- Offline analysis and fault signature databases are being created.
- The machine has concentrated windings or significant spatial harmonic content even in healthy operation.
- Phase-specific fault localization is required.
- Moderate sampling rates (1–5 kHz) are acceptable for the diagnostic system.

The abc VBR approach is less suitable for ultra-high-speed real-time control loops or extremely resource-constrained embedded systems where dq models remain superior.

3.2.4 Model Selection Guidelines

Having examined both dq-axis and abc reference frame VBR lumped parameter models in detail, it is essential to provide comparative guidance on when to use each approach. The optimal choice depends on multiple factors including the fault types of interest, computational resources available, whether the application is real-time or offline, and the required accuracy of fault signature prediction. Table 3.1 summarizes the key differences between these modeling approaches.

Recommended Application Scenarios:

Use dq-axis models when:

1. Implementing real-time model-based observers on resource-constrained embedded hardware.
2. Integrating diagnostics into existing field-oriented control systems.
3. Monitoring global parameter changes (overall demagnetization, average resistance increase).

Table 3.1 Comparison of dq-axis and abc VBR modeling approaches for PM machine fault diagnosis.

Criterion	dq-axis model	abc VBR model
State variables	2 (i_d, i_q)	3 (i_a, i_b, i_c)
Parameter variation	Constant	Position-dependent
Computational speed	Very fast (10–20 kHz)	Moderate (1–5 kHz)
Spatial harmonics	Not represented	Explicitly captured
Asymmetric faults	Limited/averaged	Fully represented
Back-EMF harmonics	Typically ignored	Can be included
Control integration	Seamless (FOC)	Requires transformation
Parameter identification	Simple (4–5 parameters)	Complex (matrix functions)
Physical intuition	High (d = flux, q = torque)	Moderate (phase quantities)
Real-time capability	Excellent	Good
Fault signature accuracy	Fair to good	Good to excellent

4. The machine has distributed windings with low harmonic content.
5. Computational efficiency is the primary concern.
6. The faults of interest primarily manifest as parameter variations rather than harmonic signatures.

Use abc VBR models when:

1. Predicting or analyzing harmonic fault signatures for diagnostic algorithm development.
2. Detecting asymmetric faults (static eccentricity, partial demagnetization, single-phase faults).
3. Working with machines having concentrated windings or significant spatial harmonics.
4. Building fault signature databases through simulation.
5. Conducting offline analysis where computational speed is less critical.
6. Phase-specific fault localization is required.
7. Interfacing with power electronics models or studying motor-drive interactions.

Hybrid Approaches: In many practical fault diagnosis systems, a hybrid strategy proves most effective:

- Use **dq-axis models** for the primary control system and high-rate monitoring of global parameters.
- Simultaneously run **abc VBR models** at reduced sampling rates (e.g., 1 kHz vs. 10 kHz for control) specifically for harmonic signature extraction and asymmetric fault detection.
- Transform between frames as needed: monitor dq parameters continuously, but periodically analyze abc current harmonics using FFT or other spectral techniques.

This hybrid approach leverages the strengths of each modeling framework: dq for efficient real-time control and basic monitoring, abc for detailed fault signature analysis. Modern motor drive processors have sufficient computational capacity to support both models running concurrently.

Model Selection Guidelines:

1. **Real-time execution at >10 kHz required:** If yes, dq-axis models are preferred due to their computational efficiency.
2. **Spatial harmonics or asymmetries critical for fault detection:** If yes, abc VBR models are preferred as they capture these effects.

3. **Seamless FOC integration required:** If yes, dq-axis models are preferred as they align with field-oriented control frameworks.
4. **Phase-specific fault localization needed:** If yes, abc VBR models are preferred as they retain phase-specific information.
5. **Offline analysis acceptable:** If yes, abc VBR models enable higher accuracy for fault diagnosis studies.
6. **Parameter identification effort a major constraint:** If yes, dq-axis models are simpler to identify with fewer parameters.

Ultimately, lumped parameter models, whether in dq or abc form, represent the most practical modeling approach for real-time fault diagnosis in permanent magnet machines. Their computational efficiency, relative ease of parameter identification, and direct connection to measurable electrical quantities make them indispensable tools. Understanding when each formulation is appropriate, and how to potentially combine them, enables engineers to design robust, accurate fault diagnosis systems optimized for their specific application requirements and computational resources.

3.3 Finite Element Method

3.3.1 Introduction

The FEM stands as the gold standard for high-fidelity electromagnetic modeling of permanent magnet machines, offering unparalleled accuracy in predicting machine behavior under both healthy and faulty operating conditions. Unlike lumped parameter models that use globally averaged quantities, FEM discretizes the machine's geometry into thousands or millions of small elements and solves Maxwell's equations numerically across this detailed spatial representation. This approach enables FEM to capture complex three-dimensional (3D) field distributions, magnetic saturation, slotting effects, end-winding leakage, eddy current losses, and virtually any geometric or material property variation, including those introduced by fault conditions.

The fundamental principle of finite element analysis is the discretization of a continuous domain into a finite number of subdomains called elements. In electromagnetic problems, the machine's cross-section (for 2D analysis) or complete 3D geometry (for 3D analysis) is subdivided into triangular or tetrahedral elements forming a mesh. Within each element, the field quantities such as magnetic vector potential are approximated using interpolation functions, typically low-order polynomials. Maxwell's equations are then formulated in a weak or variational form and solved simultaneously across all elements subject to appropriate boundary conditions, yielding a large system of algebraic equations. The solution provides detailed spatial distributions of magnetic flux density, electric field, current density, and other electromagnetic quantities at every point in the computational domain.

For permanent magnet machine analysis, FEM typically solves the magnetic vector potential formulation of Maxwell's equations [9, 10]. In 2D problems assuming axial symmetry or negligible end effects, the governing equation reduces to:

$$\nabla \times \left(\frac{1}{\mu}\nabla \times \mathbf{A}\right) = \mathbf{J} - \nabla \times \mathbf{M} \tag{3.15}$$

where $\mathbf{A}$ is the magnetic vector potential, μ is the magnetic permeability (which may be nonlinear and spatially varying to capture saturation), $\mathbf{J}$ is the current density in the windings, and $\mathbf{M}$ is the magnetization of the permanent magnets. 3D formulations extend this to account for end effects, skewing, and axially varying phenomena. Time-stepping FEM solves these equations at successive time instants to capture transient behavior and dynamic faults.

The workflow for applying FEM to permanent magnet machine fault analysis typically follows the six steps illustrated in Figure 3.3: geometry creation, mesh generation, material property assignment, boundary conditions and excitation definition, numerical solution, and post-processing to extract fault signatures.

Modern commercial FEM software packages such as ANSYS Maxwell [11], COMSOL Multiphysics, JMAG [12], and open-source tools like FEMM [13] provide sophisticated environments for this workflow, handling much of the mathematical complexity behind user-friendly interfaces. Figure 3.4 illustrates a typical FEM analysis of a V-shaped interior permanent magnet synchronous machine (IPMSM) created using ANSYS Maxwell, showing both the triangular mesh discretization with refined elements in the air gap region and the resulting magnetic flux line distribution under no-load conditions.

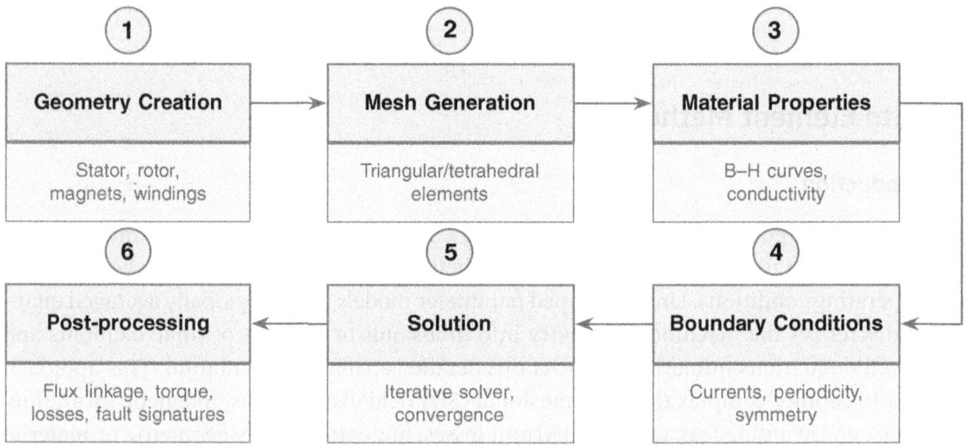

Figure 3.3 Six-step workflow for finite element analysis of permanent magnet machines, from geometry creation through post-processing of fault signatures.

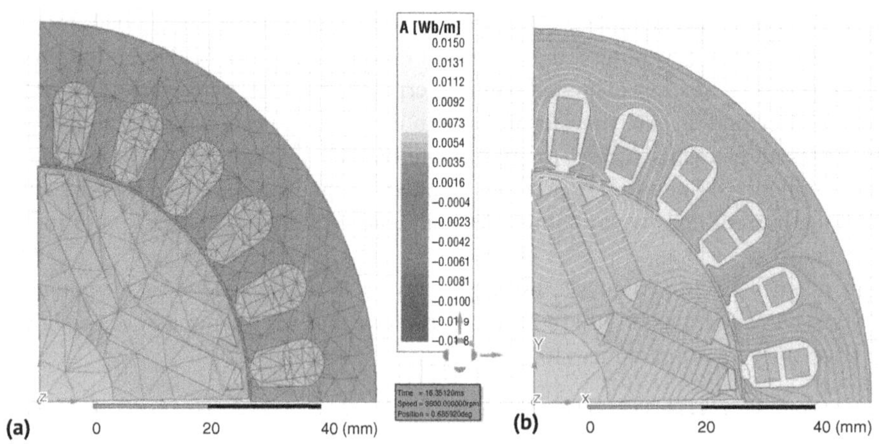

Figure 3.4 Finite element analysis of a V-shaped IPMSM cross-section: (a) triangular mesh discretization showing refined elements in the air gap region, and (b) magnetic flux line distribution under no-load conditions.

However, the computational cost remains substantial. A single 2D time-stepping simulation of one electrical cycle at one operating point might require solving systems with 50,000–500,000 degrees of freedom at hundreds of time steps, taking minutes to hours depending on mesh refinement and convergence criteria. 3D analyses or parametric studies exploring multiple operating conditions or fault severities can extend this to days of computation, even on high-performance workstations.

This computational burden represents the fundamental trade-off that defines FEM's role in fault diagnosis: unmatched accuracy and capability at the cost of execution time that makes real-time monitoring impractical. FEM serves as a research tool, a design validation method, and increasingly as a means to generate training data for faster reduced-order models or machine learning approaches, rather than as an online diagnostic technique itself.

3.3.2 Application to Fault Modeling

The unique capabilities of finite element analysis make it indispensable for understanding, characterizing, and predicting the electromagnetic signatures of faults in permanent magnet machines. FEM's ability to model arbitrary geometric and material variations with high fidelity enables it to capture fault effects that simpler models cannot represent accurately or at all.

Comprehensive Fault Coverage: FEM can model virtually any fault type that affects the electromagnetic behavior of the machine. Demagnetization faults are modeled by reducing the remanent flux density or coercivity of affected magnet regions, allowing investigation of uniform, partial, or localized demagnetization patterns and their impact on back-EMF waveforms and torque production [14, 15]. Eccentricity faults, whether static (misalignment between stator and rotor centers), dynamic (rotor orbiting around stator center), or mixed, are incorporated by modifying the air-gap geometry to reflect the spatially and/or temporally varying air-gap length, enabling prediction of the characteristic harmonic patterns in flux density and current spectra that serve as diagnostic signatures [16]. Inter-turn and phase-to-phase short circuits are represented by introducing conducting paths between affected turns with appropriate short-circuit resistance, allowing simulation of the resulting localized heating, circulating currents, and asymmetric magnetic field distributions [17, 18]. Broken rotor bars or end-ring faults in induction machines, while less common in PM machines, can be modeled by removing conductor material from the rotor cage. Even combined or cascading faults where one fault condition leads to or exacerbates another can be investigated.

High-fidelity Fault Signature Prediction: For fault diagnosis algorithm development, knowing precisely what signatures a given fault produces in measurable quantities is essential. FEM provides this ground truth information with excellent accuracy. For instance, eccentricity fault detection often relies on specific sideband frequencies appearing in the current spectrum around the fundamental supply frequency. FEM simulations can predict the exact frequencies, amplitudes, and phases of these sidebands as functions of eccentricity severity and type, providing target signatures for diagnostic algorithms to detect [19]. Similarly, demagnetization produces characteristic changes in back-EMF waveforms, harmonic content, and average torque that FEM can quantify, enabling development of detection thresholds and fault severity estimation algorithms. This capability has made FEM the primary tool for generating labeled fault signature databases used to train and validate both model-based and data-driven diagnostic systems [20].

Validation of Simplified Models: One of the most valuable applications of FEM in fault diagnosis research is validating the accuracy of faster reduced-order models. Lumped parameter, magnetic equivalent circuit, or analytical models make simplifying assumptions about field distributions and geometric effects. FEM provides a reference against which these simpler models can be benchmarked. For example, does an abc reference frame VBR model with a modified inductance matrix accurately predict the current harmonics produced by static eccentricity? FEM can answer this by computing the same scenario with full geometric fidelity. If the simplified model agrees well with FEM, confidence in using the fast model for real-time diagnosis increases. If not, the discrepancies reveal where model improvements are needed. This validation process is particularly important because experimental validation of fault conditions is expensive, potentially destructive, and limited in the range of fault severities that can be safely tested.

Understanding Fault Physics: Beyond prediction and validation, FEM provides deep physical insight into fault mechanisms. Visualizing magnetic flux density distributions, current density patterns, and loss distributions reveals how faults alter the electromagnetic environment within the machine. For instance, FEM can show how localized demagnetization creates flux redistribution that increases saturation in neighboring regions or how static eccentricity concentrates magnetic pull force in specific stator sectors, potentially leading to mechanical vibration and bearing damage. This understanding guides both diagnostic strategy development (identifying the most sensitive indicators) and fault mitigation approaches (design modifications to reduce fault sensitivity).

Parametric Studies and Sensitivity Analysis: FEM enables systematic parametric studies that would be impractical experimentally. How does the amplitude of eccentricity-related current harmonics vary with eccentricity severity from 1% to 50% of nominal air gap? How much does demagnetization need to progress before it becomes detectable above measurement noise? How do these fault signatures change with load level, speed, or temperature? FEM can answer such questions through automated parametric sweeps, providing comprehensive fault signature maps that inform diagnostic system design decisions such as sensor selection, detection thresholds, and operating condition dependencies.

Recent research has leveraged FEM extensively for fault analysis in permanent magnet machines. Studies on demagnetization faults use FEM to analyze the mechanisms of irreversible demagnetization due to thermal stress, armature reaction, or combined effects, and to predict the resulting changes in flux linkage and back-EMF harmonics [14, 15]. Eccentricity fault investigations employ FEM coupled with analytical models to predict unbalanced magnetic pull forces and characteristic current spectrum sidebands, with experimental validation confirming the accuracy of FEM predictions [16, 19]. Inter-turn short circuit research uses FEM to compute the circulating currents, local heating, and asymmetric magnetic fields, often coupling electromagnetic FEM with thermal finite element analysis to predict fault progression [17, 21]. The consistency with which FEM appears as the modeling tool in high-quality fault diagnosis research underscores its position as the reference standard in this field.

3.3.3 Advantages and Limitations

3.3.3.1 Advantages

The widespread adoption of finite element analysis for permanent magnet machine fault research stems from several compelling advantages that no other modeling technique fully matches:

Highest Modeling Accuracy: FEM provides the most accurate predictions of electromagnetic behavior available from purely computational methods. By solving Maxwell's equations with minimal simplifying assumptions across a detailed geometric representation, FEM captures effects that other models must approximate or ignore entirely. Predicted quantities such as torque, back-EMF, inductance, and losses typically agree with experimental measurements to within a few percent under both healthy and faulty conditions, assuming accurate material properties and geometry are provided. This accuracy makes FEM the natural choice when validating new diagnostic approaches or investigating subtle fault effects where high precision is essential.

Rigorous Treatment of Geometry and Saturation: Real permanent magnet machines have complex geometries with intricate slot shapes, magnet configurations, and structural features that significantly influence electromagnetic performance. FEM represents this geometry explicitly, down to fine details such as slot opening widths, magnet chamfers, and rotor rib thicknesses. Moreover, ferromagnetic materials in the stator and rotor laminations exhibit nonlinear magnetic saturation, where permeability decreases significantly at high flux densities. FEM handles saturation rigorously through iterative solution methods that update material properties based on the local magnetic field, capturing the impact of saturation on inductances, back-EMF harmonics, and fault signatures. Faults that alter saturation patterns such as demagnetization (which can change local flux density) or eccentricity (which affects air-gap reluctance distribution) are therefore modeled with high fidelity.

Comprehensive Harmonic Content: Permanent magnet machines inherently produce spatial and temporal harmonics due to slotting, winding distribution, magnet shape, and saturation effects. Faults often manifest primarily through their impact on these harmonics, introducing new harmonic components or amplifying existing ones. FEM naturally captures the complete harmonic spectrum in computed quantities without requiring assumptions about which harmonics are significant. Post-processing of FEM results through Fourier analysis reveals the full harmonic content of back-EMF, cogging torque, torque ripple, and current waveforms, providing the detailed frequency-domain information needed for harmonic-based fault diagnosis techniques such as motor current signature analysis (MCSA) [22].

Arbitrary Fault Condition Modeling: The flexibility of FEM to model arbitrary variations in geometry, material properties, and excitation conditions means it can represent virtually any conceivable fault scenario. Want to model partial demagnetization affecting only the trailing edge of every third magnet pole? Simply modify the remanent flux density of those specific finite elements. Need to investigate mixed eccentricity with a specific static offset combined with dynamic whirling? Adjust the air-gap geometry accordingly at each rotor position. This flexibility contrasts sharply with analytical models that often require significant mathematical rework to accommodate new fault types, and with lumped parameter models that may be fundamentally incapable of representing certain spatial phenomena.

Coupling with Other Physics: Modern multiphysics FEM platforms enable coupling electromagnetic field solutions with thermal, structural, and fluid dynamics analyses. This capability is particularly valuable for understanding fault progression [21]. For example, inter-turn short circuits produce localized heating that can be computed by electromagnetic FEM (calculating losses), then used as input to thermal FEM (predicting temperature rise), which in turn affects material properties such as magnet coercivity (feeding back to electromagnetic FEM), and may cause mechanical stress (structural FEM). Such coupled analyses reveal fault progression mechanisms and cascading failure modes that single-physics models cannot capture.

Visual Insight: FEM provides rich visualization of field distributions, flux lines, current patterns, and loss densities throughout the machine. These visual representations offer an intuitive understanding of fault effects that complement quantitative predictions. Seeing how flux lines distort around a demagnetized magnet region or how current density concentrates in a shorted turn provides insights that guide both diagnostic algorithm development and fault-tolerant design strategies.

3.3.3.2 Limitations

Despite its advantages, finite element analysis faces significant practical limitations that constrain its role in fault diagnosis systems:

Very High Computational Cost: The most severe limitation of FEM is its computational expense. A 2D time-stepping electromagnetic simulation of a permanent magnet machine over one electrical cycle at one operating condition typically requires minutes to hours, depending on mesh density, time step size, and convergence criteria. 3D models, necessary to capture end effects, skewing, and truly 3D fault phenomena, increase this to hours or days per operating point. Parametric studies sweeping fault severity, load level, and speed can require hundreds or thousands of simulation runs, translating to weeks of computation even on powerful workstations. This computational burden makes FEM fundamentally unsuitable for real-time fault monitoring, where diagnostic algorithms must execute in milliseconds. FEM is strictly an offline analysis tool.

Impractical for Real-time Diagnosis: Real-time fault diagnosis systems embedded in motor drives must process sensor data and make diagnostic decisions at rates of 1–20 kHz to track rapidly changing operating conditions and detect faults quickly. Even the fastest FEM solvers cannot approach these execution rates. Consequently, FEM cannot be deployed as an online diagnostic model. Its role is limited to offline activities: design-stage fault sensitivity analysis, generation of fault signature databases for training faster models, validation of real-time diagnostic algorithms, and investigation of fault mechanisms after detection by other means.

Requires Detailed Geometry and Material Data: FEM demands comprehensive information about machine geometry (slot dimensions, magnet sizes, air-gap length, stack length, etc.) and material properties (B–H curves for lamination steel, magnet remanent flux density and coercivity, winding conductivity, etc.). Obtaining this data can be challenging, particularly for commercial off-the-shelf machines where manufacturers may not provide detailed internal geometry. Magnetic material properties vary with manufacturing processes, temperature, and stress, introducing uncertainty. Lack of accurate input data degrades FEM accuracy, partially negating its theoretical advantage. This contrasts with lumped parameter models, where a few global parameters (resistance, inductance, flux linkage) can often be measured experimentally without detailed geometric knowledge.

Specialized Expertise Required: Effective use of FEM requires significant expertise in electromagnetic theory, numerical methods, and the specific FEM software package. Proper mesh design, appropriate boundary conditions, convergence criteria selection, and interpretation of results all demand experience. Poorly constructed FEM models can produce misleading results, such as mesh-dependent torque ripple predictions or nonconverged solutions. Organizations without in-house FEM expertise face a barrier to using these tools effectively, whereas simpler lumped parameter models can be implemented and used with more modest technical backgrounds.

Difficulty with Uncertainty Quantification: While FEM produces highly accurate results for specified input parameters, real machines have manufacturing tolerances, material variability, and operating condition uncertainties. Quantifying how these uncertainties propagate through the FEM solution to affect predictions (uncertainty quantification) requires running large numbers of FEM simulations with perturbed parameters, exacerbating the computational cost issue. Consequently, FEM typically provides deterministic predictions for nominal parameter values rather than probabilistic bounds accounting for uncertainty, which can be a limitation when developing robust diagnostic algorithms that must perform reliably across manufacturing variations.

Limited Representation of Auxiliary Effects: While FEM excels at electromagnetic modeling, it may not fully capture all phenomena relevant to fault diagnosis. Mechanical vibrations, acoustic noise, bearing dynamics, and thermal transients involve complex physics that may require specialized models beyond electromagnetic FEM. Although multiphysics coupling addresses some of these issues, the computational cost increases dramatically, and coupling models from different physics domains introduce additional complexity and potential sources of error.

Despite these limitations, FEM remains an indispensable tool in the permanent magnet machine fault diagnosis research and development workflow. Its role is not as a real-time diagnostic method but as a high-fidelity reference for understanding fault physics, validating faster models, generating training data, and conducting parametric studies that inform diagnostic system design. Understanding where FEM fits in the broader modeling ecosystem and how to leverage its strengths while mitigating its weaknesses through hybrid approaches represents a key skill for engineers developing advanced fault diagnosis systems.

3.3.4 Role in Fault Diagnosis Workflow

Given the complementary strengths and weaknesses of FEM relative to other modeling approaches, an effective fault diagnosis development strategy typically employs FEM at specific stages rather than throughout the entire workflow:

Design Stage Fault Sensitivity Analysis: During machine design, FEM identifies which faults the design is most sensitive to and how geometric or material choices affect fault signatures. For instance, concentrated windings versus distributed windings significantly influence harmonic content, affecting the visibility of certain fault signatures. FEM parametric studies guide design choices to enhance fault detectability or reduce fault susceptibility.

Fault Signature Database Generation: FEM simulations across ranges of fault types, severities, and operating conditions generate comprehensive fault signature databases. These databases characterize how each fault type affects measurable quantities (current harmonics, voltage asymmetries, torque ripple, vibration spectra) under various loads and speeds. These signatures then inform the design of real-time diagnostic algorithms and set detection thresholds. In data-driven approaches, FEM-generated data can augment limited experimental fault data for training machine learning classifiers.

Reduced-order Model Validation: FEM serves as ground truth for validating faster lumped parameter, magnetic equivalent circuit, or analytical models. By comparing predictions from a simplified model against FEM results for the same fault scenarios, engineers quantify the accuracy trade-offs and identify operating regions where simplified models are sufficiently accurate versus where they fail. This validation process builds confidence in using fast models for real-time diagnosis.

Post-detection Fault Analysis: After a fault is detected by real-time diagnostic algorithms, FEM can be used offline to analyze the fault in detail, estimate fault severity more accurately, predict fault progression, and assess the remaining useful life of the machine. For example, if online diagnostics detect signatures consistent with partial demagnetization, detailed FEM simulations can estimate the extent and location of demagnetization and predict when it will progress to unacceptable performance degradation.

Hybrid Modeling Approaches: Emerging research combines FEM with other techniques in hybrid frameworks. For instance, FEM might be used offline to generate lookup tables of inductances as functions of rotor position and fault parameters, which are then interpolated during real-time simulation in a lumped parameter model. Or FEM generates training data for machine learning models that learn to predict fault signatures orders of magnitude faster than FEM, enabling near-real-time high-fidelity predictions [20]. Physics-informed neural networks represent another hybrid approach, using FEM-like physics equations as constraints during neural network training to create fast data-driven models that respect electromagnetic principles [23].

By strategically deploying FEM where its accuracy is most valuable, while relying on faster models for real-time execution, engineers can develop fault diagnosis systems that are both accurate and practically implementable. This multi-fidelity modeling strategy represents the current state of the art in permanent magnet machine fault diagnosis research and development.

3.4 Field Reconstruction Method

3.4.1 Introduction

The FRM represents an innovative semi-analytical approach to electromagnetic modeling of permanent magnet machines that bridges the gap between the computational efficiency of lumped parameter models and the high fidelity of finite element methods [24, 25]. Developed to address the fundamental tension between modeling accuracy and computational speed, FRM applies the superposition principle to electromagnetic fields, enabling rapid prediction of machine behavior under arbitrary operating conditions and fault scenarios from a minimal set of pre-computed basis functions. This approach has demonstrated the ability to achieve computation times 30–150 times faster than time-stepping finite element analysis while maintaining accuracy comparable to FEM for many applications [26, 27], making it particularly attractive for fault diagnosis applications that require parametric studies, optimization, or moderate-rate online monitoring.

The core innovation of FRM lies in its recognition that for machines operating in the linear magnetic regime or with controlled saturation, the electromagnetic field distribution can be decomposed into independent contributions from different excitation sources through superposition. Specifically, the total magnetic field in the air gap of a permanent magnet machine results from the superposition of fields produced by the permanent magnets and fields produced by currents in each phase winding. Rather than solving Maxwell's equations repeatedly for every combination of rotor position and phase currents, FRM performs a small number of magnetostatic finite element analyses to compute and store these basis field distributions once, then reconstructs fields for arbitrary operating conditions through weighted superposition. This pre-computation and storage strategy transforms the problem from repeated nonlinear PDE solutions to simple algebraic operations, dramatically reducing computational cost.

For fault diagnosis applications, FRM provides several valuable capabilities. The method naturally handles geometric faults such as eccentricity by including the fault condition in the pre-computed basis functions, enabling rapid prediction of fault-induced field distortions and their effects on electrical signatures. Multiple fault conditions can be analyzed quickly by storing separate basis function libraries for healthy and various faulty configurations, then reconstructing behavior under any operating point in seconds rather than hours. This computational efficiency enables comprehensive fault signature databases covering wide ranges of fault severities, load levels, and speeds to be generated practically. Furthermore, FRM has been extended to transient analysis and combined fault scenarios, including the simultaneous presence of inter-turn short circuits and magnet defects, expanding its applicability to realistic fault diagnosis challenges.

The following sections examine the theoretical foundation of FRM, its implementation methodology, applications to fault modeling, validation results demonstrating its accuracy, and comparative assessment of its advantages and limitations within the broader modeling ecosystem for permanent magnet machine fault diagnosis.

3.4.2 Mathematical Foundation

The Field Reconstruction Method is founded on the principle of electromagnetic field superposition, which is valid when the magnetic system can be treated as linear or when nonlinearities can be handled through iterative updates. The fundamental mathematical premise states that the total magnetic field in a machine's air gap is the vector sum of fields produced by individual excitation sources acting independently [24, 28].

For a permanent magnet synchronous machine, the primary excitation sources are the rotor permanent magnets and the three-phase stator currents. The radial and tangential components of the magnetic flux density in the air gap can be expressed through superposition as:

$$B_n(\theta, t) = B_{n,\mathrm{pm}}(\theta, \theta_r) + B_{n,s}(\theta, \theta_r, i_a, i_b, i_c) \tag{3.16}$$

$$B_t(\theta, t) = B_{t,\mathrm{pm}}(\theta, \theta_r) + B_{t,s}(\theta, \theta_r, i_a, i_b, i_c) \tag{3.17}$$

where $B_{n,\mathrm{pm}}$ and $B_{t,\mathrm{pm}}$ are the radial and tangential flux density components produced solely by the permanent magnets with no stator current, and $B_{n,s}$ and $B_{t,s}$ are the components produced by the stator phase currents with magnets hypothetically removed. The subscripts n and t denote normal (radial) and tangential components respectively, θ is the spatial angle around the stator bore, and θ_r is the rotor electrical position.

Under the linearity assumption, the stator current contribution can be further decomposed into contributions from individual phase currents:

$$B_{n,s}(\theta, \theta_r, i_a, i_b, i_c) = i_a \cdot K_{n,a}(\theta, \theta_r) + i_b \cdot K_{n,b}(\theta, \theta_r) + i_c \cdot K_{n,c}(\theta, \theta_r) \tag{3.18}$$

$$B_{t,s}(\theta, \theta_r, i_a, i_b, i_c) = i_a \cdot K_{t,a}(\theta, \theta_r) + i_b \cdot K_{t,b}(\theta, \theta_r) + i_c \cdot K_{t,c}(\theta, \theta_r) \tag{3.19}$$

Here, $K_{n,a}$, $K_{n,b}$, $K_{n,c}$ and $K_{t,a}$, $K_{t,b}$, $K_{t,c}$ are the basis functions representing the radial and tangential flux density distributions produced by unit current (typically 1 A) in phases a, b, and c respectively. These basis functions depend on rotor position but not on current magnitude, capturing the geometry-dependent magnetic circuit response.

The validity of this superposition approach rests on several key assumptions that define FRM's applicability and limitations [24, 27]:

Linear Magnetic Structure: The magnetic materials in both stator and rotor cores are assumed to operate in the linear region of their B–H curves, where permeability remains approximately constant. This assumption is justified for surface-mounted permanent magnet machines with relatively large air gaps, where the air-gap reluctance dominates and magnetic saturation effects are minimal underrated load conditions. For interior PM machines or surface PM machines under heavy overload, saturation can be significant, requiring iterative FRM implementations or acceptance of reduced accuracy.

Negligible Structural Deformation: The method assumes that radial and tangential electromagnetic forces do not cause appreciable deformation of stator teeth or permanent magnets that would alter the magnetic geometry. For typical operating conditions, this is valid, but under severe fault conditions such as large eccentricity producing significant unbalanced magnetic pull, structural deflections could theoretically affect the field distribution. In practice, such deformations are usually small compared to air-gap dimensions.

Independence of Permanent Magnet and Stator Fields: The magnetic field produced by permanent magnets is assumed to remain unaltered by the presence of stator currents. This is equivalent to assuming that armature reaction does not significantly demagnetize the magnets during the time scale of interest. For properly designed permanent magnet machines operating within ratings, this assumption holds. However, for modeling progressive demagnetization faults where armature reaction contributes to irreversible magnet degradation, this assumption may need revisiting.

Negligible Axial Flux Variation (2D Formulation): In the most common 2D FRM implementations, magnetic flux density in the axial direction is assumed zero, effectively ignoring end effects. This simplification is acceptable for machines with relatively long stack lengths compared to diameter, but introduces errors for short-stack machines or when modeling phenomena with significant axial variation. 3D FRM extensions exist that relax this assumption at increased computational cost.

Neglect of Hysteresis and Eddy Currents: The standard FRM formulation uses magnetostatic finite element solutions to compute basis functions, neglecting magnetic hysteresis in laminations and eddy current effects. For fault diagnosis applications focused on fundamental and low-order harmonic electrical signatures at typical operating frequencies (less than a few kilohertz), this is generally acceptable. For high-frequency analysis or detailed loss prediction, these effects may need consideration through post-processing corrections or time-harmonic FRM extensions.

When these assumptions are satisfied, FRM provides an accurate and efficient means to predict electromagnetic behavior. Understanding these assumptions helps identify scenarios where FRM is most appropriate and where alternative methods such as nonlinear time-stepping FEM may be necessary.

3.4.3 Implementation Methodology

The practical implementation of the Field Reconstruction Method follows a systematic workflow that separates offline precomputation from online field reconstruction. This separation is the key to FRM's computational efficiency, as the computationally expensive FEM solutions are performed once during pre-processing, while the rapid reconstruction phase can be repeated for arbitrary operating conditions.

3.4.3.1 Step 1: Compute Permanent Magnet Basis Functions

The first pre-computation step uses magnetostatic finite element analysis to compute the flux density distribution produced solely by the rotor permanent magnets with zero stator current. Using a finite element software package (e.g., ANSYS Maxwell, COMSOL, JMAG, or FEMM), the machine geometry is meshed and material properties assigned. The permanent magnets are specified with their remanent flux density B_{rem} and relative permeability, while stator and rotor iron are assigned appropriate B–H curves. With all phase currents set to zero, the magnetostatic field equation:

$$\nabla \times \left(\frac{1}{\mu}\nabla \times \mathbf{A}\right) = -\nabla \times \mathbf{M} \tag{3.20}$$

is solved, where $\mathbf{M}$ is the permanent magnet magnetization. From the resulting magnetic vector potential distribution, the radial and tangential flux density components $B_{n,\mathrm{pm}}(\theta, \theta_r)$ and $B_{t,\mathrm{pm}}(\theta, \theta_r)$ are extracted along a contour in the middle of the air gap at one or more rotor positions as needed to cover one electrical period.

3.4.3.2 Step 2: Remove Magnets and Compute Stator Current Basis Functions

In this step, all permanent magnets are removed from the FEM model (setting $\mathbf{M} = 0$), converting the machine into an electrically excited structure. A base current (conventionally 1 A) is injected into one phase winding, with the other phases carrying zero current. The magnetostatic FEM solution is computed, now solving:

$$\nabla \times \left(\frac{1}{\mu}\nabla \times \mathbf{A}\right) = \mathbf{J}_a \tag{3.21}$$

where $\mathbf{J}_a$ represents the current density distribution in phase a. The air-gap flux density components $K_{n,a}(\theta, \theta_r)$ and $K_{t,a}(\theta, \theta_r)$ are extracted. This process is repeated for phases b and c, yielding six basis function arrays: $K_{n,a}$, $K_{t,a}$, $K_{n,b}$, $K_{t,b}$, $K_{n,c}$, $K_{t,c}$. For machines without rotor saliency (surface-mounted PM machines), these basis functions may be independent of rotor position, reducing storage requirements. For interior PM machines with saliency, basis functions must be stored at multiple rotor positions or fitted to Fourier series.

3.4.3.3 Step 3: Store Basis Functions

The basis function data, typically consisting of flux density values at several hundred to several thousand spatial points around the air-gap circumference, are stored in lookup tables or fitted to Fourier series representations. Fourier fitting reduces storage requirements and enables analytical differentiation but may introduce fitting errors if harmonic truncation is too aggressive. The choice between tabulated and Fourier-represented basis functions trades memory usage against computational speed and accuracy.

3.4.3.4 Step 4: Reconstruct Fields for Arbitrary Operating Conditions

Given a specific operating condition defined by rotor position θ_r and phase currents i_a, i_b, i_c, the air-gap flux density is reconstructed through weighted superposition:

$$B_n(\theta, \theta_r) = B_{n,\mathrm{pm}}(\theta, \theta_r) + i_a K_{n,a}(\theta, \theta_r) + i_b K_{n,b}(\theta, \theta_r) + i_c K_{n,c}(\theta, \theta_r) \tag{3.22}$$

$$B_t(\theta, \theta_r) = B_{t,\mathrm{pm}}(\theta, \theta_r) + i_a K_{t,a}(\theta, \theta_r) + i_b K_{t,b}(\theta, \theta_r) + i_c K_{t,c}(\theta, \theta_r) \tag{3.23}$$

This reconstruction involves only lookup operations (or Fourier series evaluation) and scalar multiplications, executing in milliseconds. The reconstructed flux density can then be used to compute performance quantities as described below.

3.4.3.5 Step 5: Compute Electromagnetic Forces and Torque

The Maxwell stress tensor (MST) method is commonly employed to compute electromagnetic forces from the reconstructed air-gap flux density [24]. The normal and tangential force densities per unit area are:

$$f_n(\theta) = \frac{1}{2\mu_0}(B_n^2 - B_t^2) \tag{3.24}$$

$$f_t(\theta) = \frac{1}{\mu_0} B_n B_t \tag{3.25}$$

where μ_0 is the permeability of free space. Integration of these force densities around the air-gap contour and over the axial length yields the total normal force (radial component at any angular position), tangential force (relevant for torque), and unbalanced magnetic pull (vector sum of radial forces):

$$F_n = l_s r \int_0^{2\pi} f_n(\theta) \cos(\theta - \phi_s) d\theta \tag{3.26}$$

$$F_t = l_s r \int_0^{2\pi} f_t(\theta) d\theta \tag{3.27}$$

where l_s is the stack length, r is the radius at which forces are computed, and ϕ_s is the spatial angle for radial force component calculation. Electromagnetic torque is obtained from the tangential force:

$$T_e = r \cdot F_t \tag{3.28}$$

These integrations are performed numerically using the reconstructed flux density values, enabling torque and force prediction for any combination of position and currents without additional FEM solutions.

3.4.3.6 Step 6: Compute Flux Linkages and Inductances

Flux linkage of each phase winding is calculated by integrating the flux density over the surfaces enclosed by the winding conductors. In practice, this is implemented using the magnetic vector potential or by summing flux contributions from each coil of the winding:

$$\lambda_a = \sum_{\text{coils}} \int_{A_{\text{coil}}} \mathbf{B} \cdot d\mathbf{A} \tag{3.29}$$

The flux linkage can be separated into permanent magnet and stator current contributions:

$$\lambda_a = \lambda_{\text{pm},a}(\theta_r) + L_{\text{aa}}(\theta_r) i_a + L_{\text{ab}}(\theta_r) i_b + L_{\text{ac}}(\theta_r) i_c \tag{3.30}$$

where $\lambda_{\text{pm},a}$ is computed from the permanent magnet basis functions, and the inductances are computed from the stator current basis functions:

$$L_{\text{aa}}(\theta_r) = \frac{\lambda_{a,1A} - \lambda_{a,0A}}{1A} \tag{3.31}$$

with $\lambda_{a,1A}$ being the flux linkage with 1 A in phase a (and zero in other phases) and $\lambda_{a,0A}$ the linkage with no current. These inductances form the position-dependent inductance matrix needed for electrical dynamic simulations or abc reference frame models discussed earlier.

This workflow enables FRM to serve as a bridge between high-fidelity FEM geometry and rapid electromechanical system simulation, making it valuable for fault diagnosis applications requiring many simulations across varying operating conditions. The complete FRM implementation procedure is summarized in Algorithm 1.

Figures 3.5 and 3.6 illustrate examples of the tangential and radial flux density distributions in the air gap obtained from FEM simulations for SPM and IPM machines, respectively. These distributions are computed with zero stator current, representing the permanent magnet basis functions $B_{t,\text{pm}}$ and $B_{n,\text{pm}}$ that capture the rotor magnet contribution to the air gap field. These flux density waveforms serve as essential basis functions for field reconstruction, capturing the spatial harmonic content and geometric effects required for accurate fault signature prediction.

Algorithm 1 Field Reconstruction Method for PM Machines

Offline Precomputation Phase:

Compute PM basis functions $B_{n,\text{pm}}(\theta, \theta_r)$, $B_{t,\text{pm}}(\theta, \theta_r)$ via magnetostatic FEM with $\mathbf{i}_{\text{abc}} = 0$

Remove magnets from FEM model ($\mathbf{M} = 0$)

Compute stator current basis functions $K_{n,\phi}(\theta, \theta_r)$, $K_{t,\phi}(\theta, \theta_r)$ for $\phi \in \{a, b, c\}$ with $i_{\phi,\text{base}} = 1$ A

Store basis functions in lookup tables or fit to Fourier series

Online Reconstruction Phase:

Require: Phase currents i_a, i_b, i_c; initial rotor position θ_{m0}; angular step $\Delta\theta_m$

Ensure: Reconstructed flux density, torque, and flux linkages over one electrical cycle

Initialize rotor position: $\theta_m \leftarrow \theta_{m0}$

repeat

  Retrieve PM basis functions at current θ_m

  Retrieve stator current basis functions at current θ_m

  Reconstruct radial flux density:

  $B_n(\theta) = B_{n,\text{pm}}(\theta, \theta_m) + i_a K_{n,a} + i_b K_{n,b} + i_c K_{n,c}$

  Reconstruct tangential flux density:

  $B_t(\theta) = B_{t,\text{pm}}(\theta, \theta_m) + i_a K_{t,a} + i_b K_{t,b} + i_c K_{t,c}$

  Compute Maxwell stress tensor components: f_n, f_t

  Calculate electromagnetic torque: $T_e = r \cdot l_s \int_0^{2\pi} f_t(\theta) d\theta$

  Calculate phase flux linkages: λ_a, λ_b, λ_c

  Update rotor position: $\theta_m \leftarrow \theta_m + \Delta\theta_m$

until $\theta_m = 2\pi$

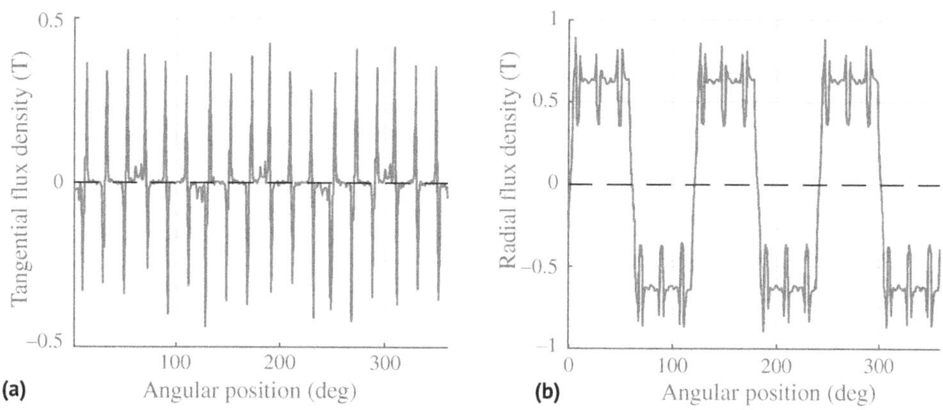

Figure 3.5 Air gap flux density distribution from FEM simulation of SPM machine with zero stator current: (a) tangential flux density, (b) radial flux density.

Similarly, Figures 3.7 and 3.8 present the tangential and radial flux density distributions produced by stator phase current at base current with the permanent magnets removed. These distributions represent the stator current basis functions $K_{t,\phi}$ and $K_{n,\phi}$ that quantify the armature reaction field contribution to the total air gap flux density.

The specifications of the SPM and IPM machines used for the FRM and MEC simulations throughout this chapter are summarized in Table 3.2. Both machines operate at a rated speed of 1000 rpm with a fundamental frequency of 50 Hz.

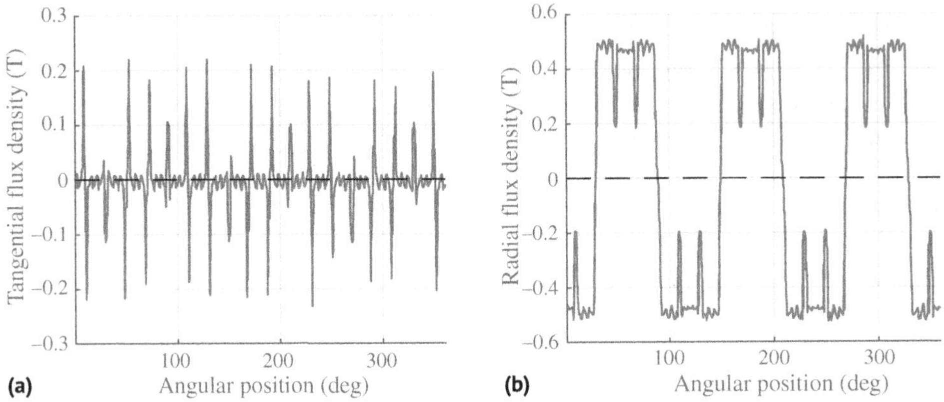

Figure 3.6 Air gap flux density distribution from FEM simulation of IPM machine with zero stator current: (a) tangential flux density, (b) radial flux density.

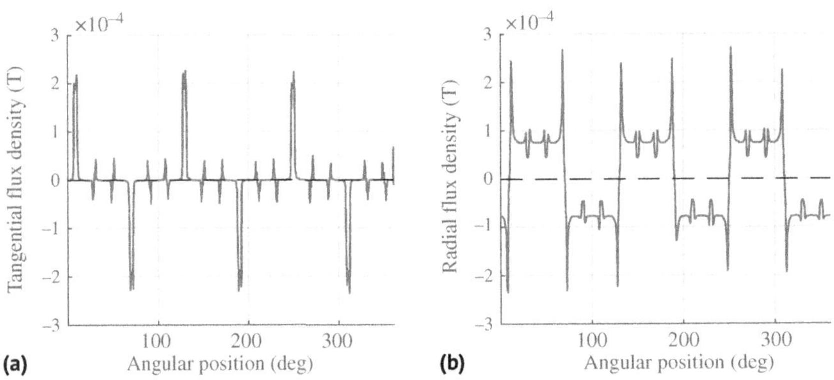

Figure 3.7 Air gap flux density distribution from FEM simulation of SPM machine with base stator current and magnets removed: (a) tangential flux density, (b) radial flux density.

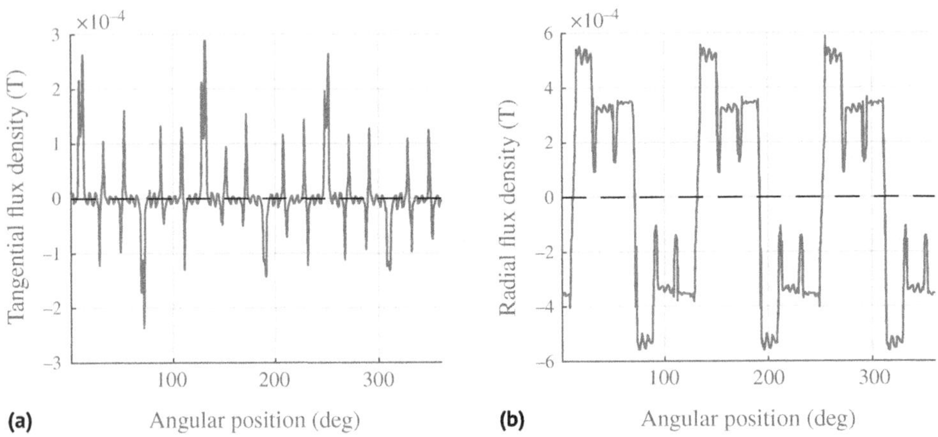

Figure 3.8 Air gap flux density distribution from FEM simulation of IPM machine with base stator current and magnets removed: (a) tangential flux density, (b) radial flux density.

Table 3.2 Specifications of SPM and IPM machines used for FRM and MEC simulations.

Parameter	SPM	IPM
Stator pole pairs, P	3	3
Stator turns per coil	20	20
Stack length (mm)	120	120
PM thickness (mm)	10	10
PM width (deg.)	148	36
Stator inner/outer diameter (mm)	150/270	150/270
Rotor inner/outer diameter (mm)	35/149	35/149
Air gap length (mm)	1	1
Slot area (mm^2)	380	380
Current density (A/mm^2)	5	4
Torque (Nm)	72	58
Rated power (kW)	7.5	6.5
Rated current (A)	24	21
PM remanence, B_{rem} (T)	1.12	1.12

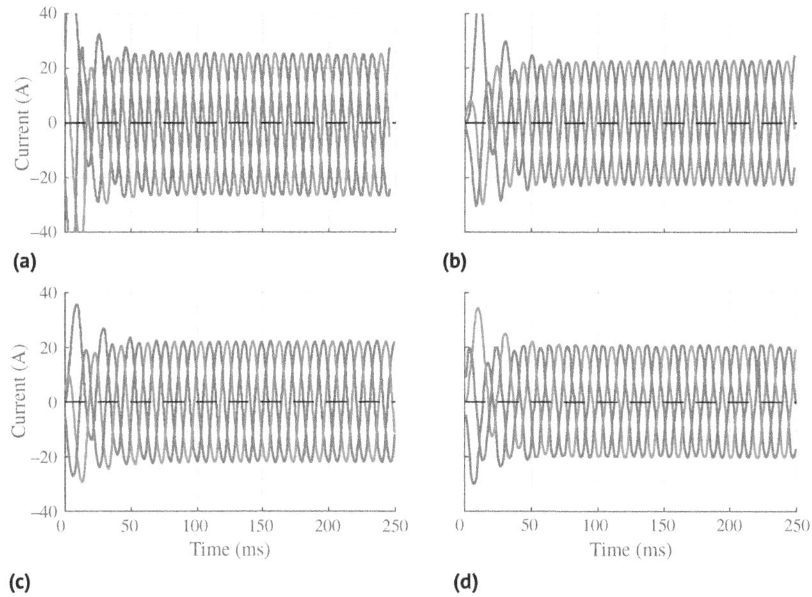

Figure 3.9 Comparison of three-phase stator currents from FRM and FEM simulations: (a) FRM results for SPM machine, (b) FRM results for IPM machine, (c) FEM results for SPM machine, (d) FEM results for IPM machine.

To validate the accuracy of the Field Reconstruction Method, Figure 3.9 presents a comparison of three-phase stator currents obtained from FRM and FEM simulations for both SPM and IPM machines under identical operating conditions. The close agreement between FRM and FEM results demonstrates that the superposition-based field reconstruction accurately captures the electromagnetic behavior, with FRM achieving computational speeds 30–150 times faster than

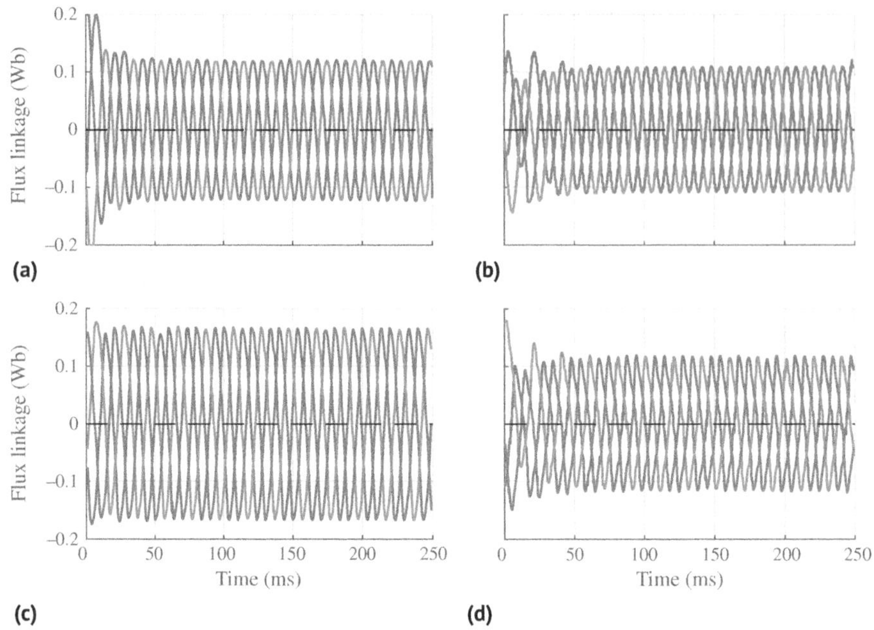

Figure 3.10 Comparison of three-phase flux linkage from FRM and FEM simulations: (a) FRM results for SPM machine, (b) FRM results for IPM machine, (c) FEM results for SPM machine, (d) FEM results for IPM machine.

time-stepping FEM while maintaining comparable accuracy in predicting current waveforms and their harmonic content.

Similarly, Figure 3.10 compares the three-phase flux linkage waveforms obtained from FRM and FEM simulations. The flux linkage, which is computed by integrating the reconstructed air gap flux density over each phase winding, shows excellent agreement between the two methods for both SPM and IPM machine topologies. This validation confirms that FRM accurately reconstructs not only the instantaneous electromagnetic quantities but also the integral quantities essential for machine performance prediction and fault signature analysis.

3.4.4 Application to Fault Modeling

The Field Reconstruction Method's rapid computation and flexibility make it particularly well-suited for modeling faults in permanent magnet machines. By incorporating fault conditions into the basis function pre-computation step, FRM can predict fault-induced changes in electromagnetic behavior with speed that enables comprehensive parametric studies. Recent research has successfully applied FRM to several important fault types.

3.4.4.1 Eccentricity Fault Modeling

Eccentricity faults, where rotor and stator axes are misaligned, introduce spatially varying air-gap length. FRM accommodates eccentricity by including the nonuniform air gap in the finite element geometry when computing basis functions [27, 29]. For static eccentricity, the air-gap variation pattern remains fixed in space but varies around the circumference. The FEM mesh is constructed with this geometric distortion, and the standard FRM workflow proceeds, generating basis functions that inherently capture the eccentricity-induced field distortions.

For dynamic eccentricity, the air-gap minimum rotates with the rotor. This requires computing basis functions at multiple rotor positions that span not only the electrical period but also a complete mechanical revolution. While this increases the number of pre-computation FEM solutions from typically 8–12 rotor positions to potentially 36–72 positions, this remains orders of magnitude faster than time-stepping FEM simulation of dynamic eccentricity over multiple revolutions. Once these basis functions are stored, reconstruction for arbitrary current waveforms and rotor trajectories proceeds at the same computational cost as the healthy machine case.

The characteristic signatures of eccentricity, particularly the sideband harmonics in phase currents and the unbalanced magnetic pull force, can be extracted from FRM predictions by Fourier analysis of the reconstructed currents and by computing the vector sum of radial forces at each time step. This enables rapid generation of fault signature databases correlating eccentricity severity with measurable electrical and mechanical quantities.

3.4.4.2 Demagnetization Fault Modeling

Permanent magnet demagnetization, whether uniform, partial, or localized, is naturally represented in FRM by modifying the remanent flux density B_{rem} of affected magnet elements in the FEM model used for basis function computation [27, 30]. Uniform demagnetization affecting all magnets equally is modeled by reducing B_{rem} proportionally throughout. Partial demagnetization of specific poles is modeled by reducing B_{rem} only in those pole regions. Localized demagnetization at magnet edges or surfaces experiencing high armature reaction fields is captured by spatially varying B_{rem} within individual magnet elements.

After computing the permanent magnet basis functions with the demagnetization included, the reconstruction process proceeds identically to the healthy case. The resulting back-EMF waveforms, flux linkages, and torque characteristics automatically reflect the demagnetization, showing reduced average values and altered harmonic content. By generating separate basis function sets for various demagnetization severities, a comprehensive fault signature library can be created with computational effort proportional to the number of severity levels times the cost of magnetostatic FEM (hours to days of computation), rather than the multiple weeks that time-stepping FEM for the same library would require.

3.4.4.3 Inter-turn Short Circuit Faults

Inter-turn short circuits present a more complex modeling challenge because the fault introduces a new current path (the shorted turns with their fault current) that must be represented. Recent research has extended FRM to handle inter-turn faults by treating the fault current as an additional independent excitation source [31]. The methodology computes basis functions for the healthy turns of the affected phase carrying normal current, and separate basis functions for the shorted turns carrying unit fault current. The field reconstruction then superposes contributions from all phases' normal currents plus the fault current contribution, with the fault current magnitude determined by the coupled electrical equations accounting for the short-circuit resistance.

This approach enables FRM to predict the circulating fault current, the resulting asymmetric magnetic field distribution, and the changes in terminal electrical quantities. The computational advantage compared to time-stepping FEM is particularly pronounced for inter-turn fault analysis because many different fault locations (which coil group is affected) and severities (number of shorted turns) can be explored using the basis function library without requiring new FEM solutions for each case, as long as the set of basis functions includes the relevant winding subsections.

3.4.4.4 Combined and Transient Fault Scenarios

A powerful recent extension of FRM enables modeling of combined fault conditions, such as the simultaneous presence of inter-turn short circuits and magnet defects, as demonstrated in recent research on PMSMs in transient states [31, 32]. The methodology develops separate magnet basis function libraries for different magnet defect patterns and combines these with inter-turn fault modeling to simulate multiple concurrent faults. This capability is valuable for fault diagnosis algorithm development because real machines may exhibit multiple degradation mechanisms simultaneously, and diagnostic systems must distinguish between single and combined faults.

Time-stepping FRM for transient analysis couples the field reconstruction with electrical circuit equations for the stator and any fault paths, solving the coupled system over time [26]. At each time step, the field reconstruction updates based on current rotor position and instantaneous phase currents, then electromagnetic torque is computed and used to update mechanical dynamics (speed and position), which feed back to the electrical equations for the next time step. This time-stepping FRM simulation remains 30–150 times faster than equivalent time-stepping FEM while capturing transient phenomena such as fault inception, short-circuit transients, and dynamic eccentricity during motor starting [25, 26].

3.4.5 Advantages and Limitations

The Field Reconstruction Method offers a compelling combination of advantages that have driven its adoption for permanent magnet machine fault modeling, while also exhibiting limitations that define its appropriate application domain.

3.4.5.1 Primary Advantages

Dramatic Computational Speed-up: FRM achieves 30–150 times faster execution than time-stepping finite element analysis for parametric studies and transient simulations. This transforms problems that would require weeks of FEM computation (e.g., simulating fault progression over thousands of operating cycles) into problems solvable in hours or days. For fault diagnosis research requiring extensive parametric sweeps over fault severity, load level, and speed, this speed advantage is often the difference between practical feasibility and intractability.

Comparable Accuracy to FEM: When operating assumptions are satisfied (linear magnetic conditions or mild saturation, negligible structural deformation), FRM accuracy approaches that of finite element analysis, typically within 2–10% for key electromagnetic quantities. This accuracy level is sufficient for fault signature prediction, diagnostic algorithm development, and most design studies. The method captures spatial harmonics, slotting effects, and geometric details through the basis functions, providing much higher fidelity than lumped parameter models.

Flexibility for Fault Scenarios: By pre-computing and storing basis function sets for various fault conditions (multiple eccentricity severities, demagnetization patterns, etc.), FRM enables rapid switching between healthy and faulty machine models during simulation. This supports fault scenario exploration, combined fault analysis, and sensitivity studies that inform diagnostic system design.

Intermediate-rate Online Capability: While too slow for ultra-high-bandwidth control loops (10–20 kHz), FRM can potentially execute at rates of 1–5 kHz on modern embedded processors for machines with moderate complexity. This opens possibilities for model-based online diagnostics using FRM rather than lumped parameter models, achieving higher accuracy in fault signature prediction while still running in real time.

Seamless Integration with System Simulation: FRM can be directly coupled with electrical circuit models and mechanical dynamic equations, enabling electromechanical system simulation that captures machine-level detail. This facilitates studies of machine-drive interactions, fault effects on overall system performance, and control algorithm evaluation under fault conditions.

3.4.5.2 Key Limitations

Restricted to Linear Magnetic Regime: The superposition principle underlying FRM is valid only when magnetic materials operate linearly or when saturation can be treated through iterative updates. For heavily saturated machines (interior PMSMs under high d-axis current, machines operating in field-weakening), FRM accuracy degrades. Iterative FRM approaches that update permeability based on reconstructed flux density can mitigate this but increase complexity and computational cost, partially negating FRM's speed advantage. For highly nonlinear problems, time-stepping FEM may ultimately be more reliable.

Precomputation Overhead: FRM requires an initial investment in finite element modeling to generate basis functions. For each machine geometry, material configuration, and fault type of interest, separate FEM analyses must be performed. While these are one-time costs, they can be substantial when exploring large fault spaces or many machine variants. In contrast, lumped parameter models require only identification of a few global parameters, which can often be measured experimentally.

Limited to Modeled Fault Conditions: FRM can only predict faults for which basis functions have been pre-computed. Unexpected or novel fault modes not included in the basis function library cannot be analyzed without returning to FEM to generate new basis functions. This contrasts with adaptive data-driven methods that might generalize to unforeseen conditions, though FRM's physics-based foundation provides more reliability when interpolating between modeled fault severities.

Storage Requirements: Basis function data for multiple rotor positions, multiple spatial points, and multiple fault conditions can require significant memory storage, particularly for 3D implementations. While modern computer memory capacities make this manageable, embedded implementation of FRM in motor drives must consider memory constraints. Fourier series fitting can reduce storage at the cost of potential fitting errors.

End-effect Approximation in 2D: Standard 2D FRM implementations neglect end effects, which can introduce errors for short-stack machines or when end-winding asymmetries contribute to faults. 3D FRM extensions address this but with substantially increased computational cost and storage requirements, reducing the speed advantage over 3D FEM.

3.4.6 Role in Fault Diagnosis Workflow

The Field Reconstruction Method occupies a valuable niche between lumped parameter models and full finite element analysis. It is most appropriately deployed in the following scenarios within permanent magnet machine fault diagnosis workflows:

Parametric Fault Studies and Signature Database Generation: When comprehensive fault signature databases covering wide ranges of fault severities and operating conditions are needed for diagnostic algorithm development, FRM provides an optimal balance. It is fast enough to generate thousands of operating points across multidimensional parameter spaces in practical time frames, yet accurate enough to capture spatial harmonics and fault-induced asymmetries that lumped parameter models miss.

Design-stage Fault Sensitivity Analysis: During machine design, FRM enables rapid evaluation of how design choices affect fault sensitivity and fault signature strength, guiding selection of winding configurations, pole/slot combinations, and magnet arrangements to enhance diagnosability or reduce fault susceptibility.

Model-based Diagnostics Requiring Moderate Update Rates: For diagnostic algorithms running at sampling rates of 1–5 kHz, where lumped parameter model accuracy is insufficient but FEM is far too slow, FRM may provide a viable compromise, delivering field-based model fidelity at computationally feasible rates on modern processors.

Validation of Simplified Models: FRM serves as an intermediate validation reference between simple lumped parameter models (easy to implement but low fidelity) and time-stepping FEM (high fidelity but expensive). Simplified models can be validated against FRM across many operating points quickly, with selective FEM validation at a few critical points to validate the FRM itself.

Hybrid Multi-fidelity Modeling Systems: In advanced diagnostic systems, FRM can serve as the middle layer in a three-tier hierarchy: lumped parameter models for real-time monitoring and control, FRM for moderate-rate diagnostic analysis and fault classification, and FEM reserved for detailed offline forensic analysis after fault detection. Each modeling level is invoked according to the required accuracy and available computational resources.

FRM is not a universal replacement for either lumped parameter models or FEM. For applications requiring maximum computational efficiency (embedded real-time control), lumped parameter models remain superior. For applications demanding absolute maximum accuracy or involving severe nonlinearity, time-stepping FEM is more reliable. However, for the large class of fault diagnosis applications in the intermediate regime, requiring better-than-lumped accuracy at better-than-FEM speed, the Field Reconstruction Method represents a powerful and increasingly adopted solution.

3.5 Winding Function Models

3.5.1 Introduction

Winding function theory represents a classical analytical approach to electric machine modeling that has found renewed importance in fault diagnosis applications. Originally developed in the 1960s and later refined by numerous researchers including Lipo, Toliyat, and others [33–35], this method provides closed-form analytical expressions for calculating inductances, flux linkages, and electromagnetic torque by explicitly incorporating the spatial distribution of windings and the geometry-dependent magnetic permeance of the air gap. Unlike lumped parameter models that use constant or position-dependent parameters obtained empirically or through finite element precomputation, winding function models derive these parameters directly from machine geometry through analytical integration, offering valuable physical insight into how geometric features and faults affect machine performance.

The fundamental premise of winding function theory is elegantly simple yet powerful: the inductance between any two windings can be calculated by integrating the product of their respective winding distribution functions and the air-gap permeance function over the stator bore. This formulation naturally incorporates spatial effects such as winding distribution, slot openings, and most importantly for fault diagnosis, geometric asymmetries like eccentricity or localized demagnetization. The analytical nature of the approach enables parametric studies that would be computationally prohibitive with finite element methods, while the explicit use of geometric quantities provides interpretability that data-driven models lack.

For permanent magnet machine fault diagnosis, winding function models occupy a valuable middle ground in the modeling hierarchy. They are more detailed than simple dq-axis lumped parameter models, explicitly capturing spatial harmonics and position-dependent phenomena, yet far more computationally efficient than finite element analysis. This combination makes them particularly well-suited for offline fault signature prediction, diagnostic algorithm development, and understanding the physical mechanisms by which geometric faults such as eccentricity produce characteristic electrical signatures. Recent extensions called the Modified Winding Function Approach (MWFA) have adapted the classical theory to handle nonuniform air gaps and permanent magnet machines [36–38], expanding its applicability to modern fault diagnosis challenges.

3.5.2 Mathematical Foundation

The winding function approach is built upon two fundamental distribution functions that characterize the spatial arrangement of windings around the stator bore: the **turns function** and the **winding function** [33, 34]. Understanding the distinction between these two functions is essential, as it forms the basis for both the classical winding function theory and its modified extension for nonuniform air gaps.

3.5.2.1 Turns Function

The **turns function** $n_k(\theta)$ describes the cumulative number of turns of winding k that enclose the magnetic axis at mechanical angle θ around the stator bore. For a single full-pitch coil with N turns placed in slots at angles θ_1 and $\theta_1 + \pi$ (for a 2-pole machine), the turns function is a step function that increases by N at the "go" conductor and decreases by N at the "return" conductor:

$$n_k(\theta) = \begin{cases} N & \theta_1 \leq \theta < \theta_1 + \pi \\ 0 & \text{elsewhere} \end{cases} \tag{3.32}$$

For distributed windings with multiple coils, the turns function is the superposition of individual coil contributions, resulting in a staircase waveform that approximates a sinusoidal distribution as the number of slots per pole per phase increases.

3.5.2.2 Winding Function for Uniform Air Gap

The **winding function** $N_k(\theta)$ is derived from the turns function by subtracting its spatial average. For a machine with uniform air gap, this relationship is:

$$N_k(\theta) = n_k(\theta) - \langle n_k(\theta) \rangle \tag{3.33}$$

where the angle brackets denote the spatial average over one complete revolution:

$$\langle n_k(\theta) \rangle = \frac{1}{2\pi} \int_0^{2\pi} n_k(\theta) \, d\theta \tag{3.34}$$

This subtraction ensures that the winding function has zero mean, which is a physical requirement arising from Ampère's law: the net magnetomotive force (MMF) around any closed path encircling the machine must be zero under balanced conditions. The winding function thus represents the spatial MMF distribution produced by unit current in the winding. Figure 3.11 illustrates both

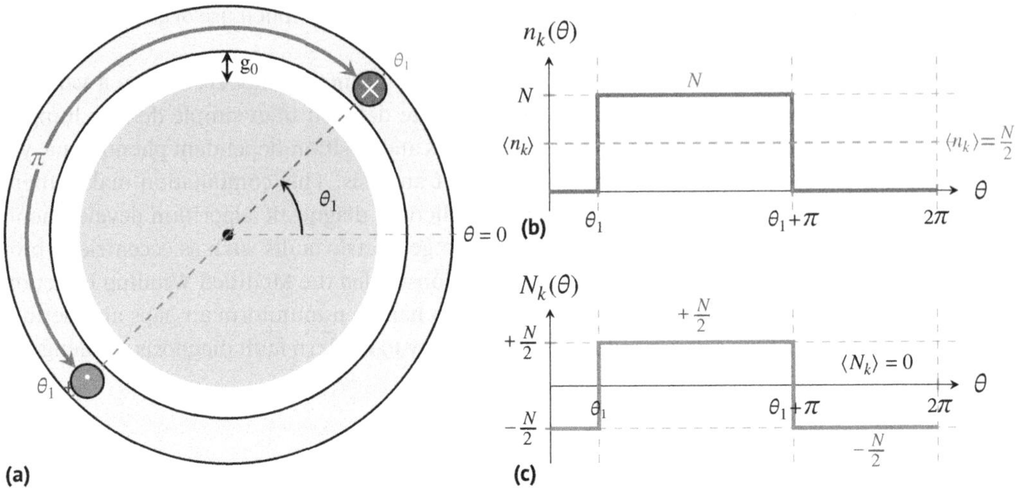

Figure 3.11 Turns function and winding function for a single coil: (a) stator cross-section with coil conductors at angles θ_1 and $\theta_1 + \beta$, (b) turns function $n_k(\theta)$ showing the cumulative turn count, and (c) winding function $N_k(\theta) = n_k(\theta) - \langle n_k \rangle$ with zero mean.

the turns function and winding function for a single coil, showing how the spatial distribution of conductors translates into these step-function representations.

For distributed windings with multiple coils per phase, the total winding function is the sum of individual coil contributions. Figure 3.12 shows a three-phase distributed winding with $q = 2$ slots per pole per phase and $p = 2$ pole pairs. The resulting winding functions for phases A, B, and C are shifted by 120 electrical degrees, and the harmonic spectrum reveals that the fundamental spatial harmonic occurs at order $p = 2$, with higher harmonics at odd multiples ($3p$, $5p$, $7p$, etc.) due to the discrete slot distribution.

3.5.2.3 Modified Winding Function for Nonuniform Air Gap

The classical winding function theory described above assumes a uniform air gap, which limits its applicability to healthy machines or those with faults that do not affect air-gap geometry. Al-Nuaim and Toliyat [36] introduced the **modified winding function approach (MWFA)** to extend the theory to machines with nonuniform air gaps, such as those experiencing eccentricity faults.

The key insight is that when the air gap varies spatially, the magnetic flux does not distribute uniformly around the machine. Regions with smaller air gaps carry proportionally more flux than regions with larger gaps. To account for this, the spatial average in the classical winding function definition must be weighted by the inverse air-gap function $g^{-1}(\theta, \theta_r)$, which represents the local permeance. The **modified winding function** is defined as:

$$M_k(\theta, \theta_r) = n_k(\theta) - \frac{\langle n_k(\theta) \cdot g^{-1}(\theta, \theta_r) \rangle}{\langle g^{-1}(\theta, \theta_r) \rangle} \tag{3.35}$$

where the weighted averages are computed as:

$$\langle n_k(\theta) \cdot g^{-1}(\theta, \theta_r) \rangle = \frac{1}{2\pi} \int_0^{2\pi} n_k(\theta) \cdot g^{-1}(\theta, \theta_r) \, d\theta \tag{3.36}$$

$$\langle g^{-1}(\theta, \theta_r) \rangle = \frac{1}{2\pi} \int_0^{2\pi} g^{-1}(\theta, \theta_r) \, d\theta \tag{3.37}$$

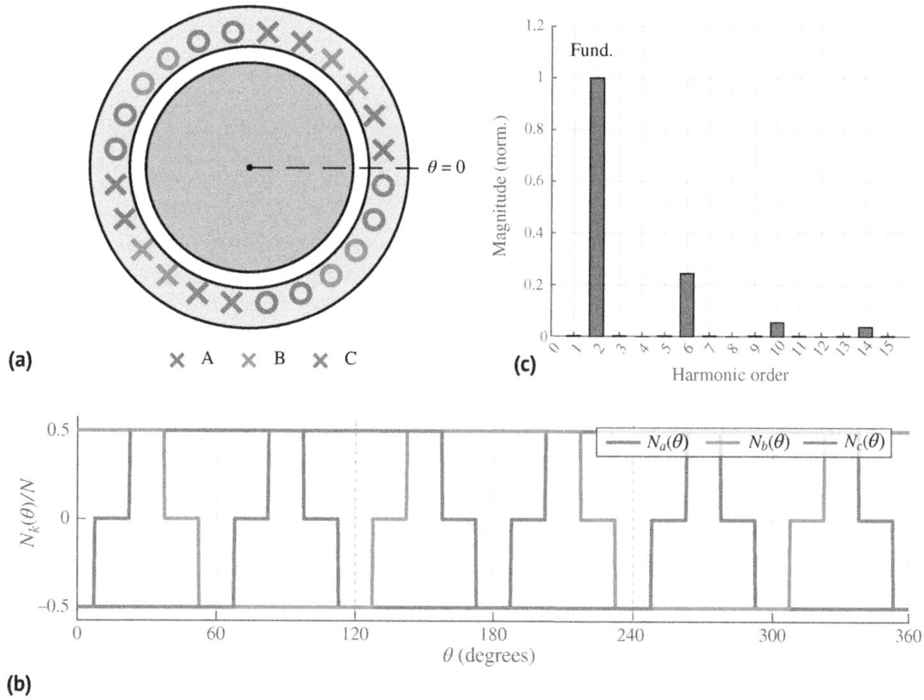

Figure 3.12 Distributed three-phase winding function: (a) stator cross-section showing conductor distribution for a 24-slot, 4-pole machine with $q = 2$ slots per pole per phase, (b) three-phase winding functions $N_a(\theta)$, $N_b(\theta)$, $N_c(\theta)$ displaced by 120 electrical degrees, and (c) spatial harmonic spectrum showing the fundamental at harmonic order $p = 2$.

The physical interpretation of this equation is that the correction term (the second term on the right-hand side) is no longer a simple spatial average but a flux-weighted average that accounts for how the nonuniform air gap redistributes the magnetic flux. When the air gap is uniform, $g^{-1}(\theta, \theta_r) = g_0^{-1}$ becomes constant and cancels from the ratio, reducing the modified winding function to the classical winding function $N_k(\theta) = n_k(\theta) - \langle n_k(\theta) \rangle$.

For eccentricity faults, the air-gap function varies as described in the fault modeling section, and the modified winding function captures how this geometric asymmetry alters the effective MMF distribution and, consequently, the machine inductances. This formulation has been widely adopted for modeling static, dynamic, and mixed eccentricity in both induction and permanent magnet machines [37, 38].

3.5.2.4 Air-gap Permeance Function

The **air-gap permeance function** $\Lambda(\theta, \theta_r)$ characterizes the magnetic conductance of the air gap as a function of both stator position θ and rotor position θ_r. For a healthy machine with uniform air gap g_0, slotting effects, and rotor saliency:

$$\Lambda(\theta, \theta_r) = \Lambda_0 + \sum_{n=1}^{\infty} \Lambda_n \cos(nN_s\theta) + \sum_{m=1}^{\infty} \Lambda_m \cos(mp(\theta - \theta_r)) \tag{3.38}$$

where $\Lambda_0 = \mu_0/g_0$ is the average permeance, N_s is the number of stator slots, and p is the number of pole pairs. The first harmonic series captures slotting effects, while the second captures

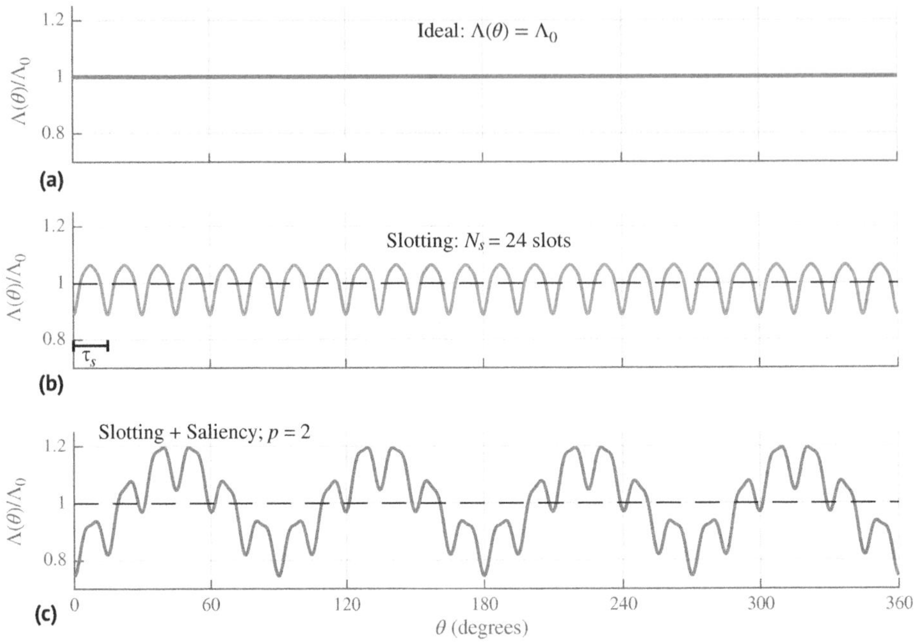

Figure 3.13 Air-gap permeance function: (a) ideal uniform permeance Λ_0, (b) permeance with slotting harmonics showing high-frequency ripple from $N_s = 24$ stator slots, and (c) permeance with both slotting and rotor saliency effects for $p = 2$ pole pairs.

rotor saliency. Figure 3.13 illustrates this progression: panel (a) shows the ideal uniform permeance, panel (b) demonstrates the high-frequency ripple introduced by stator slotting with $N_s = 24$ slots, and panel (c) reveals the combined effect of slotting and rotor saliency for a $p = 2$ pole-pair machine, where the saliency introduces a lower-frequency modulation superimposed on the slotting harmonics. Under eccentricity or other geometric faults, this function is further modified to reflect the varying air-gap length, and the modified winding function $M_k(\theta, \theta_r)$ must be used in inductance calculations.

3.5.2.5 Inductance Calculation

The **mutual inductance** between two windings j and k is calculated by integrating the product of their winding functions (or modified winding functions for nonuniform air gaps) with the inverse air-gap function. For the general case with nonuniform air gap, the inductance formula using the modified winding function approach is:

$$L_{jk}(\theta_r) = \mu_0 rl \int_0^{2\pi} M_j(\theta, \theta_r) \cdot n_k(\theta) \cdot g^{-1}(\theta, \theta_r)\, d\theta \tag{3.39}$$

where μ_0 is the permeability of free space, r is the mean air-gap radius, and l is the axial stack length. For uniform air-gap machines, this reduces to the classical winding function formulation:

$$L_{jk}(\theta_r) = \mu_0 rl \int_0^{2\pi} N_j(\theta) N_k(\theta) \Lambda(\theta, \theta_r) d\theta \tag{3.40}$$

This elegant integral formulation encapsulates the entire geometric contribution to inductance. Self-inductances L_{jj} are computed similarly by setting $j = k$. For a three-phase machine, this produces the position-dependent inductance matrix $\mathbf{L}_{\text{abc}}(\theta_r)$ needed for the abc reference frame models discussed earlier.

The **flux linkage** contribution from permanent magnets can be expressed as:

$$\lambda_{\text{pm},k}(\theta_r) = \mu_0 r l \int_0^{2\pi} N_k(\theta) B_{\text{pm}}(\theta, \theta_r) \Lambda(\theta, \theta_r) d\theta \tag{3.41}$$

where $B_{\text{pm}}(\theta, \theta_r)$ is the permanent magnet flux density distribution in the air gap. For machines with nonuniform air gaps, the modified winding function $M_k(\theta, \theta_r)$ replaces $N_k(\theta)$ in this expression. This formulation naturally captures how winding distribution affects the fundamental and harmonic content of the back-EMF.

3.5.3 Advantages and Limitations

3.5.3.1 Advantages

Winding function models provide several critical capabilities that make them valuable for permanent magnet machine fault diagnosis:

Computationally Efficient Analytical Predictions: The closed-form integral formulations enable extremely fast computation compared to finite element methods. Once the winding and permeance functions are defined, inductance calculations involve straightforward numerical integration that can be performed in milliseconds, even for complex winding arrangements. This computational efficiency enables rapid parametric studies exploring how fault severity affects electrical signatures across wide operating ranges, generating comprehensive fault signature databases that would be impractical with FEM.

Explicit Incorporation of Machine Geometry: Unlike lumped parameter models where inductances are treated as empirical parameters, winding function theory derives inductances directly from geometric quantities: number of turns, coil pitch, air-gap length, slot geometry, and rotor position. This explicit geometric dependence means that when a geometric fault occurs, such as eccentricity causing air-gap variation, its effect on inductances and flux linkages can be predicted analytically by modifying the permeance function. This direct physics-based connection between fault geometry and electrical signatures provides interpretability and confidence that empirical models cannot match.

Natural Capture of Spatial Harmonics: The winding and permeance functions inherently contain harmonic content due to discrete winding distributions and slotting effects. When these functions are integrated to compute inductances and flux linkages, the resulting quantities naturally include spatial harmonics. This is particularly valuable for fault diagnosis because many faults manifest primarily through their harmonic signatures. Eccentricity introduces specific permeance harmonics, inter-turn shorts alter winding functions, and demagnetization modifies permanent magnet flux distribution, all of which affect the harmonic content of inductances and back-EMF in ways that winding function theory can predict analytically.

Prediction of Inductance Variation and Torque Ripple: Position-dependent inductances and their derivatives directly determine cogging torque and reluctance torque components. Winding function models compute $L_{jk}(\theta_r)$ as explicit functions of rotor position, enabling analytical prediction of how faults affect torque ripple. For example, eccentricity modifies the permeance function harmonics, which propagate through the inductance calculation to alter torque ripple patterns. This capability is valuable for diagnosing faults through vibration analysis or torque quality monitoring.

Analytical Incorporation of Faults: Because faults often manifest as changes in geometry (eccentricity), winding configuration (inter-turn shorts), or field distribution (demagnetization), they can be incorporated into winding function models by modifying the appropriate functions. This analytical fault modeling enables exploration of fault physics: which harmonics increase with fault severity, which electrical quantities are most sensitive to particular faults, and how faults in different machine designs produce different signatures. These insights guide diagnostic algorithm development and sensor selection.

Validation of Simplified Models: Winding function models serve as an intermediate validation reference between simple lumped parameter models and high-fidelity FEM. When developing fast abc or dq models for real-time diagnosis, comparing their predictions against winding function theory helps identify whether simplifications are valid for the machine and fault types of interest, without requiring expensive FEM computations for every validation case.

3.5.4 Application to Fault Modeling

Winding function theory has been successfully applied to model several important fault types in permanent magnet machines, with eccentricity and inter-turn short circuits being the most extensively studied [36, 38–40].

3.5.4.1 Eccentricity Fault Modeling

Eccentricity faults, where the rotor and stator centers are misaligned, represent one of the most natural applications of modified winding function theory [36, 37]. The fault manifests as a spatially varying air gap, which directly affects the permeance function.

For **static eccentricity**, where the minimum air-gap position remains fixed in space, the air-gap length varies as:

$$g(\theta) = g_0(1 - \varepsilon_s \cos(\theta - \theta_{ecc})) \tag{3.42}$$

where g_0 is the nominal air-gap length, ε_s is the static eccentricity ratio (0 to 1), and θ_{ecc} is the angular position of minimum air gap. The permeance function, being inversely proportional to air gap length, becomes:

$$\Lambda(\theta) \approx \frac{\Lambda_0}{1 - \varepsilon_s \cos(\theta - \theta_{ecc})} \approx \Lambda_0[1 + \varepsilon_s \cos(\theta - \theta_{ecc}) + \varepsilon_s^2 \cos^2(\theta - \theta_{ecc}) + \cdots] \tag{3.43}$$

This modified permeance, when integrated with winding functions, produces inductances containing additional harmonic components not present in healthy machines. These harmonics propagate to phase currents, producing the characteristic sideband frequencies that serve as eccentricity diagnostic signatures.

For **dynamic eccentricity**, where the rotor whirls around the stator center, the air gap varies as:

$$g(\theta, \theta_r) = g_0(1 - \varepsilon_d \cos(\theta - \theta_r - \theta_{ecc})) \tag{3.44}$$

where ε_d is the dynamic eccentricity ratio. The permeance now depends on both stator and rotor positions, introducing time-varying inductances with different harmonic patterns than static eccentricity, enabling fault type discrimination [38].

Recent analytical work has developed exact Fourier series solutions to the modified winding function equations under eccentricity, eliminating the need for numerical integration and explicitly revealing the harmonic content of inductances as functions of eccentricity severity [41]. These closed-form solutions enable extremely fast parametric studies and provide clear physical insight into which harmonics grow with increasing eccentricity.

3.5.4.2 Inter-turn Short Circuit Modeling

Inter-turn short circuits in a phase winding can be modeled by splitting the winding function into healthy and faulted portions [39]. If phase a has N_s shorted turns out of N_a total turns, the winding function is modified to:

$$N_a(\theta) = N_{a,\text{healthy}}(\theta) + N_{a,\text{shorted}}(\theta) \tag{3.45}$$

where $N_{a,\text{healthy}}$ represents $(N_a - N_s)$ turns carrying normal current, and $N_{a,\text{shorted}}$ represents the N_s shorted turns carrying fault current determined by the short-circuit impedance. This split allows computing inductances between healthy and faulted portions, leading to a model that predicts the circulating current in the short circuit and its effect on terminal currents and electromagnetic torque.

The analytical nature of the winding function approach enables studies of how fault location (which coils are affected) and severity (number of shorted turns) affect the diagnostic signatures, providing insights difficult to obtain from purely empirical or FEM-based approaches.

3.5.4.3 Partial Demagnetization Modeling

Localized demagnetization of permanent magnets can be incorporated by modifying the permanent magnet flux density function $B_{\text{pm}}(\theta, \theta_r)$ to reduce the remanent flux density in affected angular regions. When this modified function is integrated with winding functions to compute flux linkage, the resulting back-EMF waveforms contain additional harmonic components and reduced fundamental amplitude, matching experimental observations of demagnetization faults. The analytical framework enables prediction of how demagnetization extent and location affect back-EMF harmonics, guiding detection algorithm design.

3.5.4.4 Limitations

Despite its advantages, winding function theory has fundamental limitations that constrain its accuracy and applicability to certain machine types and fault conditions:

Assumption of Radial Flux and Parallel Field Lines: The classical winding function formulation assumes that magnetic flux crosses the air gap radially and that field lines are parallel to the radial direction [33]. This assumption is reasonable for machines with small air gaps relative to rotor radius, but breaks down for large air-gap machines, axial flux machines, or regions near slot openings where fringing fields are significant. Faults that produce strong nonradial flux components may not be accurately captured.

Linear Magnetic Conditions: Standard winding function theory assumes linear magnetic materials with constant permeability. Real permanent magnet machines exhibit significant magnetic saturation in stator and rotor iron, especially at high loads. Saturation alters the effective permeability and can significantly affect inductances, particularly the d-axis inductance in interior PM machines. Extensions incorporating saturation effects exist [40, 42] but significantly increase model complexity and reduce the analytical tractability that makes winding function theory attractive. For fault diagnosis, this limitation means that predicted fault signatures may be inaccurate in highly saturated operating conditions unless saturation corrections are applied.

No 3D Effects: Winding function theory typically formulates machines in two dimensions, assuming infinite axial length or that end effects can be accounted for through simple correction factors. 3D effects such as end-winding leakage, axial flux variations, and skewing are not naturally represented. Some faults may produce localized 3D field distortions that 2D winding function models cannot capture. Extensions to 3D exist [43] but lose much of the computational advantage over FEM.

Lower Accuracy Than Numerical Methods: Even when its assumptions are valid, winding function theory provides lower accuracy than well-executed finite element analysis. Typical agreement with FEM or experiments is in the range of 5–15% for inductance calculations, compared to 1–3% for FEM [40, 42]. For fault diagnosis applications where quantitative accuracy is critical, such as severity estimation based on harmonic amplitudes, this accuracy limitation must be considered. Winding function models are best suited for qualitative fault signature prediction, algorithmic development, and parametric studies rather than precise quantitative predictions.

Limited Fault Type Coverage: While eccentricity and inter-turn shorts are well-suited to winding function modeling because they map naturally to permeance and winding function modifications, other faults are more challenging. Bearing faults, rotor mechanical defects, power electronics faults, and thermal degradation may not have clear representations in terms of winding and permeance functions. The method is most applicable to faults that directly affect the electromagnetic circuit geometry or winding configuration.

Complexity for Concentrated Windings and Fractional-slot Designs: Modern permanent magnet machines increasingly use concentrated windings and fractional-slot designs for manufacturing simplicity and fault tolerance [35]. The winding functions for these configurations are more complex than distributed windings, and the permeance function contains richer harmonic content due to large slot openings. While winding function theory can handle these cases, the analytical tractability decreases, and numerical computation of the integrals becomes necessary, reducing the computational advantage over other methods.

3.5.5 Role in Fault Diagnosis Workflow

Given its strengths and limitations, winding function theory serves specific valuable purposes in permanent magnet machine fault diagnosis:

Rapid Fault Signature Generation: When developing diagnostic algorithms, engineers need to understand which electrical signatures each fault type produces and how signatures vary with fault severity and operating conditions. Winding function models can generate these fault signature maps orders of magnitude faster than FEM, enabling comprehensive parametric studies that guide algorithm design, threshold setting, and operating range specification.

Physical Insight and Algorithm Validation: The analytical nature and explicit geometric dependence of winding function theory provide physical understanding of fault mechanisms. Why does static eccentricity produce current sidebands at $f_s \pm f_r$? The modified permeance function harmonics provide the answer. This insight helps validate whether diagnostic algorithms are detecting the correct physical phenomena rather than spurious correlations.

Trade-off Analysis for Machine Design: During machine design, winding function models can quickly assess how design choices (distributed vs. concentrated windings, number of poles, slot/pole combinations) affect fault sensitivity and diagnosability. Designs can be compared for their ability to produce strong, distinguishable fault signatures before committing to detailed FEM optimization.

Educational and Research Tool: For students and researchers learning permanent magnet machine fault diagnosis, winding function theory provides an accessible entry point that reveals the connection between geometry and electrical signatures without requiring extensive FEM expertise or computational resources.

Complement to Other Methods: Winding function models work best as part of a multi-fidelity modeling strategy. Use FEM to validate the winding function model for a few key operating points and fault severities, establishing trust. Then use the validated winding function model for extensive parametric studies and algorithm development. Finally, implement simplified lumped parameter models for real-time diagnosis, with confidence from the winding function and FEM validation.

Winding function theory occupies a valuable niche in the permanent magnet machine fault diagnosis modeling ecosystem. It is too slow and detailed for real-time diagnosis but too fast and insightful to ignore in favor of FEM alone. When applied to appropriate fault types, particularly geometric faults like eccentricity, it provides an optimal balance of physical insight, computational efficiency, and accuracy for offline diagnostic algorithm development and validation.

The computational workflow for winding function-based fault modeling follows a systematic sequence, as illustrated in Figure 3.14. Starting from machine geometry specifications, the process involves computing winding functions for each phase, defining the air-gap permeance function (modified appropriately for the fault condition of interest), integrating to obtain position-dependent inductances and PM flux linkages, and finally substituting these into the coupled circuit equations for time-domain simulation and fault signature analysis.

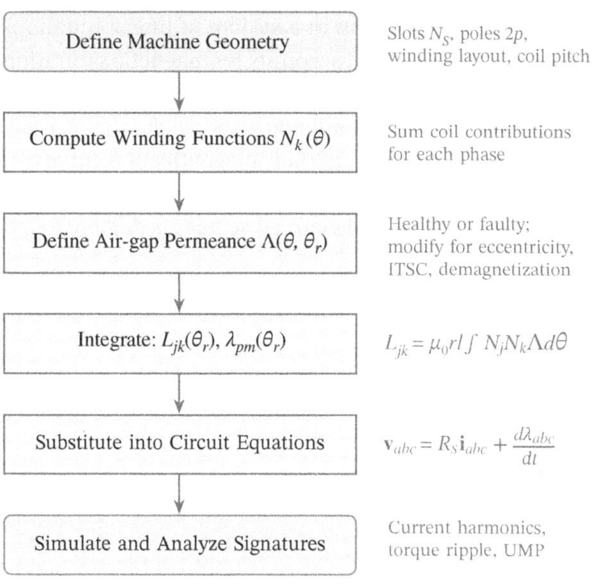

Figure 3.14 Winding function modeling workflow for fault analysis.

3.6 Magnetic Equivalent Circuit

3.6.1 Introduction

The MEC method represents an electromagnetic machine's magnetic structure as a network of magnetic circuit elements, analogous to how electrical circuits model current flow. In this approach, the complex 3D magnetic field distribution within a PM motor is abstracted into a network of reluctances (magnetic resistances), magnetomotive forces (MMF sources), and permeances (magnetic conductances). This conceptual mapping leverages the mathematical similarity between electrical and magnetic phenomena: just as Ohm's law relates voltage, current, and resistance ($V = IR$), magnetic circuits relate MMF, flux, and reluctance ($F = \Phi R$). Table 3.3 summarizes the correspondence between electrical and magnetic circuit quantities.

The fundamental premise of MEC modeling is that electromagnetic systems exhibit physical coupling between electric and magnetic components. The electric part of the system is modeled using conventional electrical elements such as voltage and current sources, resistance, inductance, and capacitance. The magnetic part is represented using magnetic elements where the magnetic field intensity is influenced by the winding magnetomotive force (MMF), which serves as the source of magnetic flux. The linear relationship between MMF and magnetic flux gives rise to reluctance [44].

In a typical MEC model, each segment of the magnetic path through the stator core, rotor core, permanent magnets, and air gap is represented by a reluctance element whose value depends on the material's permeability, the path's geometry, and (for ferromagnetic materials) the local flux density. Permanent magnets are modeled as series combinations of MMF sources and reluctances, accounting for their finite permeability and magnetizing orientation. The winding MMFs serve as excitation sources in the network, with values determined by the number of turns and phase currents.

A key distinction exists between linear and nonlinear MEC formulations. Linear MEC assumes constant permeability throughout the magnetic circuit, resulting in a system of linear equations that can be solved directly using matrix methods. Nonlinear MEC accounts for magnetic saturation by treating core reluctances as functions of local flux density, requiring iterative solution techniques such as the Newton–Raphson or Gauss–Seidel methods. This nonlinearity is crucial for accurate modeling of PM motors operating near or beyond the knee point of their magnetization curves.

Recent research continues to validate MEC's relevance in modern motor analysis. A 2024 study on permanent magnet linear synchronous motors with arbitrarily curved segments demonstrated

Table 3.3 Electric-magnetic circuit analogy showing corresponding quantities and governing laws in both domains.

Electrical domain	Magnetic domain
Voltage, V (V)	Magnetomotive force, F (A-turns)
Current, I (A)	Magnetic flux, Φ (Wb)
Resistance, R (Ω)	Reluctance, $\mathcal{R}$ (A-turns/Wb)
Conductance, G (S)	Permeance, P (Wb/A-turns)
Ohm's law: $V = IR$	Hopkinson's law: $F = \Phi\mathcal{R}$
KCL: $\sum I = 0$	Gauss's law: $\sum \Phi = 0$
KVL: $\sum V = 0$	Ampère's law: $\sum F = 0$

that MEC modeling provides accurate and computationally efficient predictions, paving the way for real-time model-based control, observers, and estimation strategies. The technique is particularly valuable for capturing geometric asymmetries and local phenomena that would require extremely fine meshes in FEM.

The MEC method occupies a strategic position in the modeling hierarchy for PM motor fault diagnosis, offering a unique balance between physical insight, computational efficiency, and accuracy. This positioning makes it particularly valuable for several critical aspects of fault analysis and diagnosis system development.

Computational Efficiency with Physical Fidelity: Unlike FEM, which can require hours to simulate a single operating condition, MEC models typically execute in seconds to minutes while maintaining acceptable accuracy. A 2024 study reports that MEC gives fast results (minutes) with acceptably accurate outcomes, representing a computation time reduction of 30–100 times compared to FEM. This speed advantage makes MEC practical for applications requiring multiple operating point analyses, parametric studies, or preliminary design optimization tasks that would be prohibitive with FEM.

Natural Incorporation of Saturation Effects: One of MEC's most significant advantages for fault diagnosis is its straightforward handling of magnetic saturation. Since reluctance network elements can be directly formulated as functions of local flux density through B–H curve relationships, saturation effects emerge naturally from the model without requiring specialized formulation techniques. This capability is crucial because many fault conditions (particularly demagnetization, inter-turn shorts, and eccentricity) alter the machine's saturation distribution in ways that affect diagnostic signatures. Research has demonstrated that MEC methods can successfully incorporate saturation effects through B–H curve fitting using novel nonlinear functions.

Fault Modeling Flexibility: The flexibility of MEC with adjustable accuracy allows analysis of both healthy and faulty motors through a unique model, thanks to shorter processing time compared to FEM. This adaptability is particularly valuable for fault diagnosis research and algorithm development. Several fault types can be incorporated into MEC models:

- **Inter-turn Short Circuits:** Modeled by modifying winding MMF sources and adding additional current paths to represent the shorted turns.
- **Partial Demagnetization:** Represented by reducing the MMF or permeability of affected permanent magnet elements. Studies demonstrate that demagnetization fault detection is achievable using flexible magnetic equivalent circuit methods with adjustable accuracy for various motor topologies.
- **Eccentricity Faults:** Captured by modifying air gap permeances as functions of rotor position to reflect nonuniform air gap length.
- **Phase-to-phase Faults:** Incorporated through modified winding connection topologies and MMF distributions.

Diagnostic Signature Generation: MEC models provide detailed predictions of magnetic flux distributions throughout the motor structure, enabling calculation of diagnostic signals that would be difficult to obtain from simpler models. These include localized flux densities, tooth-specific flux patterns, and spatial harmonic content. A significant finding is that some faults, such as demagnetization, are not readily detectable through stator current analysis alone; MEC models can predict the benefits of additional sensing approaches such as pole-specific search coils.

Algorithm Development and Validation: MEC serves as an excellent middle-ground tool for fault diagnosis algorithm development. It is detailed enough to capture relevant fault phenomena and generate realistic synthetic data for training and testing diagnostic algorithms, yet fast enough to enable extensive parametric studies exploring fault severity ranges and operating conditions. This makes MEC particularly valuable for developing and validating model-based diagnostic techniques before experimental implementation.

3.6.2 Mathematical Foundation

The implementation of MEC modeling for PM motors involves several systematic steps, from initial network topology definition through equation formulation and solution. This section details the practical methodology, distinguishing between linear and nonlinear solution approaches.

3.6.2.1 Network Topology and Element Definition

The first step in MEC modeling is defining the network topology by identifying all significant magnetic flux paths and representing each as a reluctance or permeance element. The modeling process relies on a standard magnetic branch concept shown in Figure 3.15. Each branch n has a series MMF source with branch reluctance in parallel with the MMF source of the branch, where the input flux is denoted as φ_{bn} and MMF as F_{bn} [44].

The network formation considers the physical structure of the motor:

Stator Magnetic Circuit: The stator region is divided into distinct sections: yoke, leg, tooth tip, and slot, each with characteristic permeances calculated from geometry and material properties. The distribution of magnetic field for surface-mounted PM (SPM) and interior PM (IPM) machines is shown in Figure 3.16, while the corresponding general MEC network structures are illustrated in Figure 3.17. The MMF sources are determined based on winding coil turns N_c and phase current i as:

$$F_{\mathrm{as}} = N_c \times i \tag{3.46}$$

According to the geometric subdivision, permeances for stator elements are calculated from the machine geometry. The stator yoke permeance, slot permeance, and tooth tip permeance are given by:

$$P_{\mathrm{sy}} = \mu_c \frac{l_{\mathrm{stk}}}{\theta_{\mathrm{sv}}} \ln \frac{R_{\mathrm{syo}}}{R_{\mathrm{svi}}}, \quad P_{\mathrm{ls}} = \mu_0 l_{\mathrm{stk}} \frac{2 l_{\mathrm{st}}}{3(W_s + W_l)}, \quad P_{\mathrm{st}} = \mu_c \frac{W_{\mathrm{st}} l_{\mathrm{stk}}}{l_{\mathrm{st}}} \tag{3.47}$$

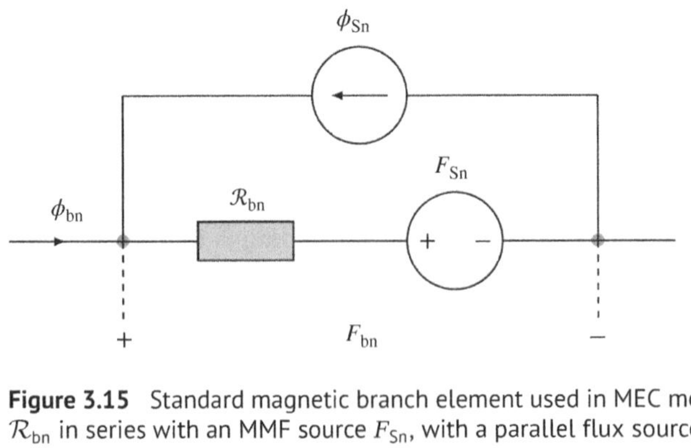

Figure 3.15 Standard magnetic branch element used in MEC modeling. The branch consists of a reluctance $\mathcal{R}_{\mathrm{bn}}$ in series with an MMF source F_{Sn}, with a parallel flux source ϕ_{Sn}. The branch MMF F_{bn} is measured between the input (+) and output (-) nodes.

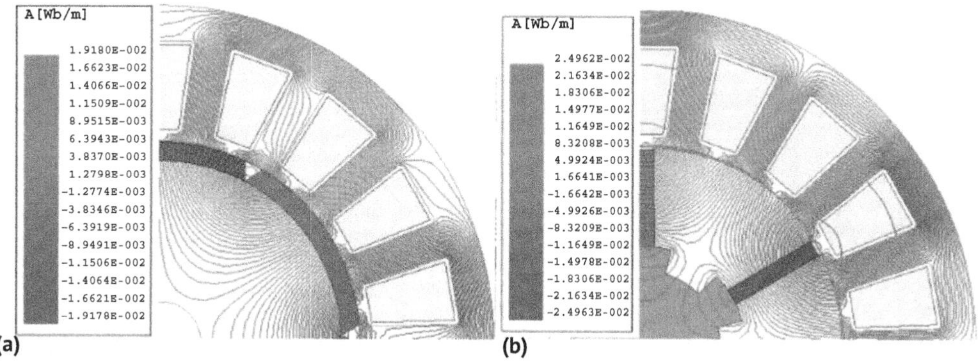

Figure 3.16 Distribution of magnetic field in PM motors: (a) SPM configuration and (b) IPM configuration.

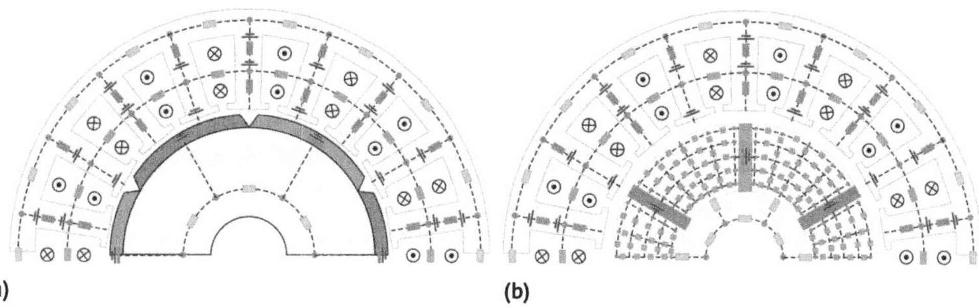

Figure 3.17 General view of MEC network topology: (a) SPM network and (b) IPM network.

where l_{stk} is the stack length, θ_{sv} is the angular span of the stator yoke segment, R_{syo} and R_{svi} are the outer and inner radii of the stator yoke, l_{st} is the stator tooth length, W_s and W_l are the slot and leg widths, and W_{st} is the tooth tip width. Additionally, vertical permeance (P_{vt}) and horizontal permeance (P_{ht}) in each square mesh element are defined to capture local flux paths:

$$P_{ht} = \mu_c l_{stk} \frac{2b}{a}, \quad P_{vt} = \mu_c l_{stk} \frac{2a}{b} \tag{3.48}$$

where a and b are the mesh element dimensions in the radial and tangential directions, respectively [44].

The detailed stator MEC network is shown in Figure 3.18, which illustrates how the stator components are interconnected through permeance elements.

Rotor Magnetic Circuit: For surface-mounted PM machines, the rotor circuit includes permanent magnet elements, rotor yoke, and connecting regions. Permanent magnets (typically N35 grade NdFeB or similar) are modeled considering their constant or field-dependent permeability. The PM elements are represented as series connections of MMF sources and permeances, accounting for PM thickness (h_{PM}), width (W_{PM}), and magnetizing orientation:

$$F_{PM} = H_{PM} h_{PM} = \frac{B_{rem} h_{PM}}{\mu_{PM}}, \quad P_{PM} = \mu_{PM} \frac{W_{PM} l_{stk}}{h_{PM}} \tag{3.49}$$

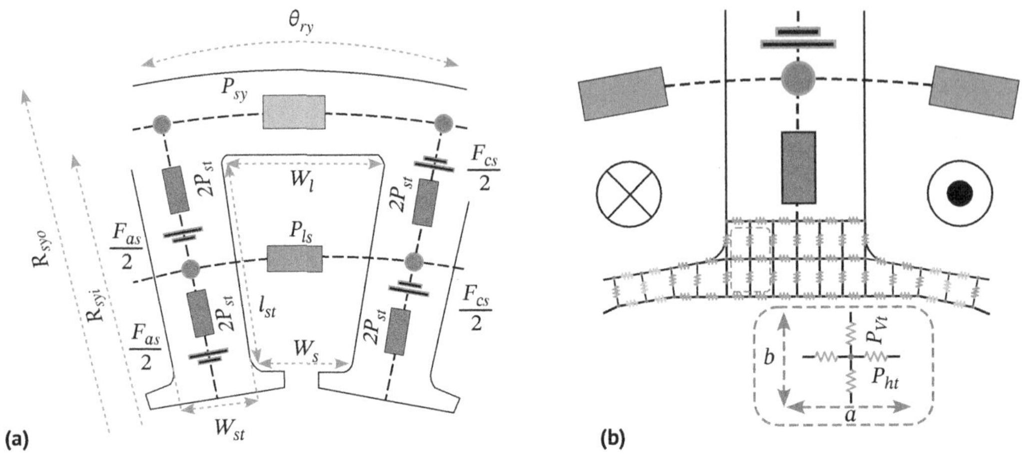

Figure 3.18 Detailed stator MEC network showing permeance distribution: (a) complete stator network topology and (b) mesh element structure.

where B_{rem} is the remanent flux density, μ_{PM} is the PM permeability, h_{PM} is the magnet thickness, and W_{PM} is the magnet width. The rotor yoke permeance (P_{ry}) and the permeance connecting the core and yoke (P_{rm}) are calculated as:

$$P_{\text{ry}} = \mu_c \frac{l_{\text{stk}}}{\theta_{\text{ry}}} \ln \frac{R_{\text{ryo}}}{R_{\text{ryi}}}, \quad P_{\text{rm}} = \mu_c \frac{l_{\text{stk}}\theta_{\text{ry}}}{\ln \frac{r_{\text{rm}}}{R_{\text{ryo}}}} \tag{3.50}$$

where θ_{ry} is the angular span of the rotor yoke segment, R_{ryo} and R_{ryi} are the outer and inner radii of the rotor yoke, and r_{rm} is the radius of the core-yoke connection region. The vertical and horizontal permeances in each rotor square mesh element follow the same form as the stator mesh elements [44]:

$$P_{\text{hr}} = \mu_c l_{\text{stk}} \frac{2b}{a}, \quad P_{\text{vr}} = \mu_c l_{\text{stk}} \frac{2a}{b} \tag{3.51}$$

Air Gap Coupling: One of the most sophisticated aspects of MEC modeling is properly representing the varying magnetic coupling between stator and rotor as the rotor rotates. This requires position-dependent permeances that connect stator boundary nodes to rotor boundary nodes based on their geometric overlap.

The coupling methodology works as follows: Angular regions are determined for every stator and rotor boundary node (the nodes closest to the air gap). As the rotor rotates and overlaps with these angular regions, corresponding nodes are connected by a permeance, with this connection persisting as long as overlapping exists. For a typical motor geometry, each rotor boundary node may overlap with 30–40 stator boundary nodes, requiring that many varying permeances for complete connection representation [44].

The number of varying permeances (N_{vp}) is generally determined as:

$$N_{\text{vp}} = n_{\text{HN}} \times N_{\text{HN}} \tag{3.52}$$

where n_{HN} is the number of boundary nodes for whichever (stator or rotor) has more boundary nodes, and N_{HN} is the number for whichever has fewer [44].

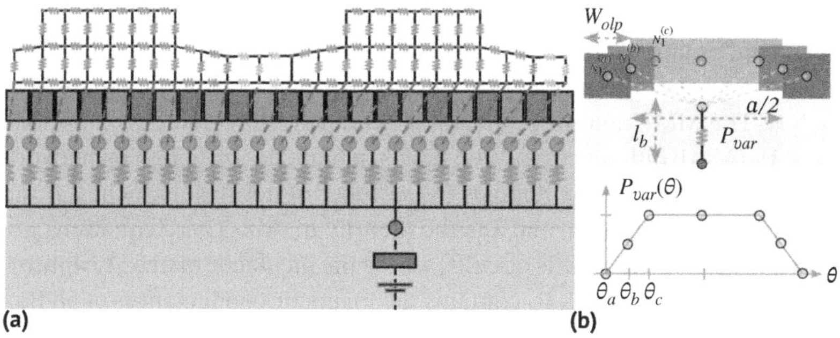

Figure 3.19 Air gap MEC network: (a) connection mechanism showing moving permeances and varying air gap coupling and (b) typical air gap permeance variation curve with rotor displacement.

The varying permeances change continuously with rotor position based on the degree of overlap between node regions. This continuous variation (rather than discrete switching) is essential for producing smooth output waveforms and avoiding the numerical spikes that represent a basic challenge in MEC modeling. The varying and moving permeances are calculated as:

$$P_{\text{var}}(\theta) = \mu_0 l_{\text{stk}} \frac{|\theta|}{\ln \frac{R_{\text{so}}+l_b}{R_{\text{so}}}}, \quad |\theta| \leq \theta_c = \frac{\pi D_{\text{ave}}}{n} = \frac{W_{\text{olp}}}{R_{\text{so}} + l_b} \tag{3.53}$$

where θ is the angular overlap between stator and rotor nodes, R_{so} is the stator bore radius, l_b is the air gap length, D_{ave} is the average air gap diameter, n is the number of stator boundary nodes, and W_{olp} is the overlap width. The permeance value increases with larger overlap angles and decreases with larger air gap lengths [44]. The air gap MEC network connection mechanism and typical permeance variation curve are illustrated in Figure 3.19.

3.6.2.2 Linear MEC Analysis

For linear analysis, the magnetic circuit is assumed to have constant permeability throughout, resulting in a system of linear algebraic equations. The objective is to determine both node voltages (magnetic potentials) and branch currents (fluxes). Since magnetic flux in each branch is more physically significant than MMF at each node, the formulation establishes a direct relationship between MMF sources and branch fluxes as the main unknowns.

Using node analysis principles analogous to electrical circuits, the fundamental constraints are:

$$\mathbf{A}^{n\times m}\boldsymbol{\varphi}_b^{m\times 1} = 0, \quad \mathbf{F}_b^{m\times 1} = \mathbf{A}^T\mathbf{F}_n^{n\times 1} \tag{3.54}$$

where the first equation represents flux conservation (Gauss's law) and the second relates branch MMF drops to node potentials. The branch fluxes can be expressed in terms of the MMF sources through the network permeance matrix **G**:

$$\boldsymbol{\varphi}_b = \mathbf{G}\mathbf{F}_s \tag{3.55}$$

where the network permeance matrix is defined as:

$$\mathbf{G} = \mathbf{P}_b\left[\mathbf{A}^T(\mathbf{A}\mathbf{P}_b\mathbf{A}^T)^{-1}\mathbf{A}\mathbf{P}_b - \mathbf{I}\right] \tag{3.56}$$

In these equations, $\boldsymbol{\varphi}_b$ denotes the branches' magnetic flux column matrix ($m \times 1$), $\mathbf{P}_b$ represents the branches' diagonal permeance matrix ($m \times m$) with the i-th entry on the main diagonal equal to the i-th branch permeance, and $\mathbf{A}$ is the incidence matrix ($n \times m$) describing the connection between nodes and branches. The MMF quantities include $\mathbf{F}_s$ (branch MMF sources), $\mathbf{F}_n$ (node MMF potentials, $n \times 1$), and $\mathbf{F}_b$ (MMF difference between nodes in branches, $m \times 1$). The network permeance matrix $\mathbf{G}$ has dimensions $m \times m$, m denotes the number of network branches, n the number of nodes excluding the reference node, and $\mathbf{I}$ is the identity matrix. These equations are derived using Kirchhoff's laws applied to magnetic circuits, where the incidence matrix $\mathbf{A}$ captures the network topology and the permeance matrix $\mathbf{P}_b$ contains the magnetic conductances of all flux paths. Once the system is solved for node MMFs, branch fluxes can be directly calculated [44].

3.6.2.3 Nonlinear MEC Analysis

For accurate representation of PM motors operating under typical load conditions, magnetic saturation in the stator and rotor cores must be considered. This renders core permeances nonlinear functions of their local flux densities, necessitating iterative solution methods. Research has demonstrated that systems based on loop equations exhibit higher convergence speed compared to systems based on node equations [44].

Two primary iterative methods are employed:

Gauss–Seidel Method: This straightforward approach solves the nonlinear system $\mathbf{AX} = \mathbf{B}$ by iteratively updating the solution based on results from the previous iteration. While simple to implement and not requiring derivative calculations, the Gauss-Seidel method's major drawback is requiring a significant number of iterations to achieve convergence [44].

Newton–Raphson Method: This technique exhibits faster convergence due to its reliance on equation derivatives. However, for node-based MEC equations, the standard Newton-Raphson method can encounter result divergence issues. A practical solution is to approximate the Jacobian matrix as $\mathbf{J} \approx \mathbf{A}$, enabling rapid convergence without adding undue computational complexity [44].

Gauss Elimination with Slight Rotation: This widely used technique for solving nonlinear systems offers superior numerical stability and faster operation even under adverse matrix coefficient conditions. The network equation is reformulated as:

$$\mathbf{A}^T\mathbf{P}_b\mathbf{A}\mathbf{F}_{\text{cal}}^{(k)} = \mathbf{A}^T\mathbf{P}_b\mathbf{F}_s \tag{3.57}$$

The iterative solution proceeds as:

$$\mathbf{F}_{\text{cal}}^{(k)} = \mathbf{F}_{\text{cal}}^{(k-1)} + \alpha\left(\mathbf{F}_{\text{act}}^{(k)} - \mathbf{F}_{\text{cal}}^{(k-1)}\right) \tag{3.58}$$

where α is a relaxation parameter that must remain below 1 for convergence. Values of α close to 1 achieve convergence with fewer iterations, while values closer to zero require more iterations but may provide better stability [44].

Saturation Modeling: To incorporate magnetic saturation effects, core permeability must be determined as a function of local magnetic field intensity. The magnetic permeability relationship can be expressed as:

$$\mu_b(H_b) = \mu_0 + \text{sign}(H_b)\sum_{K=1}^{\infty}\frac{m_K}{h_K}\frac{1}{1+\left(\frac{H_b}{h_K}\right)^{n_K}} \tag{3.59}$$

where m_K, n_K, and h_K are material-specific parameter vectors determined from ferromagnetic material properties. This formulation provides a smooth approximation to the B–H characteristic that is suitable for iterative calculations. For practical implementations, the magnetization characteristic of standard lamination materials (such as M470-50A) is typically used [44, 45].

Solution Algorithm: The complete nonlinear solution algorithm is presented in Algorithm 2.

Algorithm 2 Nonlinear MEC Solution with Saturation

Require: Initial permeabilities μ, rotor position θ, MMF estimates $\mathbf{F}_{\text{cal}}^{(0)}$
Ensure: Converged flux distribution and updated rotor position
Initialize permeabilities μ, rotor position θ, MMF $\mathbf{F}_{\text{cal}}^{(0)}$
repeat
  Update permeances $\mathbf{P}^{(k-1)}$ based on current permeability distribution
  Solve system using iterative method to obtain $\mathbf{F}_{\text{cal}}^{(k)}$
  if $|\mathbf{F}_{\text{cal}}^{(k)} - \mathbf{F}_{\text{cal}}^{(k-1)}| > \delta$ **then** ▷ $\delta \approx 10^{-5}$
    Calculate branch fluxes: $\varphi_b = \mathbf{P}_b(\mathbf{F}_s - \mathbf{A}\mathbf{F}_b)$
    Determine flux densities: $\mathbf{B}_b = \varphi_b / S_{\text{eff}}$
    Obtain $\mathbf{H}_b$ from B–H curve
    Update permeabilities μ_b from $\mathbf{H}_b$ and $\mathbf{B}_b$
    if $|\mu_b^{(k)} - \mu_b^{(k-1)}| > \varepsilon$ **then**
      Update permeances $\mathbf{P}_{\text{new}}$ with $\mu_{b,\text{new}}$
    end if
  end if
until MMF and permeability convergence achieved
Store results and advance rotor position by $\Delta\theta$

This iterative process ensures that the permeability distribution throughout the motor is consistent with the actual flux density distribution, properly capturing saturation effects [44].

3.6.2.4 Simulation Results

To validate the MEC modeling approach, time-domain simulations were performed for both SPM and IPM machines, with results compared against FEM analysis. Figure 3.20 presents the three-phase stator currents obtained from both methods during steady-state operation. The MEC results demonstrate excellent agreement with the FEM reference, accurately capturing the amplitude, phase relationships, and waveform characteristics of all three phases. The close correspondence between MEC and FEM predictions confirms that the magnetic equivalent circuit approach provides sufficient accuracy for fault diagnosis applications while maintaining significant computational advantages over full electromagnetic field solutions.

Similarly, Figure 3.21 compares the three-phase flux linkage waveforms obtained from both modeling approaches. The flux linkage predictions from MEC closely match the FEM results for both machine topologies, further validating the accuracy of the magnetic equivalent circuit method for electromagnetic analysis of PM machines.

3.6.3 Advantages and Limitations

The MEC approach offers distinct benefits and faces specific challenges when applied to PM motor modeling for fault diagnosis applications.

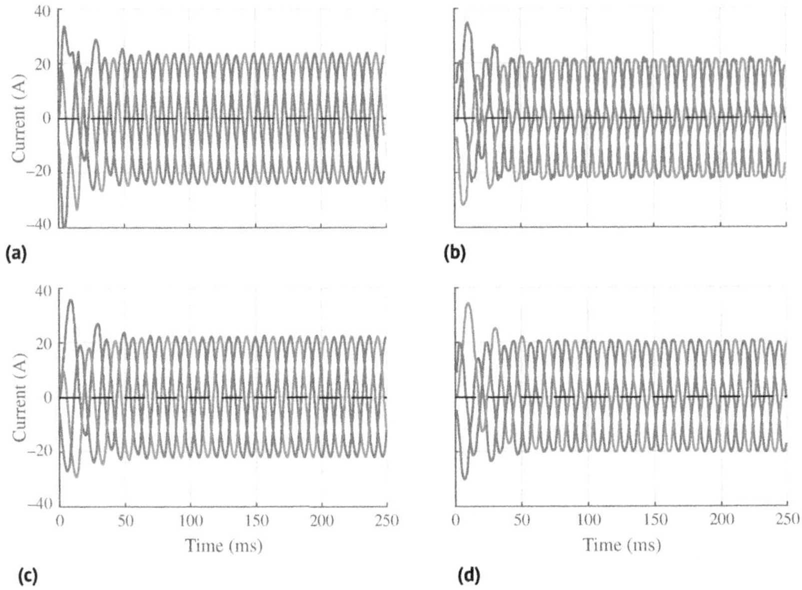

Figure 3.20 Comparison of three-phase stator currents from MEC and FEM simulations: (a) MEC results for SPM machine, (b) MEC results for IPM machine, (c) FEM results for SPM machine, (d) FEM results for IPM machine.

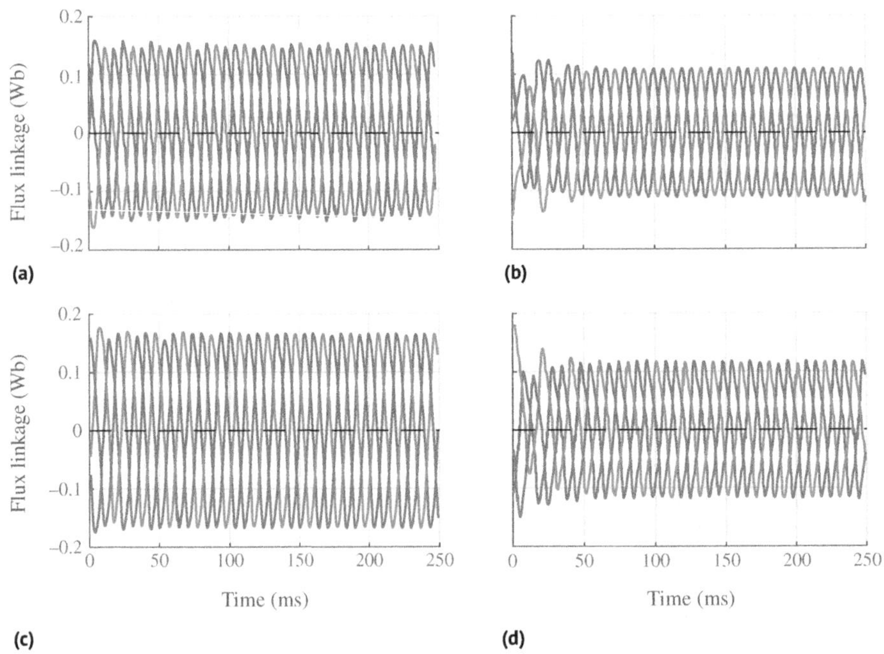

Figure 3.21 Comparison of three-phase flux linkage from MEC and FEM simulations: (a) MEC results for SPM machine, (b) MEC results for IPM machine, (c) FEM results for SPM machine, (d) FEM results for IPM machine.

3.6.3.1 Advantages

Computational Efficiency: MEC models execute 30–100 times faster than equivalent FEM models while maintaining acceptable accuracy. Recent 2024 research confirms that MEC provides fast results (minutes) with acceptably accurate predictions, making it practical for applications requiring multiple analyses, parametric studies, or semi-real-time evaluation.

Physical Insight: The network structure of MEC models provides an intuitive understanding of magnetic flux paths, bottlenecks, and saturation locations. Engineers can directly visualize how faults affect flux distribution and identify which circuit elements are most influenced by specific fault types. This transparency aids in diagnostic algorithm development and signature interpretation.

Natural Saturation Handling: Unlike some analytical methods that struggle with nonlinearity, MEC naturally incorporates saturation through permeability-flux density relationships. The iterative solution framework seamlessly integrates material B–H curves, providing accurate predictions even in heavily saturated regions.

Flexibility for Fault Modeling: MEC's network structure makes it straightforward to modify models for various fault conditions. Demagnetization is represented by changing PM MMF sources, inter-turn shorts by modifying winding MMF distribution and adding short-circuit paths, and eccentricity by adjusting air gap permeances. A single flexible model can represent both healthy and multiple faulty configurations through parameter variations.

Geometric Adaptability: MEC can handle complex motor geometries including nonuniform air gaps, skewing, and asymmetric structures that are relevant to fault conditions. The method accommodates arbitrary pole numbers, magnet configurations, and slot numbers, providing versatility across motor topologies.

Intermediate Fidelity: MEC occupies a valuable middle ground between simple lumped parameter models (too simplified for many fault phenomena) and FEM (too computationally expensive for parametric studies). This positioning makes MEC ideal for offline diagnostic algorithm development, training data generation, and validation studies.

3.6.3.2 Limitations

Mesh Complexity and Setup Time: Constructing a detailed MEC model requires significant effort in defining network topology, identifying all relevant flux paths, determining node positions, and calculating geometric parameters for each permeance element. This setup phase can be time-consuming, particularly for complex motor geometries or when modeling faults that require topology modifications.

End-winding Effects: Standard 2D MEC models cannot directly represent 3D phenomena such as end-winding leakage flux and axial components of the magnetic field. While approximate methods (such as additional lumped permeances for end effects) can be incorporated, these lack the rigor of full 3D field solution methods.

Air Gap Discretization Challenges: Achieving smooth output waveforms requires careful handling of varying air gap permeances. Coarse discretization can produce numerical artifacts and harmonic content that does not reflect physical phenomena. Ensuring permeance variations are continuous rather than discrete is critical but can complicate the implementation [44].

Convergence Issues: Nonlinear MEC solutions can face convergence difficulties, particularly for heavily saturated machines or when using inappropriate relaxation parameters. The Newton-Raphson method may diverge for node-based formulations, requiring modified approaches or alternative solution methods [44].

Limited High-frequency Phenomena: MEC is fundamentally a quasi-static approach that does not naturally incorporate eddy current effects, skin effect, or other high-frequency electromagnetic phenomena. For fault diagnosis techniques relying on high-frequency signal injection or transient analysis, MEC may not provide sufficient accuracy.

Validation Requirements: MEC models typically require validation against FEM or experimental data to ensure accuracy, particularly when applied to novel motor topologies or fault conditions. Parameter values (particularly for air gap and leakage permeances) may need tuning based on comparison with higher-fidelity methods.

Difficulty with Distributed Faults: While MEC handles localized faults (such as single-magnet demagnetization or single-phase shorts) effectively, representing spatially distributed or progressive faults may require extensive model modifications and can reduce computational advantages.

3.6.4 Application to Fault Modeling

MEC modeling has been successfully applied to various fault diagnosis scenarios in PM motors:

Demagnetization Fault Detection: Studies demonstrate that demagnetization faults can be modeled and detected using flexible MEC methods. The research indicates that demagnetization is not readily detectable through stator current analysis alone, motivating the evaluation of additional sensing approaches such as pole-specific search coils. MEC models can predict the magnetic signatures that these search coils would measure, enabling diagnostic algorithm design.

Inter-turn Short Circuit Analysis: MEC approaches enable inter-turn and demagnetization fault analysis in surface-mounted PM synchronous machines. The method can represent the altered MMF distribution caused by shorted turns and predict resulting changes in flux distribution, terminal quantities, and diagnostic signal harmonics.

Eccentricity fault modeling: MEC is particularly well-suited for eccentricity fault analysis because it naturally represents the varying air gap through position-dependent permeances. Research specifically addresses magnetic equivalent circuit modeling for interior PM synchronous machines under eccentricity faults, demonstrating the method's capability to predict unbalanced magnetic pull and current signature changes.

Multiple Fault Discrimination: The flexibility of MEC enables simultaneous representation of multiple fault types, which is valuable for developing diagnostic algorithms that can distinguish between demagnetization, eccentricity, and inter-turn faults. Studies address diagnosis and discernment between eccentricity and demagnetization faults using modeling approaches that incorporate MEC principles.

Saturation-dependent Fault Signatures: Because MEC naturally incorporates saturation, it can predict how fault signatures change with loading conditions. This is particularly important for faults such as inter-turn shorts where the short-circuit current (and thus the fault severity) depends on the saturation state of the magnetic circuit.

3.6.5 Role in Fault Diagnosis Workflow

Based on the characteristics and demonstrated applications of MEC modeling, the following recommendations guide its effective use in PM motor fault diagnosis:

When to use MEC:

- Developing fault diagnostic algorithms requiring multiple simulation runs across operating conditions.

- Analyzing faults with significant geometric or magnetic asymmetry (eccentricity, local demagnetization).
- Studying saturation-dependent fault phenomena.
- Conducting parametric studies to understand fault severity effects.
- Generating synthetic training data for machine learning-based diagnostic algorithms.
- Preliminary design studies where FEM would be too slow but lumped models too simplified.

When to consider alternatives:

- Real-time online diagnosis applications (use lumped parameter models).
- Highest accuracy requirements for research validation (use FEM).
- Faults involving significant high-frequency or eddy current effects (use FEM with appropriate formulations).
- Very simple preliminary studies (use analytical methods or lumped models).

Best practices:

- Validate MEC models against FEM or experimental data for the specific motor topology.
- Use appropriate mesh refinement near air gap and in saturation-prone regions.
- Ensure continuous variation of air gap permeances to avoid numerical artifacts.
- Select relaxation parameters carefully to balance convergence speed and stability.
- Document assumptions regarding end effects and leakage flux paths.
- Consider hybrid approaches combining MEC for magnetic field with external circuit equations for electrical dynamics.

MEC modeling represents a mature yet continually evolving approach that strikes a practical balance for fault diagnosis applications. While not displacing FEM as the reference standard nor matching lumped models' computational speed, MEC occupies a productive niche that makes it an essential tool in the fault diagnosis modeling toolkit.

3.7 Machine Learning-based Modeling

3.7.1 Introduction

Machine learning (ML) based modeling represents a fundamentally different paradigm from the physics-based approaches discussed in previous sections. Rather than deriving models from first principles such as Maxwell's equations or circuit theory, ML approaches learn input-output relationships directly from data. This chapter focuses specifically on ML techniques that replicate the physical behavior of PM motors as surrogate models, not on diagnostic ML methods for fault classification (covered in Chapter 4).

The key distinction is important: physics-based ML modeling creates fast, accurate surrogate models that predict electromagnetic quantities (flux linkage, torque, current waveforms) under various operating conditions, effectively complementing traditional simulation methods. This differs from diagnostic ML, which uses extracted features to identify fault patterns through classification or regression.

Traditional high-fidelity methods like FEM provide excellent accuracy but require hours per operating point. Simplified analytical models execute quickly but may miss critical nonlinear phenomena. ML-based modeling offers a compelling middle ground: after training on FEM simulations or experimental measurements, ML models achieve near-FEM accuracy at speeds comparable to analytical models.

The applications of ML-based motor modeling include rapid design optimization (evaluating thousands of variants), real-time digital twins for predictive maintenance, hardware-in-the-loop simulation for controller testing, multi-physics integration across electromagnetic, thermal, and mechanical domains, and efficient fault scenario analysis.

3.7.2 Key Modeling Approaches

Recent advances in physics-informed machine learning have proven particularly effective for PM motor simulation. The following approaches represent the current state of the art:

3.7.2.1 Physics-informed Neural Networks (PINNs)

Physics-informed Neural Networks represent a breakthrough in combining data-driven learning with physical laws. Unlike standard neural networks that learn purely from input-output pairs, PINNs incorporate partial differential equations (PDEs) directly into the training process through the loss function, enabling them to learn solutions that are consistent with underlying physics.

For PM motor modeling, a PINN takes spatial coordinates (x, y), rotor angular position θ_r, time t, and current components (i_d, i_q) as inputs. The network outputs electromagnetic quantities such as the magnetic vector potential $\mathbf{A}$ and flux density $\mathbf{B}$. The architecture, illustrated in Figure 3.22, consists of a multi-layer perceptron (MLP) with fully connected hidden layers that learns to approximate the electromagnetic field distribution.

The key innovation of PINNs lies in their physics-informed loss function, which combines data fidelity with physics compliance:

$$\mathcal{L}_{\text{total}} = \mathcal{L}_{\text{data}} + \lambda_{\text{pde}}\mathcal{L}_{\text{pde}} + \lambda_{\text{bc}}\mathcal{L}_{\text{boundary}} \tag{3.60}$$

The data loss $\mathcal{L}_{\text{data}}$ measures the error between network predictions and available training data from FEM simulations or experimental measurements. The PDE loss $\mathcal{L}_{\text{pde}}$ penalizes violations of the governing Maxwell's equations, specifically the magnetostatic formulation $\nabla \times (\nu \nabla \times \mathbf{A}) = \mathbf{J}$, where ν is reluctivity and $\mathbf{J}$ is current density. This residual is evaluated at collocation points

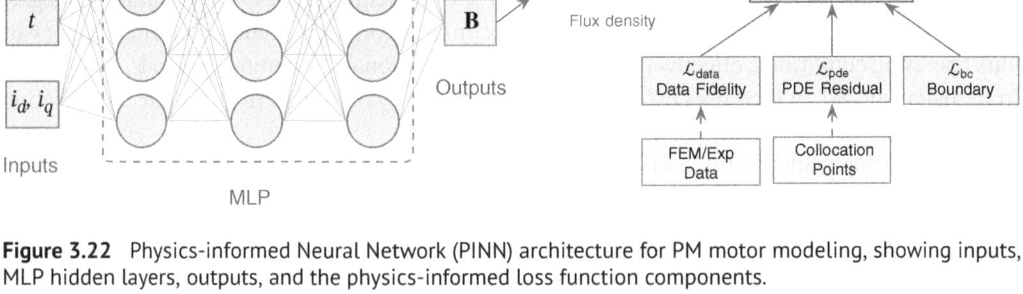

Figure 3.22 Physics-informed Neural Network (PINN) architecture for PM motor modeling, showing inputs, MLP hidden layers, outputs, and the physics-informed loss function components.

distributed throughout the computational domain. The boundary loss $\mathcal{L}_{\text{boundary}}$ enforces boundary conditions at material interfaces and domain boundaries. The weighting coefficients λ_{pde} and λ_{bc} balance the relative importance of each loss component.

A critical enabler of PINNs is automatic differentiation, which computes exact derivatives of the network outputs with respect to inputs. This allows the PDE residuals to be evaluated without numerical differentiation errors. During training, backpropagation propagates the gradient of the total loss through the network, updating the MLP weights to simultaneously minimize data error and physics violations.

Recent research has demonstrated PINNs specifically designed for PMSM modeling [46]. These implementations incorporate several key innovations to achieve high-fidelity motor simulation. Rotational coordinate transformation allows three PDEs with coordinate transformation to supervise the network during training with limited data. Domain decomposition employs separate networks for rotor and stator domains due to their distinct electromagnetic characteristics, while an interface loss term compensates for discontinuity between separate networks at the air gap interface. After training, PINNs achieve ultra-fast inference with evaluation times of 1–10 ms compared to minutes or hours for FEM, while maintaining similar accuracy with reported deviations of approximately 1% compared to reference FEM simulations for inductance and coupling calculations.

The advantages of PINNs for motor modeling are significant. They enforce physical consistency including energy conservation, symmetry, and causality, while requiring less training data than pure data-driven methods. This physics-informed approach enables better generalization to unseen operating conditions and provides interpretable predictions aligned with established physical principles.

However, PINNs also present certain challenges. The training complexity requires expertise in both machine learning and electromagnetics, and hyperparameter tuning involving λ weights and network architecture can be difficult. Additionally, PINNs may struggle with very complex geometries or highly nonlinear materials where the governing equations become particularly difficult to incorporate effectively.

3.7.2.2 Hybrid Physics-data Models

Hybrid approaches strategically combine physics-based models with data-driven corrections, leveraging the strengths of both paradigms.

Architecture:

$$y_{\text{total}} = y_{\text{physics}} + \text{NN}_{\text{correction}}(x, \theta) \tag{3.61}$$

where, y_{physics} comes from a simplified analytical or lumped parameter model - $\text{NN}_{\text{correction}}$ learns to compensate for modeling errors and assumptions - θ represents neural network parameters.

Example for PM motors:

Start with dq-axis lumped parameter model (fast, interpretable):

$$\begin{aligned} \psi_d &= L_d i_d + \psi_{\text{PM}} \\ \psi_q &= L_q i_q \end{aligned} \tag{3.62}$$

Neural networks can be trained to learn corrections for saturation effects such as $L_d(i_d, i_q)$ and $L_q(i_d, i_q)$, cross-coupling terms $\psi_d(i_d, i_q)$ and $\psi_q(i_d, i_q)$, additional spatial harmonic components, and temperature dependencies reflecting parameter variations with thermal state.

Recent multi-physics digital twin research has demonstrated that combining physics-based models with neural network refinement achieves temperature prediction errors under 2.5°C with execution times of 256.4 μs on microcontrollers, making them suitable for real-time control. These hybrid approaches have been successfully deployed in high-voltage EV motors, demonstrating their practical viability for demanding automotive applications.

3.7.3 Neural Network Architectures for Sequential Motor Behavior

PM motor behavior inherently involves temporal dynamics: current responses to voltage inputs, flux linkage variations with rotor position, and transient phenomena during faults. Specialized neural network architectures can capture these sequential dependencies.

3.7.3.1 Recurrent Neural Networks (RNNs)

What they are:

Recurrent Neural Networks (RNN) are specialized architectures introduced in the 1980s for time series modeling [47]. Unlike feedforward networks, RNNs include connections with time delays, allowing the network to maintain internal state and discover temporal correlations between events.

Architecture:

The RNN structure consists of an unrolled network with input, hidden, and output layers at each time step. At each time step t, the RNN updates its hidden state and produces an output:

$$\begin{aligned} h(t) &= \sigma(W_h h(t-1) + Ux(t) + b_h) \\ y(t) &= W_y h(t) + b_y \end{aligned} \tag{3.63}$$

where, $x(t)$ is input vector (e.g., applied voltage) - $h(t)$ is hidden state vector (internal memory) - $y(t)$ is output vector (e.g., predicted current) - W, U are weight matrices - σ is activation function (e.g., tanh, sigmoid)

RNNs can model the dynamic relationship between terminal voltage (input) and stator current (output), capturing transient responses to voltage changes, position-dependent back-EMF effects, dynamic inductance variations, and fault-induced asymmetries in current waveforms.

Training via Backpropagation Through Time (BPTT):

The network is unrolled through time and standard backpropagation is applied [48, 49]. Gradients are computed recursively backward in time to update all parameters.

Limitations:

Standard RNNs suffer from vanishing and exploding gradient problems during training, making them difficult to train for long sequences and limiting their ability to capture long-term dependencies [48].

3.7.3.2 Long Short-term Memory (LSTM) Networks

What they are:

Long Short-Term Memory networks address the fundamental limitations of RNNs through specialized memory cell structures [50]. LSTMs can effectively learn and remember information over long time sequences, making them particularly suitable for modeling PM motor dynamics.

Architecture:
The LSTM architecture comprises a memory block with three gates controlling information flow. The memory block consists of a memory cell that stores temporal state information, an input gate that controls how much new information flows into the cell, a forget gate that controls how much past information to retain, and an output gate that controls how much cell information influences the output.

Mathematical formulation (Vanilla LSTM with peephole connections) [51–53]:
Input block:

$$z_t = g(W_z x_t + R_z y_{t-1} + b_z) \tag{3.64}$$

Input gate:

$$i_t = \sigma(W_i x_t + R_i y_{t-1} + P_i c_{t-1} + b_i) \tag{3.65}$$

Forget gate:

$$f_t = \sigma(W_f x_t + R_f y_{t-1} + P_f c_{t-1} + b_f) \tag{3.66}$$

Cell state:

$$c_t = i_t \odot z_t + f_t \odot c_{t-1} \tag{3.67}$$

Output gate:

$$o_t = \sigma(W_o x_t + R_o y_{t-1} + P_o c_t + b_o) \tag{3.68}$$

Block output:

$$y_t = o_t \odot h(c_t) \tag{3.69}$$

where $\odot$ denotes element-wise multiplication, σ is the sigmoid activation function defined as $\sigma(x) = 1/(1 + e^{-x})$, g and h are typically tanh functions, and P matrices represent peephole connection weights from the cell state to the gates.

LSTMs excel at motor modeling for several reasons. They can capture long-term dependencies by remembering patterns over extended sequences spanning multiple electrical cycles. The gated structure avoids gradient problems by preventing the vanishing and exploding gradients that plague standard RNNs [54]. Selective memory through the forget gate allows the network to discard irrelevant past information while retaining critical state information. Furthermore, LSTMs have a proven track record of successful application to diverse problems including handwriting recognition, speech synthesis, machine translation, and increasingly to motor modeling tasks.

Application to PM motor simulation:
LSTMs have been demonstrated for PMSM modeling where terminal voltage serves as input and stator current as output [55, 56]. The network learns to predict three-phase current waveforms from applied voltages, flux linkage variations with rotor position and current magnitude, torque production under various operating points, and transient responses during load changes or fault events.

The supervised machine learning workflow (Algorithm 3) outlines the systematic process from data collection through validation.

Algorithm 3 Supervised Machine Learning for PM Motor Surrogate Modeling

Input: Motor specifications, target predictions (current, flux, torque)
Output: Trained surrogate model for PM motor behavior
Phase 1: Data Generation
Define input variables: terminal voltages (v_a, v_b, v_c), rotor position θ_r, speed ω
Define output variables: phase currents (i_a, i_b, i_c), flux linkages (λ_a, λ_b, λ_c), torque T_e
Generate training data via FEM simulations or experimental measurements
Ensure coverage of operating range: various speeds, loads, and rotor positions
Phase 2: Data Preprocessing
Clean data: remove outliers and measurement artifacts
Normalize features using standardization or tanh estimator
Split data into training, validation, and test sets
Phase 3: Model Training
Select network architecture (e.g., LSTM)
Define loss function (MSE or MAE for regression)
repeat
    Train model using optimization algorithm (e.g., Adam)
    Evaluate on validation set: compute prediction error for currents, flux, torque
    if validation error above threshold **then**
        Adjust hyperparameters (learning rate, hidden units, regularization)
    end if
until validation metrics satisfied
Verify on test set with unseen operating conditions
return Trained PM motor surrogate model

The training process begins with data collection, generating a training dataset from FEM simulations or experimental measurements covering a wide operating range including various speeds, loads, and rotor positions. Data normalization follows, typically applying the hyperbolic tangent estimator or other normalization methods [57, 58]:

$$z = 0.5\left[\tanh\left(0.01\frac{a-\mu}{\sigma}\right) + 1\right] \tag{3.70}$$

where μ represents the mean and σ is the standard deviation. Architecture selection determines the number of LSTM layers and hidden units per layer, followed by optimization using the Adam optimizer [59], which combines the advantages of adaptive learning rates with momentum. Finally, validation tests the trained model on a separate validation set not seen during training.

Demonstrated Results for PM Motors:
Figures 3.23 and 3.24 show comparisons between LSTM and FEM predictions for three-phase current and flux linkage waveforms in SPM and IPM motors. The LSTM-predicted currents for SPM (Figure 3.23a) and IPM (Figure 3.23b) motors closely match the FEM-generated reference waveforms (Figure 3.23c, d). Similarly, the LSTM flux linkage predictions (Figure 3.24a, b) show excellent agreement with FEM results (Figure 3.24c, d) for both motor types.

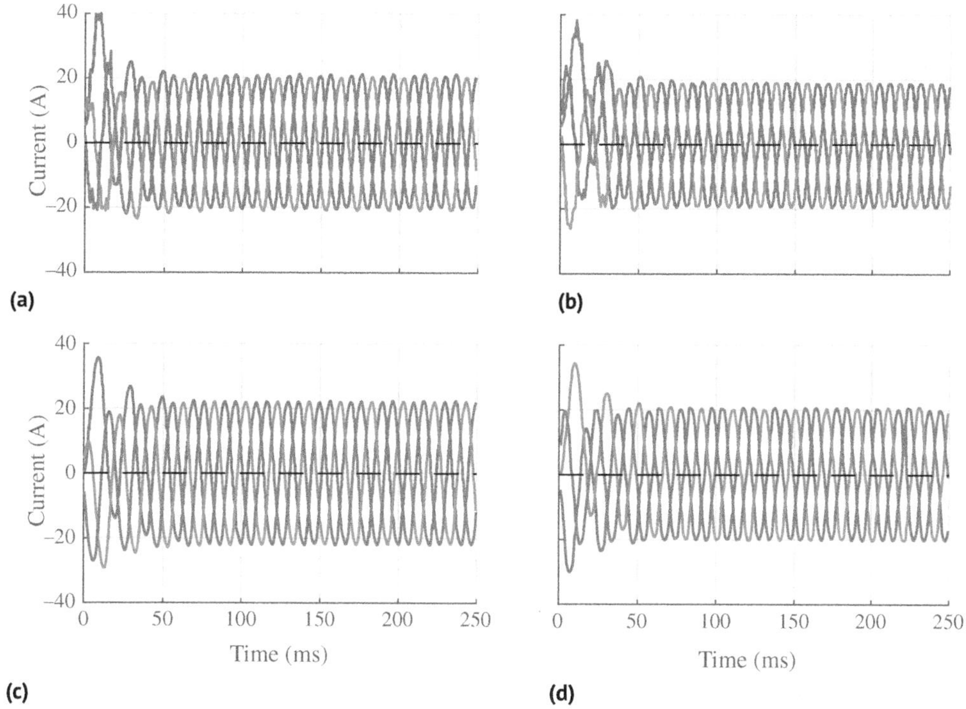

Figure 3.23 Three-phase stator currents comparing LSTM predictions with FEM simulation: (a) LSTM-SPM motor, (b) LSTM-IPM motor, (c) FEM-SPM motor, (d) FEM-IPM motor.

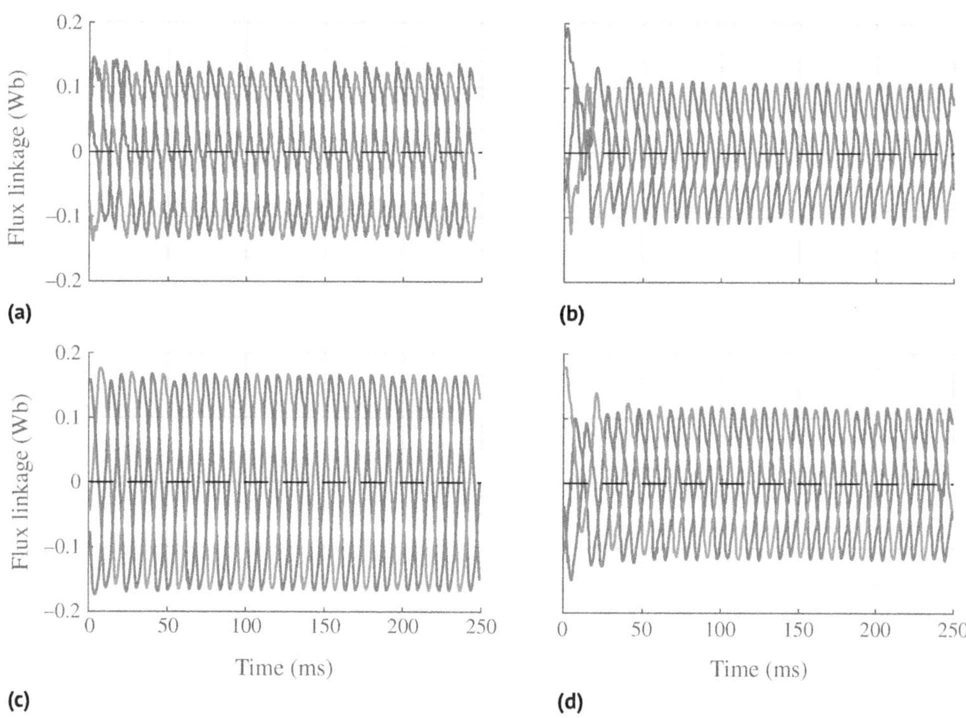

Figure 3.24 Three-phase flux linkages comparing LSTM predictions with FEM simulation: (a) LSTM-SPM motor, (b) LSTM-IPM motor, (c) FEM-SPM motor, (d) FEM-IPM motor.

The close agreement between LSTM predictions and FEM results in Figures 3.23 and 3.24 demonstrates that properly trained LSTMs can serve as accurate surrogate models for PM motor electromagnetic behavior.

3.7.3.3 Deep Learning for Temperature and Torque Prediction

Beyond electromagnetic modeling, recent research has applied deep learning to multi-physics predictions:

Bidirectional LSTM (BiLSTM) architectures have emerged as particularly effective for PMSM modeling. These networks process sequences in both forward and backward directions, capturing dependencies from both past and future time steps. BiLSTM implementations have achieved $R^2 = 0.99$ for both temperature and torque prediction, effectively modeling temporal dynamics and complex parameter relationships. While feedforward neural networks with two hidden layers also achieve $R^2 = 0.99$ for torque and temperature predictions, BiLSTM networks demonstrate superior capability for capturing the temporal evolution of thermal and mechanical states.

3.7.4 Deep Operator Networks for Multi-physics Modeling

Recent advances have introduced multiphysics-informed deep operator networks (MPI-DON) specifically for PM motor digital twins. Unlike standard neural networks that map inputs to outputs for fixed partial differential equations, operator networks learn mappings between function spaces, enabling them to handle varying operating conditions, geometries, or material properties. These networks incorporate electromagnetic, thermal, and mechanical physics in a unified framework, demonstrating accuracy, robustness, and generality in systematic comparisons against both FEM and experimental results. MPI-DON enables virtual sensing for quantities difficult to measure directly and has proven effective for bearing fault detection in field applications. Notably, these networks achieve approximately $10\times$ speedup over FEM while maintaining comparable accuracy.

3.7.5 Training Methodologies

3.7.5.1 Supervised Learning for Motor Modeling

Supervised learning is the most common approach for training neural networks to model motor behavior [60]. The process begins by determining the training sample type, which may include steady-state operating points, transient trajectories, or fault scenarios. Training data is then collected from FEM simulations, analytical models, or experimental measurements. The data must be represented appropriately through feature engineering, normalization, and sequence formatting. During training, network parameters are adjusted to minimize prediction errors on the training set, followed by validation to evaluate performance on an independent test set.

For regression tasks predicting continuous outputs such as current, flux, and torque, input features typically include voltage commands, rotor position, speed, and load torque. Output targets comprise phase currents, flux linkages, and electromagnetic torque, with loss functions based on mean squared error (MSE) or mean absolute error (MAE).

3.7.5.2 Reinforcement Learning for Control-oriented Models

While less common for physical modeling per se, reinforcement learning (RL) has emerged for model-based control of PMSMs [61]. In this approach, an agent learns optimal control policies through interaction with the system. The state representation includes motor speed, currents, and rotor position, while actions consist of voltage commands or current references. Rewards are based on tracking error, efficiency, and constraint satisfaction, with learning proceeding by updating the policy to maximize cumulative reward.

Recent comprehensive evaluations conducted of Deep Reinforcement Learning for PMSM control have examined approaches including deep Q-network (DQN), proximal policy optimization (PPO), and advantage actor-critic (A3C). These model-free, data-driven approaches offer transformative solutions for PMSM control without requiring explicit motor models, though they fall outside the scope of physical modeling for simulation purposes.

3.7.6 Training Considerations

Effective training of neural networks for motor modeling requires attention to optimization algorithms and data preprocessing. The Adam (adaptive moment estimation) optimizer [59] has proven particularly effective, combining adaptive learning rates with momentum methods to achieve fast convergence and good generalization. Alternative optimizers such as RMSprop and Nadam may be preferred for specific architectures, particularly recurrent networks.

Data normalization significantly impacts training effectiveness. For motor time series data, the hyperbolic tangent estimator has proven most effective [58], though standard z-score normalization also performs well. Proper normalization ensures all features are at comparable scales, accelerating training convergence and improving model generalization to unseen operating conditions.

3.7.7 Digital Twin Concept

Machine learning-based models form the computational core of digital twin systems for PM motors. A digital twin is a virtual replica that mirrors the physical motor in real-time, enabling several critical capabilities. Real-time monitoring provides continuously updated state estimation, while predictive maintenance forecasts remaining useful life and identifies incipient faults. Performance optimization identifies opportunities for efficiency improvements, and what-if analysis evaluates scenarios without requiring physical testing. Fault detection compares expected versus actual behavior to identify anomalies.

The implementation architecture follows a sequential data flow: sensors mounted on the physical motor capture operational data, which passes through a data acquisition system for conditioning and digitization. This processed data feeds into an ML model that has been trained offline on historical or simulated datasets. The model generates real-time predictions of motor states, fault indicators, or performance metrics, which are then delivered to the monitoring or control system for decision-making and actuation. Recent comprehensive studies on digital twin modeling for PMSMs have examined three distinct approaches. Physics-based digital twins offer high fidelity but are computationally intensive. Data-driven approaches provide fast execution but require extensive training data. Hybrid approaches balance accuracy, speed, and data requirements, combining simplified physics models with ML corrections. This hybrid paradigm shows the most promise for real-time applications with limited computational resources.

3.7.8 Advantages and Limitations

3.7.8.1 Advantages

ML-based modeling offers substantial computational speed advantages, with inference times of 1–10 ms for PINNs and less than 1 ms for trained LSTMs, enabling real-time simulation and rapid design exploration. ML models achieve near-FEM precision when trained on sufficient high-quality data, capturing complex nonlinear phenomena including saturation, spatial harmonics, and cross-coupling effects. The flexibility of ML approaches allows representation of multi-physics interactions and adaptation to different motor designs through transfer learning. ML methods also excel at learning relationships that are difficult to model analytically, discovering patterns in complex fault signatures, and bridging gaps in material properties or manufacturing variations.

3.7.8.2 Limitations

Despite their advantages, ML-based approaches face significant challenges. Data requirements are substantial, with supervised learning demanding large, representative training datasets that can be expensive to generate. The black-box nature of neural networks limits physical interpretability and makes it difficult to verify predictions outside the training distribution. Model performance degrades for operating conditions far from the training data, and extrapolation is unreliable. Training complexity requires significant expertise in hyperparameter tuning, with constant risk of overfitting.

3.7.8.3 Mitigating Limitations

Many limitations can be addressed through physics-informed and hybrid approaches. PINNs embed governing PDEs to ensure physical consistency, reducing data requirements and improving generalization. Hybrid physics-data models combine simplified analytical models with ML corrections, providing graceful degradation and reduced data needs. Transfer learning enables models pretrained on one motor design to be fine-tuned for new variants with limited additional data.

3.7.9 Role in Fault Diagnosis Workflow

ML-based modeling is most appropriate for fault diagnosis when speed is critical and accuracy must be maintained. Key applications include real-time digital twins for online monitoring, hardware-in-the-loop simulation requiring sub-millisecond updates, and rapid design optimization. ML approaches are also well-suited when complex nonlinear phenomena must be captured, including severe magnetic saturation, motors with complex geometries, and multi-physics interactions.

ML-based modeling becomes viable when high-quality training data is available from extensive FEM simulation databases or experimental measurements across wide operating ranges. The approach works best when physics-informed or hybrid methods can be employed.

When to Use Alternative Approaches: Consider physics-based methods (FEM, MEC, analytical) when training data is scarce, physical interpretability is paramount for certification requirements, operating conditions may vary widely beyond the training range, or computational resources for training are limited.

Best Practices for Fault Diagnosis Applications: Training data should span the full operating range and include relevant fault scenarios when fault modeling is an objective. LSTMs or BiLSTMs work well for sequence-to-sequence motor behavior, while PINNs should be considered when physics compliance is critical. Validation on independent test sets representing real deployment conditions is essential, and physics violations should be checked systematically. For real-time applications, thorough testing on experimental data should precede deployment, with online monitoring of prediction residuals to detect model degradation.

For fault diagnosis specifically, ML-based physical modeling enables rapid generation of fault signature databases for algorithm training and validation, real-time observers comparing model predictions to measurements for anomaly detection, digital twins that track motor health and predict remaining useful life, and efficient multi-scenario analysis to understand fault propagation and severity. These capabilities make ML-based modeling an essential tool for next-generation condition monitoring and predictive maintenance systems.

ML-based modeling is not a universal replacement for traditional physics-based methods but rather complements the modeling toolkit. FEM remains the gold standard for research validation, analytical and MEC models provide physical insight, while ML models excel when speed is critical and training data is available. The most promising direction combines these strengths through physics-informed machine learning and hybrid physics-data models.

3.8 Simulation Validation and Verification

The credibility of any simulation model depends critically on rigorous validation and verification. Verification ensures that the model is implemented correctly and solves the governing equations accurately, while validation confirms that the model represents the physical system with sufficient fidelity for its intended purpose. For PM machine modeling and fault diagnosis applications, this distinction is particularly important because simulation results often guide critical decisions about diagnostic algorithms, maintenance strategies, and safety protocols.

3.8.1 Purpose of Validation and Verification

Verification addresses the question: "Are we solving the equations correctly?" This involves mathematical and computational checks that ensure: - Numerical solutions converge to the true mathematical solution - Discretization errors are within acceptable bounds - Conservation laws (energy, momentum) are satisfied - Boundary conditions are correctly imposed

Validation addresses the question: "Are we solving the right equations?" This requires comparing model predictions against experimental measurements or higher-fidelity references to confirm: - Physical phenomena are accurately represented - Model assumptions are justified for the application - Predictions match reality within quantified uncertainty bounds - The model is fit for its intended diagnostic purpose

The distinction is crucial: a model can be perfectly verified (correctly implemented) yet invalid (does not match reality), or poorly verified (numerical errors) yet still useful for qualitative trend analysis. Both aspects must be rigorously addressed.

3.8.2 Verification of Models

Verification establishes confidence in the numerical implementation before attempting validation against experiments.

3.8.2.1 Sanity Checks

Basic physics principles provide the first line of verification:

Energy Conservation: For electromagnetic systems, input electrical power should equal the sum of output mechanical power, stored field energy changes, and losses:

$$P_{\text{electrical}} = P_{\text{mechanical}} + \frac{d}{dt} W_{\text{field}} + P_{\text{losses}} \tag{3.71}$$

Any violation indicates implementation errors. FEM models should demonstrate energy balance within numerical precision (typically < 0.1% error) [9].

Unit Consistency: All equations must maintain dimensional consistency. Automated dimensional analysis tools can detect errors where, for example, inductance values are accidentally treated as resistances.

Models should reproduce known analytical solutions in simplified limiting cases. For example, under no-load conditions, torque should vanish when $i_q = 0$ in dq models. During symmetric operation, three-phase currents should sum to zero. In the high-frequency limit, inductive reactance should dominate resistance.

3.8.2.2 Benchmark Comparisons

Comparing multiple modeling approaches on the same machine geometry provides powerful verification. If FEM, analytical, and MEC methods all predict similar flux density distributions and torque values, confidence in each implementation increases.

Recent benchmark initiatives provide standardized test cases for this purpose. The CREATOR research project has released complete design specifications and experimental measurements for reference PMSM and induction machines specifically for validation and benchmarking purposes [62]. These datasets include complete geometric parameters and material properties, low-frequency equivalent circuit parameters, transient drive cycle measurements covering diverse operating conditions, and manufacturing tolerances with uncertainty bounds. Such standardized benchmarks enable objective comparison of modeling approaches and expose implementation errors that might not be evident from single-method analysis.

3.8.2.3 Code Verification Techniques

Systematic code verification employs several complementary approaches. Mesh convergence studies for FEM models progressively refine the mesh until predicted quantities such as torque, flux linkage, and loss stabilize. Convergence should follow theoretical rates, such as second-order spatial convergence for quad elements. Time-step sensitivity analysis ensures that transient simulations use sufficiently small time steps, with results independent of further time-step reduction beyond a certain threshold.

Symmetry exploitation provides another verification avenue. If the geometry possesses symmetry such as pole-pair periodicity, models exploiting this symmetry should yield identical results to full-geometry models. The method-of-manufactured-solutions constructs a problem with a known analytical solution by specifying source terms that produce the desired solution, then verifies that the code recovers this solution accurately.

3.8.3 Validation Against Experiments

After verification establishes implementation correctness, validation confirms physical accuracy by comparing predictions to experimental measurements.

3.8.3.1 Comparison with Measurements

Key measurable quantities for model validation include:

Back-EMF Waveforms: No-load voltage measurements under constant-speed rotation provide direct validation of flux linkage modeling without complications from resistive drops or controller dynamics. Peak voltage amplitude, harmonic content, and position dependence must all match predictions [63].

Inductance Parameters: Both d-axis and q-axis inductances can be measured via standstill tests or high-frequency injection. Validation should cover the full current range to verify saturation modeling [64].

Torque Characteristics: Dynamometer measurements of electromagnetic torque versus current provide direct validation of torque production models. Both average torque and ripple components should be compared.

Efficiency Maps: Measured efficiency over the torque-speed plane integrates validation of loss predictions (copper, iron, mechanical). Discrepancies often reveal missing loss mechanisms.

Thermal Behavior: Temperature rise measurements validate coupled electromagnetic-thermal models, particularly important for understanding fault-induced heating.

3.8.3.2 Fault Scenario Validation

For fault diagnosis applications, validation must extend to faulty machine conditions. The landmark work of Haddad and Strangas [19] demonstrated comprehensive FEM validation for combined eccentricity and demagnetization faults, showing:

- Current harmonic signatures matched between FEM and experiments within 5% error.
- Fault detection methods maintained accuracy across 25% and 50% eccentricity levels.
- Fault separation algorithms achieved 95% classification accuracy when validated against experimental data.

This level of validation is essential because fault conditions often involve coupled nonlinear phenomena (local saturation, unbalanced magnetic pull, harmonic coupling) that may be poorly represented if the model is validated only under healthy operation.

Recent multi-physics validation studies have extended this rigor to electromagnetic-vibration coupling [65]. Using anechoic chamber testing, researchers verified that coupled FEM-structural-acoustic models predicted both vibration amplitude and noise spectra within measurement uncertainty, validating the complete diagnostic chain from electromagnetic force to radiated noise.

3.8.3.3 Validation Across Operating Conditions

Models validated at a single operating point may fail elsewhere. Comprehensive validation requires coverage of the full speed range from standstill to maximum speed, the full load range including both motoring and generating modes, transient conditions such as acceleration, load changes, and fault inception, and environmental variations including temperature and humidity effects for insulation modeling.

3.8.4 Validation for Fault Diagnosis

Fault diagnosis applications impose unique validation requirements beyond standard motor performance modeling.

3.8.4.1 Validate Variables Used for Detection

Diagnostic algorithms extract fault signatures from specific measured or estimated quantities. These quantities must be validated with particular care:

Current Harmonics: Motor current signature analysis (MCSA) relies on spectral components at specific frequencies. Validation must confirm not just fundamental current magnitude but the amplitude and phase of diagnostic harmonics (e.g., eccentricity sidebands at $f \pm f_r$, slot harmonics) [66].

Zero-sequence Components: Winding fault detection often uses zero-sequence voltage or current. Models must accurately represent the neutral point behavior and common-mode impedances.

Vibration Signatures: If vibration-based diagnosis is employed, the model must predict radial force harmonics and their coupling to structural modes, not just electromagnetic torque.

Flux Leakage: External search coil or Hall sensor methods detect faults via leakage flux. FEM validation should compare predicted and measured flux density at sensor locations under both healthy and faulty conditions.

3.8.4.2 Cross-validation Between Model Fidelities

A practical validation strategy uses FEM as an intermediate reference between analytical models and experiments through a hierarchical approach. First, FEM is validated against experiments for representative healthy and faulty cases. Next, faster models including analytical, MEC, and ML approaches are validated against FEM across a broader range of conditions. Finally, faster models are selectively validated against experiments to confirm critical predictions.

This hierarchical approach leverages the strengths of each method while managing experimental costs. For example, magnetic equivalent circuit models validated against FEM for multiple fault severities provide confidence for rapid fault library generation, with targeted experimental validation at key operating points [66].

3.8.4.3 Using FEM as Proxy for Experiments

When physical fault implementation is difficult, such as for progressive demagnetization or mixed eccentricity, validated FEM models serve as virtual experiments. This approach requires initial FEM validation under controlled, reproducible fault conditions, quantified uncertainty bounds on FEM predictions for extrapolation scenarios, and periodic experimental validation as new fault types are explored.

Recent studies on demagnetization fault diagnosis have successfully employed this strategy, establishing FEM as a reliable proxy by validating against uniform demagnetization tests, then using FEM to explore partial demagnetization patterns that are difficult to reproduce experimentally [64].

3.8.5 Verification and Validation Plan

A systematic verification and validation plan for fault diagnosis modeling should follow a structured approach. Component testing begins by verifying material models such as B–H curves and loss coefficients against datasheet values, validating winding impedance measurements against lumped parameter calculations, and confirming that geometric dimensions match CAD specifications within tolerances.

Healthy condition validation compares no-load back-EMF waveforms including amplitude, harmonics, and position dependence. Locked-rotor and short-circuit tests validate impedance parameters, while torque-speed characteristics must be confirmed across the operating range and efficiency predictions verified at the rated operating point.

Fault condition validation implements controlled faults such as known demagnetization percentages or measured eccentricity levels. Fault signatures in diagnostic variables including current harmonics, vibration, and flux must be validated, with fault severity indicators confirmed to track measured fault levels. Diagnostic algorithm performance should be tested on experimental fault data.

The process concludes with iterative refinement. Discrepancies between model and experiment must be documented, missing physics identified such as inter-lamination eddy currents or end-winding effects, and model complexity refined as needed to achieve target accuracy. Re-validation follows each refinement iteration.

3.8.6 Documentation and Uncertainty Quantification

Credible validation requires transparent documentation of model accuracy and limitations.

Model accuracy bounds should be quantified using appropriate metrics. Root mean square error (RMSE) applies to continuous quantities, peak error characterizes worst-case scenarios, correlation coefficient assesses waveform comparisons, and confusion matrices evaluate fault classification accuracy. For example, stating that "back-EMF peak voltage predictions exhibit RMSE < 2% and maximum error < 5% across the speed range 500–5000 rpm" provides actionable uncertainty information.

Assumption documentation requires explicitly stating and justifying all modeling assumptions. These include linearity assumptions where applied despite actual saturation, 2D versus 3D approximations and their validity ranges, neglected phenomena such as stray capacitance or bearing impedance, and material property sources with their associated uncertainties.

Modern uncertainty estimation approaches propagate parameter uncertainties through models to quantify output uncertainty. Monte Carlo sampling or response surface methods applied to uncertain inputs including magnet remanence, air gap tolerance, and iron permeability produce statistical distributions of predictions rather than single point values [67]. Recent work on robust design optimization demonstrates that accounting for manufacturing tolerances and material property variations changes optimal designs by 5–10%, highlighting the importance of uncertainty quantification for reliable predictions [68]. Such comprehensive documentation enables users to assess model suitability for their specific application and interpret predictions with appropriate confidence levels.

3.9 Summary

This chapter has presented six major modeling techniques for permanent magnet machines, each offering distinct advantages for fault diagnosis applications. Table 3.4 provides a comprehensive comparison to guide method selection.

The selection of modeling approach depends on application requirements. For real-time embedded diagnostics requiring microsecond execution rates, dq-axis lumped parameter models remain the practical choice despite their limited ability to capture spatial fault effects. When detailed fault signature prediction is needed for algorithm development, FEM provides the necessary fidelity at the cost of computational time. For parametric studies requiring thousands of simulations, FRM and MEC offer attractive middle-ground solutions with reasonable accuracy and manageable computation times.

A multi-fidelity modeling strategy represents the current best practice for comprehensive fault diagnosis system development. This approach uses FEM for initial validation and understanding of fault mechanisms, semi-analytical methods such as FRM, winding function theory, or MEC for extensive parametric studies and fault signature database generation, and lumped parameter or ML-based models for real-time implementation. Each modeling tier is validated against the higher-fidelity methods, building confidence that real-time diagnostic predictions reflect accurate underlying physics.

Table 3.4 Comparison of PM machine modeling techniques for fault diagnosis applications.

Method	Speed	Accuracy	Real-time capable	Best fault applications
Lumped parameter (dq)	Fastest (μs)	Moderate	Yes	Demagnetization, uniform degradation
Lumped parameter (abc VBR)	Fast (ms)	Good	Limited	Eccentricity, asymmetric faults
FEM	Slowest (hours)	Highest	No	All faults (reference standard)
FRM	Moderate (seconds)	High	Limited	Eccentricity, demagnetization
Winding function	Fast (ms)	Moderate	Yes	Eccentricity, winding faults
MEC	Moderate (minutes)	Good	No	All faults with saturation effects
ML-based	Very fast (ms)	Variable	Yes	Complex multi-physics, digital twins

The emergence of machine learning-based approaches, particularly physics-informed neural networks and hybrid physics-data models, promises to bridge the gap between accuracy and speed more effectively than traditional methods alone. As computational capabilities continue to advance and validation methodologies mature, these data-driven approaches will likely play an increasingly central role in PM machine fault diagnosis.

Ultimately, effective fault diagnosis requires matching modeling capabilities to specific diagnostic objectives. The techniques presented in this chapter provide engineers with a comprehensive toolkit for developing accurate, efficient, and reliable fault diagnosis systems for permanent magnet machines across diverse application domains.

References

1 Colm J. O'Rourke, Mohammad M. Qasim, Matthew R. Overlin, and James L. Kirtley. A geometric interpretation of reference frames and transformations: dq0, clarke, and park. *IEEE Transactions on Energy Conversion*, 34(4):2070–2083, 2019. https://doi.org/10.1109/TEC.2019.2941175.

2 Florence Meier and Juliette Soulard. dq theory applied to a permanent magnet synchronous machine with concentrated windings. In *4th IET International Conference on Power Electronics, Machines and Drives (PEMD 2008)*, pages 194–198, 2008. https://doi.org/10.1049/cp:20080510.

3 Sergio Zarate, Gaizka Almandoz, Gaizka Ugalde, Javier Poza, and Ana Julia Escalada. Extended dq model of a permanent magnet synchronous machine by including magnetic saturation and torque ripple effects. In *2017 IEEE International Workshop of Electronics, Control, Measurement, Signals and their Application to Mechatronics (ECMSM)*, pages 1–6, May 2017. https://doi.org/10.1109/ECMSM.2017.7945881.

4 Navid Amiri, Seyyedmilad Ebrahimi, Mehrdad Chapariha, Juri Jatskevich, and Hermann W. Dommel. Voltage-behind-reactance model of six-phase synchronous machines considering stator mutual leakage inductance and main flux saturation. *Electric Power Systems Research*, 138:155–164, September 2016. https://doi.org/10.1016/j.epsr.2016.03.036.

5 José M. Aller, José A. Restrepo, Juan C. Viola, and Jorge R. Mayor. Simplified voltage behind reactance model for the six-phase open-end salient pole PMSM. In *2020 International Conference on Electrical Machines (ICEM)*, pages 2216–2221, August 2020. https://doi.org/10.1109/ICEM49940.2020.9270701.

6 Ramón Pérez, Jérôme Cros, and Mathieu Picard. Real-time modeling of static, dynamic and mixed eccentricity in permanent magnet synchronous machines. *Machines*, 13(2):120, February 2025. https://doi.org/10.3390/machines13020120.

7 MathWorks. Permanent magnet synchronous machine - three-phase permanent magnet synchronous machine with sinusoidal or trapezoidal back electromotive force. MATLAB Simscape Documentation, 2024. https://www.mathworks.com/help/sps/powersys/ref/permanentmagnetsynchronousmachine.html.

8 Zhichao Chen, Zhe Liang, Deliang Liang, and Shaofeng Jia. Partial demagnetization fault analysis and diagnosis for fractional slot concentrated winding PMSMs based on dq-axis components. *IEEE Transactions on Energy Conversion*, 40(2):1062–1072, June 2025. https://doi.org/10.1109/TEC.2024.3505255.

9 Peter Monk. *Finite Element Methods for Maxwell's Equations*. Numerical Analysis and Scientific Computation Series. Oxford University Press, Oxford, UK, 2003. ISBN 9780198508885. https://doi.org/10.1093/acprof:oso/9780198508885.001.0001.

10 M. V. K. Chari and Sheppard J. Salon. *Numerical Methods in Electromagnetism*. Academic Press, 2000. ISBN 9780126157604. https://doi.org/10.1016/B978-012615760-4/50000-6.

11 ANSYS Inc. Ansys maxwell - low-frequency electromagnetic field simulation. Commercial FEM software, 2024. https://www.ansys.com/products/electronics/ansys-maxwell.

12 JSOL Corporation. JMAG - simulation technology for electromechanical design. Commercial FEM software, 2024. https://www.jmag-international.com/.

13 David Meeker. Finite element method magnetics (FEMM). Open-source software, 2024. URL https://www.femm.info/.

14 Praveen Kumar and P. Balakrishnan. EWT implementation for examining demagnetization fault in PMSM using FEM. In *2022 IEEE International Conference on Power Electronics, Drives and Energy Systems (PEDES)*, pages 1–6, Jaipur, India, December 2022. https://doi.org/10.1109/PEDES56012.2022.10080132.

15 Jawad Faiz and Ehsan Mazaheri-Tehrani. Demagnetization modeling and fault diagnosing techniques in permanent magnet machines under stationary and nonstationary conditions: An overview. *IEEE Transactions on Industry Applications*, 53(3):2772–2785, May 2017. https://doi.org/10.1109/TIA.2016.2608950.

16 Behrooz M. Ebrahimi, Javad Faiz, and Mohammad J. Roshtkhari. Static-, dynamic-, and mixed-eccentricity fault diagnoses in permanent-magnet synchronous motors. *IEEE Transactions on Industrial Electronics*, 56(11):4727–4739, 2009. https://doi.org/10.1109/TIE.2009.2029577.

17 Abolfazl Sarikhani and Osama A. Mohammed. Inter-turn fault detection in pm synchronous machines by physics-based back electromotive force estimation. *IEEE Transactions on Industrial Electronics*, 60(8):3472–3484, 2013. https://doi.org/10.1109/TIE.2012.2229678.

18 Jun Hang, Jinquan Zhang, Ming Cheng, and Jian Huang. Online interturn fault diagnosis of permanent magnet synchronous machine using zero-sequence components. *IEEE Transactions on Power Electronics*, 30(12):6731–6741, 2015. https://doi.org/10.1109/TPEL.2015.2388453.

19 Ramy Z. Haddad and Elias G. Strangas. On the accuracy of fault detection and separation in permanent magnet synchronous machines using MCSA/MVSA and LDA. *IEEE Transactions on Energy Conversion*, 31(3):924–934, 2016. https://doi.org/10.1109/TEC.2016.2558183.

20 Mohamed Krichen, El Houssin Elbouchikhi, Nabil Benhadj, Mohamed Chaieb, Mohamed Benbouzid, and Riadh Neji. Motor current signature analysis-based inter-turn short circuit fault diagnosis in induction machines. *Energies*, 13(14):3719, 2020. https://doi.org/10.3390/en13143719.

21 Yonghoon Park, Daniel Fernandez, Sang B. Lee, Dongil Hyun, Minho Jeong, S. K. Kommuri, Changhwan Cho, Daniel D. Reigosa, and Fernando Briz. Online detection of rotor eccentricity and

demagnetization faults in PMSMs based on hall-effect field sensor measurements. *IEEE Transactions on Industry Applications*, 55(3):2499–2509, 2019. https://doi.org/10.1109/TIA.2018.2886772.

22 Gang Niu, Xun Dong, and Yuejian Chen. Motor fault diagnostics based on current signatures: A review. *IEEE Transactions on Instrumentation and Measurement*, 72:1–19, 2023. https://doi.org/10.1109/TIM.2023.3285999.

23 Arslan Ahmed Amin, Ansa Mubarak, and Saba Waseem. Application of physics-informed neural networks in fault diagnosis and fault-tolerant control design for electric vehicles: A review. *Measurement*, 246:116728, March 2025. https://doi.org/10.1016/j.measurement.2025.116728.

24 Wei Zhu, Babak Fahimi, and Steven D. Pekarek. A field reconstruction method for optimal excitation of permanent magnet synchronous machines. *IEEE Transactions on Energy Conversion*, 21(2):305–313, June 2006. https://doi.org/10.1109/TEC.2005.859979.

25 Dezheng Wu, Steven D. Pekarek, and Babak Fahimi. A field reconstruction technique for efficient modeling of the fields and forces within induction machines. *IEEE Transactions on Energy Conversion*, 24(2):366–374, June 2009. https://doi.org/10.1109/TEC.2008.2001439.

26 Marcio L. Magri Kimpara, Ruben Barros Godoy, Pedro E. M. Justino Ribeiro, Luiz E. Borges da Silva, Babak Fahimi, and João Onofre Pereira Pinto. A new synchronous machine modeling using the field reconstruction method. *Journal of Control, Automation and Electrical Systems*, 25(4):481–492, 2014. https://doi.org/10.1007/s40313-014-0124-8.

27 Dimitri Torregrossa, Amir Khoobroo, and Babak Fahimi. Prediction of acoustic noise and torque pulsation in PM synchronous machines with static eccentricity and partial demagnetization using field reconstruction method. *IEEE Transactions on Industrial Electronics*, 59(2):934–944, February 2012. https://doi.org/10.1109/TIE.2011.2151810.

28 Amir Khoobroo, Babak Fahimi, and Steven D. Pekarek. A new field reconstruction method for permanent magnet synchronous machines. In *2008 34th Annual Conference of IEEE Industrial Electronics*, pages 2009–2013, November 2008. https://doi.org/10.1109/IECON.2008.4758265.

29 Amir Khoobroo and Babak Fahimi. Detection of static rotor eccentricity in permanent magnet synchronous drives using field reconstruction method. In *2010 IEEE Vehicle Power and Propulsion Conference*, pages 1–7, September 2010. https://doi.org/10.1109/VPPC.2010.5729044.

30 Amir Khoobroo and Babak Fahimi. A novel method for permanent magnet demagnetization fault detection and treatment in permanent magnet synchronous machines. In *Twenty-Fifth Annual IEEE Applied Power Electronics Conference and Exposition (APEC)*, pages 2231–2237, February 2010.

31 Sveinung Attestog, Huynh Van Khang, and Kjell G. Robbersmyr. Field reconstruction for modeling multiple faults in permanent magnet synchronous motors in transient states. *IEEE Access*, 9:127131–127140, 2021. https://doi.org/10.1109/ACCESS.2021.3112224.

32 Ehsan Ajily, Mohammad Ardebili, and Karim Abbaszadeh. Magnet defect and rotor eccentricity modeling in axial-flux permanent-magnet machines via 3-d field reconstruction method. *IEEE Transactions on Energy Conversion*, 31(2):486–495, June 2016. https://doi.org/10.1109/TEC.2015.2506819.

33 Thomas A. Lipo. *Analysis of Synchronous Machines*. CRC Press, Boca Raton, FL, second edition, 2017. ISBN 9781439880678.

34 Xiaodong Luo, Yung Liao, Hamid A. Toliyat, Ahmed El-Antably, and Thomas A. Lipo. Multiple coupled circuit modeling of induction machines. *IEEE Transactions on Industry Applications*, 31(2):311–318, March/April 1995. https://doi.org/10.1109/28.370279.

35 Hamid A. Toliyat, Thomas A. Lipo, and John C. White. Analysis of a concentrated winding induction machine for adjustable speed drive applications part 1 (motor analysis). *IEEE Transactions on Energy Conversion*, 6(4):679–683, December 1991. https://doi.org/10.1109/60.103642.

36 Nasser A. Al-Nuaim and Hamid A. Toliyat. A novel method for modeling dynamic air-gap eccentricity in synchronous machines based on modified winding function theory. *IEEE Transactions on Energy Conversion*, 13(2):156–162, June 1998. https://doi.org/10.1109/60.678979.

37 Javad Faiz and Iraj Tabatabaei. Extension of winding function theory for nonuniform air gap in electric machinery. *IEEE Transactions on Magnetics*, 38(6):3654–3657, November 2002. https://doi.org/10.1109/TMAG.2002.804805.

38 Javad Faiz and Mehdi Ojaghi. Unified winding function approach for dynamic simulation of different kinds of eccentricity faults in cage induction machines. *IET Electric Power Applications*, 3(5):461–470, September 2009. https://doi.org/10.1049/iet-epa.2008.0206.

39 Behrooz M. Ebrahimi, Mohammad J. Roshtkhari, Javad Faiz, and Seyed V. Khatami. Modeling and detecting the stator winding inter turn fault of permanent magnet synchronous motors using stator current signature analysis. *ISA Transactions*, 80:359–371, September 2018. https://doi.org/10.1016/j.isatra.2018.04.020.

40 Peng Liang, Yulong Pei, Feng Chai, and Kai Zhao. Analytical calculation of d- and q-axis inductance for interior permanent magnet motors based on winding function theory. *Energies*, 9(8):580, 2016. https://doi.org/10.3390/en9080580.

41 Carsten Klein, Mira Pinter, Marco Palmieri, Matthias Nienhaus, and Emanuele Grasso. An exact solution to the modified winding function for eccentric permanent magnet synchronous machines. *Mathematics*, 10(3):345, January 2022. https://doi.org/10.3390/math10030345.

42 Mehrdad Gholami, Om-Kolsoom Shahryari, and Hadi Tarimoradi. An extended winding function model for induction machine modelling considering saturation effect. *IET Electric Power Applications*, 15(1):79–91, 2021. https://doi.org/10.1049/elp2.12006.

43 Farhad Rezaee-Alam, Behrooz Rezaeealam, and Vahid Naeini. An improved winding function theory for accurate modeling of small and large air-gap electric machines. *IEEE Transactions on Magnetics*, 57(5):1–13, May 2021. https://doi.org/10.1109/TMAG.2021.3064402.

44 Hamza W. Derbas, Joshua M. Williams, Andreas C. Koenig, and Steven D. Pekarek. A comparison of nodal- and mesh-based magnetic equivalent circuit models. *IEEE Transactions on Energy Conversion*, 24(2):388–396, June 2009. https://doi.org/10.1109/tec.2008.2002037.

45 S. D. Sudhoff and F. Shane. Pmmtoolbox: power magnetic material toolbox (version 1.0) [matlab toolbox]. Available online at https://engineering.purdue.edu/ECE/Research/Areas/PES/Software/power_magnetic_material_toolbox,2018

46 Seho Son, Hyunseung Lee, Dayeon Jeong, Ki-Yong Oh, and Kyung Ho Sun. A novel physics-informed neural network for modeling electromagnetism of a permanent magnet synchronous motor. *Advanced Engineering Informatics*, 57:102035, 2023. ISSN 1474-0346. https://doi.org/10.1016/j.aei.2023.102035.

47 Xiaofeng Yuan, Lin Li, and Yalin Wang. Nonlinear dynamic soft sensor modeling with supervised long short-term memory network. *IEEE Transactions on Industrial Informatics*, 16(5):3168–3176, 2020. https://doi.org/10.1109/TII.2019.2902129.

48 Razvan Pascanu, Tomas Mikolov, and Yoshua Bengio. On the difficulty of training recurrent neural networks. In Sanjoy Dasgupta and David McAllester, editors, *Proceedings of the 30th International Conference on Machine Learning*, volume 28 of *Proceedings of Machine Learning Research*, pages 1310–1318, Atlanta, Georgia, USA, 17–19 Jun 2013. PMLR. https://proceedings.mlr.press/v28/pascanu13.html.

49 Ronald J. Williams and Jing Peng. An efficient gradient-based algorithm for on-line training of recurrent network trajectories. *Neural Computation*, 2(4):490–501, December 1990. ISSN 0899-7667. https://doi.org/10.1162/neco.1990.2.4.490.

50 Klaus Greff, Rupesh K. Srivastava, Jan Koutník, Bas R. Steunebrink, and Jürgen Schmidhuber. LSTM: A search space odyssey. *IEEE Transactions on Neural Networks and Learning Systems*, 28(10):2222–2232, 2017. https://doi.org/10.1109/TNNLS.2016.2582924.

51 Felix A. Gers, Jürgen Schmidhuber, and Fred Cummins. Learning to forget: Continual prediction with LSTM. *Neural Computation*, 12(10):2451–2471, 2000. https://doi.org/10.1162/089976600300015015.

52 Felix A. Gers and Jürgen Schmidhuber. Recurrent nets that time and count. In *Proceedings of the IEEE-INNS-ENNS International Joint Conference on Neural Networks. IJCNN 2000. Neural Computing: New Challenges and Perspectives for the New Millennium*, volume 3, pages 189–194, 2000. https://doi.org/10.1109/IJCNN.2000.861302.

53 Alex Graves. Long short-term memory. In *Supervised Sequence Labelling with Recurrent Neural Networks*, pages 37–45. Springer Berlin Heidelberg, Berlin, Heidelberg, 2012. ISBN 978-3-642-24797-2. https://doi.org/10.1007/978-3-642-24797-2̇4.

54 Yuxiu Hua, Zhifeng Zhao, Rongpeng Li, Xianfu Chen, Zhiming Liu, and Honggang Zhang. Deep learning with long short-term memory for time series prediction. *IEEE Communications Magazine*, 57(6):114–119, 2019. https://doi.org/10.1109/MCOM.2019.1800155.

55 Davide Bianchi, Emilio Camponogara, Piero Miotto, Pietro De Rossi, Antonio Concilio, Giorgio Previati, and Massimiliano Gobbi. Learning-based model predictive current control for synchronous machines: An LSTM approach. *Control Engineering Practice*, 124:105163, 2022. ISSN 0947-3580. https://doi.org/10.1016/j.conengprac.2022.105163.

56 Suleyman Ali Baran, Mehmet Yildirim, and Berat Berat Yildiz. Machine learning techniques for vector control of permanent magnet synchronous motor drives. *Cogent Engineering*, 11(1):2323813, 2024. https://doi.org/10.1080/23311916.2024.2323813.

57 Dalwinder Singh and Birmohan Singh. Investigating the impact of data normalization on classification performance. *Applied Soft Computing*, 97:105524, 2020. ISSN 1568-4946. https://doi.org/10.1016/j.asoc.2019.105524.

58 Einoshin Suzuki and Tadahiro Tanigaki. Impact of data normalization on deep neural network for time series forecasting. *arXiv preprint arXiv:1812.05519*, 2018.

59 Diederik P. Kingma and Jimmy Ba. Adam: A method for stochastic optimization. May 2015. http://arxiv.org/abs/1412.6980.

60 Mrutyunjaya Biswal, Soumya Ranjan Das, Prakash Kumar Hota, and Asit Mohanty. Design of machine learning-based controllers for speed control of PMSM drive. *Scientific Reports*, 15(1):2140, 2025. https://doi.org/10.1038/s41598-025-02396-y.

61 Xuejian Zhou, Xinyang Liu, Dongsheng Yang, Yuxiang Song, Haotian Chen, Liang Feng, and Lijun Diao. Comprehensive evaluation of deep reinforcement learning for permanent magnet synchronous motor current tracking and speed control applications. *Engineering Applications of Artificial Intelligence*, 143:110551, 2025. ISSN 0952-1976. https://doi.org/10.1016/j.engappai.2025.110551.

62 Annette Mütze, Kourosh Heidarikani, Pawan Kumar Dhakal, Roland Seebacher, and Sebastian Schöps. "Creator case": PMSM and IM electric machine data for validation and benchmarking of simulation and modeling approaches. *arXiv preprint*, 2025. https://arxiv.org/html/2501.15921v1.

63 Ahmed Belkhadir, Remus Pusca, Driss Belkhayat, Raphaël Romary, and Youssef Zidani. Analytical modeling, analysis and diagnosis of external rotor PMSM with stator winding unbalance fault. *Energies*, 16(7):3198, April 2023. https://doi.org/10.3390/en16073198.

64 Anmol Aggarwal and Elias G. Strangas. Review of detection methods of static eccentricity for interior permanent magnet synchronous machine. *Energies*, 12(21):4105, October 2019. https://doi.org/10.3390/en12214105.

65 Jie Xu, Lijun Zhang, De Meng, and Hui Su. Simulation, verification and optimization design of electromagnetic vibration and noise of permanent magnet synchronous motor for vehicle. *Energies*, 15(16):5808, August 2022. https://doi.org/10.3390/en15165808.

66 P. Naderi and Azita Fathi. Fault diagnosis/separation of surface mounted permanent magnet synchronous machine by current and its homopolar orders analysis. *IEEE Transactions on Energy Conversion*, 38(2):924–934, June 2023. https://doi.org/10.1109/TEC.2022.3216686.

67 Vahid Rafiee and Jawad Faiz. Uncertainty quantification of permanent magnet motor using response surface surrogate modeling, pages 1–6, February 2020. https://doi.org/10.1109/PEDSTC49159.2020.9088480.

68 Mohamed Ibrahim, Vladimir Rjabtšikov, Anton Rassõlkin, Toomas Vaimann, and Anton Kallaste. Validation of an EV-permanent magnet synchronous motor model based on analytical dynamic approach, pages 240–246, September 2022. https://doi.org/10.1109/ICEM51905.2022.9910755.

4

Fault Diagnosis Techniques

4.1 Part A: Introduction and Foundations

4.1.1 Introduction

Fault diagnosis in permanent magnet machines represents a critical aspect of modern condition monitoring and predictive maintenance strategies. This chapter presents a comprehensive overview of diagnostic techniques employed to detect, identify, and characterize various fault conditions in PM machines.

Methodological Classification: Fault diagnosis techniques can be fundamentally categorized into two distinct approaches:

- **Signal-based methods** operate as pattern recognition pipelines comprising: **signal acquisition → preprocessing → feature extraction → decision logic → fault classification and severity estimation** (Figure 4.1). These methods identify characteristic signatures (spectral patterns, trajectory deformations, envelope statistics) in measured signals that correlate with specific fault conditions. The implementation of each pipeline stage differs across techniques based on the physical signals measured, the fault phenomena targeted, and the available instrumentation.
- **Model-based methods** follow a different paradigm: they solve inverse problems to estimate physical machine parameters (R_s, L_d, L_q, λ_{pm}) from measured electrical quantities. Faults are detected as deviations of estimated parameters from expected healthy values. Rather than searching for patterns in signal features, these methods provide quantitative, physically interpretable metrics directly tied to component degradation.

The techniques can also be classified by the physical phenomena they exploit:

- **Electromagnetic methods** analyze electrical signals (currents, voltages, flux) to detect faults that alter electromagnetic behavior.
- **Mechanical methods** monitor vibrations and acoustic emissions resulting from mechanical anomalies.
- **Hybrid approaches** combine multiple signal types or integrate physics-based models with data-driven analytics.

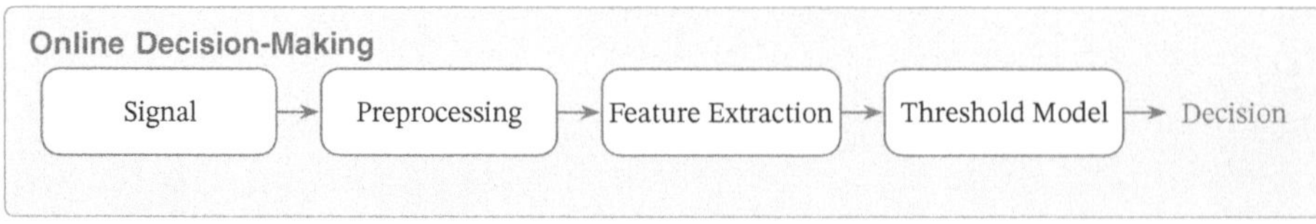

Figure 4.1 Signal-based fault diagnosis as a pattern recognition pipeline showing the sequential stages from signal acquisition through preprocessing, feature extraction, threshold model evaluation, to final decision.

This chapter is organized into three parts to eliminate redundancy and improve clarity. **Part A** consolidates the foundational concepts, signal processing techniques, preprocessing methods, confounding factors, calibration procedures, decision logic frameworks, validation methodologies, and implementation considerations that are common across multiple diagnostic techniques. **Part B** examines each specific diagnostic technique, focusing exclusively on what makes each method unique, including its distinctive physical principles, specific signal requirements, characteristic features, and particular strengths and limitations. **Part C** provides comparative analysis through tables and selection guidelines to help practitioners choose appropriate techniques for their applications.

This chapter presents the diagnostic techniques in a fault-agnostic manner, emphasizing their fundamental principles, signal requirements, and general capabilities. Detailed information on specific fault types (eccentricity, inter-turn shorts, bearing faults, demagnetization, etc.), their characteristic signatures, and the application of these diagnostic methods to detect and characterize each fault condition is provided in the subsequent fault-specific chapters (Chapters 5–9). Each fault chapter demonstrates how the techniques described here are applied in practice to diagnose that particular fault type, including fault-specific frequency formulas, signature interpretation, and case studies.

4.1.2 Preprocessing Pipeline

Before applying diagnostic signal processing techniques, raw sensor data must undergo systematic preprocessing to enhance fault signatures while suppressing noise and artifacts. The preprocessing pipeline transforms raw measurements into conditioned signals suitable for reliable fault feature extraction. Each preprocessing step addresses specific signal quality issues that could otherwise obscure fault indicators or introduce false alarms.

Synchronization: Multi-channel measurements must be precisely time-aligned to preserve phase relationships critical for coordinate transformations and power calculations. Inter-channel synchronization within 1–10 μs is typical.

Resampling and Antialias Filtering: Uniform sampling at appropriate rates (5–50 kHz depending on application) captures fault signatures while managing computational load. Analog antialiasing filters (4th–6th order Butterworth) prevent high-frequency content from aliasing.

Detrending: DC offsets and slow thermal drifts are removed via mean subtraction, polynomial detrending, or high-pass filtering (0.1–1 Hz cutoff).

Baseline Normalization: Operating condition variations are accommodated through amplitude scaling, operating-point lookup tables, or functional models capturing parameter dependencies on speed, load, and temperature.

Demodulation: Amplitude and frequency demodulation via Hilbert transform or envelope detection reveals modulation components at fault characteristic frequencies.

Order Tracking: Variable-speed operation requires angular resampling to create order-domain signals where fault harmonics appear at fixed orders regardless of speed variations.

4.1.3 Signal Processing Foundations

Modern signal-based fault diagnosis techniques rely on a common toolkit of signal processing methods to extract fault signatures from measured signals. These methods form the core of the **feature extraction** stage in the pattern recognition pipeline, transforming raw time-domain measurements into informative representations, such as frequency spectra, time-frequency maps, coordinate-transformed quantities, or statistical descriptors, that reveal characteristic fault patterns. This section consolidates the mathematical foundations and practical considerations that apply broadly across multiple signal-based diagnostic methods.

4.1.3.1 Fourier-based Methods

Fast Fourier transform (FFT) converts time-domain signals into the frequency domain, revealing harmonic content and characteristic fault frequencies [1, 2]:

$$X[k] = \sum_{n=0}^{N-1} x[n]e^{-j2\pi kn/N} \tag{4.1}$$

where $x[n]$ is the time-domain signal, $X[k]$ is the frequency-domain representation, N is the number of samples, and k is the frequency bin index. The frequency resolution is $\Delta f = f_s/N$ where f_s is the sampling frequency.

Practical Considerations:

- Window length determines frequency resolution; longer windows provide finer resolution but increase latency.
- Windowing functions (Hanning, Hamming, Blackman) reduce spectral leakage [2].
- Zero-padding interpolates the spectrum but does not improve true frequency resolution.
- Overlapping windows (typically 50% overlap) increase temporal resolution.

4.1.3.2 Time–Frequency Analysis

Rationale for Time–Frequency Methods: Standard FFT-based frequency analysis assumes signal stationarity, that statistical properties remain constant over the analysis window. This assumption breaks down when machine operating conditions vary (speed changes, load transients, startup/shutdown sequences). Time–frequency methods address this limitation by revealing how spectral content evolves with time, enabling fault detection during variable-speed operation, transient events, and fault progression monitoring where frequency characteristics change temporally. Note that time-frequency methods generally incur higher computational complexity compared to standard FFT, with the notable exception of the discrete wavelet transform (DWT), which offers reasonable computational efficiency through fast filter-bank implementations.

Short-time Fourier transform (STFT) computes FFT over successive windowed segments for time-frequency analysis [3]:

$$\text{STFT}(t, f) = \int_{-\infty}^{\infty} x(\tau)w(\tau - t)e^{-j2\pi f\tau}d\tau \tag{4.2}$$

STFT enables tracking of how frequency content evolves during transient conditions or fault progression.

Wavelet transform provides multi-resolution analysis with frequency-dependent time resolution [4, 5]. The continuous wavelet transform (CWT):

$$\mathrm{CWT}(a, b) = \frac{1}{\sqrt{|a|}} \int_{-\infty}^{\infty} x(t)\psi^*\left(\frac{t-b}{a}\right)dt \tag{4.3}$$

The **Discrete wavelet transform (DWT)** decomposes signals into approximation and detail coefficients at multiple scales, commonly using Daubechies (db4, db6) or Symlet wavelets [6].

Cohen's class of time–frequency distributions provides a unified framework for bilinear time-frequency representations [7]:

$$C_x(t, f) = \iiint A(\theta, \tau)x\left(u + \frac{\tau}{2}\right)x^*\left(u - \frac{\tau}{2}\right)e^{-j2\pi\theta t}e^{-j2\pi f\tau}e^{j2\pi\theta u}\, du\, d\tau\, d\theta \tag{4.4}$$

where $A(\theta, \tau)$ is a kernel function defining specific distributions. Common members include the **Wigner–Ville distribution (WVD), Choi–Williams distribution (CWD),** and **smoothed pseudo Wigner–Ville distribution (SPWVD)**. These methods offer superior time–frequency resolution compared to STFT but suffer from cross-term interference when analyzing multi-component signals. Their high computational complexity limits real-time implementation in embedded systems.

Hilbert–Huang transform (HHT) combines empirical mode decomposition (EMD) with Hilbert spectral analysis to adaptively decompose nonstationary and nonlinear signals [8]. EMD decomposes a signal into intrinsic mode functions (IMFs):

$$x(t) = \sum_{i=1}^{n} c_i(t) + r_n(t) \tag{4.5}$$

where $c_i(t)$ are IMFs and $r_n(t)$ is the residual. The Hilbert transform is then applied to each IMF to extract instantaneous frequency and amplitude, forming the Hilbert spectrum. HHT is particularly effective for fault detection in machines with nonlinear dynamics and time-varying operating conditions, as it requires no predefined basis functions and adapts to the signal characteristics. However, mode mixing and end effects can complicate interpretation.

Teager–Kaiser energy operator (TKEO) enhances detection of amplitude and frequency modulations [9]:

$$\Psi[x(n)] = x^2(n) - x(n+1)x(n-1) \tag{4.6}$$

The three-sample nonlinear operation instantaneously captures the energy required to generate the signal, making it highly sensitive to coupled amplitude-frequency variations that characterize bearing impacts and early-stage faults. The quadratic term $x^2(n)$ amplifies transient events, while the product $x(n+1)x(n-1)$ provides temporal context, enabling real-time fault detection with minimal computational overhead.

4.1.3.3 Coordinate Transformations

Clarke transformation converts three-phase quantities (a, b, c) to two-phase stationary reference frame (α, β) [10]:

$$\begin{bmatrix} x_\alpha \\ x_\beta \end{bmatrix} = \frac{2}{3}\begin{bmatrix} 1 & -\frac{1}{2} & -\frac{1}{2} \\ 0 & \frac{\sqrt{3}}{2} & -\frac{\sqrt{3}}{2} \end{bmatrix}\begin{bmatrix} x_a \\ x_b \\ x_c \end{bmatrix} \tag{4.7}$$

Park transformation rotates the $\alpha\beta$ frame to align with the rotor, producing d-q components in the synchronously rotating reference frame [11, 12]:

$$\begin{bmatrix} x_d \\ x_q \end{bmatrix} = \begin{bmatrix} \cos\theta & \sin\theta \\ -\sin\theta & \cos\theta \end{bmatrix} \begin{bmatrix} x_\alpha \\ x_\beta \end{bmatrix} \tag{4.8}$$

In the dq frame, fundamental frequency components appear as DC quantities under steady-state conditions, while faults introduce AC ripple.

Space vector representation treats α and β components as real and imaginary parts of a complex vector, enabling geometric fault pattern analysis.

4.1.3.4 Envelope Analysis and Demodulation

Hilbert transform extracts the analytic signal enabling envelope and instantaneous frequency estimation. The envelope $A(t) = |x(t) + j\tilde{x}(t)|$ reveals amplitude modulation patterns characteristic of bearing defects and eccentricity.

Demodulation techniques extract fault signatures from carrier signals modulated by fault phenomena through band-pass filtering followed by envelope extraction.

4.1.3.5 Parametric Spectral Methods

Autoregressive (AR) modeling represents the signal as $x(n) = \sum_{k=1}^{p} a_k x(n-k) + w(n)$, where a_k are AR coefficients and $w(n)$ is white noise. The power spectral density is:

$$P_{\mathrm{AR}}(f) = \frac{\sigma_w^2}{\left|1 + \sum_{k=1}^{p} a_k e^{-j2\pi fk}\right|^2} \tag{4.9}$$

providing sharp spectral peaks for resonant modes.

MUSIC (multiple signal classification) decomposes the correlation matrix into signal and noise subspaces, with spectrum [13]:

$$P_{\mathrm{MUSIC}}(f) = \frac{1}{\mathbf{a}^H(f)\mathbf{U}_n\mathbf{U}_n^H\mathbf{a}(f)} \tag{4.10}$$

where $\mathbf{U}_n$ contains noise eigenvectors and $\mathbf{a}(f)$ is the steering vector. Sharp peaks indicate sinusoidal frequencies.

ESPRIT (estimation of signal parameters via rotational invariance techniques) exploits rotational invariance in signal subspace [14]:

$$\mathbf{U}_s^{(2)} = \mathbf{U}_s^{(1)}\mathbf{\Phi} \tag{4.11}$$

where $\mathbf{\Phi}$ contains phase shifts $e^{j2\pi f_i}$ yielding frequency estimates:

$$f_i = \frac{1}{2\pi}\angle(\lambda_i) \tag{4.12}$$

from eigenvalues λ_i of $\mathbf{\Phi}$.

These methods provide superior frequency resolution for sinusoidal components compared to FFT, valuable when fault signatures comprise narrowband tones embedded in noise.

4.1.3.6 Cepstrum Analysis

The **cepstrum** applies a second Fourier transform to the logarithm of the power spectrum [15]:

$$C(q) = \left|\mathcal{F}^{-1}\{\log(|X(f)|^2)\}\right|^2 \tag{4.13}$$

where q denotes quefrency (time-like domain). The logarithmic operation converts multiplicative harmonics into additive components, while the second transform maps periodic patterns in the frequency domain to distinct peaks in the quefrency domain. This makes cepstrum analysis particularly effective at detecting bearing faults and gear defects that generate families of harmonics or sidebands with regular frequency spacing.

4.1.4 Decision Logic Frameworks

Once diagnostic features have been extracted and conditioned, the critical step of converting measurements into actionable fault decisions remains. Decision logic frameworks provide the algorithms and statistical tools to classify operating states, detect anomalies, and estimate fault severity. The choice of decision framework fundamentally shapes diagnostic performance: simple threshold methods offer transparency and computational efficiency but struggle with overlapping fault signatures, while sophisticated machine learning classifiers can handle complex decision boundaries at the cost of interpretability and training data requirements. Modern industrial systems increasingly employ hierarchical decision architectures that combine multiple frameworks, using fast threshold checks for initial screening followed by more computationally intensive classification when anomalies are detected.

Threshold-based Methods: Fixed or adaptive thresholds compare features against limits. Adaptive thresholds adjust based on operating point using lookup tables or functional models. Threshold selection via statistical analysis (2-3σ margins), ROC curve analysis, or domain knowledge.

Statistical Hypothesis Testing: GLRT, SPRT, Chi-Square, and Hotelling T^2 tests provide rigorous frameworks with explicit false alarm rate control through significance level α.

One-class and Anomaly Detection: OC-SVM [16], Isolation Forest [17], and Autoencoders learn healthy operation boundaries without requiring fault data, flagging deviations as anomalies.

Supervised Classification: SVM [18], Decision Trees/Random Forests [19], Neural Networks, and Gaussian Process Classifiers enable fault type identification and severity estimation when labeled data is available. Requires 60–70% training, 15–20% validation, 15–20% holdout testing.

4.1.5 Confounding Factors

Even the most sophisticated diagnostic algorithms face a common challenge: distinguishing genuine fault signatures from benign operational variations. Load transients, temperature swings, manufacturing tolerances, and sensor imperfections all introduce signal variations that can mimic or mask fault indicators. Successful fault diagnosis requires not only detecting anomalies but also attributing them correctly to either faults or confounding factors. Understanding these confounding influences and implementing appropriate mitigation strategies is essential for achieving low false-alarm rates and high detection sensitivity in real-world deployments.

Load Variations and Transients: Steady-state load changes alter current magnitudes, saturation effects, and temperature. Transient load steps temporarily overwhelm fault signatures. Mitigation: suspend diagnostics during transients or use adaptive sensitivity.

Speed Ripple: Rotor speed fluctuations from cogging torque or load oscillations modulate signals and blur spectral lines. Mitigation: order tracking or instantaneous frequency estimation.

PWM Artifacts and Inverter Nonlinearities: Switching harmonics (5–20 kHz), dead-time effects, and voltage drops contaminate measurements. Mitigation: hardware and digital filtering, differential measurements, inverter compensation models.

Magnetic Saturation: Nonlinear saturation causes inductances to vary with current, introducing legitimate parameter variations. Mitigation: multi-dimensional parameter maps as functions of i_d and i_q.

Temperature Drift: Resistance increases $\sim 0.4\%/°C$ (typical for copper and aluminum windings), magnet flux decreases $\sim 0.11\%/°C$ (typical for NdFeB magnets). A $\pm 30°C$ variation produces $\pm 12\%$ resistance change. Mitigation: direct temperature measurement, thermal modeling, temperature-indexed baselines.

Sensor Bias and Noise: Current sensor offsets, gain errors, and encoder quantization corrupt measurements. Mitigation: online offset estimation, periodic calibration, redundant measurements.

Structural Resonances: Mechanical resonances couple into electrical signals through speed oscillations and air-gap modulation. Mitigation: system identification during commissioning, frequency discrimination.

Comparative Impact on Diagnostic Approaches: Signal-based methods (MCSA, vibration analysis) prove particularly vulnerable to PWM artifacts, structural resonances, and speed ripple, as these contaminate the spectral content directly and can obscure fault harmonics. However, signal-based techniques demonstrate relative robustness to model uncertainties and manufacturing tolerances. Model-based approaches (parameter estimation, observers) face their greatest challenges from magnetic saturation and temperature drift, which introduce legitimate parameter variations that violate model assumptions and trigger false alarms. Load transients severely impact both paradigms, though model-based methods can partially compensate through adaptive baseline adjustment. Sensor bias affects model-based techniques more critically, since parameter estimation algorithms interpret systematic offsets as parameter changes, whereas signal-based methods using spectral ratios or relative indicators naturally cancel common-mode sensor errors. Hybrid strategies combining physics-based models with data-driven adaptation show the most promise for handling the full spectrum of confounding factors, achieving 15–30% lower false-alarm rates compared to single-paradigm approaches in industrial trials.

4.1.6 Baseline Formation and Calibration

4.1.6.1 Machine-specific Fingerprinting

Baseline data collection during verified healthy operation captures unique electromagnetic and mechanical characteristics. Measurements should span representative speed-load combinations (5–10 points minimum) at thermal steady-state.

4.1.6.2 Speed and Load Normalization

- **Speed:** Order-based indexing or angle-domain representations provide speed normalization.
- **Load:** Discrete operating point baselines, functional modeling, or adaptive baseline tracking.

4.1.6.3 Per-unit Scaling

Normalization by rated values creates dimensionless quantities facilitating cross-machine comparison:

$$L_d^* = \frac{L_d}{L_{\text{base}}}, \quad \lambda_m^* = \frac{\lambda_m}{\lambda_{\text{base}}}, \quad R_s^* = \frac{R_s}{Z_{\text{base}}} \tag{4.14}$$

4.1.6.3.1 Temperature Compensation

Temperature variations affect both signal-based and model-based diagnostic methods. For **signal-based methods**, temperature compensation normalizes feature baselines (e.g., current amplitudes, spectral magnitudes) to account for thermally-induced signal variations. For **model-based methods**, temperature compensation is critical for parameter estimation, as resistance increases approximately 0.4%/°C and PM flux decreases approximately 0.11%/°C. Common compensation approaches include: - Direct temperature measurement with correction factors: $R_s(T) = R_s(T_{\text{ref}})[1 + 0.004(T - T_{\text{ref}})]$ - Thermal modeling using machine thermal time constants - Temperature-indexed baseline lookup tables spanning the expected operating temperature range

4.1.7 Validation Methodology

A diagnostic algorithm that performs flawlessly in simulation may fail catastrophically when deployed on real hardware facing manufacturing variations, environmental disturbances, and unanticipated operating conditions. Rigorous validation requires a multi-stage progression from controlled simulation environments through hardware-in-the-loop testing, laboratory bench experiments with seeded faults, and ultimately field deployment with long-term monitoring. Each validation stage exposes different classes of implementation issues: simulation reveals algorithmic soundness and parameter sensitivity, HIL testing uncovers real-time constraints and numerical precision limitations, bench testing validates performance under known fault conditions with measurement noise, and field trials confirm practical utility under genuine operating diversity. Skipping validation stages or relying solely on in-house data risks deploying algorithms with hidden failure modes that manifest only after costly field failures.

Simulation Validation: FEM (2D/3D), circuit-coupled models, and analytical models enable controlled fault injection at known severities. Simulation provides a cost-effective initial validation stage where fault parameters can be precisely controlled and varied systematically, allowing systematic exploration of fault severity ranges and operating conditions that may be difficult or dangerous to replicate experimentally. Monte Carlo simulation assesses robustness by introducing parameter uncertainties, measurement noise, and operating condition variations to evaluate algorithm performance under realistic uncertainty.

Hardware-in-the-loop (HIL) Testing: Real-time platforms (dSPACE, Typhoon HIL, OPAL-RT) execute algorithms on target hardware with simulated machine, validating timing, fixed-point effects, and interrupt handling. HIL bridges the gap between pure simulation and physical testing by running the diagnostic algorithm on actual embedded hardware while the machine dynamics are simulated in real-time, exposing implementation issues such as computational delays, numerical precision limitations, and real-time scheduling constraints. This approach enables rapid iteration and comprehensive testing across fault scenarios without requiring multiple physical machines or risking damage from severe fault conditions.

Bench Testing: Dynamometer-coupled machines with seeded faults provide ground-truth validation under controlled laboratory conditions where fault severity is known and environmental factors can be regulated. Testing multiple machines of the same design validates algorithm generalizability and ensures robustness to manufacturing tolerances, while testing across different machine designs confirms broader applicability. Data discipline: temporal splits (train on early data, test on later data to simulate deployment), hold-out sets (reserve machines never seen during development), cross-machine validation (train on subset of machines, test on others).

Field Validation: Operational deployment exposes algorithms to real operating conditions including environmental variations, duty cycle diversity, and operational transients that cannot be fully replicated in laboratory settings. Field validation represents the ultimate test of diagnostic robustness and practical utility. Long-term logging (months to years) with maintenance correlation confirms diagnostic accuracy by comparing algorithm predictions with actual maintenance findings and failure events, establishing true positive rates, false alarm rates, and prognostic horizon accuracy.

Cross-machine and External Validation: Testing across different designs, ratings, and manufacturers ensures diagnostic techniques generalize beyond the specific machines used during development and prevents overfitting to particular machine characteristics or test conditions. Cross-validation across machine families (surface-mounted vs. interior PM, radial vs. axial flux, different slot-pole combinations) demonstrates fundamental applicability of the diagnostic approach. External testbeds and benchmark datasets enable unbiased validation by independent researchers, facilitating reproducible comparisons and preventing optimistic bias from validating only on in-house data.

4.1.8 Real-time Implementation Considerations

Transitioning a diagnostic algorithm from offline post-processing to real-time embedded execution imposes severe constraints on latency, computational complexity, memory footprint, and numerical precision. A sophisticated machine learning classifier validated offline with unlimited processing time may prove completely impractical when constrained to execute within 1 ms on a cost-optimized microcontroller with 64 KB of RAM and fixed-point arithmetic. Real-time implementation demands careful architectural choices: multi-rate processing hierarchies that match computational intensity to fault dynamics, efficient algorithms that exploit hardware acceleration (SIMD instructions, DSP cores, FPGAs), fixed-point numerical implementations that preserve diagnostic accuracy despite quantization, and memory-efficient data structures that minimize RAM usage while maintaining sufficient history for statistical reliability. The implementation challenge intensifies for integrated motor drives where diagnostic algorithms must coexist with control loops, communication protocols, and safety functions while sharing limited processor cycles and memory resources.

Latency Requirements: Different fault types demand vastly different detection speeds based on their progression rates and consequence severity: - **Electrical faults:** 1–10 ms detection for protective shutdown - **Magnetic faults:** 100–1000 ms sufficient - **Mechanical faults:** seconds to minutes adequate.

Multi-rate architectures accommodate these diverse requirements: Fast layer (5–20 kHz) for electrical protection, medium layer (100–1000 Hz) for parameter tracking, slow layer (1–100 Hz) for trend analysis.

Computational Cost: Algorithm complexity directly determines processor requirements and energy consumption: - Threshold monitoring: 10–20k operations, 100–200 μs - FFT (2048-point): 50–500k operations, 1–5 ms - Extended Kalman Filter: 100–500k operations, 100–500 μs - Neural networks: 100k–1M operations, may need accelerators.

Memory Requirements: RAM availability constrains buffering depth, model complexity, and feature storage: - Lightweight: 20–100 KB (microcontroller RAM) - Comprehensive: 200–500 KB to several MB.

Sample Rates: Acquisition rates must satisfy Nyquist criterion for target fault frequencies: - Electrical faults: 5–20 kHz - Mechanical faults: 10–50 kHz - Acoustic: up to 100 kHz.

Window Lengths: Temporal averaging introduces fundamental trade-off between noise rejection and detection latency: - Short windows (1–10 ms): fast detection, high variance - Long windows (100–200 ms): low noise, high latency - Overlapping windows (50–75%) improve update rate.

4.2 Part B: Diagnostic Techniques

4.2.1 Motor Current Signature Analysis

Motor current signature analysis (MCSA) is **the most common fault diagnosis technique** for permanent magnet machines, owing to its noninvasive nature, zero incremental hardware cost, and ability to detect multiple fault types using existing current sensors [20, 21].

A modern MCSA diagnostic architecture for PM machine drives is illustrated in Figure 4.2. The system comprises the PM motor fed by an inverter/motor drive, which contains built-in current sensors for the three-phase stator currents (i_a, i_b, i_c). These current signals, already available for field-oriented control, are routed through signal conditioning circuitry to a DSP/MCU for digital signal processing. The DSP performs FFT-based spectrum analysis to extract frequency-domain features, which are then fed to the fault diagnosis algorithm. The diagnostic results are displayed on an HMI/monitoring system for operator awareness and decision-making. Position feedback from the encoder returns to the inverter for motor control purposes. This architecture leverages the existing drive infrastructure, eliminating the need for external current transformers or dedicated spectrum analyzers that characterized earlier MCSA implementations in induction motor applications.

4.2.1.1 Physical Principle

MCSA is distinguished by its focus on **characteristic sideband harmonics** in stator current spectra. Physical changes or faults modify electromagnetic conditions, creating specific harmonic patterns around the fundamental supply frequency and its multiples. The primary diagnostic feature is the identification and analysis of these sidebands, whose amplitude and spacing indicate specific fault conditions. This sideband-centric approach makes MCSA particularly effective for detecting asymmetries that modulate the current waveform.

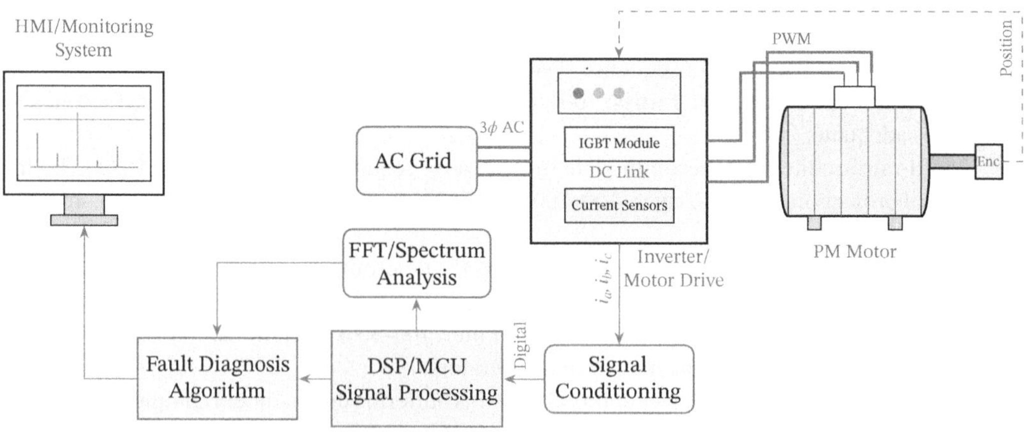

Figure 4.2 Modern MCSA diagnostic system architecture for PM machine drives with integrated current sensing, DSP-based spectrum analysis, and fault diagnosis.

4.2.1.2 Signal Requirements

Primary Observable: Three-phase stator currents.

Accessibility: Already available in typical inverter-fed PMSM drives from standard current sensors used for motor control. No additional sensors, wiring modifications, neutral access, or search coils required, making MCSA the most readily implementable technique for PM machines.

Practical Use Cases Enabled by Zero-cost Accessibility: The zero incremental hardware cost of MCSA makes it particularly valuable for: (1) **Retrofit condition monitoring** where existing machines can be upgraded with software-only diagnostics without physical modifications or production downtime; (2) **Large fleet deployments** in industrial plants, HVAC systems, and building automation where installing additional sensors across hundreds of machines would be cost-prohibitive; (3) **Cost-sensitive applications** in consumer appliances, power tools, and entry-level automotive systems where component cost margins are tight; (4) **Remote or inaccessible installations** such as offshore wind turbines, subsea pumps, or embedded building infrastructure where physical access for sensor installation is difficult or expensive; and (5) **Rapid prototyping and development** where engineers can implement diagnostic algorithms during the design phase using existing control hardware without waiting for specialized sensor procurement and integration.

4.2.1.3 Fault Features and Indicators

Under Stationary Conditions: - **Characteristic sideband harmonics:** Primary indicators appearing around fundamental and harmonics at frequencies determined by fault type - **Sub-harmonic components:** Particularly sensitive to partial demagnetization and eccentricity - **Amplitudes at specific fault-related frequency lines:** Derived via FFT analysis.

Under Nonstationary Conditions: - **Time–frequency representations:** STFT, wavelet transforms, or HHT reveal evolving spectral content during speed/load transients - **Order-domain analysis:** Angular resampling transforms time-domain signals to angle domain, where fault harmonics appear at fixed orders independent of speed variations - **Instantaneous frequency tracking:** Adaptive methods track fault characteristic frequencies as they shift with operating conditions - **Envelope modulation patterns:** Time-varying amplitude modulation of fault-related frequency components.

4.2.1.4 Strengths and Limitations

Strengths:

- Zero incremental hardware cost by leveraging existing sensors.
- Nonintrusive implementation.
- Real-time monitoring capability.
- Particularly effective for electrical faults (demagnetization, inter-turn shorts) and bearing faults.

Limitations:

- Sensitivity degrades substantially at low speeds.
- Load variation sensitivity without normalization.
- Spectral signatures from different faults can overlap.
- Uniform demagnetization across all poles produces weak asymmetric signatures.

Best-fit Applications: Industrial and automotive with stable/predictable load profiles; steady-state or slowly varying conditions; traction drives and continuous manufacturing processes.

4.2.2 Motor Flux Signature Analysis

4.2.2.1 Physical Principle

MFSA is distinguished by **direct monitoring of electromagnetic flux distributions** rather than inferring from current signatures alone. Faults cause asymmetries in flux patterns generating unique harmonic signatures. The method has demonstrated superior capability compared to MCSA in reducing false alarms and missed detections for challenging faults (inter-turn shorts, air gap eccentricity, irreversible demagnetization). Direct flux measurement captures phenomena that produce subtle current changes but distinct flux variations.

In electrical machines, magnetic flux can be classified into main flux and leakage flux. The main flux is the portion that crosses the air gap and links with both stator and rotor windings, contributing to torque production and energy conversion. Leakage flux, in contrast, does not contribute to energy conversion and includes flux components that either remain confined within the stator or rotor (tooth-tip leakage, slot leakage, end-winding leakage, pole leakage, and PM leakage) or cross the air gap as harmonic components without fully linking the windings.

While most flux components complete their paths internally, a small portion of leakage flux can escape into the stator frame and radiate outward as stray flux. Axial stray flux typically originates from end-ring currents and PM leakage fluxes, while radial stray fluxes at the machine's ends are primarily caused by end-winding leakage flux, as illustrated in Figure 4.3.

Internal fluxes undergo varying levels of attenuation before leaking into the surrounding air, leading to generally low flux density in stray flux, especially in machines with steel frames. However, when a fault occurs, the flux distribution becomes asymmetrical, resulting in increased leakage flux density and unique harmonic patterns related to the oscillatory fields of specific failure modes. Various flux leakage paths exist in typical nonsalient pole electrical machines, as shown in Figure 4.4.

4.2.2.2 Signal Requirements

Primary Observables:

- **Internal flux:** Air gap flux via sensors placed inside the machine (invasive).
- **External flux:** Stray flux leaking outside machine housing (noninvasive).

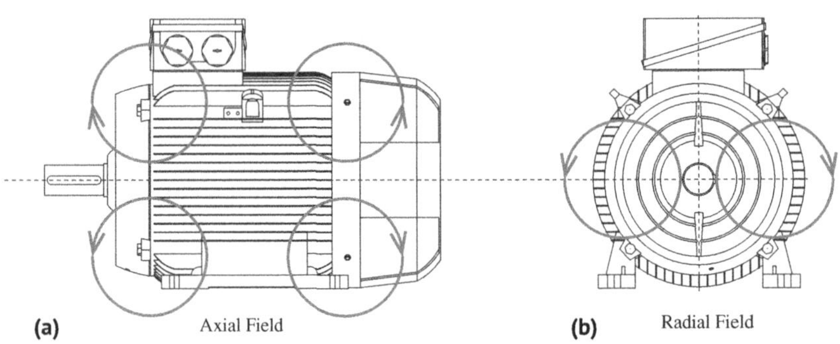

Figure 4.3 Principal flux directions in an electrical machine: (a) side view perspective showing axial flux components (b) frontal view perspective showing radial flux components.

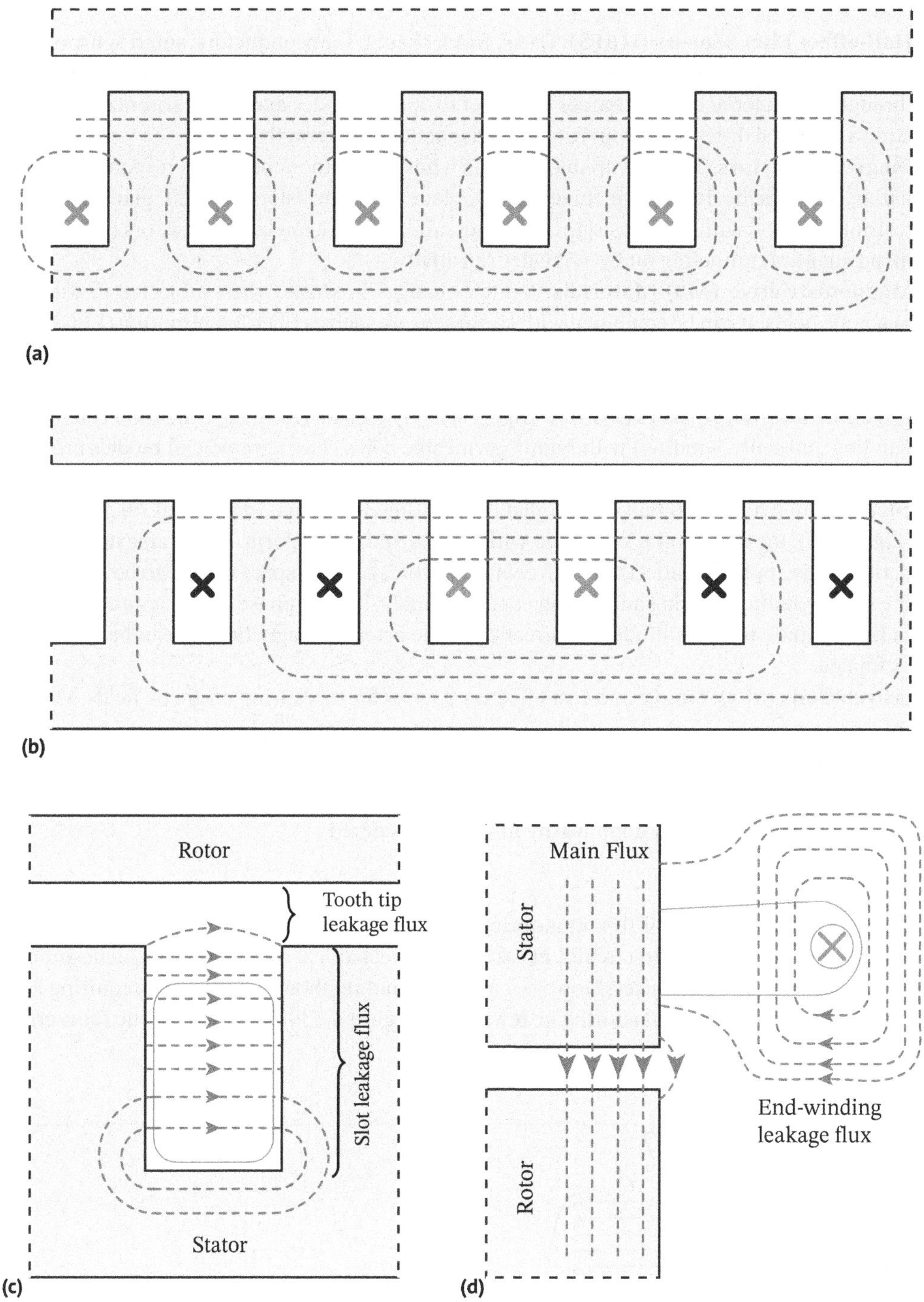

Figure 4.4 Various flux leakage paths in a typical nonsalient pole electrical machine: (a) depiction of pole and slot leakage flux paths (b) visualization of slot leakage flux path for a slot winding (c) detailed view of slot and tooth tip leakage flux paths within a single slot (d) representation of end-winding leakage flux path. *Source:* Adapted from [22].

Sensor Types:

- **Hall-effect Flux Sensors (HEFS):** GaAs, InAs, or InSb semiconductors; smart sensors capture flux in three directions. When an external magnetic field is applied, current passing through the material causes charges to deflect to opposite sides due to the Lorentz force, creating a potential difference proportional to magnetic field strength.
- **Magneto-resistors (MRs):** Variable resistors whose resistance changes in response to external magnetic fields. Its types include anisotropic magneto-resistors (AMRs), giant magneto-resistors (GMRs) with large resistance changes, and tunneling magneto-resistors (TMRs) utilizing quantum tunneling for even greater sensitivity.
- **Magnetostrictive (MS) Materials:** Exhibit changes in shape when subjected to external magnetic fields. It can be combined with strain sensors such as fiber Bragg grating (FBG) integrated with highly volumetric MS material like Terfenol-D to detect geometric deformations caused by rotating PM fields.
- **Flux-gate Magnetometers (FGMs):** Highly sensitive sensors utilizing two sets of coils (drive winding and sense winding) with highly permeable cores. Two commercial models are popular: rod-core FGMs with two rods wound in opposite directions, and ring-core FGMs with a circular core where each half is saturated in opposing directions. In a typical ring-core FGM (Figure 4.5), the drive coil is energized with a trapezoidal waveform. When an external magnetic field is applied, it affects the drive winding core, causing spike voltage to be induced in the sense winding at rising and falling edges. By analyzing the phase and magnitude of these induced spikes, the magnitude and direction of the external magnetic field can be accurately estimated.
- **Search Coils (SCs):** Simple and cost-effective devices for measuring magnetic fields. Voltage is induced when exposed to external magnetic fields: $e = -NA\frac{dB}{dt}$ where e is induced voltage, N is number of turns, A is surface area, and B is air gap flux density. Can be PCB-based, circular, rectangular, or C-shaped; air-core or iron-core. Iron cores significantly enhance induced voltage amplitude, though nonlinearity must be considered.

Installation Approaches:

Internal Positioning: Internal flux monitoring captures air gap flux with higher sensitivity to various faults (inter-turn short circuits, electrical and mechanical rotor faults, magnetic anomalies) compared to external fluxes. However, most internal methods are invasive, requiring sensor installation during manufacturing or rewinding. Figure 4.6 illustrates magnetic flux sensor positions at different internal locations:

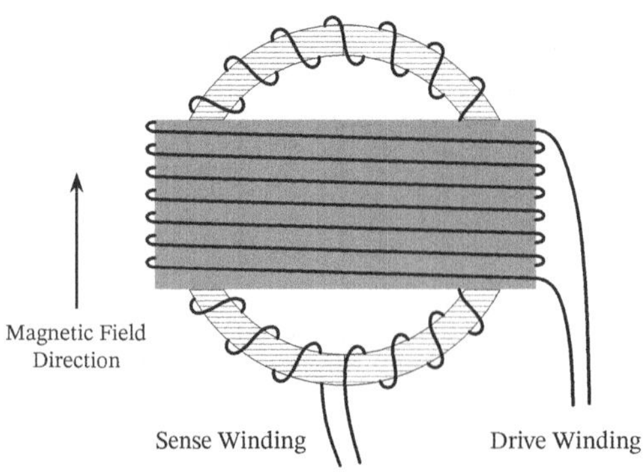

Figure 4.5 Schematic representation of a flux-gate magnetometer featuring a circular core. *Source:* Adapted from [23].

- **IPS1:** Search coil wrapped around a stator tooth in SPMMs.
- **IPS2:** Search coil with straight section placed inside slot beneath stator slot wedge, exiting through holes in stator housing.
- **IPS3:** HEFS, FGM, or TMR sensor mounted on stator tooth between coil extensions.
- **IPS4:** FBG-Terfenol-D sensor embedded in stator slot wedge.
- **IPS5:** HEFS, FGM, or TMR sensor positioned on stator bore within air gap.

External Positioning: External flux monitoring provides noninvasive alternatives with easier installation and more space for larger sensors and innovative multisensory configurations. However, challenges include defining decision boundaries due to sensitivity to sensor location and orientation, and vulnerability to environmental electromagnetic noise. Sensor placement determines whether axial, radial, or combined flux is measured:

- **EPS1:** Frontal axial flux.
- **EPS2:** Combined axial and radial flux.
- **EPS3:** Radial flux.
- **EPS4/EPS5:** C-shaped search coils for radial/axial flux.
- **EPS6:** Emerson flux probe for rear axial flux (when frontal access limited).

(a)

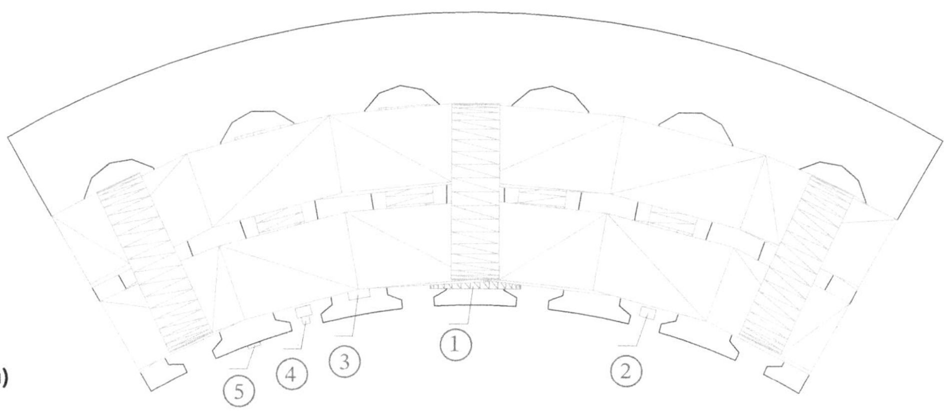

(b)

Figure 4.6 Sensor positions for capturing internal magnetic flux: (a) top view and (b) side view. 1: IPS1, 2: IPS2, 3: IPS3, 4: IPS4, 5: IPS5.

Multisensory Systems: Multiple homogeneous sensors providing spatial information crucial for distinguishing different fault types. Data fusion can occur at sensor level, feature level, or decision level. Figure 4.7 illustrates internal multisensory configurations:

- **IPM1:** Concentrated or pole-specific search coil systems within stator slots.
- **IPM2:** Search coils placed around all stator teeth or every alternate tooth.
- **IPM3:** Array of miniature HEFS sensors arranged around stator circumference inside air gap.

External multisensory systems include:

- **EPM1:** Pseudo-three-phase sensor systems (three search coils at 120° mechanically/electrically, or optimized angular separations).
- **EPM2:** Two search coils placed 180 electrical degrees apart.
- **EPM3:** Array of PCB-mounted HEFS on circular or linear path in axial direction.
- **EPM4:** 24 TMR and 8 FGM sensors uniformly distributed around machine external surface.
- **EPM5:** Three built-in HEFS arranged with 120 electrical degrees in axial direction.

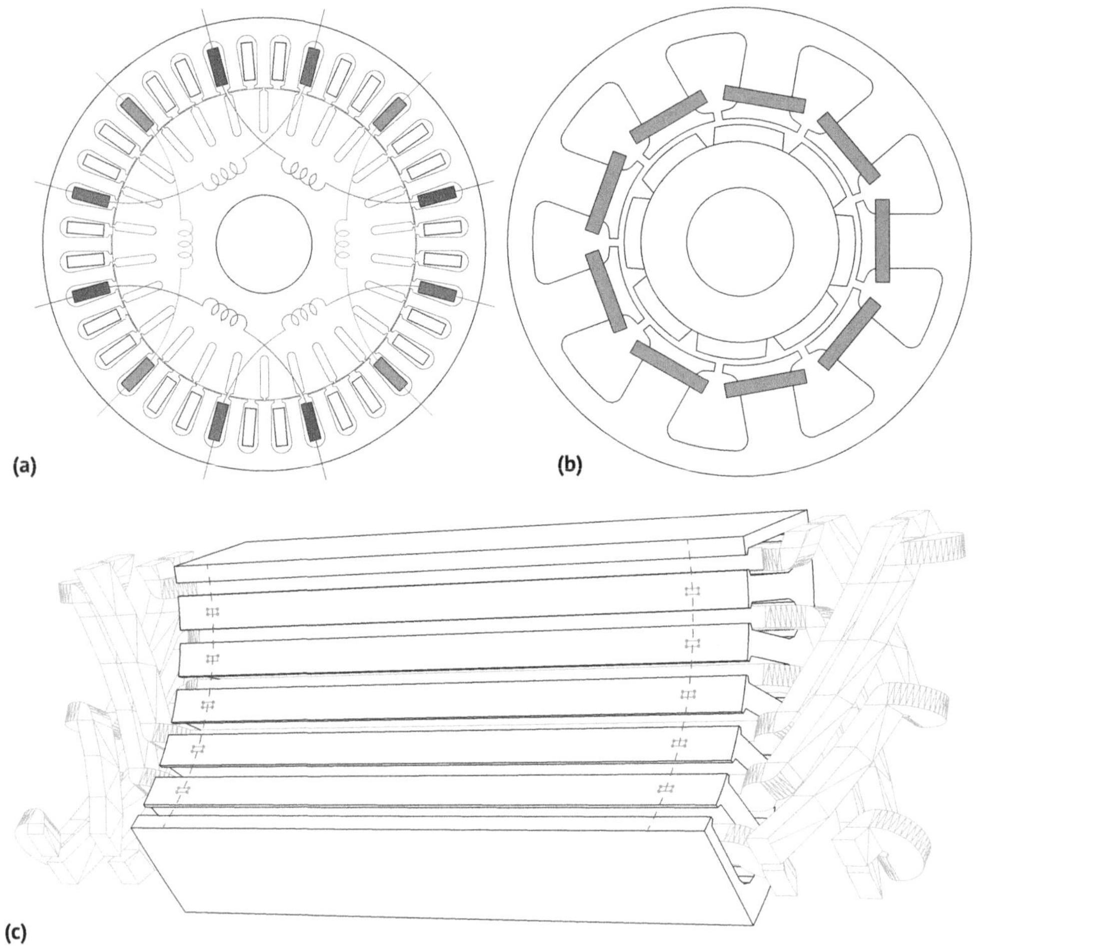

Figure 4.7 Various multisensory approaches for capturing internal magnetic flux: (a) IPM1, (b) IPM2, and (c) IPM3.

Figure 4.8 illustrates external sensor positions and orientations:

Sensorless Approaches: Flux estimation via parameter estimation methods, field reconstruction models, inverse problem solutions, MEC models, or hybrid models, avoiding physical sensors but requiring complex machine models.

4.2.2.3 Fault Features and Indicators

- Spatial flux distribution patterns revealing localized vs. uniform demagnetization.
- Harmonic content in flux linkage estimates.
- Flux-linkage magnitude reductions (% from baseline).
- Position-dependent flux variations (flux as function of rotor angle).

4.2.2.4 Strengths and Limitations

Strengths:

- Higher sensitivity for certain faults (ITSC, AGE, IDM) compared to current-only methods.
- Multisensory systems provide spatial fault localization.
- Captures voltage distortions amplified in flux but subtle in current.

Limitations:

- **Internal Methods:** Invasive, requires sensor installation during manufacturing or rewinding, not feasible for retrofit.
- **External Methods:** Lower signal strength (stray flux), sensitive to sensor placement and orientation, vulnerable to environmental EMI.
- Sensorless methods require detailed machine geometry knowledge and complex models.

Best-fit Applications: New machine designs where internal sensors can be integrated; applications where spatial fault information is critical; when MCSA alone provides insufficient sensitivity.

4.2.3 Zero Sequence Voltage Component Analysis

4.2.3.1 Physical Principle

Zero sequence voltage component (ZSVC) exploits the fact that in an ideally balanced three-phase system, **zero-sequence voltage remains at zero during healthy operation**. Any deviation provides a clear indicator of asymmetry or fault.

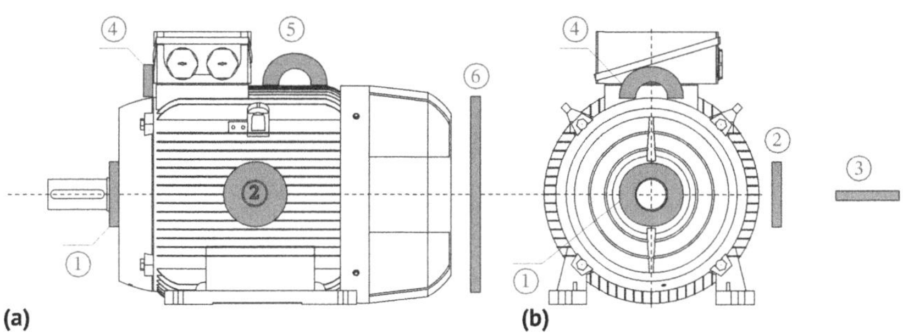

Figure 4.8 Various search coil positions and orientations for measuring external magnetic flux: (a) Side view and (b) Frontal view. 1: EPS1, 2: EPS2, 3: EPS3, 4: EPS4, 5: EPS5, and 6: EPS6.

In a three-phase system, the ZSVC is mathematically defined as the average of the instantaneous phase voltages:

$$\mathrm{ZSVC} = \frac{V_a + V_b + V_c}{3} \tag{4.15}$$

Under ideal balanced conditions with balanced three-phase supply, the sum of instantaneous phase voltages equals zero, resulting in zero ZSVC. However, when faults occur, such as eccentricity, inter-turn shorts, or high-resistance connections in specific phases or poles, asymmetries arise in the magnetic flux linkages and air gap distribution. These asymmetries disturb the balance, causing the sum of instantaneous phase voltages (and consequently the zero-sequence components) to deviate from zero. The resulting nonzero ZSVC contains significant harmonic content at fault characteristic frequencies, providing a high signal-to-noise indicator for early fault warning and classification.

4.2.3.2 Signal Requirements

Primary Observable: Zero-sequence voltage component derived from three-phase stator voltages.

Two Principal Measurement Approaches:

1. **Direct Neutral Point Measurement (Traditional):** In accessible configurations, ZSVC is directly measured at the machine neutral point. The prerequisite for accessing the physical neutral point makes this approach straightforward when available.
2. **Inverter-based Artificial Neutral Point (Zero Sequence Back EMF Component, ZSBEC):** When the neutral point is inaccessible (delta-connected machines or certain inverter configurations), an artificial neutral point can be created by measuring voltage against inverter DC bus midpoint using a balanced three-phase resistor network connected in star configuration to the phase terminals. This configuration is illustrated in Figure 4.9. The balanced resistor network ($R_{ma} = R_{mb} = R_{mc} = R$) forms a star connection whose neutral point serves as an artificial reference. The measured voltage $V_{0,m}$ between this artificial neutral and the machine neutral represents ZSVC influenced primarily by machine characteristics (third harmonics and geometric distortions), while ZSVC injected by the inverter is substantially separated. This method provides a diagnostic-oriented ZSVC parameter serving as a direct fault indicator.

Critical Limitations:

- **Neutral Access Requirement:** Traditional direct measurement requires access to the **neutral point**, which is often inaccessible in delta-connected machines. This represents a fundamental hardware constraint precluding direct measurement in certain motor designs.
- **Need for Neutral Point access Constraint:** One critical limitation for widespread ZSVC adoption is the necessity of accessing the neutral point. This can be prohibitive for machines where the neutral is not externally available. Extra sensors or wiring for neutral point access are generally required in retrofit scenarios.
- **Inverter-fed Machines Without Neutral Point:** The artificial neutral point method addresses this limitation but requires additional wiring for the balanced resistor network, though remaining minimally invasive compared to internal sensor installation.

Alternative Implementations: External stray magnetic flux sensors (Hall effect) mounted around stator yoke provide complementary information but add cost and complexity.

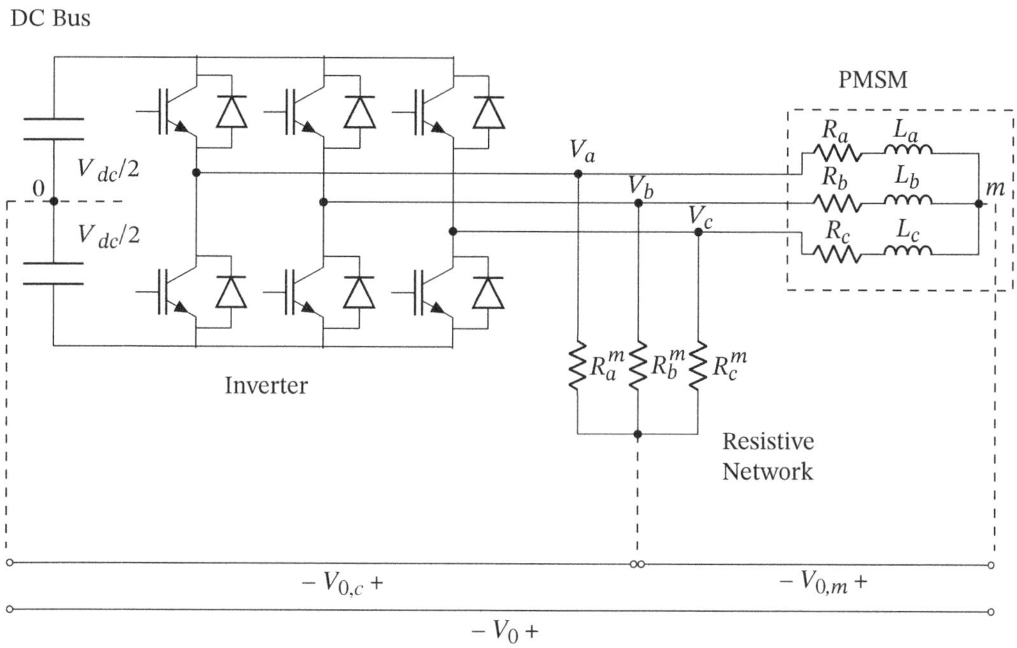

Figure 4.9 Inverter-based artificial neutral point measurement for ZSVC using balanced star-connected resistive network to measure machine-originated zero-sequence voltage $V_{0,m}$. *Source:* Adapted from [24].

4.2.3.3 Measurement Techniques and Implementation

Physical Access Requirements for Traditional ZSVC Measurement:

The prerequisite for traditional ZSVC measurement is physical access to the neutral point. This is inherently unavailable in delta-connected machines and often inaccessible in star-connected machines fed by inverters where the neutral point is not brought out externally. Accessing the neutral typically requires either:

- Direct line-to-neutral voltage measurement when the stator neutral is physically accessible.
- Retrofitting additional wiring to access the neutral point (when feasible for maintenance purposes but not always practical).

Artificial Neutral Point for Inverter-fed Machines:

For motor drives fed by inverters, when direct neutral access is not available, one common method to measure ZSVC employs a balanced three-phase resistor network. The three equal resistors R (i.e., $R_{ma} = R_{mb} = R_{mc} = R$) are connected in star configuration to the three phase terminals. The star point of this resistor network serves as an artificial neutral point.

By applying Kirchhoff's current law to the resistor network, the voltage at the artificial neutral point $V_{0,c}$ with respect to the inverter DC bus midpoint is derived as:

$$\frac{V_a - V_{0,c}}{R} + \frac{V_b - V_{0,c}}{R} + \frac{V_c - V_{0,c}}{R} = 0 \tag{4.16}$$

This leads to the neutral point voltage of the resistor network:

$$V_{0,c} = \frac{1}{3}(V_a + V_b + V_c) \tag{4.17}$$

The neutral voltage of the machine (with respect to the same machine reference) can be expressed as $V_0 = V_{0,c} + V_{0,m}$, where $V_{0,m}$ is the ZSVC primarily influenced by machine characteristics. Measurement of $V_{0,m}$ provides the diagnostic parameter primarily reflecting machine electromagnetic imbalances, as the voltage $V_{0,m}$ represents the ZSVC component attributable to machine faults with a much higher sensitivity to imbalance-related faults compared to direct terminal voltage measurements.

This technique provides a clearer separation between machine-originated ZSVC and inverter-injected ZSVC, enabling more focused fault detection purely based on isolated machine quantities. This makes the method particularly valuable for inverter-fed motors where traditional methods are impractical, and allows more reliable detection with higher signal-to-noise ratio compared to direct phase voltage analysis performed solely at fundamental frequency quantities.

Vitality and Applicability Considerations:
A critical constraint for widespread ZSVC diagnostic adoption is the necessity for accessing the neutral point. This is obligatory for machines with externally accessible neutral but not feasible for machines in enclosed designs or delta-connected stators unless an external accessible external neutral is retrofitted. For machines where neutral access is unavailable externally and retrofitting an accessible neutral is infeasible, the balanced resistor network method provides an alternative, though at the expense of additional wiring and passive components. For machines without neutral access altogether, alternative diagnostic methods (MCSA, EPVA, model-based methods) remain viable without such constraints.

4.2.3.4 Fault Features and Indicators

- **Characteristic harmonic amplitudes** at frequencies corresponding to:
  - Static eccentricity: fundamental frequency components.
  - Dynamic eccentricity: sideband harmonics.
  - Inter-turn faults: fundamental frequency with distinct phase shifts.
  - High-resistance connections: fundamental with phase angle differences.
- **Initial phase angle of ZSVC phasor:** Additional discriminator for fault classification.
- **Harmonic sideband offsets:** Relative to supply frequency.
- **ZSVC amplitude difference** between winding sets (dual three-phase machines).
- **Mathematical expression for ZSVC-based fault indicator:** A concise mathematical expression for ZSVC can serve as a direct fault indicator parameter, with specific harmonic components identifiable via FFT analysis indicating fault presence and type.

4.2.3.5 Strengths and Limitations

Strengths:

- Strong sensitivity to mechanical eccentricities with clear spectral distinction (static vs. dynamic vs. mixed).
- High sensitivity to inter-turn shorts and high-resistance connection faults.
- Clear reference point (ideal zero) simplifies anomaly detection.
- Relatively noninvasive once neutral access established.

Limitations:
- **Fundamental requirement for neutral point access** precludes use in delta-connected or certain inverter configurations.
- Reduced sensitivity to magnetic irregularities (demagnetization better detected via current/flux).
- Limited effectiveness for power-electronic faults and bearing defects.
- Performance degrades at very low speeds (small voltage magnitudes, poor SNR).
- Susceptible to PWM harmonics and common-mode voltages.

Best-fit Applications: Star-connected PMSMs with accessible neutral; medium to high-speed operation; industrial drives with relatively stable conditions; when inter-turn fault detection is primary objective.

4.2.4 Extended Park Vector Approach

4.2.4.1 Physical Principle

Extended park vector approach (EPVA) is distinguished by **geometric analysis of current space vector trajectories** in the rotating reference frame [25]. Healthy machines trace approximately circular patterns in the *d*-*q* plane under steady-state operation, while faults introduce characteristic distortions, such as ellipticity, asymmetries, or higher-order harmonic content visible as trajectory deformations. This spatial-geometric perspective provides intuitive visualization and quantitative metrics based on trajectory shape analysis.

4.2.4.2 Signal Requirements

Primary Observables:
- Three-phase stator currents (standard implementation).
- Optional: Search coils positioned circumferentially at 120° intervals (alternative implementation providing electrical isolation but requiring additional hardware).

Accessibility: Standard implementation uses existing current sensors with zero incremental hardware cost for inverter-fed machines.

4.2.4.3 Fault Features and Indicators

- **Park Vector Modulus:** Magnitude of current space vector in *d*-*q* plane with its spectral content.
- **Eccentricity of Park Vector Trajectory:** Degree of ellipticity or deviation from circular pattern.
- **Major and Minor Axis Ratios:** Quantifying trajectory shape distortion.
- **Area Deviations and Shape Distortion Metrics:** Geometric measures of asymmetry.
- **Phase Angle Evolution:** Tracking vector rotation characteristics.
- **Model Prediction Residuals:** Deviations from expected circular trajectory.

4.2.4.4 Strengths and Limitations

Strengths:
- Nonintrusive, leverages existing current sensors.
- Provides intuitive geometric visualization of fault signatures.
- Multi-fault detection capability (demagnetization, eccentricity, winding faults).
- Software-only deployment without hardware modifications.
- Computational efficiency suitable for standard embedded controllers.

Limitations:

- **Speed Dependence:** Diagnostic sensitivity deteriorates substantially at low speeds.
- **Load Variation Sensitivity:** Load changes introduce trajectory variations potentially masking/mimicking faults.
- **Baseline Data Requirements:** Typically requires healthy machine reference for comparative analysis.
- **Fault Signature Overlap:** Different fault mechanisms can produce similar spectral/geometric signatures.
- **Transient Operation Challenges:** Control system responses during transients obscure fault indicators.
- **Uniform Fault Limitations:** Symmetric faults (uniform demagnetization) produce weak asymmetric signatures.

Best-fit Applications: Medium to high-speed applications with relatively steady loading; cost-sensitive industrial applications where dedicated diagnostic hardware is prohibitive; retrofit scenarios requiring minimal machine modifications; continuous process industries with extended steady-state operation.

4.2.5 Model-based Parameter Estimation Methods

4.2.5.1 Physical Principle

Model-based parameter estimation is distinguished by **explicit computation of physical machine parameters** (R_s, L_d, L_q, λ_{pm}) through identification algorithms. Unlike signature-based methods analyzing spectral patterns, this approach solves inverse problems to estimate parameters from measured voltages, currents, and rotor position. Faults manifest as deviations in estimated parameters from expected healthy values, providing **quantitative, physically interpretable** diagnostic indicators with clear connections to underlying fault mechanisms.

4.2.5.2 Signal Requirements

Primary Observables:

- Phase currents (all three phases).
- Phase voltages (measured or reference voltages from FOC).
- Rotor position/speed.

Accessibility: All signals already available in standard motor drive systems for control purposes with zero incremental hardware cost. Option to use reference voltages (no voltage sensors) or measured voltages (improved accuracy but adds three sensors at $30–100 each).

4.2.5.3 Algorithms

4.2.5.3.1 Lumped-parameter Model-based Approaches

Classical model-based parameter estimation methods employ **lumped-parameter models** of the machine in the dq or $\alpha\beta$ reference frames. These approaches represent the machine dynamics through differential equations with concentrated parameters (R_s, L_d, L_q, λ_{pm}) that capture the overall electromagnetic behavior without explicit spatial resolution. A typical implementation integrates the parameter estimation block within the Field-Oriented Control (FOC) structure, as illustrated in Figure 4.10, where reference voltages from the current controller are used instead of direct voltage measurements, eliminating the need for additional voltage sensors.

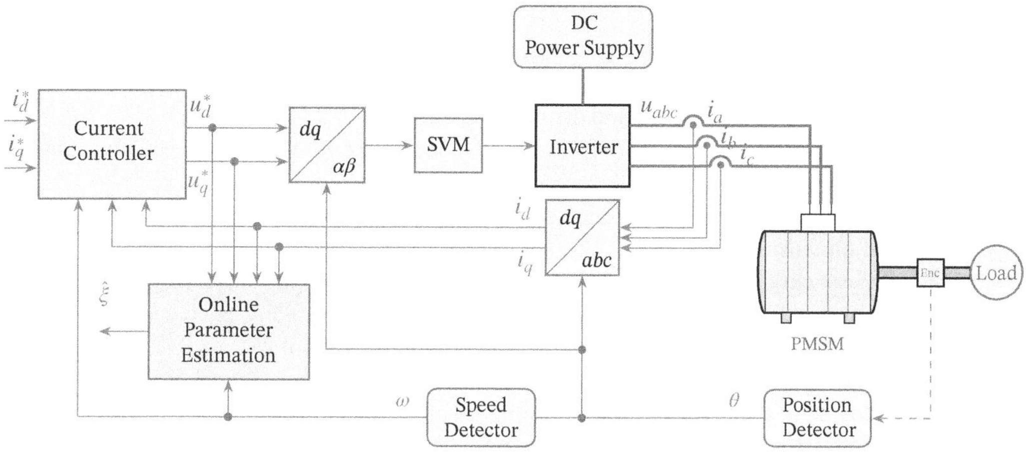

Figure 4.10 Online parameter estimation integrated with Field-oriented Control using reference voltages u_d^*, u_q^* instead of direct voltage measurements.

- **Extended Kalman Filter (EKF):** Recursive state-space estimation with covariance tracking for nonlinear systems [26].
- **Recursive Least Squares (RLS):** Adaptive parameter identification with forgetting factors for time-varying parameters.
- **Luenberger Observers:** State estimation with residual-based fault detection through output prediction error [27].

These methods excel in computational efficiency and real-time implementation but rely on simplified models that assume uniform magnetic field distribution and lumped circuit parameters. The fundamental limitation is their inability to capture spatially localized faults or geometric effects without explicit modeling extensions.

4.2.5.3.2 High-fidelity Geometry-based Inverse Problem Approaches

Apart from lumped-parameter approaches, advanced model-based methods create **high-fidelity models** of the machine using detailed geometric and electromagnetic knowledge to solve inverse problems for fault parameter estimation. These approaches leverage spatial resolution to detect and localize faults that produce localized electromagnetic effects:

Magnetic Equivalent Circuit (MEC) Models: MEC modeling represents the machine's magnetic structure as a network of reluctances (magnetic resistances) and magnetomotive force sources. The machine geometry is discretized into flux tubes, each modeled by reluctances that depend on the material permeability, cross-sectional area, and path length:

$$\mathcal{R} = \frac{l}{\mu A} \tag{4.18}$$

where $\mathcal{R}$ is reluctance, l is magnetic path length, μ is permeability, and A is cross-sectional area.

Inverse Problem Formulation: Given measured terminal electrical quantities (voltages, currents, flux linkages), the inverse problem solves for fault-related parameters such as:

- Local permeance variations (air-gap eccentricity).
- PM magnetization distribution (demagnetization faults).
- Winding turn ratios (inter-turn short circuits).

Advantages:

- Moderate computational complexity compared to finite element methods.
- Physical interpretability through circuit analogies.
- Real-time feasibility for observer and estimation strategies.
- Captures spatial fault localization (e.g., localized demagnetization, phase-specific turn faults).

Limitations:

- Requires detailed machine geometry (stator/rotor tooth dimensions, slot geometry, air-gap length, PM dimensions).
- Reluctance network topology must be predefined based on machine structure.
- Nonlinear magnetic saturation requires iterative solutions.
- Model accuracy depends on correct reluctance path identification.

Improved Winding Function Theory (IWFT):

The IWFT provides an analytical framework that precisely accounts for:

- Stator slotting effects and winding distribution.
- Magnetic flux density within permanent magnets.
- PM magnetization angle variations.
- Magnetic saturation effects.

IWFT uses conformal mapping techniques to calculate the relative complex permeance of the slotted air gap, obtaining the slotted air-gap length function. Machine inductances are computed directly from geometric data and winding layout.

Inverse Parameter Estimation: By measuring terminal quantities and rotor position, IWFT-based inverse models estimate:

- Effective air-gap function (eccentricity detection).
- Turn function variations (winding fault localization).
- Flux linkage spatial distribution (demagnetization mapping).

Advantages:

- Removes drawbacks of conventional winding function theory.
- Faster than numerical techniques (FEM) while maintaining analytical rigor.
- User-friendly for design, optimization, and fault diagnosis.
- Geometry-based formulation provides spatial fault resolution.

Limitations:

- Requires complete geometric specifications (tooth geometry, slot dimensions, winding pitch).
- Analytical complexity increases with machine structural complexity.
- Limited to 2D cross-sectional analysis (end-effects require separate treatment).

Subspace Identification Methods:

Subspace identification techniques provide data-driven black-box or gray-box state-space model identification for permanent magnet machines:

$$\begin{aligned} \mathbf{x}(k+1) &= \mathbf{A}\mathbf{x}(k) + \mathbf{B}\mathbf{u}(k) \\ \mathbf{y}(k) &= \mathbf{C}\mathbf{x}(k) + \mathbf{D}\mathbf{u}(k) \end{aligned} \tag{4.19}$$

where $\mathbf{x}$ represents internal states (fluxes, currents), $\mathbf{u}$ represents inputs (applied voltages), and $\mathbf{y}$ represents measured outputs (terminal currents, voltages).

Auxiliary Variable Closed-loop Subspace Identification: Addresses modeling difficulties in closed-loop controlled systems (typical for motor drives) by introducing auxiliary variables to decorrelate noise from input signals, enabling consistent parameter estimation.

Geometry Integration: When combined with geometric machine knowledge, subspace models can be structured such that identified state-space matrices relate to physical parameters:

- System matrix **A** elements relate to resistances and inductances.
- Input matrix **B** relates to flux linkages and winding configurations.
- Output matrix **C** encodes sensor placement and measurement geometry.

Advantages:

- Handles closed-loop operation without explicit controller modeling.
- Captures system dynamics including unmodeled resonances and nonlinearities.
- Suitable for time-varying parameter tracking (fault progression monitoring).
- Can incorporate geometric constraints to improve physical interpretability.

Limitations:

- Requires persistent excitation for accurate identification.
- Black-box nature may obscure physical parameter interpretation without structural constraints.
- Computational complexity scales with model order.
- Requires substantial input-output data for reliable identification.

Hybrid FEM-analytical Models:

Hybrid approaches combine the spatial resolution of finite element method (FEM) with the computational efficiency of analytical models:

Typical Workflow: 1. **FEM Calibration Phase:** High-fidelity 2D/3D FEM simulations characterize the machine under healthy and various fault conditions across operating points. 2. **Analytical Model Parameterization:** FEM results populate lookup tables or fit analytical model parameters (inductance maps, flux linkage functions, reluctance network values). 3. **Real-time Inverse Problem Solution:** The calibrated analytical model runs in real-time estimation algorithms, solving inverse problems to estimate fault parameters from measured signals.

Sparse Identification Techniques: Reduce the number of FEM simulations required by intelligently sampling the parameter space and using interpolation or machine learning to predict intermediate operating points.

Advantages:

- FEM provides high-accuracy parameter calibration accounting for saturation, eddy currents, and geometric complexities.
- Analytical model enables real-time execution suitable for online diagnostics.
- Time savings of 95% or more compared to full FEM at every estimation step.
- Combines spatial resolution of FEM with speed of analytical methods.

Limitations:

- Initial FEM characterization requires substantial offline computational effort.
- Geometry knowledge mandatory for FEM model construction.
- Interpolation between FEM-calibrated points introduces approximation errors.
- Model validity limited to the range of conditions simulated during calibration phase.

Permeance Network and Reluctance Network Models:

These approaches extend traditional MEC by incorporating detailed geometric discretization:

Machine Geometry Discretization: The machine is divided into multiple geometric zones (stator teeth, stator yoke, rotor yoke, permanent magnets, air gap), with each zone represented by:

- Nonlinear reluctances (for ferromagnetic regions with saturation).
- Linear reluctances (for air regions and permanent magnets).
- MMF sources (magnetization of permanent magnets, winding currents).

Inverse Problem for Fault Detection:

- **Demagnetization:** Local PM MMF reduction identified from flux distribution measurements.
- **Eccentricity:** Air-gap reluctance variations estimated from permeance function changes.
- **Winding Faults:** Equivalent MMF source variations from turn count changes.

Advanced Network Formulations: Generic reluctance network modeling tools provide flexibility to model various machine topologies and fault scenarios by adjusting network connectivity and element values.

Advantages:

- Physical transparency through circuit-magnetic analogies.
- Captures geometric effects (slot opening, tooth saturation, air-gap variations).
- Computationally efficient compared to full-field numerical methods.
- Suitable for parametric sensitivity analysis and fault signature prediction.

Limitations:

- Geometry knowledge is mandatory (all dimensions, material properties).
- Network topology construction requires expert knowledge of flux paths.
- 3D effects (end-windings, axial leakage) require specialized treatment.
- Accuracy for complex geometries may be limited without dense discretization.

4.2.5.4 Common Limitation: Geometry Knowledge Requirement

All high-fidelity geometry-based inverse problem methods share a fundamental limitation: **detailed knowledge of the machine geometry is required**. This includes:

- Stator and rotor lamination dimensions (tooth width, slot depth, yoke thickness).
- Air-gap length and variations.
- Permanent magnet dimensions, placement, and material properties.
- Winding configuration (turns per coil, slot fill factor, pitch).
- Material properties (B–H curves for laminations, PM remanence and coercivity).

Practical Implications:

- **New Machine Design:** Geometry readily available from design files, enabling integration during development.
- **Retrofit Applications:** Geometry may be unavailable or require reverse-engineering (dimensional measurements, material testing).
- **Proprietary Machines:** Manufacturers may not disclose detailed geometry, limiting applicability.
- **Manufacturing Tolerances:** Nominal geometry may differ from actual manufactured dimensions, introducing model errors.

Mitigation Strategies:

- Use lumped-parameter models when geometry is unavailable.
- Perform geometric measurements and material characterization for critical applications.
- Employ adaptive calibration to compensate for geometry uncertainties.
- Combine geometry-based and data-driven approaches (hybrid methods).

4.2.5.5 Fault Features and Indicators

Estimated Parameter Values:

- Stator resistance R_s (temperature-compensated deviations detect inter-turn shorts).

- d-q axis inductances L_d, L_q (asymmetries indicate eccentricity, magnetic degradation).
- PM flux linkage λ_{pm} (reductions indicate demagnetization).

Parameter Deviations and Residuals:

- Temperature-compensated resistance deviations.
- Flux linkage reductions below temperature-corrected baselines.
- Inductance asymmetries between phases or axes.
- Current/voltage estimation errors (model prediction residuals).

Parameter Ratios:

- Saliency ratio L_q/L_d changes.
- Normalized inductance variations across electrical cycles.
- Resistance ratios between phases.

Estimation Uncertainty:

- Parameter covariance matrix diagonal elements (estimation confidence).
- Correlation patterns between parameters (fault mechanism identification).

Temporal Evolution:

- Gradual resistance increases (winding degradation).
- Abrupt resistance changes (discrete fault events).
- Flux linkage decay rates (demagnetization progression).

4.2.5.6 Strengths and Limitations

- **Early electrical fault detection** using existing drive sensors.
- **Continuous operation** without test signals, allowing passive monitoring during normal operation.
- **Component Localization:** Distinguishes stator faults from rotor demagnetization.
- **Quantitative Degradation Metrics:** Percentage changes rather than binary decisions.
- **Natural FOC Integration:** Seamlessly leverages existing control infrastructure.
- **Physical Interpretability:** Parameters directly understandable by maintenance personnel.

Limitations:

- **Weak Coupling to Mechanical Faults:** Early bearing defects, localized demagnetization may go undetected.
- **Temperature Variations:** ±30 °C produces ±12% resistance change requiring careful compensation.
- **Model Accuracy Dependence:** Unmodeled saturation/cross-coupling limits resolution (±5–10% baseline uncertainty).
- **Load Transients:** Dynamic responses temporarily overwhelm fault signatures.
- **Multiple Simultaneous Faults:** Combined effects may not decompose uniquely.
- **Operating Condition Dependencies:** Poor observability at low-speed/light-load.

Best-fit Applications: Constant or slowly-varying loads (HVAC, pumps, conveyors, fans); existing high-quality current sensing with known machine models; safety-critical systems requiring continuous monitoring; medium-to-high power machines (>5 kW) with better SNR; predictive maintenance emphasis.

4.2.6 Instantaneous Power/Power-factor Signature Analysis

4.2.6.1 Physical Principle

Instantaneous Power/Power-Factor Signature Analysis (IPSA/IPFSA) is distinguished by analyzing **instantaneous electrical power and power factor** computed from terminal voltage and current measurements. The method exploits how faults alter electromagnetic energy distribution and flow, manifesting as characteristic changes in power signatures. The fundamental advantage over current-only analysis: power represents actual energy transfer and incorporates information from both voltage and current waveforms. Certain faults (demagnetization, static eccentricity) create subtle voltage distortions amplified in the power calculation through the nonlinear voltage-current product.

4.2.6.2 Signal Requirements

Primary Observables:

- Three-phase currents.
- Three-phase voltages (line or phase).

Computed Quantities:

$$P(t) = v_a(t)i_a(t) + v_b(t)i_b(t) + v_c(t)i_c(t) \tag{4.20}$$

Power factor: $PF(t) = P(t)/S(t)$, where $S(t) = \sqrt{V_{rms}^2 I_{rms}^2}$.

Accessibility: Both signals are already measured in typical inverter-fed PMSM drives for control and protection with near-zero incremental hardware cost.

4.2.6.3 Fault Features and Indicators

- **Harmonic amplitudes at characteristic frequencies** in power spectrum (static eccentricity: $f = f_e \pm kf_r$; dynamic eccentricity: $f = f_e[1 \pm k/p]$; inter-turn shorts: $f = 2f_e$).
- **Envelope Statistics:** RMS, kurtosis, skewness, crest factor of demodulated power signals.
- **Power Factor Ripple:** RMS of AC component or peak-to-peak amplitude.
- **Vector trajectory metrics** in d-q or α-β coordinates (eccentricity, higher-order harmonic content).
- **Observer Residuals:** Model prediction errors for power/voltage.
- **Sequence Components:** Negative-sequence to positive-sequence ratio I^-/I^+.

4.2.6.4 Strengths and Limitations

Strengths:

- Noninvasive using existing drive sensors with near-zero incremental hardware cost.
- Real-time capable with 50–100 ms latency.
- Highly sensitive to electrical asymmetries (5–10% parameter changes detectable).
- Operates at all speeds including standstill and low-speed.
- Works in both transient and steady-state regimes without signal injection.

Limitations:

- Requires clean measurements ($\geq$12-bit ADC, high CMRR > 60 dB).
- Load-dependent signatures complicate threshold setting.
- Limited fault type discrimination without multivariate analysis.
- Insensitive to purely mechanical faults unless they induce significant electrical asymmetry.
- Baseline drift from aging, temperature, saturation requires periodic recalibration.

Best-fit Applications: Servo drives and traction motors with high-bandwidth sensing; high-reliability applications (medical pumps, electric vehicles); complement to vibration monitoring (catches electrical faults with minimal mechanical signatures); inverter-fed machines.

4.2.7 Acoustic and Vibration Signature Analysis

4.2.7.1 Physical Principle

Vibration and acoustic monitoring exploit **mechanical and acoustic physical effects** rather than electromagnetic signals. Changes in mechanical vibrations and sound waves generated by moving parts (bearings, rotors, stators) reveal faults such as imbalance, misalignment, bearing defects, or gear issues. Vibration monitoring measures oscillations via accelerometers mounted on machine frame/bearings. Acoustic monitoring captures sound signatures propagating through air or structure via microphones or acoustic sensors.

4.2.7.2 Signal Requirements

Primary Observables:

- Vibration signals (oscillations, amplitude, frequency) from accelerometers.
- Acoustic signals (noise, sound pressure levels) from microphones or sound-level meters.

Accessibility: Unlike electrical signals readily available in drives, vibration and acoustic signals **require additional sensors and wiring**. No neutral access or search coils needed, but additional wiring from sensor locations to monitoring system is necessary.

Instrumentation:

- Sensor costs: low-cost MEMS accelerometers to precision piezoelectric types.
- EMI susceptibility requires proper shielding and careful routing.
- Mounting constraints: structurally rigid attachment with good signal transmission.
- Sensor placement critically affects detection sensitivity.

4.2.7.3 Fault Features and Indicators

- **Amplitudes at Characteristic Lines:** Fault-specific harmonics, sidebands around rotational frequencies, bearing fault characteristic frequencies.
- **Envelope Statistics from Demodulation:** Kurtosis, RMS, crest factor, peak-to-peak values (enhance impulsive fault detection).
- **Vector Eccentricity Indicators:** From complex vibration vector loci (differentiate misalignment types).
- **Comprehensive Time-domain and Statistical Features:** Mean, variance, skewness, kurtosis extracted via wavelets or STFT.

4.2.7.4 Strengths and Limitations

Strengths:

- Rich, multi-dimensional fault information capturing diverse characteristics.
- Early fault detection before catastrophic failure.
- Well-established for mechanical faults (imbalance, misalignment, bearing defects, rotor issues).
- Noncontact capability (acoustic monitoring).
- Effective integration with advanced signal processing and machine learning.
- Data fusion (vibration + acoustic) improves diagnostic accuracy and robustness.

Limitations:

- Sensors need precise placement close to fault source for reliable detection.
- Acoustic signals corrupted by environmental noise.
- Operating condition variability complicates consistent feature extraction.
- Additional sensors introduce costs, wiring complexity, EMI susceptibility.
- Signal processing and model complexity increase computational cost.

Best-fit Applications: Mechanical fault detection is primary concern with accessible sensor mounting points; noncontact sensing needed or vibration sensor placement impractical; complex/variable operating conditions benefiting from multi-source fusion.

Fault Sensitivity:

- **High Sensitivity:** Mechanical eccentricities, imbalances, bearing faults, rotor shaft deformations, magnetic irregularities manifesting mechanically.
- **Moderate Sensitivity:** Electrical asymmetries (stator faults, inter-turn shorts) producing indirect mechanical effects.
- **Low Sensitivity:** Power electronic or sensor issues.

4.2.8 Data-driven and Hybrid Analytics

4.2.8.1 Overview and Distinction from Signal-based Methods

Data-driven diagnostic methods represent a paradigm shift from traditional signal-based approaches by leveraging machine learning algorithms to automatically learn discriminative fault patterns directly from data. While the signal-based techniques described previously (MCSA, MFSA, ZSVC, EPVA, vibration analysis) follow an explicit pattern recognition pipeline, signal acquisition, preprocessing, feature extraction, decision logic, and fault classification, data-driven approaches can automate or entirely replace portions of this pipeline through learning-based algorithms, as illustrated in Figure 4.11. These methods are particularly valuable when: (1) fault mechanisms are poorly understood or analytically intractable, (2) systems exhibit highly nonlinear behavior with unknown models, (3) large volumes of labeled operational data are available, or (4) fusion of multiple heterogeneous signal types is required.

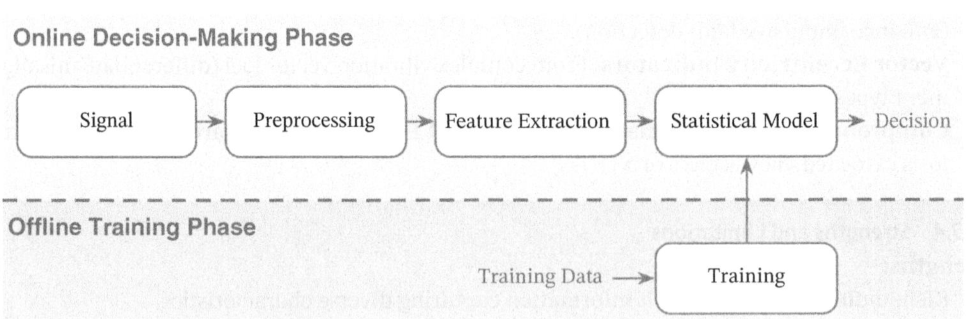

Figure 4.11 Machine learning diagnostic pipeline with offline training phase and online decision-making phase for fault detection.

4.2.8.2 Statistical Classification at the Decision Layer

As discussed in the pattern recognition pipeline framework (Part A), traditional signal-based methods extract handcrafted features (spectral amplitudes, harmonic ratios, statistical descriptors) and apply threshold-based decision logic or simple statistical tests. Data-driven approaches enhance this decision layer by formulating fault diagnosis as a **supervised classification problem**, where a statistical classifier learns to map extracted features to fault classes using labeled training data.

Common Classification Algorithms:

- **Support Vector Machines (SVM):** Construct optimal hyperplanes separating fault classes in feature space, effective for high-dimensional problems with clear margins between classes. Kernel functions (RBF, polynomial) enable nonlinear decision boundaries. The genetic algorithm-support vector machine (GA-SVM) variant uses genetic algorithms for hyperparameter optimization, improving classification accuracy for demagnetization fault detection.
- **Decision Trees and Random Forests:** Hierarchical rule-based classifiers providing interpretable decision logic. Random Forests ensemble multiple trees to improve robustness and generalization, particularly effective when dealing with mixed feature types (time-domain statistics, frequency-domain amplitudes, coordinate-transformed quantities).
- **Naive Bayes Classifiers:** Probabilistic classifiers assuming feature independence, computationally efficient for real-time implementation despite simplifying assumptions.
- **Neural Network Classifiers:** Multi-layer perceptrons (MLPs) with nonlinear activation functions learn complex nonlinear mappings from features to fault classes. Suitable when decision boundaries are highly irregular or when feature interactions are complex.

These classifiers operate on features extracted by traditional signal processing methods (FFT, wavelet transforms, Park transforms, envelope analysis), effectively replacing simple threshold comparisons with learned decision functions trained on historical fault data. This approach maintains the interpretability of engineered features while leveraging machine learning's ability to discover complex decision boundaries.

4.2.8.3 Physics-informed Machine Learning for Parameter Estimation

Model-based parameter estimation methods traditionally solve inverse problems using observers and filters (EKF, RLS, Luenberger observers) to estimate physical machine parameters (R_s, L_d, L_q, λ_{pm}) from measured electrical quantities. **Physics-informed machine learning (PIML)** enhances these methods by integrating governing physical laws directly into neural network training, providing a hybrid approach that combines data-driven flexibility with physics-based constraints.

Physics-informed Neural Networks (PINNs) embed governing differential equations (voltage equations, flux linkage equations, electromagnetic torque equations) as soft constraints in the loss function [28]:

$$\mathcal{L}_{\text{total}} = \mathcal{L}_{\text{data}} + \lambda \mathcal{L}_{\text{physics}} \tag{4.21}$$

where $\mathcal{L}_{\text{data}}$ measures prediction error on measured data (currents, voltages) and $\mathcal{L}_{\text{physics}}$ penalizes violations of physical laws (e.g., Kirchhoff's voltage law, electromechanical energy balance). The regularization parameter λ balances data fitting with physics compliance.

Advantages for Inverse Parameter Estimation:

- **Reduced Training Data Requirements:** Physics constraints regularize the solution space, enabling accurate parameter estimation with limited training data compared to purely data-driven models.
- **Physical Plausibility:** Estimated parameters respect fundamental physical relationships (energy conservation, electromagnetic coupling), preventing unphysical predictions that purely empirical models may produce.
- **Improved Generalization:** Physics-informed models extrapolate more reliably to operating conditions outside the training distribution (speed/load ranges, temperature variations) by enforcing universal physical principles.
- **Application to Complex Inverse Problems:** PINNs effectively solve parameter identification problems where analytical observers struggle due to model complexity, nonlinearity, or coupled multi-physics phenomena. Examples include simultaneous estimation of temperature-dependent resistance and flux linkage from terminal measurements under time-varying operating conditions.

Applications in Fault Diagnosis: PIML has been demonstrated for parameter estimation in DC-DC converters and permanent magnet synchronous motors, where the neural network learns inverse models to identify circuit parameters or machine parameters from input-output measurements. For fault diagnosis, PINNs can estimate degradation-sensitive parameters (winding resistance increases from inter-turn shorts, flux linkage reductions from demagnetization) while maintaining physical consistency, providing quantitative fault severity metrics.

4.2.8.4 Deep Learning End-to-end Models

Traditional signal-based diagnostics decompose the problem into sequential stages: preprocessing, feature extraction, and classification. **End-to-end deep learning models** bypass explicit feature engineering by learning hierarchical representations directly from raw or minimally processed signals, collapsing the pattern recognition pipeline into a single trainable neural network, as illustrated in Figure 4.12.

Convolutional Neural Networks (CNNs): Automatically extract spatial and spectral features from raw time-domain signals or time-frequency representations (spectrograms, wavelet scalograms). Multi-layer convolutional architectures learn hierarchical features: early layers detect low-level patterns (transients, oscillations), while deeper layers combine these into fault-specific

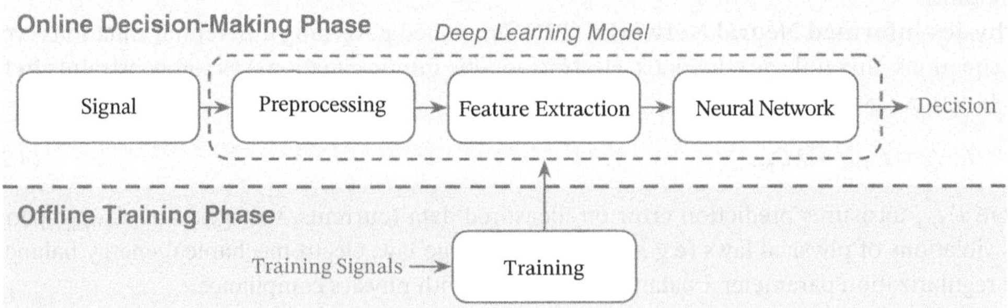

Figure 4.12 End-to-end deep learning approach that collapses all stages of pattern recognition framework into a single trainable model.

signatures. CNNs applied to motor current signals or vibration data achieve high classification accuracy (>99% for bearing, rotor, and stator faults) without manual feature design.

Recurrent Neural Networks (RNNs) and Long Short-term Memory (LSTM): Capture temporal dependencies in sequential data, effective for monitoring fault progression or diagnosing faults during dynamic operating conditions (speed transients, load variations). LSTM networks address vanishing gradient problems [29], enabling learning of long-term temporal patterns relevant to gradual degradation processes.

Transformer Networks and Attention Mechanisms: Recent architectures apply self-attention mechanisms to automatically weigh the importance of different temporal segments or frequency bands, dynamically focusing on fault-relevant signal portions. Transformer-based ensemble networks have been developed for multi-modal time-series data (current, vibration, acoustic, temperature), achieving robust fault classification by learning to integrate complementary information across signal types.

Time–Frequency Hybrid Architectures: Combining convolutional layers with recurrent layers (CNN-LSTM hybrids) extracts both spatial features from time-frequency representations and temporal dynamics from sequential patterns, particularly effective for nonstationary signals encountered during variable-speed operation.

Advantages:

- **Automated Feature Learning:** Eliminates labor-intensive feature engineering and domain expertise requirements for feature selection.
- **Unified Architecture:** Single model handles preprocessing, feature extraction, and classification, simplifying deployment.
- **Robustness to Raw Signal Variations:** Deep networks learn invariant representations robust to noise, operating point variations, and sensor placement differences.
- **Practical Cost-effectiveness:** Motor current-signal-based deep learning approaches are particularly attractive since current signals are readily available from inverters without additional accelerometer installation costs.

Limitations:

- **Data Hunger:** Requires large labeled datasets (thousands of samples per fault class) to avoid overfitting.
- **Interpretability:** Black-box nature complicates root-cause analysis and regulatory approval in safety-critical applications.
- **Computational Demands:** Training requires GPU acceleration; inference may need specialized hardware for real-time embedded deployment.

4.2.8.5 Multi-signal Fusion for Comprehensive Diagnostics

No single measurement modality captures all fault types with equal sensitivity. Mechanical faults (bearing defects, eccentricity) manifest strongly in vibration but weakly in current; electrical faults (inter-turn shorts, demagnetization) produce clear current signatures but subtle vibration changes. **Multi-sensor data fusion** integrates information from heterogeneous sources, current, voltage, vibration, acoustic, temperature, flux to achieve comprehensive fault coverage and improved diagnostic reliability.

Fusion Architectures:

1. **Data-level Fusion:** Concatenates raw signals from multiple sensors into a unified input representation fed to a single classifier. Example: stacking three-phase current waveforms with triaxial accelerometer readings as a multi-channel input to a CNN. Preserves maximum information but increases dimensionality and computational cost.

2. **Feature-level Fusion:** Extracts features independently from each signal type (current spectra, vibration envelope statistics, temperature trends) and concatenates these engineered features into a combined feature vector for classification. Reduces dimensionality compared to data-level fusion while maintaining interpretability. Short-time Fourier transform (STFT) applied to both current and vibration generates time-frequency features for weighted probability ensemble deep learning (WPEDL) classifiers.
3. **Decision-level Fusion:** Trains separate classifiers for each sensor modality, then combines their outputs (posterior probabilities, confidence scores) through voting, averaging, or learned fusion rules. Provides fault classification even if individual sensors fail; enables incremental sensor addition without retraining the entire system. Graph neural networks (GNNs) have been applied to construct heterogeneous multi-sensor fusion architectures, where graph edges represent relationships between different sensor modalities.

Common Sensor Combinations:

- **Current + Vibration:** Most common pairing, captures both electromagnetic and mechanical fault signatures.
- **Current + Voltage + Flux:** Comprehensive electromagnetic monitoring, particularly effective for electrical faults.
- **Current + Vibration + Acoustic:** Adds noncontact acoustic sensing for bearing and gear diagnostics.
- **Multi-Modal Industrial Systems:** Advanced systems integrate current, vibration, device sound, ambient sound, and temperature for holistic condition monitoring.

Benefits:

- **Improved Accuracy and Reliability:** Redundant and complementary information reduces false alarms; multi-class fault diagnosis accuracy exceeding 99% has been demonstrated with WPEDL on fused current-vibration features.
- **Comprehensive Fault Coverage:** Different modalities sensitive to different fault types; fusion ensures no critical failure mode is missed.
- **Robustness to Sensor Degradation:** Redundancy maintains diagnostic capability if individual sensors fail or degrade.

Challenges:

- **Sensor Synchronization:** Precise time alignment (1–10 μs) required to preserve temporal relationships.
- **Feature Heterogeneity:** Balancing features with different physical units, dynamic ranges, and statistical properties; normalization and feature scaling critical.
- **Increased Complexity:** Additional sensors introduce cost, wiring, mounting constraints, and electromagnetic interference concerns.
- **Data Fusion Algorithm Complexity:** Learning optimal fusion strategies requires substantial computational resources and labeled multi-modal training data.

4.2.8.6 Hybrid Approaches: Combining Physics-based and Data-driven Methods

Hybrid diagnostic frameworks synergistically combine model-based methods with data-driven learning to leverage the strengths of both paradigms:

- **Physics-based Feature Engineering with ML Classification:** Use analytical models or simulations (FEM, MEC) to generate fault-indicative features (parameter deviations, flux distribution asymmetries), then apply machine learning classifiers to these physically meaningful features. Provides interpretability while exploiting ML's classification power.

- **Model Residual-based Learning:** Train data-driven classifiers on residuals (differences between model predictions and measurements) rather than raw signals. Residuals amplify fault signatures by removing nominal behavior explained by the physics model.
- **Ensemble Methods:** Combine outputs from multiple diagnostic techniques (MCSA, EPVA, model-based parameter estimation, vibration analysis) using ensemble learning (random forests, boosting, stacking) to form a meta-classifier with superior performance to individual methods.
- **Transfer Learning with Physics Priors:** Pre-train neural networks on simulated fault data generated from high-fidelity physics models (FEM), then fine-tune on limited experimental data. Reduces experimental data requirements while maintaining physical realism.

4.2.8.7 Implementation Considerations

Training Data Requirements: Supervised learning requires balanced datasets with representative samples of all fault types and severities across operating conditions. Data augmentation (synthetic fault injection via simulation, signal transformations) can supplement limited experimental data.

Real-time Constraints: Deep learning inference times vary: lightweight CNNs (10–50 ms), complex Transformers (100–500 ms). Edge computing platforms (GPU-equipped industrial PCs, FPGA accelerators) enable real-time deployment, while cloud-based processing suits noncritical predictive maintenance applications.

Validation and Trust: Black-box models require rigorous validation through cross-machine testing, external benchmarks, and comparison with physics-based methods. Explainable AI techniques (Grad-CAM [30], SHAP values [31]) can identify which signal regions or features drive predictions, improving operator trust and regulatory acceptance.

4.3 Part C: Comparative Analysis

4.3.1 Comparative Summary of Techniques

The selection of fault diagnosis techniques for permanent magnet machines requires careful consideration of multiple factors including cost, invasiveness, sensitivity to specific faults, computational requirements, and application constraints. This comparative analysis provides a systematic framework for technique selection by examining five critical dimensions: hardware requirements, fault sensitivity, implementation complexity, advanced method characteristics, and application-specific recommendations.

The following tables and discussion synthesize the detailed information from parts A and B, enabling practitioners to make informed decisions about which techniques best suit their specific monitoring requirements. The analysis progresses from practical implementation considerations (hardware, cost, invasiveness) through diagnostic performance (fault sensitivity matrix) to computational feasibility (complexity and real-time capabilities) and finally to application-specific guidance.

Key Observations from Table 4.1:
The zero-cost techniques (MCSA, EPVA, model-based methods using existing current/voltage sensors, IPSA/IPFSA, and data-driven approaches) represent the optimal starting point for any fault diagnosis implementation. These methods leverage sensors already present in motor drive systems,

Table 4.1 Signal requirements and hardware needs for fault diagnosis techniques. The table reveals fundamental trade-offs between diagnostic capability and implementation requirements, with advanced techniques often requiring additional sensors or invasive installation procedures.

Technique	Primary signals	Additional hardware	Invasiveness
MCSA	Stator currents	None	Noninvasive
MFSA (internal)	Air gap flux	Search coils, Hall sensors, FGMs, TMRs	Highly invasive
MFSA (external)	Stray flux	External search coils, Hall sensors, FGMs	Noninvasive
MFSA (sensorless)	Currents, voltages, position	None (uses estimation)	Noninvasive
ZSVC (direct)	Zero-sequence voltage	Neutral point access wiring	Minimally invasive
ZSVC (artificial neutral)	Phase voltages	Balanced resistor network (3 resistors)	Minimally invasive
EPVA	Stator currents (or search coils)	None (or search coils if alternative)	Noninvasive (standard)
Model-based (lumped)	Currents, voltages, position	None (optional: direct voltage sensing)	Noninvasive
Model-based (geometry)	Currents, voltages, position	None	Noninvasive
IPSA/IPFSA	Currents, voltages	None	Noninvasive
Vibration/acoustic	Vibration, sound	Accelerometers, microphones	Noninvasive
Data-driven (ML)	Any of the above	None (uses existing sensors)	Noninvasive

requiring only software implementation. MCSA and EPVA can detect multiple fault types with no hardware modifications, making them universally applicable for initial condition monitoring.

Among techniques requiring additional hardware, a clear cost-capability hierarchy emerges. ZSVC with artificial neutral point represents the lowest-cost addition ($20–100) and provides exceptional sensitivity to winding faults and eccentricity. External MFSA sensors ($100–500) offer noninvasive flux monitoring with good sensitivity to electromagnetic faults. Vibration sensors ($200–2000) are essential for bearing fault diagnosis but represent a higher investment. Internal MFSA sensors ($200–1000) provide superior localization capability but require invasive installation during manufacturing or major overhauls.

The invasiveness dimension is critical for retrofit applications. Noninvasive methods (MCSA, external MFSA, EPVA, vibration with external mounting) can be deployed on operating machines without downtime. Minimally invasive methods (ZSVC requiring neutral access, artificial neutral installation) require brief electrical access but no machine disassembly. Highly invasive methods (internal MFSA sensors) are practical only for new machine designs or critical applications justifying the installation effort and cost.

Legend: Very High (primary detection method), High (reliable detection), Moderate (detectable with limitations), Low (marginal sensitivity), Very Low (generally ineffective)

Note: Geometry-based model methods (MEC, IWFT, hybrid FEM-analytical) provide spatial fault localization superior to lumped-parameter approaches but require detailed machine geometry. Data-driven/ML methods' sensitivity depends on training data quality and diversity.

Key Insights from Table 4.2:
Several patterns emerge from the fault sensitivity matrix that guide technique selection. Inter-turn short circuits, the most common electrical fault, benefit from multiple highly effective detection methods including MFSA, ZSVC, model-based approaches, and data-driven techniques. This redundancy enables multi-modal fault confirmation and reduces false alarms. In contrast, uniform demagnetization presents a detection challenge across all methods due to its symmetrical nature, with only MFSA achieving moderate sensitivity.

Eccentricity faults exhibit a distinctive signature: ZSVC and vibration methods provide very high sensitivity, while model-based approaches show variable performance depending on whether geometry information is available. Static and dynamic eccentricity produce similar diagnostic signatures across most techniques, though vibration analysis can distinguish between them through rotational frequency components.

Bearing faults demonstrate the clearest domain specialization: vibration and acoustic methods are primary detection tools (very high sensitivity), while electrical methods provide only moderate to low sensitivity. This highlights the importance of multi-domain monitoring, as relying solely on electrical techniques would miss the most common mechanical failure mode. The high effectiveness of data-driven methods for bearing faults stems from their ability to learn complex vibration patterns that correlate with specific bearing degradation stages.

Demagnetization faults reveal a critical distinction between localized and uniform degradation. Localized demagnetization (from hot spots, mechanical damage, or manufacturing defects) generates asymmetry detectable by multiple electrical methods. Uniform thermal demagnetization across all magnets produces minimal diagnostic signatures, as the motor remains balanced despite

Table 4.2 Fault sensitivity matrix for diagnosis techniques. The matrix provides guidance on which techniques are most effective for detecting specific fault types, enabling practitioners to match diagnostic methods to the failure modes most critical in their application.

Fault type	MCSA	MFSA	ZSVC	EPVA	Model (lumped)	Model (geometry)	IPSA/ IPFSA	Vib/ acoustic	Data-driven
Inter-turn shorts	High	V.High	V.High	High	V.High	V.High	High	Mod	V.High
Demagnetization (localized)	High	V.High	Mod	High	Mod	V.High	High	Mod	V.High
Demagnetization (uniform)	Low	Mod	Low	Low	Mod	Mod	Mod	Low	Mod
Static eccentricity	Mod	High	V.High	High	Low	High	Mod	V.High	High
Dynamic eccentricity	Mod	High	V.High	High	Low	High	Mod	V.High	High
Bearing faults	Mod	Low	Low	Low	V.Low	V.Low	Low	V.High	V.High
Rotor imbalance	Mod	Mod	Mod	Mod	Low	Mod	Mod	V.High	High
Phase imbalance	High	High	V.High	V.High	V.High	V.High	High	Mod	V.High
Inverter faults	Mod	Mod	Low	Mod	Mod	Mod	Mod	Low	High

reduced torque capability. Geometry-based model methods excel at localized fault spatial identification, while lumped-parameter approaches can track gradual flux linkage reduction.

Phase imbalance and inter-turn faults produce strong electrical signatures detectable by nearly all electrical techniques (high to very high ratings). Inverter faults show moderate sensitivity across electrical methods, as their signatures may be masked by control system compensation. Data-driven approaches achieve high sensitivity by learning subtle inverter fault patterns that evade traditional fixed-threshold detection.

Analysis of Computational Trade-offs:

As detailed in Table 4.3, three distinct tiers of computational requirements correspond to different hardware platforms. Lightweight techniques (EPVA at 20–100k operations, basic FFT-based MCSA at 50–500k operations, threshold-based classifiers at 10–20k operations) execute comfortably on embedded microcontrollers with limited RAM (20–100 KB) and provide excellent real-time performance with update intervals under 5 milliseconds. These methods are ideal for cost-sensitive distributed implementations and edge devices.

Moderate-complexity techniques (model-based EKF/RLS at 100–500k operations, advanced time-frequency methods like STFT at 500k–5M operations, SVM classifiers at 10–100k operations) require industrial PC-class processors with moderate memory (100–500 KB) but remain real-time feasible with update rates suitable for control integration (100 μs–100 ms). This tier balances diagnostic sophistication with deterministic execution, enabling advanced monitoring without specialized hardware.

High-complexity techniques (geometry-based models at 500k–2M operations, physics-informed machine learning at 1–10M operations, deep learning at 100k–1M operations) demand high-performance computing platforms, often requiring GPU acceleration for deep learning and PIML methods. Memory requirements escalate to 500 KB–5 MB, and real-time performance becomes challenging (10–500 ms update intervals). These methods are justified for safety-critical applications, comprehensive multi-fault diagnosis, or when spatial fault localization is essential.

A critical observation is the distinction between training cost and inference cost for data-driven methods. Deep learning models may require substantial offline training resources (GPU clusters, days of computation) but can achieve moderate inference cost (100k–1M operations) suitable for edge deployment. Simple classifiers (SVM, random forest) train quickly and execute with minimal resources (10–100k operations), making them attractive for industrial applications with limited data and computational budgets.

The total implementation cost reflects not only hardware but also development effort. Algorithms with low algorithmic complexity (threshold-based methods, FFT-based techniques) require minimal tuning and can be deployed rapidly. High algorithmic complexity methods (deep learning, geometry-based models, advanced time-frequency techniques) demand significant expertise, validation effort, and ongoing maintenance, substantially increasing total cost of ownership even when sensor costs are zero.

Understanding Advanced Method Selection:

As presented in Table 4.4, time-frequency methods address the fundamental limitation of FFT-based techniques: the inability to track evolving spectral content during transients or variable-speed operation. STFT provides a practical balance (500k–2M operations) for industrial implementation when operating conditions vary. Wavelet transforms (CWT/DWT) offer superior multi-resolution analysis (1–5M operations) with DWT being more efficient through dyadic decomposition. Cohen's class methods (WVD, CWD) achieve optimal time-frequency resolution (5–20M operations) but

Table 4.3 Implementation complexity and cost comparison of fault diagnosis techniques. Computational complexity directly impacts real-time feasibility and hardware platform requirements, guiding implementation decisions based on available computational resources.

Technique	Computational complexity	Typical operations/update	Memory requirements	Real-time feasibility	Algorithm complexity	Total implementation Cost
MCSA (FFT)	Moderate	50–500k ops, 1–5 ms	Low (20–100 KB)	Excellent	Low-moderate	Low
MCSA (Advanced TF)	High (STFT/CWT/HHT)	500k–5M ops, 10–100 ms	Moderate (100–500 KB)	Good	Moderate-high	Low
MFSA	Moderate (FFT-based)	50–500k ops, 1–5 ms	Low-moderate (50–150 KB)	Excellent	Moderate	Low-high (depends on sensors)
ZSVC	Moderate (FFT-based)	50–500k ops, 1–5 ms	Low (20–100 KB)	Excellent	Low-moderate	Low-moderate
EPVA	Low-moderate	20–100k ops, 200–500 μs	Low (20–100 KB)	Excellent	Low-moderate	Low
Model-based (EKF/RLS)	Moderate-high	100–500k ops, 100–500 μs	Moderate (50–200 KB)	Good	Moderate-high	Low-moderate
Model-based (MEC/IWFT)	High	500k–2M ops, 1–10 ms	Moderate-high (200–500 KB)	Good	High	Low-moderate (needs geometry)
Model-based (PIML)	Very high	1–10M ops, may need GPU	High (500 KB–5 MB)	Moderate	High	Moderate (training cost)
IPSA/IPFSA	Moderate	100–500k ops, 1–5 ms	Low-moderate (50–150 KB)	Excellent	Low-moderate	Low
Vib/acoustic	Moderate-high	500k–2M ops, 5–20 ms	Moderate-high (100–500 KB)	Good	Moderate-high	High
Data-driven (threshold/SVM)	Low-moderate	10–100k ops, 100–500 μs	Low-moderate (50–200 KB)	Excellent	Moderate	Low (after training)
Data-driven (deep learning)	High-very high	100k–1M ops, 10–500 ms	High (500 KB–5 MB)	Moderate-good	High	Moderate-high (training + compute)

Table 4.4 Advanced signal processing and data-driven methods for fault diagnosis. Detailed comparison of sophisticated techniques that extend beyond basic FFT-based analysis, addressing applications with variable-speed operation, transient-heavy duty cycles, spatial fault localization requirements, or limited training data.

Category	Method	Computational load	Key advantage	Key limitation	Best application
Time-frequency	STFT	Moderate-high (500k–2M ops)	Tracks evolving spectral content	Fixed time-freq. resolution trade-off	Variable-speed, transients
Time-frequency	CWT/DWT	High (1–5M ops)	Multi-resolution analysis	High complexity, DWT more efficient	Nonstationary signals, fault progression
Time-frequency	Cohen's Class	Very high (5–20M ops)	Superior resolution	Cross-term interference, not real-time	Offline analysis, research
Time-frequency	HHT (EMD + Hilbert)	High (2–10M ops)	Adaptive, no basis functions	Mode mixing, end effects	Nonlinear dynamics, adaptive decomp.
Time-frequency	TKEO	Low (10–50k ops)	Efficient modulation detection	Limited to AM/FM detection	Real-time embedded systems
Model-based	MEC models	High (500k–2M ops)	Physical interpretability, spatial faults	Requires detailed geometry	New designs with geometry available
Model-based	IWFT	High (500k–2M ops)	Analytical rigor, faster than FEM	Required, 2D limitation	Design optimization, spatial diagnosis
Model-based	Hybrid FEM-analytical	Very high (offline), mod (online)	High accuracy calibration	Initial FEM cost, geometry required	Critical applications, high-fidelity
Model-based	Subspace identification	Moderate-high (200k–1M ops)	Handles closed-loop, data-driven	Requires excitation, black-box	Unknown dynamics, time-varying params
Data-driven	SVM/random forest	Low-moderate (10–100k ops)	Interpretable features, efficient	Requires feature engineering	Real-time classification, limited data
Data-driven	PIML (PINNs)	Very high (1–10M ops, GPU)	Physics-constrained, generalizes well	Training cost, model development	Limited data, safety-critical apps
Data-driven	CNNs	High-very high (100k–1M ops)	Automated feature learning	Data hunger, black-box	Large datasets, spectrograms
Data-driven	LSTMs/RNNs	High-very high (200k–2M ops)	Temporal dependencies	Sequential processing, data hunger	Fault progression, variable-speed
Data-driven	Transformers	Very high (500k–5M ops)	Self-attention, multi-modal fusion	Extreme data/-compute requirements	Multi-sensor fusion, large-scale
Fusion	Data-level	Very high (1–10M ops)	Maximum information	Dimensionality, computational cost	GPU platforms, comprehensive diag.
Fusion	Feature-level	Moderate-high (100k–1M ops)	Reduced dimensionality, interpretable	Feature engineering required	Industrial PCs, balanced performance
Fusion	Decision-level	Low-moderate (50–200k ops)	Sensor redundancy, incremental	May lose inter-modal correlations	Distributed systems, fault-tolerant

cross-term interference and computational cost limit them to offline analysis and research applications. HHT provides adaptive data-driven decomposition (2–10M operations) without predefined basis functions, excelling with nonlinear dynamics but suffering from mode mixing artifacts. TKEO emerges as the efficiency champion (10–50k operations), extracting amplitude and frequency modulation with minimal computation, ideal for embedded real-time systems.

Geometry-based model methods leverage detailed machine specifications (winding layouts, magnet geometries, air gap profiles) to provide spatial fault localization unattainable with lumped-parameter approaches. MEC models (500k–2M operations) offer physical interpretability and moderate computational cost, enabling design optimization and spatial diagnosis when geometry is available. IWFT (500k–2M operations) provides analytical rigor and executes faster than FEM while maintaining reasonable accuracy for 2D geometries. Hybrid FEM-analytical methods achieve high-fidelity calibration through offline FEM characterization with moderate online cost, justified for critical applications requiring maximum accuracy. Subspace identification (200k–1M operations) handles closed-loop operation and time-varying parameters through data-driven state-space modeling, useful when dynamics are unknown but excitation is controllable.

Data-driven method selection depends critically on data availability and computational resources. SVM and Random Forest classifiers (10–100k operations) provide interpretable feature-based classification with efficient inference, suitable for real-time embedded systems with limited training data. PIML methods (1–10M operations, GPU-accelerated) constrain neural networks with physical laws, achieving superior generalization from limited data compared to pure black-box approaches. This physics-constrained learning is particularly valuable for safety-critical applications where extrapolation beyond training data must remain physically plausible. CNNs (100k–1M operations) automate feature extraction from spectrograms or vibration time-frequency representations, eliminating manual feature engineering but requiring substantial labeled training data. LSTMs and RNNs (200k–2M operations) capture temporal dependencies across long sequences, essential for fault progression modeling and variable-speed operation where fault signatures evolve over time. Transformers (500k–5M operations) enable multi-modal sensor fusion through self-attention mechanisms but demand extreme data and computational resources, justified only in large-scale deployments with comprehensive sensor suites.

Multi-sensor fusion architectures provide complementary advantages at different abstraction levels. Data-level fusion concatenates raw signals for end-to-end deep learning (1–10M operations), maximizing information retention but requiring GPU platforms and large training datasets. Feature-level fusion (100k–1M operations) combines engineered features from multiple modalities for classical ML classifiers, offering interpretability and moderate computational cost suitable for industrial PCs. Decision-level fusion (50–200k operations) runs separate classifiers per sensor modality with voting or weighted averaging, enabling incremental deployment, sensor redundancy, and fault tolerance at the cost of potentially missing inter-modal correlations captured by lower-level fusion.

Application-specific Insights:

As shown in Table 4.5, industrial HVAC, pump, and fan applications benefit from their predominantly steady-state operation, enabling simple FFT-based MCSA and model-based lumped-parameter estimation without complex time-frequency processing. The existing sensor infrastructure (current sensors for control) provides zero-cost monitoring capability. IPSA/IPFSA complement MCSA by providing operating-point-independent indicators, while data-driven methods can leverage large fleet data for anomaly detection and predictive maintenance scheduling.

Table 4.5 Best-fit application scenarios for fault diagnosis techniques in permanent magnet machines. Application-specific recommendations synthesize fault sensitivity, cost, computational feasibility, and practical deployment constraints for common industrial scenarios.

Application type	Recommended primary technique(s)	Complementary technique(s)	Rationale
Industrial HVAC, pumps, fans	MCSA, model-based (lumped)	IPSA/IPFSA, data-driven	Steady-state operation, existing sensors, cost-effective
Electric vehicle traction	EPVA, model-based (lumped), data-driven	MCSA, IPSA/IPFSA	Dynamic operation, existing control infrastructure, real-time constraints
High-speed spindles	MCSA, EPVA	Vibration, data-driven	Good electrical frequency content, mechanical critical
Low-speed direct drives	Model-based (geometry), vibration	MFSA	Low electrical frequency challenges MCSA/EPVA
Safety-critical (aerospace, medical)	Model-based (geometry), MFSA (internal)	MCSA, ZSVC, multi-sensor fusion	Redundancy, multiple physical domains, high reliability
Retrofit condition monitoring	MCSA, IPSA/IPFSA, data-driven (if historical data)	None initially	Zero hardware addition, software-only
New machine design	MFSA (internal), model-based (geometry)	MCSA, vibration, multi-sensor fusion	Sensors integrated during manufacturing, geometry available
Bearing-critical applications	Vibration/acoustic, data-driven (CNN/LSTM)	MCSA (secondary)	Vibration primary for bearing faults, ML for complex patterns
Winding-critical applications	ZSVC, model-based (lumped/geometry)	MFSA	Strong inter-turn short sensitivity
Large fleet deployments	MCSA, data-driven (cloud-based)	IPSA/IPFSA	Cost-effective scaling, centralized ML training
Variable-speed/ transient-heavy	Model-based (lumped), IPSA/IPFSA, data-driven (LSTM)	MCSA (with order tracking)	Methods robust to operating point variations
Unknown fault mechanisms	Data-driven (deep learning), multi-sensor fusion	Model-based (for interpretability)	ML learns patterns without explicit models

Electric vehicle traction drives present unique challenges: high dynamic operation, wide speed range, frequent transients, and stringent real-time constraints. EPVA excels here due to its computational efficiency (20–100k operations) and integration with existing sensorless control observers. Model-based lumped-parameter methods track parameter variations across the operating envelope, while LSTM-based data-driven approaches capture temporal fault progression patterns during drive cycles. MCSA and IPSA/IPFSA provide complementary fault indicators without additional sensors.

High-speed spindle applications (machine tools, turbomachinery) generate rich electrical frequency content that favors MCSA and EPVA. However, mechanical integrity is critical at high speeds, making vibration monitoring essential for bearing fault detection and rotor balance verification. Data-driven vibration analysis (CNN-based) can identify subtle bearing degradation patterns before catastrophic failure.

Low-speed direct drives (<500 RPM) challenge electrical signature-based methods due to low fundamental frequency and poor signal-to-noise ratio in the fault frequency bands. Geometry-based model methods enable spatial fault mapping without relying on high-frequency harmonics. Vibration analysis remains effective at low speeds, as bearing fault characteristic frequencies scale with shaft speed. MFSA provides direct flux monitoring less affected by low-speed operation than current-based techniques.

Safety-critical applications in aerospace and medical devices demand multi-modal redundancy and spatial fault localization. Geometry-based model methods provide detailed fault characterization, while internal MFSA sensors enable direct flux monitoring independent of electrical faults. MCSA, ZSVC, and multi-sensor fusion with decision-level voting ensure no single-point failure mode compromises diagnostic capability. The higher cost of redundant sensors and sophisticated algorithms is justified by the consequences of undetected faults.

Retrofit condition monitoring scenarios prioritize noninvasive, zero-hardware-cost methods. MCSA, IPSA/IPFSA, and data-driven approaches using existing current/voltage sensors enable immediate monitoring capability with software-only deployment. If historical operational data exists, supervised learning provides fault classification without explicit fault models. This phased approach allows validation before committing to additional sensor investments.

New machine designs present an opportunity to integrate advanced diagnostics from the outset. Internal MFSA sensors can be installed during manufacturing without the retrofit cost penalty. Geometry-based model methods leverage design-phase specifications (FEM models, detailed drawings) to enable spatial fault localization. Multi-sensor fusion architectures combine electrical, mechanical, and thermal monitoring for comprehensive coverage, with decision-level or feature-level fusion providing sensor redundancy and improved fault classification accuracy.

Bearing-critical applications (high-speed, heavy-load, or applications where bearing replacement is expensive or difficult) justify dedicated vibration monitoring. Accelerometers with advanced signal processing (envelope analysis, kurtosis, cepstral analysis) provide early bearing fault detection. CNN and LSTM models trained on large bearing fault databases achieve superior classification accuracy compared to traditional threshold-based methods, identifying specific fault types (inner race, outer race, ball/roller, cage) for targeted maintenance planning.

Winding-critical applications (high-voltage, frequent start-stop cycles, thermal cycling, or machines with insulation aging concerns) benefit from ZSVC's exceptional inter-turn short sensitivity. Model-based parameter estimation tracks insulation resistance and turn-to-turn capacitance degradation. MFSA complements ZSVC with direct flux measurement, providing independent confirmation of winding asymmetry.

Large fleet deployments enable cloud-based data-driven approaches that would be infeasible for single machines. Centralized ML training on aggregated fleet data improves anomaly detection through cross-machine learning. MCSA and IPSA/IPFSA data from each machine stream to cloud infrastructure for analysis, with local edge processing providing real-time fault alerts while comprehensive pattern mining occurs offline. Transfer learning allows models trained on one machine type to bootstrap learning for new deployments with limited initial data.

Variable-speed and transient-heavy applications (elevator drives, wind turbines, industrial robots) require techniques robust to operating point variations. Model-based methods with adaptive parameter tracking maintain accuracy across speed ranges. IPSA/IPFSA provide operating-point-independent fault indicators, avoiding the need for speed-dependent threshold tables. LSTM networks learn fault signatures that evolve with operating conditions, while MCSA with order tracking or time-frequency methods (STFT, HHT) maintain spectral resolution during speed transients.

Unknown fault mechanism scenarios benefit from data-driven methods that discover patterns without explicit physical models. Deep learning (CNNs for spectrograms, LSTMs for temporal sequences) identifies fault signatures missed by traditional model-based approaches. Multi-sensor fusion increases the likelihood of capturing unknown fault manifestations. However, model-based methods should complement purely data-driven approaches to provide interpretability, enable extrapolation beyond training data, and ensure physically plausible predictions in safety-critical contexts.

4.3.2 Selection Guidelines

The following principles provide a systematic framework for selecting and implementing fault diagnosis techniques across the full range of permanent magnet machine applications.

4.3.3 Key Selection Principles

The following eleven principles form a decision framework that progresses from immediate, low-cost implementations through progressively sophisticated techniques as application requirements, budget, and operational experience justify additional investment.

1. **Start with Zero-cost Methods:** MCSA, EPVA, model-based (lumped-parameter), and IPSA/IPFSA leverage existing sensors. Begin with these for initial condition monitoring capability without hardware modifications or additional costs.

   This principle recognizes that nearly all motor drive systems already contain current and voltage sensors for control purposes. Implementing software-based fault diagnosis using these existing signals represents the lowest-risk, fastest-deployment pathway to condition monitoring. MCSA provides broad fault coverage through spectral analysis of stator currents. EPVA offers computational efficiency suitable for real-time embedded implementation. Model-based lumped-parameter methods enable quantitative parameter tracking and incipient fault detection through gradual parameter drift. IPSA/IPFSA provide operating-point-independent indicators that simplify threshold setting across variable-speed applications. Beginning with these zero-cost methods establishes baseline diagnostic capability, validates the business case for condition monitoring, and informs decisions about which additional sensors and techniques would provide the greatest incremental value.
2. **Layer by Fault Criticality:** Add sensors and techniques based on the most critical failure modes for your application:
   - **Winding Failures Critical:** ZSVC (if neutral accessible), model-based (lumped or geometry-based), MFSA.
   - **Bearing Failures Critical:** Vibration/Acoustic primary, data-driven (CNN/LSTM) for pattern recognition.
   - **Demagnetization Critical:** MFSA (internal for localization), model-based (geometry for spatial faults), data-driven.
   - **Eccentricity Critical:** ZSVC (highest sensitivity), vibration, MFSA.

Not all fault types pose equal risk in every application. A systematic failure modes and effects analysis (FMEA) identifies which faults have the highest consequence (safety, downtime cost, repair expense) and probability. Prioritize diagnostic investment toward these critical failure modes. For applications where winding insulation failures dominate (high-voltage, thermal cycling, aggressive duty cycles), ZSVC provides exceptional inter-turn short sensitivity at low cost ($20–200). Bearing-critical applications (high-speed, heavy-load, contaminated environments) justify dedicated vibration monitoring ($200–2000) as electrical methods show only moderate bearing fault sensitivity. Demagnetization concerns in high-temperature or shock-load applications benefit from MFSA spatial mapping or geometry-based parameter tracking. Eccentricity-prone applications (large air gaps, flexible shafts, installation misalignment) achieve optimal detection with ZSVC and vibration analysis. This fault-prioritized approach ensures limited resources address the most consequential failure modes first.

3. **Consider Operating Envelope:**
   - **Low-speed Applications (<500 RPM):** Model-based (lumped/geometry), vibration, IPSA/IPFSA (MCSA/EPVA degrade).
   - **High-speed Applications (>3000 RPM):** All electrical methods effective, MCSA particularly robust.
   - **Variable-speed/transient-heavy:** Model-based with adaptive tracking, IPSA/IPFSA, data-driven (LSTM for temporal patterns), MCSA with order tracking or time-frequency methods (STFT, HHT).
   - **Steady-state Dominant:** All methods applicable, FFT-based techniques most efficient.

Operating conditions fundamentally influence diagnostic technique effectiveness. Low-speed direct drives challenge spectral methods due to low fundamental frequency (5–10 Hz at 100–200 RPM), pushing fault harmonics into the baseband where noise and measurement resolution are problematic. Model-based approaches and IPSA/IPFSA maintain effectiveness at low speeds by not relying on high-frequency spectral content. Vibration analysis benefits from reduced electromagnetic noise at low speeds, though bearing fault frequencies also scale down proportionally. High-speed applications generate rich harmonic content well-separated from the fundamental, enabling MCSA to achieve excellent signal-to-noise ratio for fault detection. Variable-speed and transient-heavy duty cycles require adaptive techniques: model-based methods with time-varying parameter tracking, IPSA/IPFSA normalization that removes operating-point dependence, LSTM networks that learn fault evolution across operating conditions, or time-frequency methods (STFT, HHT) that track spectral content during speed transients. Steady-state dominant applications enable the simplest and most efficient implementations using standard FFT-based spectral analysis with fixed thresholds.

4. **Geometry Knowledge Availability:**
   - **Geometry Available (New Designs, Well-documented Machines):** Model-based geometry methods (MEC, IWFT, hybrid FEM-analytical) provide superior spatial fault localization.
   - **Geometry Unavailable (Proprietary, Retrofit, Reverse-engineering Infeasible):** Lumped-parameter model-based methods, MCSA, EPVA, IPSA/IPFSA, data-driven approaches.

The availability of detailed machine geometry (winding layouts, slot/pole configurations, magnet dimensions and placement, air gap profiles, rotor and stator lamination geometries) divides model-based approaches into two capability tiers. Geometry-based methods (MEC models, improved winding function theory, hybrid FEM-analytical approaches) leverage this spatial information to provide localized fault identification, answering questions like "which magnet is demagnetized?" or "where is the eccentricity maximum?" rather than merely detecting that a fault exists. These methods excel during new machine design when geometry is inherently available, or for well-documented industrial machines with complete technical specifications. Conversely, proprietary commercial machines, retrofit applications without manufacturer cooperation, or applications where reverse-engineering is impractical must rely on lumped-parameter models that aggregate spatial effects into equivalent circuit parameters. While lumped-parameter approaches cannot localize faults spatially, they maintain quantitative parameter tracking capability (resistance, inductance, flux linkage) using only terminal measurements, making them universally applicable regardless of geometry knowledge availability.

5. **Data Availability and ML Considerations:**
   - **Large Labeled Historical Fault Data Available:** Data-driven (supervised ML, deep learning) highly effective for pattern recognition and multi-class classification.
   - **Limited Fault Data:** Physics-informed ML (PIML) combines limited data with physical laws for improved generalization.
   - **No Fault Data (Healthy-only Baseline):** Anomaly detection methods (OC-SVM, autoencoders), traditional signal-based methods with threshold-based decision logic.
   - **Transfer Learning Opportunity:** Pre-train on simulations (FEM-generated data), fine-tune on limited experimental data.

   Data-driven methods promise powerful pattern recognition but their effectiveness depends critically on training data characteristics. Large fleets with comprehensive fault history databases enable supervised learning (CNNs, LSTMs, transformers) that can achieve superior classification accuracy compared to hand-engineered feature-based methods. However, most industrial applications face the "rare fault" problem: operational data is predominantly healthy-state, with infrequent labeled fault examples. Physics-informed machine learning addresses this limitation by constraining neural networks with physical laws (electromagnetic equations, thermal models, mechanical dynamics), enabling plausible extrapolation beyond sparse training data. When no fault data exists, anomaly detection approaches (one-class SVM, autoencoders trained on healthy data) flag deviations from normal operation without requiring labeled fault examples. Transfer learning provides another pathway: FEM simulations can generate diverse fault scenarios under controlled conditions, pre-training models that are then fine-tuned on limited experimental data from the actual machine. This hybrid simulation-experimental approach combines the data abundance of virtual environments with the realism of physical measurements.

6. **Multi-modal Fusion Advantages:** Combining techniques from different physical domains (electrical + mechanical + thermal) provides comprehensive coverage:
   - **Data-level Fusion:** Raw signal concatenation for deep learning (CNNs, transformers) when computational resources available.
   - **Feature-level Fusion:** Engineered features from multiple modalities for SVM, Random Forest classifiers.
   - **Decision-level Fusion:** Separate classifiers per modality with voting/weighted averaging for incremental deployment and sensor redundancy.

   No single sensing modality detects all fault types optimally. Multi-modal fusion leverages complementary strengths: electrical methods excel at winding faults and demagnetization, mechanical methods at bearing faults and eccentricity, thermal methods at overload and cooling system failures. The fusion architecture choice involves fundamental trade-offs. Data-level fusion feeds raw signals from all sensors into end-to-end deep learning models, maximizing information retention and enabling discovery of cross-modal correlations. However, this approach demands GPU-class computational resources and large multi-sensor training datasets. Feature-level fusion extracts engineered features from each modality (MCSA spectral peaks, vibration envelope kurtosis, thermal gradients) and concatenates them for classical ML classifiers, offering interpretability and moderate computational cost on industrial PC platforms. Decision-level fusion runs independent classifiers on each modality and combines predictions through voting or weighted averaging, enabling incremental sensor deployment, providing fault-tolerant redundancy, and supporting distributed processing. The architecture choice depends on available computational resources, deployment constraints, and whether discovering novel cross-modal fault signatures justifies the complexity of data-level fusion.

7. **Computational Resource Constraints:**
   - **Embedded Microcontrollers (limited RAM/Flash):** Lightweight methods: Threshold-based (10–20k ops), EPVA (20–100k ops), FFT-based MCSA (50–500k ops).
   - **Industrial PCs/Edge Computing:** Model-based (EKF/RLS 100–500k ops), MCSA with advanced time-frequency (STFT 500k–5M ops), SVM classifiers (10–100k ops).
   - **GPU-equipped Platforms or Cloud:** Deep learning (CNNs/LSTMs 100k–1M ops), PIML (1–10M ops), multi-sensor fusion with transformers.

The available computing platform establishes hard constraints on implementable techniques. Embedded microcontrollers common in distributed motor drives (ARM Cortex-M series with 64–256 KB RAM, 100–200 MHz clock) support lightweight methods: simple threshold-based detection (10–20k operations), EPVA (20–100k operations), and basic FFT-based MCSA (50–500k operations) all execute within millisecond update intervals on these constrained platforms. Industrial PC or edge computing platforms (×86/ARM with gigabytes of RAM, multi-core GHz processors) enable moderate-complexity techniques: model-based parameter estimation with EKF or RLS (100–500k operations), advanced time-frequency analysis like STFT (500k–5M operations), and classical ML classifiers like SVM (10–100k operations inference). High-complexity methods require GPU acceleration or cloud computing: deep learning models (CNNs/LSTMs at 100k–1M operations per inference), physics-informed machine learning (1–10M operations), and multi-sensor fusion with transformer architectures demand specialized hardware. A practical deployment strategy matches algorithm complexity to available platforms: edge devices run lightweight real-time monitoring, streaming data to cloud infrastructure for offline deep learning-based comprehensive analysis.

8. **Temperature Compensation Requirements:**
   - All model-based parameter estimation methods require temperature compensation ($\pm 30\,°C \rightarrow \pm 12\%$ resistance change).
   - Signal-based methods benefit from temperature-indexed baselines for improved accuracy.
   - Direct temperature measurement ($20–50 sensors) or thermal modeling recommended for quantitative parameter estimation.

Temperature variations present a fundamental challenge for quantitative fault diagnosis. Winding resistance changes approximately 0.4%/°C for copper (temperature coefficient ~ 3900 ppm/K), meaning a 30 °C temperature swing produces 12% resistance variation that can easily mask or mimic fault signatures. Model-based parameter estimation methods are particularly sensitive: attempting to track 5–10% resistance increase from an inter-turn short becomes problematic when normal thermal cycling produces comparable variations. Three compensation strategies address this issue. Direct temperature measurement using thermocouples or RTDs ($20–50 per sensor) embedded in windings or mounted on the stator provides ground truth for compensation, though sensor installation may be invasive. Thermal modeling estimates winding temperature from operating history (load current, ambient temperature, duty cycle) using lumped-parameter or FEM-based thermal networks, avoiding additional sensors at the cost of model accuracy limitations. Temperature-indexed baselines maintain separate fault detection thresholds or reference signatures for different temperature ranges (e.g., 20–40, 40–60, 60–80 °C), ensuring comparisons occur under similar thermal conditions. Signal-based methods like MCSA and ZSVC, while less affected than resistance-based parameter estimation, still benefit from temperature compensation to reduce false alarm rates and improve detection sensitivity.

9. **Validation Methodology:**
   - **Simulation → HIL → Bench → Field:** Progress through validation hierarchy.
   - More complex techniques (geometry-based models, deep learning) require extensive validation across multiple machines and operating conditions.
   - Cross-machine validation essential to prevent overfitting to specific machine characteristics.
   - External benchmark datasets enable unbiased performance comparison.

   Rigorous validation is essential before deploying fault diagnosis algorithms in production, particularly for safety-critical applications. A staged validation hierarchy manages risk and development cost effectively. Simulation validation using FEM or circuit-based models enables rapid iteration, fault scenario exploration, and initial algorithm development under controlled conditions without physical test hardware. However, simulations inevitably simplify reality (perfect materials, idealized geometries, simplified nonlinearities), necessitating hardware validation. Hardware-in-the-loop (HIL) testing injects simulated machine models into physical drive hardware, validating algorithm integration with real control systems, communication protocols, and timing constraints. Bench testing with physical machines and controllable fault injection (resistive shorts, selective magnet removal, adjustable eccentricity) provides ground truth for algorithm calibration and sensitivity characterization. Field testing under actual operating conditions validates robustness to environmental factors, load variations, and unanticipated disturbances absent in laboratory environments. More complex techniques (geometry-based models requiring accurate geometry, deep learning models with high parameter counts) demand extensive validation across multiple machines and operating conditions to prevent overfitting to specific test configurations. Cross-machine validation ensures algorithms generalize across manufacturing variations, different machine designs within a product family, and variations in installation and maintenance practices. Public benchmark datasets (if available) enable unbiased performance comparison against published methods, accelerating research progress and establishing performance expectations.

10. **Implementation Progression:** Phase implementation over time to manage costs and risks:
    - **Phase 1 (Immediate):** Software-only electrical methods (MCSA, EPVA, model-based lumped, IPSA/IPFSA) using existing sensors.
    - **Phase 2 (Low-cost Additions):** External search coils for MFSA ($100–500), neutral access for ZSVC ($20–200), artificial neutral point for ZSVC ($20–100).

- **Phase 3 (Mechanical Monitoring):** Vibration/acoustic sensors ($200–2000), particularly for bearing-critical applications.
- **Phase 4 (Advanced Analytics):** Integrate data fusion architectures, deploy ML classifiers trained on accumulated operational data, implement PIML for physics-constrained parameter estimation.
- **Phase 5 (Spatial Resolution):** For new designs or critical applications: internal MFSA sensors ($200–1000), geometry-based model methods requiring detailed machine specifications.

A phased implementation strategy balances immediate monitoring capability with long-term sophistication, managing both capital expenditure and organizational learning curve. Phase 1 leverages existing infrastructure for immediate value, requiring only software development and algorithm integration with existing control systems. This initial deployment establishes baseline performance data, identifies which fault types occur in practice, and validates the business case for expanded investment. Phase 2 targets low-cost hardware additions ($20–500) that significantly enhance fault detection capability: external MFSA sensors provide flux monitoring without machine invasiveness, while ZSVC implementation (through neutral access or artificial neutral) dramatically improves inter-turn short and eccentricity sensitivity at minimal cost. Phase 3 adds mechanical monitoring for comprehensive fault coverage, particularly essential if Phase 1 operational data reveal bearing issues or if application characteristics (high-speed, heavy-load) indicate bearing criticality. Phase 4 leverages accumulated operational data for advanced analytics: multi-sensor fusion combines electrical and mechanical signals, supervised ML classifiers train on real fault examples captured during Phases 1–3, and physics-informed ML provides physicsconstrained parameter estimation with improved generalization. Phase 5 applies to new machine designs where sensors can be integrated during manufacturing, or to critical existing machines where spatial fault localization justifies invasive sensor installation and detailed geometry-based modeling. This staged approach distributes costs over time, demonstrates value at each stage before expanding investment, and aligns technical sophistication with organizational maturity in condition monitoring practices.

11. **Practical Deployment Considerations:**
    - **Retrofit Applications:** Prioritize noninvasive, zero-hardware-cost methods (MCSA, EPVA, IPSA/IPFSA, Data-driven on existing signals).
    - **Large fleet Deployments:** Centralized cloud-based data-driven approaches enable cross-fleet learning and anomaly detection.
    - **Safety-critical Systems:** Multi-modal redundancy (electrical + mechanical + thermal) with decision-level fusion ensures no single-point failure.
    - **Cost-sensitive Applications:** MCSA and EPVA provide multi-fault coverage at zero incremental cost.
    - **Unknown Failure Modes:** Data-driven deep learning discovers patterns without explicit fault models, complemented by model-based methods for interpretability.

Real-world deployment contexts impose constraints beyond pure technical capability. Retrofit applications face installed base constraints: machines already in service cannot be easily modified for invasive sensors, downtime for sensor installation may be prohibitively expensive, and capital budgets may be limited for mature assets. These constraints favor noninvasive, zero-hardware-cost methods that deliver monitoring capability through software updates alone. Large fleet deployments transform economics: centralized cloud infrastructure amortizes across thousands of machines, enabling sophisticated analytics infeasible for single assets. Cross-fleet learning identifies anomalies by comparing individual machine behavior against fleet-wide statistical distributions, detecting incipient faults before traditional threshold-based methods. Transfer learning accelerates new machine type deployment by leveraging models trained on existing fleet data. Safety-critical systems in aerospace, medical, or nuclear applications demand multi-modal redundancy: independent electrical, mechanical, and thermal monitoring with decision-level fusion ensures no single sensor failure or algorithm bug compromises diagnostic capability. Cost-sensitive applications in commodity markets (residential HVAC, low-margin industrial processes) cannot justify expensive sensor suites, making zero-cost MCSA and EPVA attractive despite limitations. Unknown failure mode scenarios, particularly with new machine designs or novel operating conditions, benefit from data-driven methods that discover patterns without explicit physical models, though interpretable model-based methods should complement black-box approaches for root cause analysis and operator trust in safety-critical contexts. Each deployment context emphasizes different aspects of the comprehensive technique portfolio, reinforcing the importance of maintaining diverse diagnostic capabilities rather than over-relying on any single approach.

References

1 James W. Cooley and John W. Tukey. An algorithm for the machine calculation of complex Fourier series. *Mathematics of Computation*, 19(90):297–301, April 1965. https://doi.org/10.1090/S0025-5718-1965-0178586-1.

2 Alan V. Oppenheim and Ronald W. Schafer. *Discrete-Time Signal Processing*. Pearson, Upper Saddle River, NJ, 3rd edition, 2010. ISBN 978-0-13-198842-2.

3 Jont B. Allen and Lawrence R. Rabiner. A unified approach to short-time Fourier analysis and synthesis. *Proceedings of the IEEE*, 65(11):1558–1564, November 1977. https://doi.org/10.1109/PROC.1977.10770.

4 Stéphane G. Mallat. Multiresolution approximations and wavelet orthogonal bases of $L^2(\mathbb{R})$. *Transactions of the American Mathematical Society*, 315(1):69–87, 1989. https://doi.org/10.2307/2001373.

5 Ingrid Daubechies. The wavelet transform, time-frequency localization and signal analysis. *IEEE Transactions on Information Theory*, 36(5):961–1005, September 1990. https://doi.org/10.1109/18.57199.

6 Ingrid Daubechies. Orthonormal bases of compactly supported wavelets. *Communications on Pure and Applied Mathematics*, 41(7):909–996, 1988. https://doi.org/10.1002/cpa.3160410705.

7 Leon Cohen. Time-frequency distributions—A review. *Proceedings of the IEEE*, 77(7):941–981, July 1989. https://doi.org/10.1109/5.30749.

8 Norden E. Huang, Zheng Shen, Steven R. Long, Manli C. Wu, Hsing H. Shih, Quanan Zheng, Nai-Chyuan Yen, Chi Chao Tung, and Henry H. Liu. The empirical mode decomposition and the Hilbert spectrum for nonlinear and non-stationary time series analysis. *Proceedings of the Royal Society of London. Series A: Mathematical, Physical and Engineering Sciences*, 454(1971):903–995, March 1998. https://doi.org/10.1098/rspa.1998.0193.

9 James F. Kaiser. On a simple algorithm to calculate the 'energy' of a signal. *Proceedings of the IEEE International Conference on Acoustics, Speech, and Signal Processing (ICASSP)*, 1:381–384, 1990. https://doi.org/10.1109/ICASSP.1990.115702.

10 Edith Clarke. *Circuit Analysis of A-C Power Systems*, volume 1. John Wiley & Sons, New York, 1943.

11 Robert H. Park. Two-reaction theory of synchronous machines—Generalized method of analysis—Part I. *Transactions of the American Institute of Electrical Engineers*, 48(3):716–727, July 1929. https://doi.org/10.1109/T-AIEE.1929.5055275.

12 Paul C. Krause, Oleg Wasynczuk, Scott D. Sudhoff, and Steven D. Pekarek. *Analysis of Electric Machinery and Drive Systems*. Wiley-IEEE Press, 3rd edition, 2013. ISBN 978-1-118-02429-4. https://doi.org/10.1002/9781118524336.

13 Ralph O. Schmidt. Multiple emitter location and signal parameter estimation. *IEEE Transactions on Antennas and Propagation*, 34(3):276–280, March 1986. https://doi.org/10.1109/TAP.1986.1143830.

14 Richard Roy and Thomas Kailath. ESPRIT—Estimation of signal parameters via rotational invariance techniques. *IEEE Transactions on Acoustics, Speech, and Signal Processing*, 37(7):984–995, July 1989. https://doi.org/10.1109/29.32276.

15 Bruce P. Bogert, M. J. R. Healy, and John W. Tukey. The quefrency analysis of time series for echoes: Cepstrum, pseudo-autocovariance, cross-cepstrum and saphe cracking. *Proceedings of the Symposium on Time Series Analysis*, pages 209–243, 1963.

16 Bernhard Schölkopf, John C. Platt, John Shawe-Taylor, Alex J. Smola, and Robert C. Williamson. Estimating the support of a high-dimensional distribution. *Neural Computation*, 13(7):1443–1471, July 2001. https://doi.org/10.1162/089976601750264965.

17 Fei Tony Liu, Kai Ming Ting, and Zhi-Hua Zhou. Isolation forest. *Proceedings of the 8th IEEE International Conference on Data Mining (ICDM)*, pages 413–422, 2008. https://doi.org/10.1109/ICDM.2008.17.

18 Vladimir N. Vapnik. *The Nature of Statistical Learning Theory*. 1995. https://doi.org/10.1007/978-1-4757-2440-0.

19 Leo Breiman. Random Forests. *Machine Learning*, 45(1):5–32, October 2001. https://doi.org/10.1023/A:1010933404324.

20 Mohamed El H. Benbouzid. A review of induction motors signature analysis as a medium for faults detection. *IEEE Transactions on Industrial Electronics*, 47(5):984–993, October 2000. https://doi.org/10.1109/41.873206.

21 William T. Thomson and Mark Fenger. Current signature analysis to detect induction motor faults. *IEEE Industry Applications Magazine*, 7(4):26–34, July/August 2001. https://doi.org/10.1109/2943.930988.

22 Juha Pyrhonen, Tapani Jokinen, and Valeria Hrabovcova. *Design of Rotating Electrical Machines*. John Wiley & Sons, 2013.

23 Aleksi Riitala. Estimating the speed of a permanent magnet synchronous machine using magnetic leakage flux. Master's thesis, Lappeenranta University of Technology, 2018.

24 Julio-César Urresty, Jordi-Roger Riba, Miguel Delgado, and Luís Romeral. Detection of demagnetization faults in surface-mounted permanent magnet synchronous motors by means of the zero-sequence voltage component. *IEEE Transactions on Energy Conversion*, 27(1):42–51, March 2012. ISSN 0885-8969, 1558-0059. https://doi.org/10.1109/TEC.2011.2176127.

25 Sergio M. A. Cruz and A. J. Marques Cardoso. Stator winding fault diagnosis in three-phase synchronous and asynchronous motors, by the extended Park's vector approach. *IEEE Transactions on Industry Applications*, 37(5):1227–1233, September/October 2001. https://doi.org/10.1109/28.952496.

26 Rudolph E. Kalman. A new approach to linear filtering and prediction problems. *Journal of Basic Engineering*, 82(1):35–45, March 1960. https://doi.org/10.1115/1.3662552.

27 David G. Luenberger. An introduction to observers. *IEEE Transactions on Automatic Control*, 16(6):596–602, December 1971. https://doi.org/10.1109/TAC.1971.1099826.

28 Maziar Raissi, Paris Perdikaris, and George E. Karniadakis. Physics-informed neural networks: A deep learning framework for solving forward and inverse problems involving nonlinear partial differential equations. *Journal of Computational Physics*, 378:686–707, February 2019. https://doi.org/10.1016/j.jcp.2018.10.045.

29 Sepp Hochreiter and Jürgen Schmidhuber. Long short-term memory. *Neural Computation*, 9(8):1735–1780, November 1997. https://doi.org/10.1162/neco.1997.9.8.1735.

30 Ramprasaath R. Selvaraju, Michael Cogswell, Abhishek Das, Ramakrishna Vedantam, Devi Parikh, and Dhruv Batra. Grad-CAM: Visual explanations from deep networks via gradient-based localization. In *Proceedings of the IEEE International Conference on Computer Vision (ICCV)*, pages 618–626, 2017. https://doi.org/10.1109/ICCV.2017.74.

31 Scott M. Lundberg and Su-In Lee. A unified approach to interpreting model predictions. In *Proceedings of the 31st International Conference on Neural Information Processing Systems (NIPS)*, pages 4768–4777, Red Hook, NY, 2017. Curran Associates Inc.

5

Eccentricity Faults

5.1 Introduction

Eccentricity faults represent one of the most prevalent mechanical–electromagnetic failure modes in permanent magnet (PM) machines, arising from air gap asymmetries between the rotor and stator. These faults compromise the electromagnetic symmetry of the machine, leading to distorted magnetic flux distribution, unbalanced magnetic forces, increased vibration, bearing wear, and ultimately, catastrophic failure if left undetected. Unlike purely mechanical or purely electrical faults, eccentricity faults exhibit coupled electromechanical characteristics, making their diagnosis and mitigation particularly challenging in PM machine applications.

The significance of eccentricity faults in PM machines stems from their progressive nature. Initial eccentricity, whether caused by manufacturing tolerances, assembly errors, or bearing degradation, generates unbalanced magnetic pull (UMP) that exerts additional radial forces on the rotor. These forces accelerate bearing wear, which in turn exacerbates the eccentricity, creating a positive feedback loop that can rapidly lead to rotor-stator contact and catastrophic damage. This self-reinforcing mechanism distinguishes eccentricity faults from many other fault types and necessitates early detection strategies.

This chapter provides a comprehensive treatment of eccentricity faults in PM machines, organized into three main sections. Section 5.2 examines the failure modes and mechanisms, including the classification of eccentricity types (static, dynamic, and mixed), their root causes, and the physics of UMP. Section 7.3 addresses modeling and analysis techniques, covering analytical methods, finite element analysis, air gap field perturbation theory, and unbalanced force calculations. Section 5.4 focuses on fault diagnosis, presenting motor current signature analysis (MCSA), vibration analysis, flux monitoring, external magnetic field analysis, and multi-signal fusion approaches specific to eccentricity detection in PM machines.

5.2 Failure Modes and Mechanisms

5.2.1 Introduction to Air Gap Eccentricity

Air gap asymmetries are inherent in all types of electrical motors, causing distorted distribution of magnetic flux density in the air gap. This distortion results in unbalanced magnetic forces on the rotor and stator, known as unbalanced magnetic pull (UMP). UMP can cause various mechanical faults in electrical motors, particularly bearing wear and progressive eccentricity development, leading to side effects such as noise, reduced efficiency, hot spots, and vibrations. The physical mechanisms and effects of UMP are detailed later in this chapter.

5.2.1.1 Classification Framework

Rotor eccentricity refers to an air gap asymmetry caused by the eccentric displacement of the rotor from the stator's center in cylindrical dimensions, while maintaining parallel rotor and stator axes. This results in a nonconstant air gap length profile relative to the stator reference angle. In a general rotor eccentricity scenario, a cylindrical rotor rotates with an angular velocity (ω_r) around its symmetry center (O_r), which is displaced from the stator symmetry center (O_s). Additionally, the rotor's symmetry center whirls on a whirling radius (r_w) with the symmetry center O_w and whirling angular velocity ω_w. These symmetry centers are located in different positions. This type of rotor eccentricity is called mixed eccentricity [1].

This general form can be categorized into two extreme forms (see Figure 5.1):

1. **Dynamic Eccentricity:** The whirling center coincides with the stator symmetry center ($O_s = O_w$).
2. **Static Eccentricity:** The whirling center coincides with the rotor symmetry center ($O_r = O_w$).

The latter is termed "static" because there is no whirling motion involved ($\omega_w = 0$), and the air gap profile remains unchanged. In dynamic eccentricity, the whirling angular velocity ω_w equals the rotational angular velocity ω_r. In all types, the resultant force on the rotor points toward the minimum air gap [1].

The general air gap length for mixed eccentricity can be expressed as:

$$g(\theta, t) = g_0[1 - \varepsilon_s \cos(\theta - \theta_s) - \varepsilon_d \cos(\theta - \omega_r t - \theta_d)] \tag{5.1}$$

where g_0 is the nominal air gap length, $\varepsilon_s = \delta_s/g_0$ and $\varepsilon_d = \delta_d/g_0$ are the static and dynamic eccentricity ratios respectively, θ_s and θ_d are the angular positions of minimum air gap for each component, and ω_r is the rotor angular velocity. Static eccentricity corresponds to $\varepsilon_d = 0$, dynamic eccentricity to $\varepsilon_s = 0$, and mixed eccentricity includes both components.

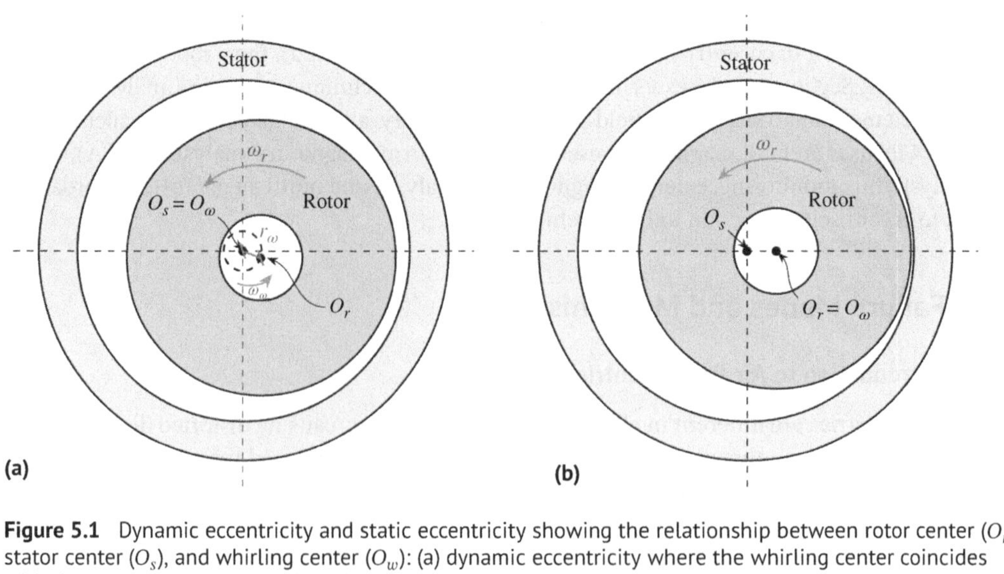

Figure 5.1 Dynamic eccentricity and static eccentricity showing the relationship between rotor center (O_r), stator center (O_s), and whirling center (O_w): (a) dynamic eccentricity where the whirling center coincides with the stator center ($O_s = O_w$), and (b) static eccentricity where the whirling center coincides with the rotor center ($O_r = O_w$).

5.2.1.2 PM Machine-specific Considerations

PM machines with alternate permanent magnet and consequent steel magnetic poles tend to exhibit higher UMP compared to surface-mounted PM (SPM) machines. This difference arises because steel poles concentrate the flux in the center, while SPM machines allow the flux to freely spread around the point of minimum air gap. Consequently, machines with smaller air gaps experience larger UMP [1]. This characteristic makes PM machines, particularly those designed with minimal air gaps for high power density, especially susceptible to the detrimental effects of eccentricity.

5.2.2 Static Eccentricity Faults

Static eccentricity occurs when the rotor's geometric center is displaced from the stator bore center, but the rotor rotates about its own geometric center. In this condition, the whirling center coincides with the rotor symmetry center ($O_r = O_w$), and there is no whirling motion ($\omega_w = 0$). The air gap profile remains stationary relative to the stator reference frame, though it exhibits a nonuniform spatial distribution.

5.2.2.1 Geometric Characteristics

In static eccentricity ($\varepsilon_d = 0$ in the general equation), the minimum air gap position remains fixed in space relative to the stator. Because the air gap profile is stationary, the magnetic flux density distribution exhibits a fixed spatial pattern that does not rotate with the rotor. This creates a stationary field component that interacts with the rotating rotor field, producing characteristic force and vibration signatures.

5.2.2.2 Root Causes

Static eccentricity primarily arises from manufacturing and assembly-related issues:

Manufacturing Tolerances: In the manufacturing of electric motors, tolerances are usually allowed in the shaft, bearing, and cap areas. These tolerances are the primary cause of static eccentricity conditions [1]. Machining imperfections can result in nonconcentric stator bore and outer diameter, shaft diameter variations, housing bore misalignment, and endbell mounting surface irregularities. Each of these geometric deviations contributes to the displacement of the rotor center from the stator center, creating the foundation for static eccentricity development.

Assembly Defects: Improper assembly procedures can introduce static eccentricity even when individual components are within tolerance. Common assembly-related causes include misalignment during press-fitting of stator into housing, uneven tightening of endbell fasteners, foreign objects or debris trapped between mating surfaces, and thermal expansion mismatch during assembly. The cumulative effect of these assembly errors can produce static eccentricity that exceeds what would be expected from individual component tolerances alone.

Bearing Installation Issues: Bearing wear and fluctuating mechanical loads can contribute to static eccentricity development [1]. Specifically, bearing race ovalization in the housing or on the shaft, improper bearing preload, and bearing creep within the housing all create conditions that allow the rotor to assume an eccentric position relative to the stator bore center.

5.2.2.3 Air Gap Profile Characteristics

The stationary nature of the air gap profile in static eccentricity creates unique electromagnetic characteristics. The flux density in the minimum air gap region is higher than in the maximum air gap region, producing a radial force that pulls the rotor toward the stator at the location of minimum clearance. This force remains fixed in the stator reference frame, creating a constant radial load on the bearings.

For a PM machine, the interaction between the rotating rotor magnets and the stationary air gap asymmetry produces additional harmonic components in the air gap flux density. These harmonics have specific spatial and temporal orders that can be exploited for diagnostic purposes, as will be discussed in Section 5.4.

5.2.2.4 Effects on Electromagnetic Performance

Static eccentricity affects several aspects of PM machine electromagnetic performance. The air gap flux density becomes asymmetric, with increased flux density in the region of minimum air gap. For surface-mounted PM machines, this effect is particularly pronounced because the magnets have permeability close to that of air, providing minimal flux regulation capability. The spatially varying air gap also creates position-dependent winding inductances, and while these variations are smaller than in dynamic eccentricity, they can still affect control performance in field-oriented control schemes.

The asymmetric flux distribution produces a net radial force on the rotor, directed toward the position of minimum air gap. This force is constant in magnitude and direction in the stator reference frame, distinguishing it from the rotating force characteristic of dynamic eccentricity. Additionally, the increase in flux density in the minimum air gap region can lead to localized saturation in the stator teeth, increasing core losses and creating hot spots that may accelerate insulation degradation in those regions.

5.2.2.5 Progressive Nature and UMP Relationship

Although static eccentricity itself is a stationary condition, it is not necessarily a static fault in time. The radial force generated by static eccentricity creates additional stress on the bearing at the position aligned with the minimum air gap. Over time, this concentrated load accelerates bearing wear in a specific circumferential location, potentially leading to bearing cage damage or race spalling.

As bearing degradation progresses, the shaft may develop additional freedom of movement, allowing dynamic eccentricity components to develop. Thus, static eccentricity can serve as a precursor to mixed eccentricity conditions, making early detection crucial for preventing progressive fault development.

5.2.3 Dynamic Eccentricity Faults

Dynamic eccentricity occurs when the rotor's geometric center coincides with its rotation center, but this rotation center is offset from the stator bore center. In this condition, the whirling center coincides with the stator symmetry center ($O_s = O_w$), and the rotor center whirls around the stator center with whirling angular velocity ω_w equal to the rotational angular velocity ω_r.

5.2.3.1 Geometric Characteristics

In dynamic eccentricity ($\varepsilon_s = 0$ in the general equation), the air gap profile is time-varying in the stator reference frame. The minimum air gap position rotates synchronously with the rotor, completing one revolution per rotor rotation. This rotating air gap asymmetry creates fundamentally different electromagnetic effects compared to static eccentricity, with the minimum air gap position sweeping around the stator bore at rotational speed and producing time-varying magnetic loading on both the stator and rotor structures.

5.2.3.2 Root Causes

Dynamic eccentricity develops from different mechanisms than static eccentricity:

Bearing Wear and Degradation: As bearings wear during operation, radial clearances increase, allowing the rotor to orbit within the bearing clearance. Unlike static eccentricity, which creates localized bearing wear, dynamic eccentricity typically results from generalized bearing degradation affecting the entire bearing circumference.

Shaft Bending: Both static and dynamic shaft deflection contribute to dynamic eccentricity. Static deflection can be measured even when the motor is not in operation, while dynamic deflection is only measurable during machine operation. Static deflection occurs due to the impact of loads on the rotor shaft, whereas dynamic deflection is primarily caused by centrifugal forces [1].

Mass Imbalance: Rotor mass imbalance creates centrifugal forces that scale with the square of rotational speed, making even small imbalances significant at high speeds. The mathematical relationship between mass eccentricity and unbalanced magnetic pull (UMP) is detailed in the UMP section of this chapter.

Mechanical Shocks: External mechanical impacts or shocks can cause permanent shaft deformation or bearing damage, introducing dynamic eccentricity that persists during operation.

Thermal Effects: Differential thermal expansion between the rotor and stator during transient operation can temporarily introduce dynamic eccentricity. Asymmetric cooling or localized hot spots can create thermal bowing of the shaft.

5.2.3.3 Time-varying Air Gap Profile

The rotating air gap asymmetry in dynamic eccentricity creates a spatially and temporally varying magnetic field distribution. As the rotor rotates, the minimum air gap position sweeps around the stator circumference, creating a traveling wave of increased magnetic loading. This traveling magnetic asymmetry interacts with the stationary stator structure to produce time-varying electromagnetic forces.

The temporal variation of the air gap introduces additional harmonic components in the electromagnetic quantities (flux, current, and force) that are distinct from those produced by static eccentricity. These harmonics appear at frequencies related to the rotational speed and can be identified through spectral analysis of motor currents, vibrations, or external magnetic fields.

5.2.3.4 Speed-dependent Effects

A critical characteristic of dynamic eccentricity is its strong dependency on rotational speed. As discussed in the UMP section, the unbalanced force exhibits a quadratic relationship with speed ($F \propto \omega_r^2$), meaning that dynamic eccentricity effects increase dramatically with speed.

For high-speed PM machines, such as those used in electric vehicle traction drives, turbomachinery, or high-speed spindles, even small amounts of dynamic eccentricity can generate substantial forces. This makes dynamic eccentricity a particular concern for high-speed applications, where precise balancing and bearing design are critical.

The speed dependency also affects diagnostic approaches. Some eccentricity detection methods may be highly effective at rated speed but insensitive at low speeds. Conversely, testing at elevated speeds may amplify eccentricity signatures but also increases the risk of damage if significant eccentricity is present.

5.2.3.5 PM Motor Characteristics

PM motors exhibit several characteristics that affect eccentricity behavior compared to induction motors: smaller air gaps increase sensitivity to absolute eccentricity displacement, the absence of rotor currents eliminates electromagnetic damping effects, and interior PM motors have inherent magnetic saliency that interacts with eccentricity-induced saliency. The characteristic harmonic frequencies in motor currents also differ due to different rotor excitation mechanisms, requiring modification of eccentricity detection algorithms originally developed for induction motors.

5.2.4 Mixed Eccentricity Faults

Mixed eccentricity represents the general case where neither the rotor center nor the rotation center coincides with the stator center. In this condition, none of the symmetry centers coincide with each other ($O_r \neq O_s \neq O_w$), and the air gap profile is both nonuniform and nonstationary [1].

5.2.4.1 Combined Static–Dynamic Characteristics

Mixed eccentricity corresponds to the general air gap equation with both $\varepsilon_s \neq 0$ and $\varepsilon_d \neq 0$. The presence of both components creates a complex air gap profile exhibiting both fixed spatial asymmetry and rotating asymmetry, with electromagnetic effects combining characteristics of both static eccentricity (stationary radial force) and dynamic eccentricity (rotating force, speed-dependent effects).

The severity of mixed eccentricity depends not only on the magnitudes of ε_s and ε_d, but also on their relative phase angle. When the components are aligned (in phase), they produce maximum combined eccentricity at one point in the rotation cycle; when in opposite phase, they may partially cancel at certain rotor positions.

5.2.4.2 Complexity in Diagnosis

Mixed eccentricity presents significant diagnostic challenges due to the simultaneous presence of both static and dynamic components. The electromagnetic signatures of these components appear simultaneously in measured quantities such as motor currents and vibrations, and separating them requires careful spectral analysis and understanding of the characteristic frequencies associated with each type. The relative phase between static and dynamic components affects the total eccentricity magnitude and its variation over a rotation cycle, and without knowledge of this phase relationship, assessing the severity and risk posed by mixed eccentricity becomes difficult.

Unlike pure static or dynamic eccentricity, mixed eccentricity severity varies during each rotor revolution. The minimum and maximum air gap values change position cyclically, creating complex electromagnetic force patterns that can excite multiple vibration modes. Furthermore, mixed eccentricity often develops progressively from initial static or dynamic eccentricity. As one form of eccentricity develops, the resulting UMP and bearing stress can create conditions that introduce the other component, leading to mixed conditions that are more challenging to diagnose and mitigate.

5.2.4.3 Industrial Occurrence Scenarios

Mixed eccentricity commonly occurs in several industrial scenarios. A motor may be manufactured with slight static eccentricity due to tolerances, and during operation, bearing wear or rotor imbalance introduces dynamic eccentricity, creating mixed conditions. Advanced bearing degradation can create both increased radial clearance (allowing dynamic eccentricity) and preferential wear patterns (creating static eccentricity), resulting in a complex fault signature.

Shaft bending with misalignment represents another common scenario where a bent shaft creates dynamic eccentricity during rotation while assembly misalignment contributes static eccentricity. Asymmetric thermal expansion can create quasi-static eccentricity components, while thermally-induced shaft bowing adds dynamic components. Additionally, mechanical shocks can simultaneously damage bearings (creating play for dynamic eccentricity) and permanently deform structural components (creating static eccentricity), leading to immediate mixed eccentricity conditions.

5.2.5 Axial Flux PM Machine Eccentricity

Axial flux PM machines exhibit eccentricity characteristics distinct from radial flux machines due to their geometry (Figure 5.2). In static eccentricity, the whirling shaft (d_W) aligns with the rotor symmetry axis (d_R) but not with the stator symmetry axis (d_S), creating a nonuniform but stationary air gap profile. In dynamic eccentricity, the whirling axis coincides with the stator center ($d_S = d_W$). Mixed eccentricity occurs when none of the symmetry centers coincide, manifesting as a combination of radial displacement and angular misalignment of the rotor disc relative to the stator [1].

The axial flux geometry presents unique challenges because the air gap is distributed radially rather than circumferentially. Eccentricity manifests as a tilt of one rotor disc relative to the stator, creating a radially varying air gap that can vary with both angular position and radial position. This geometric difference requires modified analytical and diagnostic approaches compared to radial flux machines.

5.2.6 Other Air Gap Asymmetries

Other types of air gap asymmetries, such as elliptic rotor, elliptic stator, rotor misalignment, and shaft deflection, can also accelerate the progressive fault development process (Figure 5.3). These asymmetries primarily stem from manufacturing flaws. Similar to dynamic eccentricity, an elliptic rotor exhibits a time-varying air gap profile due to the rotation of the elliptic rotor. Conversely, an elliptic stator maintains a time-invariant air gap profile relative to the stator reference angle. In the case of rotor misalignment, the axis of the rotor is not parallel to the axis of the stator, resulting in a small angle (θ) between them, so the air gap profile varies not only along the stator's angular direction but also along the axial direction [1].

5.2.7 Unbalanced Magnetic Pull

UMP represents the fundamental electromagnetic consequence of air gap asymmetry, serving as both an indicator of eccentricity and a driver of progressive fault development. Understanding UMP is essential for comprehending eccentricity fault mechanisms and developing effective diagnostic strategies.

5.2.7.1 Physical Mechanism and Mathematical Formulation

UMP arises from the asymmetric distribution of magnetic flux density in an eccentric air gap. In a healthy machine with a uniform air gap, the radial magnetic forces are distributed symmetrically around the rotor circumference, resulting in zero net radial force. When eccentricity exists, the air gap length varies circumferentially, creating corresponding variations in magnetic flux density.

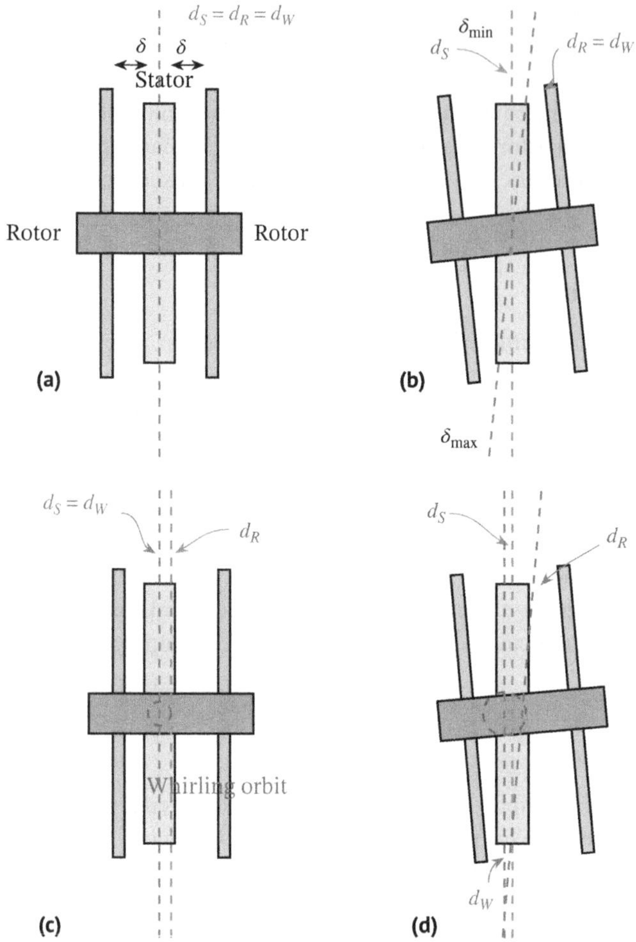

Figure 5.2 Axial flux PM machine eccentricity conditions: (a) healthy condition with aligned axes ($d_S = d_R = d_W$), (b) static eccentricity where whirling shaft aligns with rotor axis ($d_R = d_W$), (c) dynamic eccentricity where whirling axis coincides with stator center ($d_S = d_W$), and (d) mixed eccentricity with no coinciding symmetry centers.

According to the Maxwell stress tensor formulation, the magnetic pressure (force per unit area) in the air gap is proportional to the square of the magnetic flux density:

$$p(\theta) = \frac{B^2(\theta)}{2\mu_0} \tag{5.2}$$

where $B(\theta)$ is the radial component of air gap flux density at angular position θ, and μ_0 is the permeability of free space.

In an eccentric air gap, the flux density is approximately inversely proportional to the air gap length:

$$B(\theta) \propto \frac{1}{g(\theta)} \tag{5.3}$$

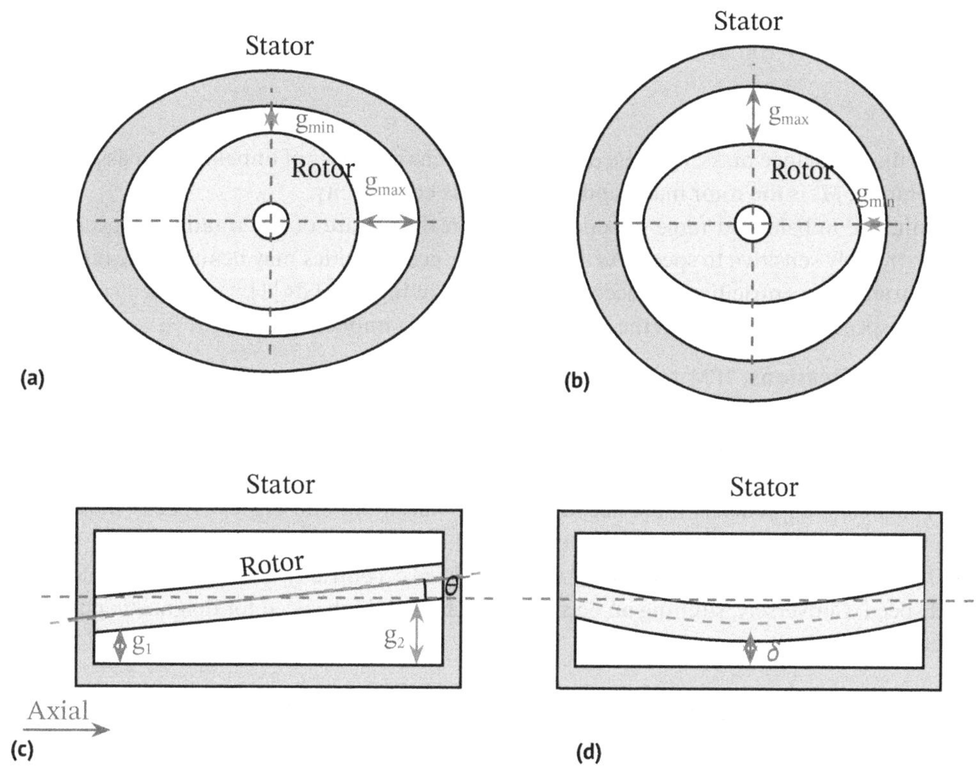

Figure 5.3 Air gap asymmetries: (a) elliptic stator showing noncircular stator bore, (b) elliptic rotor with noncircular rotor surface, (c) rotor misalignment where rotor and stator axes are not parallel with angle θ between them, and (d) shaft deflection showing bending of the rotor shaft under load.

Therefore, regions with a smaller air gap (higher flux density) experience higher magnetic stress than regions with larger air gap (lower flux density). This pressure asymmetry creates a net radial force pulling the rotor toward the stator at the position of minimum air gap.

5.2.7.2 Force Distribution Analysis

The radial force distribution can be analyzed by integrating the magnetic stress over the rotor surface. For a machine with length L and rotor radius r, the net radial force components in orthogonal directions can be expressed as follows:

$$F_x = Lr \int_0^{2\pi} p(\theta)\cos(\theta)d\theta \tag{5.4}$$

$$F_y = Lr \int_0^{2\pi} p(\theta)\sin(\theta)d\theta \tag{5.5}$$

For static eccentricity, this force is constant in magnitude and direction in the stator reference frame, creating a steady radial load on the bearings. For dynamic eccentricity, the force rotates synchronously with the rotor, creating an alternating radial load at rotational frequency.

5.2.7.3 Speed Dependency

The unbalanced force (F) on the rotor and stator due to mass eccentricity is as follows [1]:

$$F = (m_u r_u + M_r e)\omega_r^2 \tag{5.6}$$

where m_u is the unbalance mass, r_u is the distance between the center of unbalanced mass and the center of rotation, M_r is the rotor mass, and e is the mass eccentricity.

As a result, the unbalanced force depends highly on the square of the rotational speed (ω_r^2), making it extremely sensitive to speed. As a result, large eccentricities may be insignificant at low speeds, while small eccentricities can become significant at high speeds [1].

This strong speed dependency has the following important implications:

High-speed Applications: IPM machines designed for high-speed operation (electric vehicles, turbomachinery, high-speed spindles) are particularly vulnerable to eccentricity effects. Even small manufacturing tolerances can produce substantial UMP at a rated speed.

Variable-speed Operation: In variable-speed drives, UMP varies dramatically with operating speed. Diagnostic approaches must account for this variation, and critical speeds that excite structural resonances must be identified and avoided.

Testing Considerations: Low-speed testing may not reveal eccentricity that becomes problematic at rated speed. Conversely, running an eccentric machine at high speed for diagnostic purposes risks accelerating damage.

5.2.7.4 Effects on Bearing Life and Structural Integrity

UMP creates additional loading on bearings beyond the nominal radial load from rotor weight and belt/coupling forces:

Static Eccentricity Loading: The constant radial force from static eccentricity creates a steady additional load on the bearing at a specific circumferential position. This concentrated loading accelerates wear in that region, potentially leading to spalling of the bearing race at the loaded position, cage wear and damage, bearing temperature rise due to friction, and eventual bearing seizure or catastrophic failure.

Dynamic Eccentricity Loading: The rotating radial force due to dynamic eccentricity creates an alternating load on the bearings at rotational frequency. This cyclic loading causes fatigue stress in bearing components, accelerated wear of the entire bearing race circumference, excitation of structural vibration modes, and fretting damage at press-fit interfaces.

Structural Stress: Beyond bearing effects, UMP creates stress in structural components that can compromise mechanical integrity. Endbell deflection and potential cracking occur under sustained loading, while housing deformation affects air gap uniformity and exacerbates the eccentricity condition. Shaft bending increases eccentricity fault severity in a self-reinforcing manner, and mounting foot or base fatigue can lead to misalignment of the entire machine assembly.

5.2.7.5 Progressive Fault Development

Eccentricity faults exhibit progressive behavior due to a positive feedback mechanism: initial eccentricity from manufacturing tolerances or assembly errors produces UMP, which creates additional bearing stress and accelerates wear. The resulting increase in bearing clearance allows greater rotor displacement, producing higher UMP and further accelerating degradation. This feedback loop explains why eccentricity faults often progress slowly over months or years in early stages but can deteriorate rapidly over days or hours in later stages. When the magnitude of rotor eccentricity becomes significant, eventual stator and rotor rubbing leads to severe damage to both structures and the windings [1]. This progressive nature makes early detection through condition monitoring critical for preventing catastrophic failure.

5.3 Modeling and Analysis

5.3.1 Analytical Modeling of Eccentricity Fault

5.3.1.1 Introduction to Analytical Modeling Approaches

Analytical modeling of eccentricity faults in PM machines plays a crucial role in understanding fault mechanisms, predicting machine performance under faulty conditions, and developing diagnostic algorithms. Unlike dq-based methods commonly used for motor control, comprehensive fault diagnosis requires analytical approaches that capture air gap field harmonics, since diagnostic methods rely on analyzing variations in harmonic content of motor signals [2].

Several analytical approaches have been applied to model eccentric electrical machines, including magnetic equivalent circuit (MEC), winding function method (WFM), and analytical field solution methods. The theoretical foundations, mathematical formulations, and general fault modeling capabilities of these techniques are presented in Chapter 3. This section focuses specifically on their application to eccentricity fault modeling, where the key phenomena to be captured include nonuniform air gap distribution, permeance variations due to stator slotting, PM flux distribution, and electromechanical coupling under asymmetric conditions.

For IPM machines, eccentricity modeling must address unique characteristics compared to induction motors. The PM excitation creates a strong fundamental field component independent of load, the typically smaller air gaps in IPM machines increase sensitivity to eccentricity-induced perturbations, and the surface-mounted or interior magnet configurations affect the air gap permeance distribution differently than the uniform rotor structure of induction machines. Furthermore, IPM machines lack rotor conductors, eliminating certain damping effects present in induction motors and potentially amplifying vibration signatures under eccentric conditions.

5.3.1.2 Air Gap Length Distribution Under Eccentricity Fault

The air gap length function is the foundation of all analytical eccentricity models, as it directly affects permeance, flux density, inductance, and electromagnetic forces. For a machine with eccentricity, the air gap length $g(\theta, t)$ varies as a function of both spatial position θ (stator reference frame angle) and time t.

For static eccentricity, the air gap length can be expressed as follows:

$$g_{\mathrm{SE}}(\theta) = g_0(1 - \varepsilon_s \cos(\theta - \theta_s)) \tag{5.7}$$

where g_0 is the nominal air gap length, $\varepsilon_s = \delta_s/g_0$ is the static eccentricity ratio, δ_s is the absolute displacement of the rotor center from the stator center, and θ_s is the angular position of minimum air gap in the stator frame.

For dynamic eccentricity, the minimum air gap position rotates with the rotor:

$$g_{\mathrm{DE}}(\theta, t) = g_0(1 - \varepsilon_d \cos(\theta - \omega_r t - \theta_d)) \tag{5.8}$$

where $\varepsilon_d = \delta_d/g_0$ is the dynamic eccentricity ratio and θ_d is the initial phase angle.

For mixed eccentricity, both components superpose:

$$g_{\mathrm{ME}}(\theta, t) = g_0[1 - \varepsilon_s \cos(\theta - \theta_s) - \varepsilon_d \cos(\theta - \omega_r t - \theta_d)] \tag{5.9}$$

The spatial variation of air gap length in static eccentricity creates a stationary permeance distribution, while dynamic eccentricity produces a rotating permeance pattern. In a typical surface-mounted IPM machine with nominal air gap of 0.5 mm, 10% static eccentricity varies air gap length from 0.45 to 0.55 mm, while 50% eccentricity creates a more severe range from 0.25 to 0.75 mm. This nonuniformity significantly affects the magnetic circuit behavior and introduces harmonic distortions in all electromagnetic quantities.

Figure 5.4 shows the air gap length variations for a healthy motor and motors with various static eccentricity severities. The spatial variation (Figure 5.4a) represents observations from the stator reference frame over one complete rotor rotation, while the time variation (Figure 5.4b) shows the minimum air gap length as a function of time. For static eccentricity, the minimum air gap position remains fixed in space, resulting in a constant minimum air gap value over time.

When stator slotting effects are considered, the effective air gap length becomes position-dependent even in the healthy condition. The combined effect of eccentricity and slotting creates a complex permeance function:

$$\lambda(\theta, t) = \frac{g(\theta, t) + h_m/\mu_r}{g(\theta, t) + h_m/\mu_r + (\pi/2)w(\theta)} \tag{5.10}$$

where h_m is the magnet thickness, μ_r is the relative permeability of the PM, and $w(\theta)$ represents the slot opening function.

5.3.1.3 Inductance Variations Under Eccentricity Fault

The stator phase inductance consists of leakage inductance and magnetizing inductance:

$$L = L_{ls} + L_m \tag{5.11}$$

The leakage inductance L_{ls}, associated with slot leakage, end-winding leakage, and differential leakage, is largely independent of air gap length and remains essentially constant under eccentricity conditions. However, the magnetizing inductance L_m depends directly on the air gap permeance and varies with eccentricity.

For a three-phase IPM machine, the magnetizing inductance is expressed as follows:

$$L_m = \frac{3}{\pi}(qN_sk_w)^2 \times \frac{\mu_0}{g(\theta, t) + h_m/\mu_r} \times (D - g(\theta, t))l_m \tag{5.12}$$

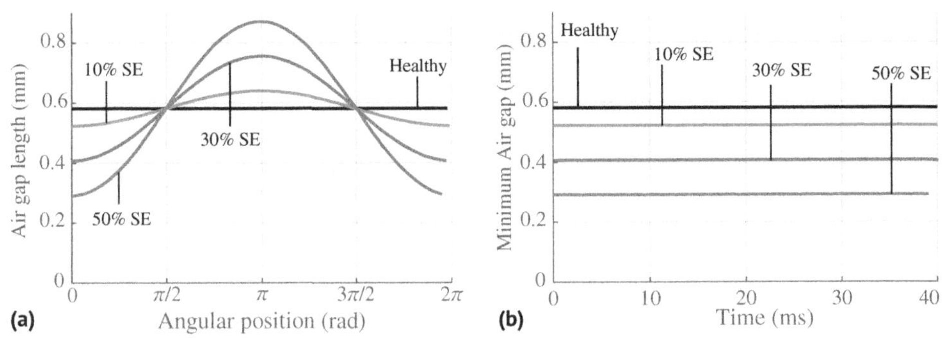

Figure 5.4 Air gap length variations under static eccentricity: (a) spatial variation showing nonuniform distribution with healthy (0.58 mm), 10% SE (0.52–0.64 mm), 30% SE (0.41–0.75 mm), and 50% SE (0.29–0.87 mm), and (b) time variation showing constant minimum air gap for each eccentricity level.

where q is the number of slots per pole per phase, N_s is the number of series turns per phase, k_w is the winding factor, D is the stator bore diameter, and l_m is the axial length.

Under static eccentricity, the magnetizing inductance exhibits spatial variation but remains constant in time at any given position. The inductance increases at the location of minimum air gap (maximum permeance) and decreases where the air gap is largest. For a machine with nominal inductance of 66 mH and nominal air gap of 0.5 mm, 10% static eccentricity causes inductance variation between 64.3 and 67.6 mH, while 50% eccentricity produces a wider range from 58.3 to 75.1 mH.

Figure 5.5 illustrates the inductance variations for healthy and static eccentricity conditions. The spatial variation (Figure 5.5a) shows nonuniform inductance distribution observed from the stator frame during one rotor revolution, while the time variation (Figure 5.5b) demonstrates that static eccentricity creates a constant maximum inductance value over time, reflecting the stationary nature of the air gap asymmetry.

Under dynamic eccentricity, the inductance variation becomes time-dependent. The minimum and maximum inductance values are similar to static eccentricity of equal severity, but these extrema rotate synchronously with the rotor. This creates pulsating inductance as observed from the stator reference frame, introducing corresponding variations in the stator current and electromagnetic torque.

The spatial-temporal variation of inductance under mixed eccentricity creates complex patterns that depend on the relative magnitudes and phases of static and dynamic components. This inductance modulation is a key mechanism by which eccentricity produces characteristic harmonic patterns in motor currents.

5.3.1.4 Air Gap Flux Density Perturbations

The air gap magnetic flux density in a IPM machine under eccentricity consists of contributions from the PMs and the armature reaction. For healthy machines with uniform air gap, the IPM flux density can be expressed using Fourier series:

$$B_{\mathrm{IPM}}(\theta, t) = \sum_{n=1,3,5,\ldots}^{\infty} B_n \cos(np\theta - np\omega_r t) \tag{5.13}$$

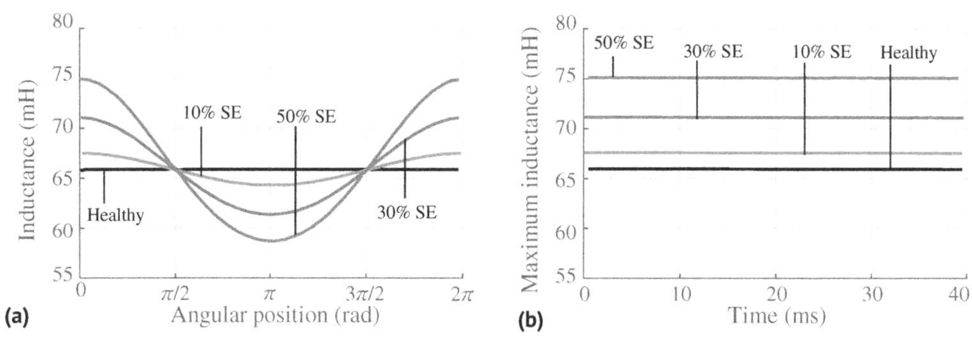

Figure 5.5 Inductance variations under static eccentricity: (a) spatial variation showing healthy motor at constant 66 mH and increasing nonuniformity with eccentricity severity (10% SE: 64.3–67.6 mH, 30% SE: 61.4–71.1 mH, 50% SE: 58.3–75.1 mH), and (b) time variation showing constant maximum inductance for static eccentricity versus time.

where p is the number of pole pairs and B_n are the harmonic amplitudes determined by magnet geometry and magnetization pattern.

When eccentricity is present, the permeance variation modulates this flux density distribution. The eccentric air gap permeance can be expanded using Taylor series and trigonometric identities as follows:

$$\Lambda(\theta, t) = \Lambda_0 \left[1 + \sum_{n=1}^{\infty} \Lambda_n^{(s)} \cos(n\theta) + \sum_{n=1}^{\infty} \Lambda_n^{(d)} \cos(n(\theta - \omega_r t)) \right] \tag{5.14}$$

where Λ_0 is the average permeance, $\Lambda_n^{(s)}$ are the static eccentricity permeance harmonics, and $\Lambda_n^{(d)}$ are the dynamic eccentricity permeance harmonics.

The flux density under eccentricity becomes:

$$B_{\text{IPM}}^{\text{ecc}}(\theta, t) = \Lambda(\theta, t) \cdot B_{\text{IPM}}^{\text{source}}(\theta, t) \tag{5.15}$$

This multiplication in the spatial-temporal domain generates sideband harmonics around the fundamental and higher-order IPM flux components. For dynamic eccentricity specifically, the interaction produces flux density components at the following frequencies:

$$f_B = \left| np \pm \frac{k}{p} \right| f_s \tag{5.16}$$

where f_s is the supply frequency, $n = 1, 3, 5, \ldots$ represents the IPM field harmonics, and $k = 1, 2, 3, \ldots$ represents the eccentricity orders.

The spatial distribution of flux density becomes nonuniform, with increased flux in the minimum air gap region and decreased flux where the air gap is maximum. For surface-mounted PM (SPM) machines, this effect is particularly pronounced because the magnet permeability is close to that of air, providing minimal flux regulation. The flux density in the minimum air gap region can increase by 10–20% for moderate eccentricity fault (30–50%), creating corresponding increases in local iron losses and potential saturation in stator teeth.

Stator slotting introduces additional complexity to the flux density distribution. The slot openings create permeance harmonics that interact with both the IPM field and the eccentricity-induced perturbations, generating a rich harmonic spectrum. The combined effect of eccentricity and slotting must be considered to accurately predict the electromagnetic signature observable in diagnostic measurements.

5.3.1.5 Harmonic Patterns in Stator Currents

The eccentricity-induced flux density perturbations create corresponding variations in the induced EMF and, through electromechanical coupling, produce characteristic harmonic patterns in the stator currents. These current harmonics form the basis for MCSA diagnostic techniques.

The stator current of an eccentric IPM machine can be expressed in compact form as follows:

$$i(t) = I_f \cos(\omega_s t - \phi_f) + \sum_{n,k} I_{n,k} \cos\left[\left(np \pm \frac{k}{p} \right) \omega_s t - \phi_{n,k} \right] \tag{5.17}$$

where I_f is the fundamental current amplitude, ϕ_f is the fundamental phase angle, and $I_{n,k}$ are the eccentricity-induced sideband amplitudes.

For dynamic eccentricity, the characteristic fault frequency patterns in the stator current are:

$$f_{\text{ecc}} = \left[(2\chi - 1) \pm \frac{\eta}{p} \right] f_s, \quad \chi, \eta = 1, 2, 3, \ldots \tag{5.18}$$

This general pattern encompasses several specific cases:

$$f_{\text{ecc}} = \left(1 \pm \frac{k}{p}\right) f_s, \quad k = 1, 2, 3, \ldots \tag{5.19}$$

The most prominent eccentricity indicators typically appear at:

$$f_{\text{ecc}} = f_s \pm f_r = f_s \pm \frac{f_s}{p} \tag{5.20}$$

where f_r is the mechanical rotational frequency.

For a 4-pole machine (p = 2) operating at 50 Hz supply frequency, the primary eccentricity sidebands appear at 25 and 75 Hz. These sidebands are distinct from other fault signatures: bearing faults produce harmonics at multiples of shaft frequency modulated by bearing geometry, while broken rotor bars (in induction motors) create sidebands at $(1 \pm 2s)f_s$ where s is slip.

Static eccentricity produces different harmonic patterns than dynamic eccentricity. Pure static eccentricity creates harmonics at frequencies:

$$f_{\text{SE}} = f_s \pm k \cdot n f_0 \tag{5.21}$$

where f_0 is a multiple of the slot-passing frequency, reflecting the interaction between stationary eccentricity and rotating slot permeance variations.

The amplitude of eccentricity-induced current harmonics depends on several factors: eccentricity severity (magnitude of ε), machine loading (affecting armature reaction), air gap length (smaller gaps increase sensitivity), and pole number (affecting harmonic frequency locations). Typically, harmonic amplitudes range from −40 to −60 dB relative to the fundamental for mild to moderate eccentricity, increasing to −30 dB or higher for severe conditions approaching rotor-stator contact.

Figure 5.6 shows the stator current waveforms for motors under 30% static and dynamic eccentricity at 25% rated load. The eccentricity introduces visible ripples in the current waveform that result from the inductance variations discussed earlier. Dynamic eccentricity (Figure 5.6b) produces more pronounced ripple compared to static eccentricity (Figure 5.6a) due to the time-varying nature of the inductance modulation. These current distortions contain the characteristic harmonic frequencies that enable eccentricity detection through motor current signature analysis.

The phase relationships between eccentricity harmonics can provide additional diagnostic information. In mixed eccentricity, the relative phase between static and dynamic components affects the total severity and its variation over each rotation cycle, creating complex modulation patterns that can be exploited for fault classification.

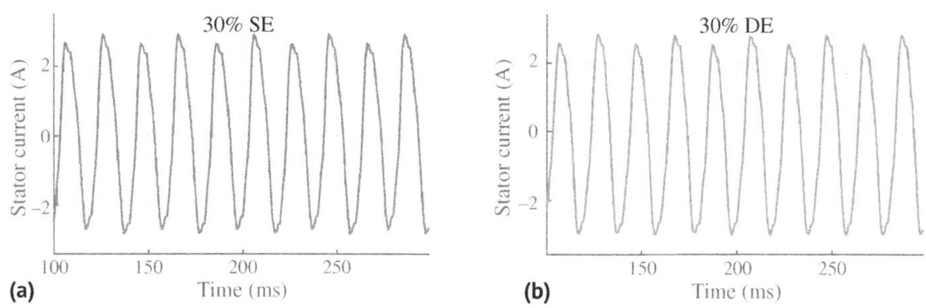

Figure 5.6 Stator current waveforms under eccentricity at 25% rated load: (a) 30% static eccentricity showing moderate current ripple from stationary air gap asymmetry, and (b) 30% dynamic eccentricity showing more pronounced ripple due to rotating air gap minimum.

5.3.1.6 Electromagnetic Torque Perturbations

Eccentricity affects the electromagnetic torque through multiple mechanisms. The air gap flux density perturbations interact with stator currents to produce torque ripple, while the inductance variations create reluctance torque components fluctuating with rotor position.

The electromagnetic torque can be calculated from the product of air gap flux density and current density:

$$T_{\mathrm{em}} = \frac{\pi D^2 l_m}{4} \int_0^{2\pi} B_{\mathrm{IPM}}(\theta, t) \cdot J(\theta, t)\, d\theta \tag{5.22}$$

where $J(\theta, t)$ is the current sheet distribution.

Under eccentricity, both B_{IPM} and J contain harmonic components that interact to produce torque pulsations at frequencies related to the eccentricity pattern and rotor speed. These torque variations can excite torsional vibration modes and contribute to acoustic noise, particularly when torque pulsation frequencies coincide with mechanical resonances.

5.3.1.7 Limitations and Validation Requirements

Analytical models of eccentric IPM machines involve several simplifying assumptions that affect accuracy:

1. **Magnetic Saturation Neglect:** Most analytical models assume linear magnetic materials. While this simplification is acceptable for steady-state diagnosis where the machine operates on the linear portion of the magnetization curve, transient analysis and heavily loaded conditions may require saturation effects to be included.
2. **End Effects:** Two-dimensional analytical models neglect axial end effects, which can be significant in short axial length machines. Three-dimensional effects become important when axial length is comparable to pole pitch.
3. **Harmonic Truncation:** Fourier series expansions must be truncated at practical harmonic orders. Typically, including harmonics up to the 11th or 15th order provides sufficient accuracy for fault diagnosis, but higher orders may be needed for precise prediction of high-frequency phenomena.
4. **Slotting Idealization:** Slot opening geometries are often simplified to rectangular profiles, while actual slots may have more complex shapes affecting local flux density distribution.

Despite these limitations, analytical models validated against experimental measurements and finite element simulations show good agreement for eccentricity diagnosis purposes. The computational efficiency of analytical models makes them invaluable for parametric studies, real-time condition monitoring algorithms, and physical understanding of fault mechanisms. Validation against both FEM results and experimental data remains essential to ensure model fidelity for the specific machine geometry and operating conditions of interest.

5.3.2 FEM Simulation of Eccentric PM Motors

Finite element method provides the highest fidelity approach for modeling eccentric PM synchronous motors (PMSMs), capturing complex geometric features, nonlinear material properties, and detailed electromagnetic field distributions. The theoretical foundations and general methodology of FEM for PM machine analysis are presented in Chapter 3. This section focuses on the specific application of time-stepping FEM (TSFEM) with voltage supply to eccentricity fault analysis, which has become the preferred approach because it directly represents actual operating conditions where voltage rather than current is the controlled quantity [3].

Eccentricity analysis presents unique challenges for FEM because the geometric symmetry exploitable in healthy machines cannot be utilized to reduce computational burden. The complete circumferential cross-section must be modeled, and for dynamic eccentricity, time-varying mesh updates are required as the rotor center orbits around the stator center. This section presents TSFEM modeling and analysis of two three-phase PMSMs (4-pole and 8-pole) under both healthy and eccentric conditions, providing validation of the analytical methods presented in the previous section.

5.3.2.1 Eccentricity Fault Implementation in FEM

The general FEM modeling procedure for PM machines, including geometric modeling, supply coupling, and load representation, is presented in Chapter 3. For eccentricity analysis, the critical adaptation involves modifying the air gap geometry to represent the fault condition [3].

Static eccentricity is implemented by displacing the rotor center from the stator bore center by a specified distance while maintaining the rotation axis coincident with the rotor geometric center. The stator and rotor meshes are adjusted to accommodate the nonuniform air gap profile, with mesh refinement in minimum air gap regions to maintain solution accuracy. Dynamic eccentricity requires time-varying mesh updates as the rotor center orbits around the stator center during rotation, with the minimum air gap position rotating synchronously with the rotor. Mixed eccentricity combines both static displacement and dynamic orbital motion, creating the most computationally demanding scenario. The eccentricity severity is specified as a percentage of nominal air gap length, typically ranging from 10% (mild) to 50% (severe) in simulation studies.

5.3.2.2 Governing Equations and Solution Approach

The TSFEM formulation couples the two-dimensional magnetic field equation with external circuit equations representing the three-phase voltage supply. The general mathematical framework for this coupled field-circuit approach is presented in Chapter 3. For eccentricity analysis, the key adaptation is that the nonuniform air gap geometry must be incorporated into the mesh at each time step, with the permeability distribution $\mu(x, y)$ reflecting the position-dependent air gap length [3].

The time-stepping solution provides the magnetic vector potential distribution $A(x, y, t)$ at each rotor position, from which all electromagnetic quantities relevant to eccentricity diagnosis can be derived: magnetic flux density ($\vec{B} = \nabla \times \vec{A}$), back-EMF, stator current, and electromagnetic torque (computed using the Maxwell stress tensor method). These quantities are analyzed in both time and frequency domains to extract the characteristic eccentricity fault signatures.

5.3.2.3 Motor Specifications

To demonstrate FEM analysis of eccentric PMSMs, two three-phase motors with different pole numbers and winding configurations are modeled. Table 5.1 presents the key specifications of both machines. Both motors operate at 50 Hz supply frequency with 36 stator slots, but differ in pole count and air gap length. The 4-pole motor features a concentrated winding with nominal air gap of 0.5 mm, while the 8-pole motor employs a distributed winding with a larger air gap of 0.8 mm. Both motors use Nd–Fe–B PMs with high remanent flux density. Notably, the relative sensitivity to eccentricity differs between these designs: for a given absolute rotor displacement, the smaller air gap motor experiences higher percentage eccentricity and consequently more pronounced fault signatures [3].

Table 5.1 Key specifications of the 4-pole and 8-pole PMSMs.

Parameter	4-Pole PMSM	8-Pole PMSM
Number of poles	4	8
Number of stator slots	36	36
Air gap length (mm)	0.5	0.8
Winding type	Single-layer concentrated	Double-layer nonconcentrated
Remanent flux density (T)	1.21	1.19
Rated voltage (V)	220	200
Rated frequency (Hz)	50	50
Magnet material	Nd–Fe–B	Nd–Fe–B

Source: Adapted from [3].

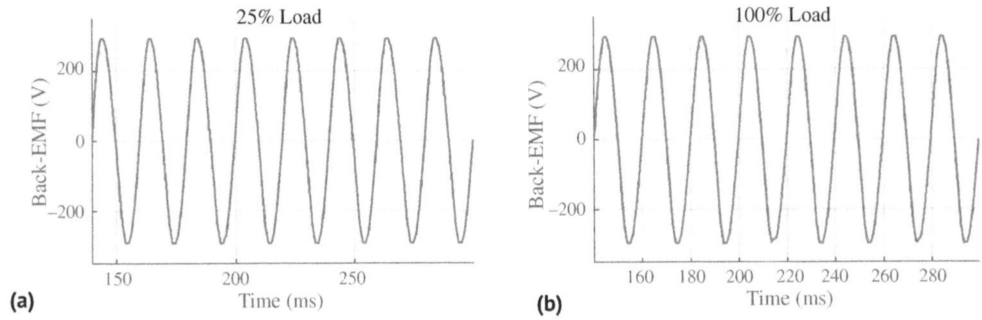

Figure 5.7 Back-EMF in healthy 4-pole PMSM showing load independence: (a) 25% rated load producing peak EMF of approximately 290 V, and (b) 100% rated load producing identical EMF waveform, confirming that PM flux linkage is unaffected by armature reaction in steady-state operation.

5.3.2.4 Baseline Healthy Motor Characteristics

Before analyzing eccentricity effects, healthy motor simulations establish baseline characteristics relevant to fault diagnosis. The back-EMF waveform is sinusoidal with peak amplitude of approximately 290 V for the 4-pole motor, independent of load level (Figure 5.7). The stator current increases proportionally with load while maintaining sinusoidal character with minimal harmonic distortion (Figure 5.8). These healthy baselines provide the reference against which eccentricity-induced changes are identified in subsequent sections [3].

5.3.2.5 Eccentricity Fault Simulation Setup

Both static and dynamic eccentricity conditions are introduced into the FEM model following the geometric definitions established in Section 5.2 (see Figure 5.1). Three severity levels are simulated: 10%, 30%, and 50% of nominal air gap. For the 4-pole motor with 0.5 mm nominal air gap, 50% eccentricity creates an air gap ranging from 0.25 mm (minimum) to 0.75 mm (maximum), representing a severe but physically realizable fault condition [3].

5.3.2.6 Impact of Eccentricity on Back-EMF

A fundamental characteristic of eccentricity faults in IPM machines is their minimal effect on back-EMF waveform, distinguishing them from faults that affect the magnetic excitation such as

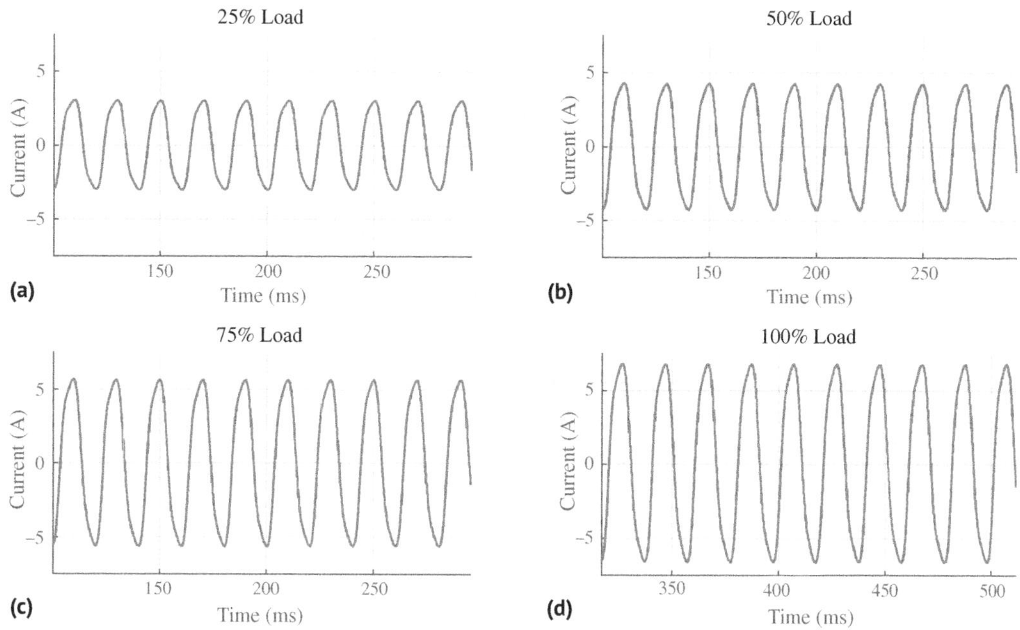

Figure 5.8 Stator current of 4-pole healthy PMSM under varying load conditions: (a) 25% rated load showing low amplitude with minimal distortion, (b) 50% rated load with moderate current amplitude, (c) 75% rated load approaching rated current, and (d) 100% rated load showing maximum current with slight harmonic content from spatial field harmonics.

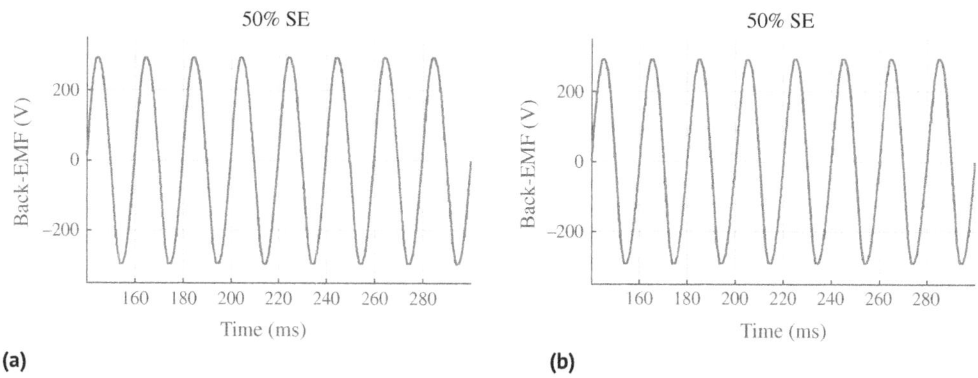

Figure 5.9 Back-EMF of 4-pole PMSM at rated load with severe eccentricity showing minimal deviation from healthy condition: (a) 50% static eccentricity producing nearly identical EMF waveform to healthy motor, and (b) 50% dynamic eccentricity also showing no significant effect on back-EMF amplitude or shape.

demagnetization. Figure 5.9 presents the back-EMF of the 4-pole PMSM operating at rated load with 50% static eccentricity (Figure 5.9a) and 50% dynamic eccentricity (Figure 5.9b). Comparison with the healthy motor back-EMF (Figure 5.7) reveals that even severe eccentricity produces negligible change in either the amplitude or waveshape of the back-EMF. The peak back-EMF remains at approximately at 290 V for both static and dynamic eccentricity conditions, identical to the healthy case [3].

This insensitivity of back-EMF to eccentricity can be explained by examining the flux linkage mechanism. The back-EMF is produced by the time rate of change of flux linkage from PMs to stator windings. While eccentricity creates spatial nonuniform distribution of air gap flux density distribution, the total flux linking a multi-turn distributed winding remains nearly constant because increased flux in the minimum air gap region is approximately compensated by decreased flux in the maximum air gap region. For surface-mounted PM (SPM) machines, the magnets act as constant flux sources with permeability close to air, so air gap length variations primarily redistribute the fixed magnet flux rather than significantly changing its total magnitude. This compensation effect makes back-EMF monitoring ineffective for eccentricity detection, in contrast to demagnetization faults where back-EMF reduction provides a clear diagnostic signature [3].

5.3.2.7 Stator Current Under Eccentricity Conditions

While back-EMF remains largely unaffected by eccentricity, the stator current exhibits clear distortion and asymmetry that serves as the primary electrical signature for eccentricity fault detection. The current distortion arises from the position-dependent inductance variations discussed in the analytical modeling section, combined with the interaction between armature reaction field and the asymmetric air gap permeance distribution. Figure 5.10 presents stator current waveforms for the 4-pole PMSM with 50% static and dynamic eccentricity at four different load conditions, demonstrating the progressive severity of current distortion as both eccentricity and loading increase [3].

For 50% static eccentricity at 25% rated load (Figure 5.10a), the current waveform shows visible asymmetry and ripple superimposed on the fundamental sinusoidal component. This distortion becomes more pronounced at 75% rated load (Figure 5.10b), where the increased armature reaction field interacts more strongly with the eccentric air gap geometry. The ripple content is most visible at lighter loads on a relative basis, though the absolute magnitude of harmonic current increases with load due to higher fundamental current and greater magnetic saturation effects. Static eccentricity creates current harmonics at specific frequencies related to slotting effects and the stationary air gap asymmetry [3].

For 50% dynamic eccentricity at 50% rated load (Figure 5.10c) and rated load (Figure 5.10d), the current waveforms exhibit even more pronounced distortion than equivalent static eccentricity cases. The dynamic eccentricity creates time-varying inductance that modulates the current at mechanical rotational frequency, producing characteristic sideband harmonics around the fundamental supply frequency. The asymmetry is clearly visible in the current peaks and zero crossings, with successive half-cycles showing different amplitudes and shapes. At rated load with 50% dynamic eccentricity (Figure 5.10d), the current distortion is substantial, with harmonic content that would be easily detectable through frequency spectrum analysis [3].

Comparison of FEM-predicted current waveforms with analytical results reveals that FEM produces more harmonics due to inclusion of spatial harmonics, slotting effects, and magnetic saturation that are typically neglected in simplified analytical models. The analytical approach tends to predict smoother waveforms because it assumes linear magnetic properties and may simplify or omit slot-related permeance variations. However, both methods agree on the fundamental characteristic that eccentricity creates asymmetry and harmonic distortion in stator current while leaving back-EMF essentially unchanged, validating the physical understanding of eccentricity fault mechanisms [3].

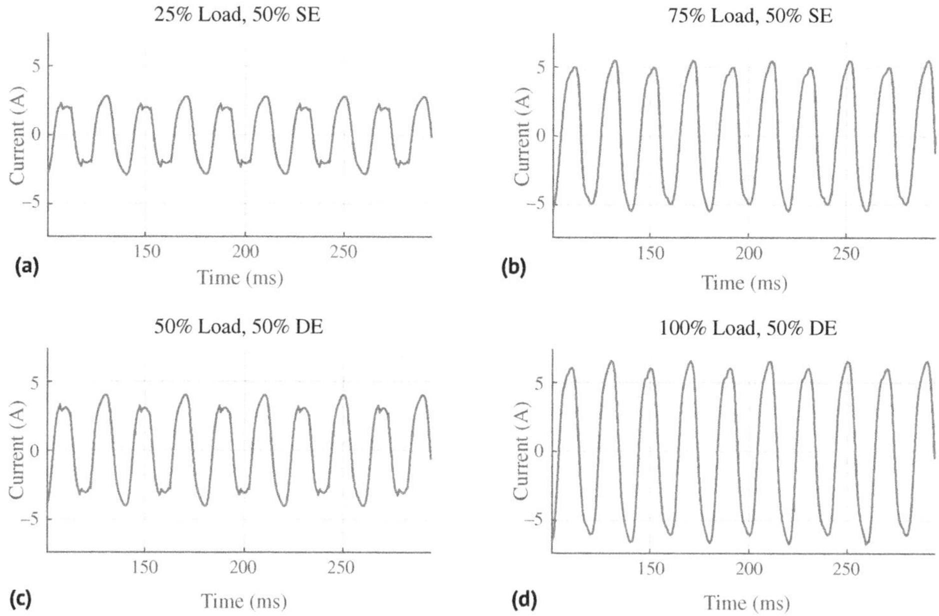

Figure 5.10 Stator current of 4-pole PMSM under 50% eccentricity showing progressive distortion: (a) 25% rated load with static eccentricity exhibiting moderate asymmetry, (b) 75% rated load with static eccentricity showing increased distortion, (c) 50% rated load with dynamic eccentricity demonstrating time-varying modulation, and (d) 100% rated load with dynamic eccentricity showing severe current distortion.

5.3.2.8 Air Gap Flux Density Under Eccentricity

The air gap magnetic flux density distribution provides direct evidence of eccentricity effects and helps explain the mechanisms by which eccentricity produces its characteristic electromagnetic signatures. Figure 5.11 presents the circumferential distribution of radial air gap flux density for the 4-pole PMSM with 50% static eccentricity (Figure 5.11a) and 50% dynamic eccentricity (Figure 5.11b). Surprisingly, comparison with the healthy motor flux density reveals relatively minor changes in the flux density magnitude and distribution pattern. The flux density maintains its fundamental spatial harmonic pattern with slot-induced modulation, and the peak flux density values do not increase substantially despite the 50% reduction in minimum air gap [3].

This apparently counterintuitive result can be explained by examining the source of air gap flux in IPM machines and the effect of magnetic saturation. The PMs act as approximately constant flux sources, and in surface-mounted PM (SPM) designs, the magnet permeability is close to that of air. When eccentricity reduces the air gap locally, the flux density at that position increases, but the increase is limited by saturation in the stator teeth, which cannot sustain arbitrarily high flux density. Additionally, the total magnet flux is conserved, so increased flux density at minimum air gap is partially compensated by decreased density at maximum air gap positions. The net result is a relatively modest change in flux density distribution despite severe geometric asymmetry [3].

This behavior distinguishes IPM machines from wound-field machines where field excitation can be adjusted or from induction machines where rotor currents respond to air gap field changes. In IPM machines, the fixed magnet MMF combined with saturation effects creates a “self-regulating” flux distribution that partially mitigates eccentricity effects on the air gap field. However, while the

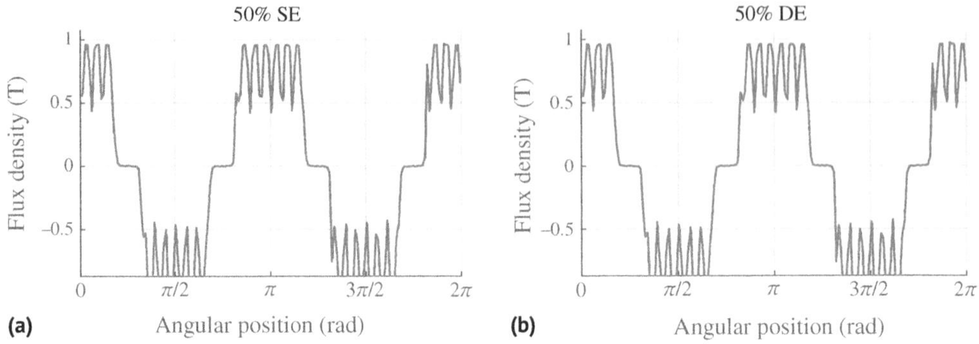

(a) (b)

Figure 5.11 Air gap magnetic flux density distribution under eccentricity showing limited deviation from healthy condition: (a) 50% static eccentricity with slightly elevated flux at minimum air gap position but maintained overall pattern, and (b) 50% dynamic eccentricity showing similar flux distribution characteristics to healthy and static eccentricity cases.

fundamental flux density may change only moderately, the eccentricity introduces important harmonic components at specific spatial and temporal frequencies that serve as diagnostic indicators, as will be explored in detail in the fault diagnosis section of this chapter [3].

5.3.2.9 Electromagnetic Torque Under Eccentricity

Eccentricity significantly affects the electromagnetic torque, introducing ripple components and increasing torque pulsation amplitude even when average torque remains relatively constant. Figure 5.12 presents torque waveforms for the 4-pole PMSM under 50% static and dynamic eccentricity at multiple load conditions. For 25% rated load with static eccentricity (Figure 5.12a), the torque ripple amplitude increases noticeably compared to the healthy condition, with pulsations arising from the interaction between rotor MMF and stationary air gap asymmetry. At 75% rated load with static eccentricity (Figure 5.12b), both average torque and ripple amplitude increase, with the ripple exhibiting complex frequency content related to pole number, slot number, and eccentricity geometry [3].

For dynamic eccentricity at 50% rated load (Figure 5.12c) and rated load (Figure 5.12d), the torque pulsations are even more pronounced, with ripple frequencies related to rotational speed creating potential for mechanical resonance excitation. The rotating air gap asymmetry produces time-varying electromagnetic forces and torque components at mechanical frequency and its harmonics, which can couple with torsional natural frequencies of the shaft system. At rated load with 50% dynamic eccentricity (Figure 5.12d), the peak-to-peak torque ripple represents a substantial fraction of average torque, creating significant vibration and acoustic noise concerns [3].

The torque ripple amplitude is generally higher at larger loads not just in absolute terms but often as a percentage of average torque, contrary to the trend observed in healthy motors. This occurs because eccentricity-induced magnetic asymmetries interact with load-dependent armature reaction fields, and magnetic saturation in the minimum air gap region creates nonlinear electromagnetic coupling that amplifies harmonic torque components. The FEM results predict greater torque ripple than analytical methods, particularly for severe eccentricity, because FEM captures local saturation effects, detailed slot geometry interactions, and higher-order spatial harmonics that analytical models typically neglect or approximate [3].

The increased torque pulsation has important practical consequences beyond diagnostic signatures. Torque ripple creates shaft stress, accelerates gearbox wear in coupled systems, generates

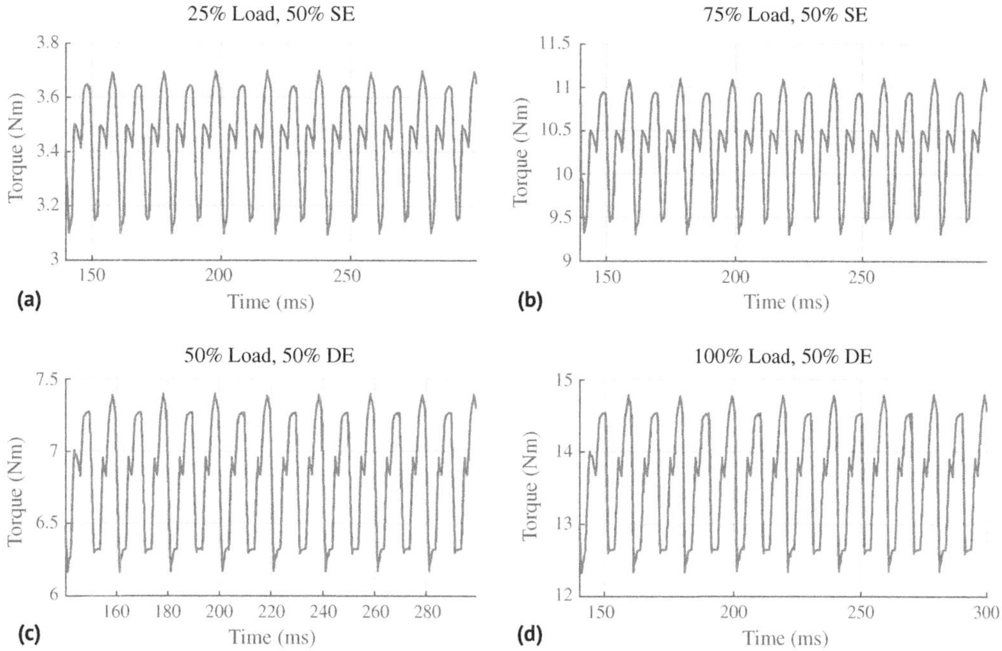

Figure 5.12 Electromagnetic torque of 4-pole PMSM under 50% eccentricity showing increased ripple: (a) 25% rated load with static eccentricity exhibiting elevated torque pulsations, (b) 75% rated load with static eccentricity showing increased average torque and ripple, (c) 50% rated load with dynamic eccentricity demonstrating time-varying torque modulation, and (d) 100% rated load with dynamic eccentricity showing maximum torque with substantial ripple content.

vibration that can excite structural resonances, and produces an acoustic noise that may be unacceptable in quiet environments. These mechanical consequences can make eccentricity detection important not only for preventing catastrophic rotor-stator contact but also for maintaining acceptable noise, vibration, and harshness (NVH) performance throughout the motor's service life [3].

5.3.2.10 Comparison of Analytical and FEM Results

Comparison of analytical and FEM results reveals consistent agreement on fundamental eccentricity characteristics: both methods predict unchanged back-EMF under eccentricity, increasing stator current distortion with fault severity, and characteristic harmonic frequency patterns in current spectra. The key differences lie in harmonic amplitudes, where FEM predicts higher values due to detailed slot geometry and saturation effects not captured in analytical models. FEM also shows baseline sideband harmonics in healthy motors from slotting and saturation, whereas analytical models predict these only under fault conditions [3].

These differences reflect the complementary roles of each approach: analytical models provide computational efficiency for real-time monitoring and physical insight into fault mechanisms, while FEM provides accuracy for detailed performance prediction and validation. Experimental measurements, presented in the fault diagnosis section, confirm the key predictions from both methods and typically fall between their amplitude predictions [3].

5.3.3 Unbalanced Magnetic Forces Modeling

UMP represents the net radial electromagnetic force acting on the rotor due to air gap asymmetry. This force creates additional mechanical loading on bearings, accelerates bearing wear, and can establish a self-reinforcing feedback mechanism where UMP-induced bearing degradation increases eccentricity, which in turn increases UMP [4].

5.3.3.1 Physical Origin of UMP

In a perfectly symmetric machine, the radial magnetic forces distributed around the air gap periphery sum to zero due to symmetry. Eccentricity breaks this symmetry by creating nonuniform air gap length, leading to nonuniform flux density distribution and consequently nonuniform magnetic stress distribution [4]. The Maxwell stress (magnetic pressure) at any point in the air gap is given by:

$$\sigma = \frac{B_n^2 - B_t^2}{2\mu_0} \tag{5.23}$$

where B_n is the normal (radial) flux density component and B_t is the tangential component. The total radial force on the rotor can be calculated by integrating the Maxwell stress over the rotor surface [5]:

$$\vec{F}_{\text{UMP}} = \int_0^{2\pi} \frac{B_n^2 - B_t^2}{2\mu_0} r l_m \, d\theta \, \hat{r}(\theta) \tag{5.24}$$

where r is the rotor radius, l_m is the axial length, and $\hat{r}(\theta)$ is the unit radial vector at angle θ.

5.3.3.2 UMP Magnitude and Direction

For small eccentricity ratios, analytical approximations show that UMP magnitude is proportional to eccentricity and increases with the square of flux density:

$$F_{\text{UMP}} \propto \varepsilon \times B_{\text{avg}}^2 \times A_{\text{rotor}} \tag{5.25}$$

where ε is the eccentricity ratio, B_{avg} is the average air gap flux density, and A_{rotor} is the rotor surface area. The UMP force acts in the direction of minimum air gap, tending to pull the rotor toward the stator at the closest approach point. This directional characteristic makes UMP a progressive fault mechanism, as the force actively works to increase the existing eccentricity.

The speed dependency of UMP differs between static and dynamic eccentricity. For static eccentricity, the UMP value remains relatively constant with speed (flux density is largely speed-independent in IPM machines), but the force direction remains fixed in the stator frame. For dynamic eccentricity, both the magnitude and direction rotate with the rotor at mechanical frequency, creating a rotating unbalanced load.

5.3.3.3 Relationship Between UMP and Mechanical Unbalance

The total unbalanced force on the rotor combines electromagnetic UMP with mechanical mass unbalance:

$$F_{\text{total}} = (m_u r_u + M_r e)\omega_r^2 + F_{\text{UMP}} \tag{5.26}$$

where m_u is the unbalanced mass, r_u is its distance from the rotation center, M_r is the rotor mass, e is the mass eccentricity, and ω_r is the rotational speed [1]. The mechanical unbalanced force increases with the square of rotational speed, dominating at high speeds, while UMP magnitude is relatively speed-independent in IPM machines, dominating at low speeds.

This speed dependency has important diagnostic implications. At low speeds, large eccentricities may produce significant UMP but modest vibration due to low mechanical unbalance force. At high speeds, even small eccentricities can generate substantial vibration due to the ω_r^2 term, making speed-dependent analysis essential for eccentricity assessment.

5.3.3.4 Impact on Bearing Life Span

UMP creates additional radial loading on bearings beyond the designed static and dynamic loads. This additional load accelerates bearing wear through increased contact stress and fatigue accumulation. While rolling element bearings have limited lifetime due to subsurface-initiated fatigue spalling even under normal loading, UMP substantially reduces this lifetime by increasing the probability of crack initiation and propagation.

The bearing life span reduction depends on the UMP magnitude relative to rated bearing load. For large IPM machines where UMP can represent a significant fraction of bearing capacity, the effect is particularly severe. Empirical studies show that larger shaft diameters experience higher bearing lifespan reduction when UMP is present, because the electromagnetic force scales with rotor surface area while bearing capacity scales less favorably with shaft diameter [1].

The progressive nature of eccentricity faults arises from the coupling between UMP and bearing degradation. Initial eccentricity (perhaps from manufacturing tolerances or assembly errors) generates UMP, which accelerates bearing wear, increasing eccentricity and consequently UMP magnitude. This positive feedback loop can lead to rapidly escalating fault severity, making early detection of incipient eccentricity critical for preventing catastrophic failure.

5.3.3.5 FEM-based Force Calculation

The electromagnetic forces calculated from FEM simulations provide accurate prediction of UMP including nonlinear effects such as local saturation [5]. The Maxwell stress tensor method integrates stress components over a closed contour in the air gap. For the radial force component:

$$F_r = \int_0^{2\pi} \frac{B_r^2 - B_\theta^2}{2\mu_0} r l_m \cos(\theta)\, d\theta \tag{5.27}$$

$$F_\theta = \int_0^{2\pi} \frac{B_r^2 - B_\theta^2}{2\mu_0} r l_m \sin(\theta)\, d\theta \tag{5.28}$$

where B_r and B_θ are the radial and tangential flux density components from the FEM solution. The resultant force magnitude and direction are as follows:

$$|\vec{F}_{\text{UMP}}| = \sqrt{F_r^2 + F_\theta^2} \tag{5.29}$$

$$\alpha = \arctan\left(\frac{F_\theta}{F_r}\right) \tag{5.30}$$

FEM analysis reveals that UMP exhibits pulsating components superimposed on a DC (constant) value, with pulsation frequencies related to pole number, slot number, and rotational speed. These pulsating force components contribute to vibration signatures and can excite structural resonances if their frequencies align with mechanical natural frequencies of the rotor-bearing-stator system.

5.4 Fault Diagnosis

5.4.1 MCSA Application for Eccentricity Diagnosis

MCSA, introduced in Chapter 4, detects eccentricity through characteristic sideband components in the stator current spectrum. The frequency patterns derived in Section 5.3.1 (see Eqs. (5.18) and (5.19)) provide the theoretical foundation for identifying eccentricity faults.

Even in healthy machines, manufacturing tolerances, magnetic saturation, and spatial harmonic content produce low-level sidebands at these characteristic frequencies. The diagnostic challenge therefore lies in establishing threshold amplitudes that reliably distinguish between acceptable manufacturing variations and developing fault conditions.

5.4.1.1 Static Eccentricity Detection Through Current Spectrum Analysis

Experimental and simulation studies have consistently demonstrated that static eccentricity produces measurable increases in the amplitude of sideband components at the characteristic frequencies. For a 4-pole PMSM operating at 50 Hz supply frequency, the most prominent sideband components appear at 25, 75, and 125 Hz, corresponding to the frequency pattern $f_s[(2\chi{-}1) \pm \eta/p]$ with specific values of η. This phenomenon is illustrated in Figure 5.13, which shows the stator current spectrum for healthy and faulty conditions with varying eccentricity severity levels.

FE simulations validated through experimental measurements reveal that the relationship between eccentricity severity and sideband amplitude is nonlinear. For static eccentricity severities up to approximately 30% of the nominal air gap, the sideband amplitudes increase monotonically with fault severity. However, at higher eccentricity severity (approaching 50%), the rate of increase

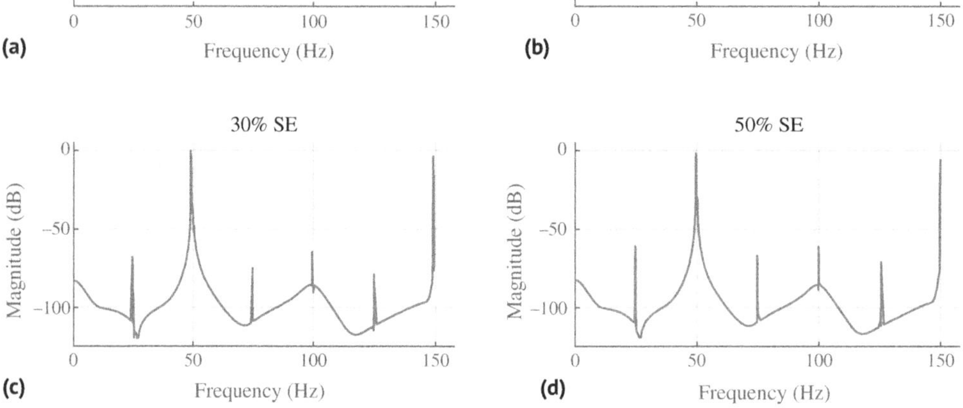

Figure 5.13 Stator current spectrum of 4-pole PMSM at: (a) healthy, (b) 10% static eccentricity, (c) 30% static eccentricity, (d) 50% static eccentricity.

diminishes due to local magnetic saturation effects. This saturation occurs because severe eccentricity brings the rotor surface extremely close to the stator bore at the point of minimum air gap, pushing the magnetic flux density into the saturation region of the iron core material.

The comparison between analytical predictions and FEM results provides important insights into the limitations and applicability of different modeling approaches. Analytical methods, which typically neglect magnetic saturation, tend to predict higher sideband amplitudes than observed in FEM simulations and experiments. This discrepancy becomes more pronounced at higher eccentricity severities where saturation effects are significant. Nevertheless, analytical models correctly predict the frequency locations and relative trends, making them valuable for preliminary analysis and diagnostic algorithm development.

For 8-pole machines, the characteristic frequency pattern remains consistent with theoretical predictions, with prominent sidebands appearing at 12.5, 25, 37.5, 62.5, 75, 87.5, 112.5, 125, and 137.5 Hz. This demonstrates that the fundamental frequency relationships scale appropriately with motor pole count, providing a consistent diagnostic framework across different machine designs. The validation of these frequency patterns through both FEM simulations and experimental measurements on physical prototypes establishes confidence in using MCSA for static eccentricity detection across a range of motor sizes and pole configurations, as shown in Figure 5.14.

5.4.1.2 Dynamic Eccentricity Fault Identification

Dynamic eccentricity, characterized by the rotation of the eccentricity position with the rotor, produces sideband components at the same characteristic frequencies as static eccentricity. However, the physical mechanism differs: dynamic eccentricity creates time-varying permeance that

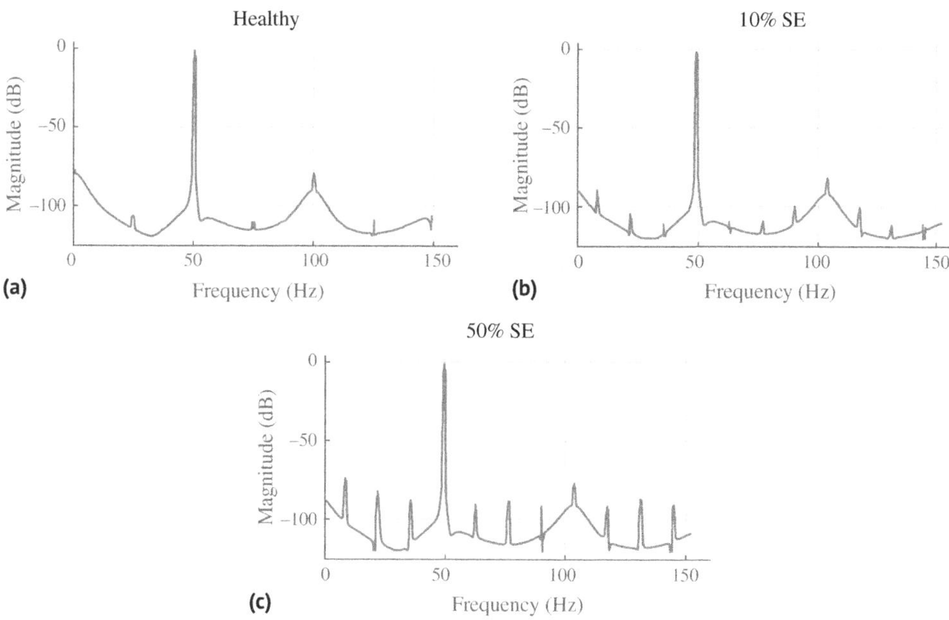

Figure 5.14 Stator current spectrum of 8-pole PMSM under static eccentricity: (a) healthy showing baseline harmonic content, (b) 10% static eccentricity with emerging sideband components, and (c) 50% static eccentricity showing pronounced sidebands at characteristic frequencies.

modulates the air gap flux density as the rotor rotates. This modulation produces current sidebands with amplitudes that depend on both the eccentricity severity and the rotational frequency.

Spectral analysis of machines with dynamic eccentricity reveals that 10% dynamic eccentricity produces measurable increases in sideband amplitudes at the characteristic frequencies. As with static eccentricity, the amplitudes increase with fault severity, providing a means to assess not only the presence but also the magnitude of the eccentricity. For both 4-pole and 8-pole machines, the frequency patterns match theoretical predictions, confirming that the analytical framework developed for static eccentricity extends to dynamic eccentricity conditions. Figure 5.15 illustrates the current spectrum characteristics for 4-pole machines under dynamic eccentricity conditions, while Figure 5.16 presents comparable results for 8-pole machines.

An interesting observation from comparative studies is that static eccentricity generally produces higher sideband amplitudes than dynamic eccentricity of equivalent magnitude. This difference arises from the different spatial distributions of the air gap asymmetry: static eccentricity creates a fixed, localized region of reduced air gap that continuously affects the flux distribution, while dynamic eccentricity distributes the perturbation around the entire air gap circumference over a complete rotation cycle. From a diagnostic perspective, this means that static eccentricity may be detected at earlier stages or lower severity levels than dynamic eccentricity using amplitude-based thresholds.

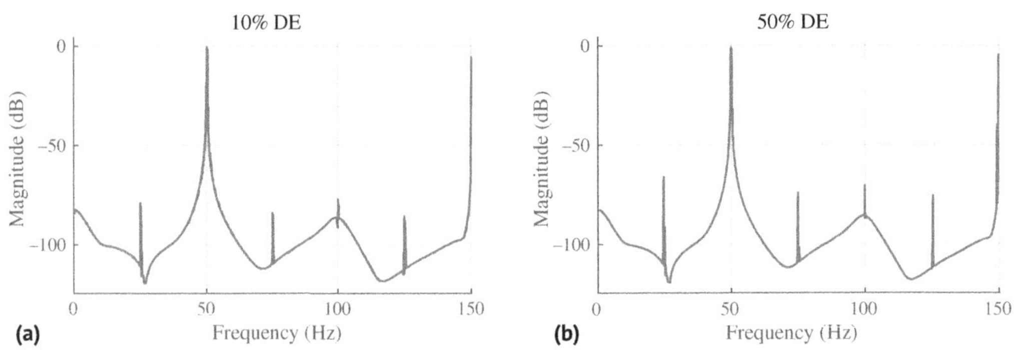

Figure 5.15 Stator current spectrum of 4-pole PMSM with dynamic eccentricity: (a) 10% dynamic eccentricity showing emerging sideband components, and (b) 50% dynamic eccentricity showing pronounced sidebands at characteristic frequencies.

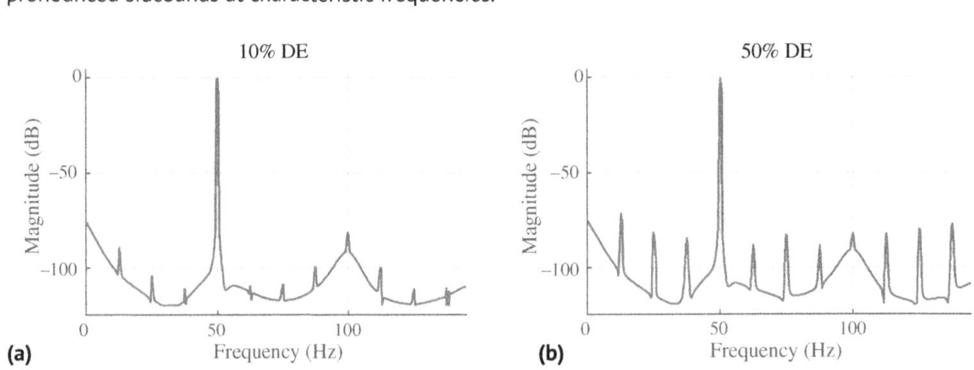

Figure 5.16 Stator current spectrum of 8-pole PMSM with dynamic eccentricity: (a) 10% dynamic eccentricity showing emerging sideband components, and (b) 50% dynamic eccentricity showing pronounced sidebands at characteristic frequencies.

5.4.1.3 Severity Assessment and Load Independence

A critical requirement for practical fault diagnosis is the ability to quantify fault severity and to distinguish between fault-induced effects and operational variations such as load changes. Studies examining the influence of load variations on eccentricity-related sideband amplitudes have demonstrated a valuable property: the amplitude of sideband components at the characteristic eccentricity frequencies remains substantially independent of load level.

This load independence represents a significant advantage for MCSA-based eccentricity detection. Unlike some other current spectrum features that vary with torque loading, the eccentricity sidebands maintain relatively constant amplitudes across the operating range. This characteristic enables the establishment of fixed threshold values for fault detection that do not require load-dependent calibration. The physical basis for this behavior lies in the geometric nature of eccentricity: the air gap asymmetry that produces the characteristic frequency components is independent of the electromagnetic loading of the machine. Figure 5.17 demonstrates this load-independent behavior for both static and dynamic eccentricity in an 8-pole PMSM, showing that sideband amplitude variations with eccentricity severity are substantially larger than variations due to load changes.

The relationship between sideband amplitude and eccentricity severity provides a quantitative measure of fault progression. By monitoring the trend of sideband amplitudes over time, incipient faults can be detected before they progress to severe levels that risk catastrophic failure. However, the nonlinear relationship between amplitude and severity, particularly at higher eccentricity fault severities, requires careful calibration and may benefit from machine-specific characterization through FEM simulations or baseline measurements.

The extensive validation through FEM simulations and experiments on both 4-pole and 8-pole machines with various eccentricity fault severities (10%, 30%, and 50%) provides strong evidence for the effectiveness of MCSA as an eccentricity diagnostic tool. The consistency between theoretical predictions, simulation results, and experimental observations, particularly regarding frequency

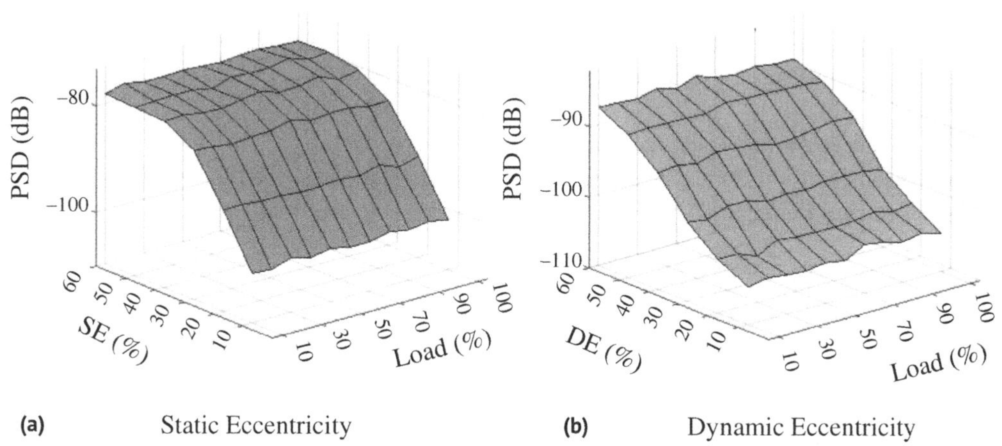

Figure 5.17 Load independence of eccentricity fault signatures showing PSD variation of characteristic sideband components: (a) static eccentricity exhibiting rapid amplitude increase at low severity levels before saturation, and (b) dynamic eccentricity showing gradual amplitude increase with fault severity. Both cases demonstrate that sideband amplitudes remain relatively constant across load variations (10–110%) while varying significantly with fault severity (10–60%).

patterns, establishes MCSA as a reliable method for eccentricity detection in PMSM. For general MCSA implementation considerations including frequency resolution, signal conditioning, and data acquisition requirements, refer to Chapter 4.

5.4.2 Vibration Analysis for Eccentricity Faults

Eccentricity manifests distinct vibration signatures due to the electromagnetic nature of the UMP that drives the vibration response. Building on the vibration monitoring fundamentals in Chapter 4, this section focuses on eccentricity-specific signatures and detection approaches.

5.4.2.1 Physical Mechanisms of Eccentricity-induced Vibration

Eccentricity generates vibration through UMP excitation, as analyzed in Section 5.3.3. Static eccentricity produces predominantly DC radial force combined with harmonics at twice the electrical frequency ($2f_e$), primarily exciting structural modes through slotting harmonic modulation. Dynamic eccentricity creates a rotating UMP component at the mechanical rotation frequency (f_r), producing characteristic synchronous whirl patterns observable in rotor orbit analysis [6]. Mixed eccentricity generates both stationary and rotating force components.

When force components coincide with structural natural frequencies, resonant amplification dramatically increases vibration amplitude even for modest eccentricity severities [7].

5.4.2.2 Characteristic Vibration Frequency Components

The vibration spectrum generated by eccentricity faults in PM machines contains specific frequency components that serve as diagnostic indicators. Understanding these characteristic frequencies enables targeted signal processing and automated fault detection algorithms. While MCSA detects eccentricity through current harmonics (as discussed in the MCSA application section), vibration analysis monitors the mechanical manifestation of the electromagnetic forces that generate those current harmonics.

The electromagnetic force frequencies that drive vibration correspond to the current harmonic frequencies but manifest through Maxwell stress in the air gap. For a PM motor with p pole pairs operating at supply frequency f_s, the mechanical rotational frequency is $f_r = f_s/p$ (in synchronous operation), and the electrical frequency is $f_e = pf_r = f_s$.

5.4.2.2.1 Static Eccentricity Fault Vibration Frequencies

For static eccentricity, the dominant electromagnetic force frequencies are DC (zero frequency) and twice the electrical frequency. The measurable vibration spectrum contains:

$$f_{\text{vib,static}} = 2kf_e = 2kpf_r, k = 1, 2, 3, \ldots \tag{5.31}$$

The DC component creates steady deflection rather than periodic vibration, while the $2f_e$ component and its harmonics produce measurable oscillation. The stator structure responds to these forces with near-DC components related to structural deflection under steady UMP, strong components at $2f_e = 2pf_r$ and higher harmonics $4f_e$, $6f_e$, and sidebands around $2kf_e$ spaced at slot passing frequency $f_{\text{slot}} = Nf_r$ (where N is the number of stator slots). Resonance peaks appear when force harmonics coincide with structural natural frequencies.

5.4.2.2.2 *Dynamic Eccentricity Vibration Frequencies*

Dynamic eccentricity produces a rotating force at the mechanical rotation frequency, creating vibration at:

$$f_{\text{vib,dynamic}} = f_r, 2f_r, 3f_r, \ldots \tag{5.32}$$

Additionally, the interaction between the rotating eccentricity and the electromagnetic field produces force components at sideband frequencies similar to those observed in current analysis (5.19):

$$f_{\text{vib,ecc}} = f_e \pm kf_r = pf_r \pm kf_r = (p \pm k)f_r, k = 1, 2, 3, \ldots \tag{5.33}$$

The most prominent dynamic eccentricity vibration indicators appear at the fundamental rotational frequency f_r, primary sidebands $f_e \pm f_r = (p-1)f_r$ and $(p+1)f_r$, and higher harmonics $2f_r$, $3f_r$ when eccentricity severity increases.

5.4.2.2.3 *Mixed Eccentricity Vibration Frequencies*

Mixed eccentricity generates the superposition of static and dynamic signatures with additional interaction terms:

$$f_{\text{vib,mixed}} = \{2kf_e, kf_r, f_e \pm kf_r\}, k = 1, 2, 3, \ldots \tag{5.34}$$

The presence of both stationary ($2kf_e$) and rotating (kf_r, $f_e \pm kf_r$) force components creates complex sideband patterns. The sum and difference frequency combinations appear as:

$$f_{\text{vib,mixed}} = 2kf_e \pm mf_r, k, m = 1, 2, 3, \ldots \tag{5.35}$$

Distinguishing mixed eccentricity from pure static or dynamic cases requires careful spectral analysis comparing the relative amplitudes of static-characteristic frequencies ($2kf_e$) and dynamic-characteristic frequencies (kf_r, $f_e \pm kf_r$) [8].

5.4.2.2.4 *PM Machines vs. Induction Motors*

In synchronous PM machines, the relationship $f_e = pf_r$ creates harmonic overlaps that complicate spectral interpretation. For example, in a 4-pole PM motor (p = 2), the static eccentricity component at $2f_e = 4f_r$ coincides with the 4th harmonic of rotational frequency. Similarly, dynamic eccentricity sidebands at $f_e \pm f_r$ become $(p \pm 1)f_r$, appearing as f_r and $3f_r$ for a 4-pole machine. This frequency overlap necessitates correlation with current signatures or multi-signal analysis to distinguish eccentricity from pure mechanical imbalance [9].

5.4.2.3 Orbit Analysis for Eccentricity Detection

Orbit analysis, a key technique for eccentricity diagnosis, examines the trajectory traced by the rotor shaft when radial vibration measurements in two orthogonal directions are plotted against each other. For healthy machines, the orbit approximates a circle centered at the static equilibrium position. Dynamic eccentricity manifests as orbit elongation in the direction of whirl, while static eccentricity shifts the orbit center from the geometric center. Mixed conditions create both shift and distortion [6].

Quantitative orbit metrics include major/minor axis lengths, ellipticity ratio, and centerpoint offset. Machine learning classifiers trained on orbit morphology can distinguish eccentricity types with accuracy exceeding 95% [9]. For general vibration signal processing techniques including FFT, envelope analysis, and order tracking, refer to Chapter 4.

5.4.2.4 Spectral Indicators for Eccentricity Detection

For eccentricity detection, spectral analysis focuses on monitoring specific frequency components: the $2f_e$ component and its harmonics for static eccentricity, and the f_r component with sidebands at $f_e \pm f_r$ for dynamic eccentricity. Distinguishing the f_r component from mechanical imbalance requires examining sideband structures or correlating with current signatures, since eccentricity produces characteristic current harmonics while pure mechanical imbalance does not affect electrical signatures [9].

Envelope analysis (see Chapter 4) proves particularly effective for eccentricity when UMP modulates higher-frequency structural resonances. The envelope spectrum should exhibit peaks at DC and $2f_e$ for static eccentricity, or f_r and $f_e \pm f_r$ for dynamic eccentricity [10].

5.4.2.5 Comparison with Current Signature Analysis

Vibration and MCSA provide complementary eccentricity detection capabilities. Vibration monitoring excels at detecting dynamic eccentricity where rotating UMP directly excites mechanical resonances, and provides spatial information through orbit analysis. However, vibration signatures lack specificity for distinguishing eccentricity from imbalance, misalignment, or bearing defects. MCSA offers superior fault discrimination through characteristic frequency patterns unique to eccentricity, requires no additional sensors, and maintains more consistent relationships to fault severity across operating speeds. Multi-signal fusion combining both modalities achieves superior diagnostic performance [9].

5.4.3 Flux-based Detection Methods

While motor current signature analysis and vibration monitoring are widely adopted diagnostic techniques for eccentricity detection, flux-based detection methods offer unique advantages for PM machines. As discussed in Chapter 4, magnetic flux monitoring provides direct observation of the air gap field perturbations caused by eccentricity faults, making these methods particularly effective for IPM motors where the PMs create a strong reference field that is directly affected by air gap variations.

Air gap asymmetries in electrical machines lead to distorted magnetic flux density distributions, resulting in unbalanced magnetic forces on both the rotor and stator. This effect manifests distinctly in the magnetic flux signals. While UMP is typically more pronounced in large electric machines, small IPM machines can also experience significant flux perturbations that are detectable through appropriate sensing techniques. Over time, UMP contributes to bearing wear, which further exacerbates air gap asymmetries and increases UMP in a continuous cycle, making early detection through flux monitoring critically important.

An air gap eccentricity fault in PM machines produces specific patterns in the stator current, air gap flux, and vibration signals [10]. The characteristic frequency signature for eccentricity-induced flux components can be expressed as:

$$f_{\text{ece}} = f_s\left(1 \pm \frac{k}{p}\right) = f_s \pm f_r \tag{5.36}$$

where f_s is the supply frequency, f_r is the rotor mechanical frequency, k is a positive integer, and p is the number of pole pairs. This frequency pattern appears in both current and flux signatures, but flux measurements often provide enhanced sensitivity and clarity for eccentricity detection.

5.4.3.1 Built-in Hall Effect Flux Sensor Applications

One practical approach for eccentricity detection leverages the built-in Hall effect flux sensors (HEFSs) already present in many PMSMs for control purposes. These sensors, typically three in number positioned with 120° spatial separation and attached to the motor end cap, can serve dual purposes for both machine control and fault diagnosis [11]. Mechanical rotor faults such as eccentricity generate distinct patterns in the axial flux measurements recorded by the HEFSs. However, these measurements can be distorted by other factors including manufacturing defects and temperature variations, necessitating robust feature extraction methods.

To address measurement sensitivity to external factors, a time-domain feature has been developed that relies on the deviation between the current diagnostic index, m_n, and its value during initial commissioning, m_{n0}. The deviation indicator is defined as:

$$e = m_n - m_{n0} \tag{5.37}$$

The index m_n comprises the normalized RMS values of the positive (subscript N) and negative (subscript S) half-cycles of the flux measurements relative to their average value, $\widehat{M}$, and is expressed as:

$$m_n = [M_{S1}\ M_{N1}\ M_{S2}\ M_{N2}\ M_{S3}\ M_{N3}]/\widehat{M} \tag{5.38}$$

This method is applicable only to IPM motors equipped with analog HEFSs. In brushless direct current (BLDC) motors with binary Hall sensors, the sensors must be replaced with analog counterparts for this diagnostic approach to be implemented. The technique provides baseline-referenced detection that compensates for manufacturing variations and initial misalignment, improving specificity to progressive eccentricity development during operation.

5.4.3.2 Search Coil-based Detection in Axial Flux PM Machines

For axial flux PMSM (AFPMSM), a specialized search coil approach has demonstrated effectiveness for static eccentricity detection [12]. This method employs three search coils positioned 120° apart around the stator teeth. A time-space representation is formed by treating these measurements as three stationary vectors, each separated by 120°, with the RMS values of the induced voltages in the respective search coils serving as the vector magnitudes.

The diagnostic principle relies on vector summation. The sum of these three vectors serves as a fault indicator. In a healthy machine with uniform air gap, this vector sum is theoretically zero due to the balanced three-phase symmetry. However, in the presence of air gap eccentricity faults, the magnitudes of vectors increase when the search coils encounter a reduced air gap region and decrease in regions with enlarged air gap. The resulting fault indicator vector exhibits two key diagnostic features:

1. **Direction:** The indicator vector points toward the location of the minimum air gap, providing spatial fault localization
2. **Magnitude:** The amplitude of the indicator vector quantifies the severity of the eccentricity fault

This approach provides both detection and localization capabilities using a minimal sensor configuration, making it particularly attractive for retrofit applications and condition monitoring systems. The search coil method is noninvasive when the coils are mounted on the stator surface and provides real-time diagnostic capability during normal motor operation.

5.4.3.3 Comparison of Flux and Current Signatures

Flux measurements directly observe air gap field perturbations, providing high sensitivity particularly for static eccentricity. In machines with symmetrical winding distributions, certain harmonic components may cancel in phase currents while remaining detectable in flux measurements. Flux monitoring is unaffected by winding connection (Y or delta) and demonstrates effectiveness independent of load and speed [13].

However, flux measurements require additional sensors positioned within the motor structure, increasing implementation complexity. Many modern IPM motors already incorporate Hall sensors for commutation, enabling dual-use diagnostic implementation with minimal additional cost. For PM machines, flux-based detection directly monitors the magnetic field produced by the PMs, providing both physical insight and diagnostic sensitivity that complement electrical monitoring.

5.4.4 EPVA Application for Eccentricity Diagnosis

EPVA, introduced in Chapter 4, detects eccentricity through pattern analysis in the dq plane. The Park transformation converts three-phase stator currents into the dq reference frame, where healthy machines trace circular trajectories. Eccentricity-induced harmonics (see (5.19)) distort this circular pattern into characteristic geometric shapes [14].

5.4.4.1 Eccentricity Signature Patterns

Static eccentricity causes elliptical distortion with the major axis orientation fixed in the stationary reference frame. The ellipticity ratio (major to minor axis) quantifies severity. Dynamic eccentricity produces pulsating ellipticity with the major axis rotating synchronously with rotor position, distinguishing it from static eccentricity [2]. Mixed eccentricity generates complex patterns combining fixed and rotating elliptical components.

5.4.4.2 Diagnostic Indicators

Key quantitative indicators for EPVA-based eccentricity detection include: (1) Park vector modulus spectral content at characteristic frequencies, where amplitudes exceeding 5% of fundamental typically indicate detectable faults [15]; (2) trajectory ellipticity ratio, where values below 0.95 suggest incipient eccentricity; and (3) total harmonic distortion exceeding 10–15%.

5.4.4.3 Challenges in PMSM Applications

EPVA sensitivity deteriorates at low speeds (below 20% rated) due to reduced signal-to-noise ratio. In closed-loop vector-controlled drives, the current controller may suppress eccentricity-induced harmonics, reducing diagnostic sensitivity [16]. Additionally, demagnetization and winding asymmetries can produce similar Park vector distortions, requiring correlation with other signals for reliable discrimination [17].

5.4.4.4 Advanced EPVA Variants

Several extensions address conventional EPVA limitations. Inverse transient complex inductance vector theory injects specific voltage sequences to analyze inductance vector trajectories, demonstrating superior discrimination between eccentricity and demagnetization at low speeds [17]. Stray flux vector space decomposition extends Park vector concepts to external flux measurements, providing reliable detection even when closed-loop control maintains sinusoidal currents [16]. Multimodal approaches combining current Park vectors with vibration time-frequency representations enhance cross-domain fault diagnosis [18].

5.4.5 Data-driven and Machine Learning Classification Approaches

Building on the ML framework established in Chapter 4, this section focuses on eccentricity-specific applications. Machine learning addresses key challenges in eccentricity diagnosis: signature overlap with other faults, sensitivity variations with operating conditions, and subtle early-stage changes obscured by noise.

5.4.5.1 Eccentricity-specific Features

For eccentricity classification, the most discriminative features include amplitudes at characteristic frequencies ($f_s \pm f_r, 2f_e$), park vector ellipticity ratio and trajectory area, orbit analysis parameters (major/minor axis lengths, centroid offset), and harmonic amplitude ratios [9, 19]. Multi-modal feature fusion combining current, vibration, and flux features achieves classification accuracies exceeding 95% [20]. Random forest feature importance ranking identifies the most discriminative eccentricity indicators, guiding sensor selection for practical implementations [21].

5.4.5.2 Deep Learning for Eccentricity Diagnosis

CNNs applied to time-frequency representations have achieved accuracies exceeding 98% for eccentricity classification [22]. A particularly effective approach combines CNNs with FEM simulation data for training augmentation, addressing the challenge of obtaining comprehensive fault datasets [23]. Domain adaptation enables transfer of detection models across motor designs through fine-tuning on limited target data [24]. LSTM-based models capture the gradual progression patterns characteristic of eccentricity faults, while autoencoders provide unsupervised anomaly detection when labeled fault data are scarce.

5.4.5.3 Advanced Techniques

Variational mode decomposition (VMD) with deep learning achieves 98.6% accuracy for simultaneous demagnetization and eccentricity detection by decomposing signals into mode-specific components before CNN classification [25]. Fuzzy neural networks provide interpretable diagnostic systems with linguistic severity descriptors, while PSO optimization reduces training data requirements [26]. Attention mechanisms and Grad-CAM interpretability techniques validate that networks learn physically meaningful eccentricity patterns.

5.4.5.4 Implementation Considerations

Physical eccentricity fault induction for training data collection is challenging, requiring motor disassembly and modified components. FEM simulation provides synthetic training data covering various severities and operating conditions [23]. Physics-informed ML combining data-driven models with UMP equations enables prognostic applications, forecasting when eccentricity-induced bearing wear will exceed operational thresholds [21].

5.4.6 Experimental Validation Summary

The experimental validation program, utilizing 4-pole and 8-pole PMSMs with controlled eccentricity introduction capabilities, establishes key findings that inform practical implementation of eccentricity diagnostic systems.

5.4.6.1 Key Validation Results

FEM simulations and experimental measurements show strong agreement for characteristic frequency patterns across both static and dynamic eccentricity conditions (Figures 5.14 and 5.16).

FEM predictions match experimental back-EMF within 5% for healthy operation and correctly predict the relative insensitivity of back-EMF to eccentricity up to 50% severity. Frequency patterns of eccentricity-induced harmonics show essentially perfect agreement, though FEM typically predicts slightly higher sideband amplitudes than measured, attributed to manufacturing tolerances and thermal effects in physical test fixtures [3, 27].

A critical finding is the presence of low-amplitude sideband components at characteristic frequencies even in nominally healthy motors, due to unavoidable manufacturing variations and bearing clearances. This implies diagnostic thresholds must be established from statistical characterization of healthy machine populations rather than theoretical zero values.

The load independence of eccentricity signatures (Figure 5.17) is convincingly validated: sideband amplitudes remain substantially constant across 25–100% rated load while load-dependent harmonics vary significantly. This enables threshold-based detection without load measurement or calibration.

Vibration validation confirms characteristic frequency patterns: prominent $2f_e$ components for static eccentricity and strong f_r components for dynamic eccentricity. Vibration shows greater sensitivity to structural resonance effects than current signatures, providing enhanced detection at resonance crossing speeds but complicating direct severity quantification.

5.4.6.2 Synthesis of Findings

The comprehensive validation program establishes several key conclusions:

1. **Physics-based Signatures Are Robust:** Characteristic frequency patterns generalize well across motor designs and operating conditions, providing reliable diagnostic foundations.
2. **Machine-specific Calibration Is Essential:** While frequency patterns generalize, amplitude thresholds require adaptation to specific motors, manufacturing quality, and installation conditions.
3. **Load Independence Simplifies Implementation:** Fixed thresholds can be established without load measurement, reducing system complexity.
4. **Multi-signal Fusion Improves Reliability:** Combining current, vibration, and flux modalities achieves superior detection compared to single-modality approaches, particularly for early fault detection and eccentricity type classification.
5. **Severity Trending Enables Prognostics:** The progressive nature of eccentricity signatures allows remaining useful life estimation and predictive maintenance scheduling.

EPVA validation confirms progressive trajectory distortion with increasing eccentricity severity, with distinct patterns enabling static/dynamic classification. EPVA demonstrates particular value during transient conditions where steady-state spectral methods lose effectiveness. Machine learning cross-validation reveals that multi-motor training datasets achieve significantly better generalization, and physics-based features (characteristic frequency amplitudes) consistently rank among the most informative, validating the complementary value of physics-informed and data-driven approaches.

5.5 Summary

This chapter has presented a treatment of eccentricity faults in permanent magnet machines, encompassing their failure mechanisms, modeling approaches, diagnostic techniques, and experimental validation. Eccentricity faults represent a particularly challenging class of failures due to their coupled electromechanical nature, progressive characteristics, and the self-reinforcing degradation mechanism driven by unbalanced magnetic pull.

The examination of failure modes established the fundamental taxonomy distinguishing static, dynamic, and mixed eccentricity conditions. Static eccentricity, with its stationary air gap profile, generates constant radial forces and characteristic vibration patterns at twice the electrical frequency. Dynamic eccentricity produces time-varying magnetic loading with synchronous vibration components. The root cause analysis revealed that eccentricity originates from manufacturing tolerances, assembly defects, bearing degradation, and shaft deflection. The progressive nature of these faults makes early detection critical, as initial eccentricity generates radial forces that accelerate bearing wear, creating a positive feedback cycle that can rapidly escalate to catastrophic failure.

The modeling and analysis section covered analytical and finite element approaches that provide complementary capabilities: analytical models offer computational efficiency and physical insight into fundamental mechanisms, while FEM delivers accurate quantitative predictions through detailed numerical solutions. Validation studies demonstrated strong agreement between the two approaches. The unbalanced magnetic pull analysis revealed the quadratic dependence of UMP on eccentricity ratio, explaining the rapid progression from mild eccentricity to catastrophic failure at higher fault severities.

The diagnostic techniques exploit different physical manifestations of air gap asymmetry. Motor current signature analysis emerged as the most widely implemented approach, with characteristic frequency patterns providing robust detection signatures that generalize across PM machine designs. The load independence of eccentricity sidebands substantially simplifies practical implementation. Vibration analysis provides complementary mechanical measurements with distinct frequency patterns, enabling fault type classification. Flux-based detection methods offer advantages for specific machine types, while Extended Park's Vector Approach provides valuable fault detection during transient conditions. Data-driven approaches leveraging machine learning demonstrate improved classification accuracy, with physics-informed feature selection consistently outperforming purely data-driven methods.

Experimental validation confirmed that physics-based diagnostic signatures are robust and reliable, while quantitative discrepancies highlight the importance of machine-specific baseline establishment and threshold calibration. The complementary strengths of different diagnostic modalities emphasize the value of multi-signal fusion, and the progression of diagnostic signatures with fault severity enables quantitative severity assessment and prognostic trending. The integration of physics-based modeling, multi-signal diagnostic techniques, and data-driven classification methods provides the foundation for effective condition monitoring systems that support the transition from reactive to predictive maintenance in critical applications.

References

1 Wei Tong. *Mechanical Design of Electric Motors*. CRC Press, 2014. ISBN 9780429145025. https://doi.org/10.1201/b16863.

2 Bashir M. Ebrahimi, Jawad Faiz, and Mehrsan J. Roshtkhari. Static-, dynamic-, and mixed-eccentricity fault diagnoses in permanent-magnet synchronous motors. *IEEE Transactions on Industrial Electronics*, 56(11):4727–4739, 2009. https://doi.org/10.1109/TIE.2009.2029577.

3 Bashir M. Ebrahimi, Jawad Faiz, M. Javan-Roshtkhari, and A. Zargham Nejhad. Static eccentricity fault diagnosis in permanent magnet synchronous motor using time stepping finite element method. *IEEE Transactions on Magnetics*, 44(11):4297–4300, November 2008. ISSN 0018-9464. https://doi.org/10.1109/TMAG.2008.2001534.

4 Xueping Xu, Qinkai Han, and Fulei Chu. Review of electromagnetic vibration in electrical machines. *Energies*, 11(7):1779, July 2018. ISSN 1996-1073. https://doi.org/10.3390/en11071779.

5 Hang Yin, Hengliang Zhang, Wei Hua, Zheng Wu, and Chen Li. Quantitative analysis of electromagnetic forces by decoupling air-gap field modulation and force modulation in rotor-permanent-magnet machines. *IEEE Transactions on Industrial Electronics*, 70(2):1310–1320, February 2023. ISSN 0278-0046. https://doi.org/10.1109/TIE.2022.3161759.

6 Zaixin Song, Chunhua Liu, Hang Zhao, and Rundong Huang. Nonlinear force and vibration analysis of an interior permanent magnet synchronous generator with eccentricity detection. *IEEE/ASME Transactions on Mechatronics*, 27(5):3154–3165, October 2022. ISSN 1083-4435. https://doi.org/10.1109/TMECH.2021.3118316.

7 Fu Lin, Shuguang Zuo, and Wenzhe Deng. Impact of rotor eccentricity on electromagnetic vibration and noise of permanent magnet synchronous motor. *Journal of Vibroengineering*, 20(2):923–935, March 2018. ISSN 1392-8716. https://doi.org/10.21595/jve.2017.18436.

8 Shu Wang and Hongfeng Li. Analysis of electromagnetic vibration of permanent magnet synchronous motor under static and dynamic eccentricity fault. *Proceedings of the 6th International Conference on Control and Robotics Engineering (ICCRE)*, pages 106–111, April 2021. https://doi.org/10.1109/ICCRE51898.2021.9435680.

9 Muhammad Sarmad Rafaq, Hyeonjun Lee, Yonghyun Park, Sang Bin Lee, Daniel Fernandez, David Diaz Reigosa, and Fernando Briz. A simple method for identifying mass unbalance using vibration measurement in permanent magnet synchronous motors. *IEEE Transactions on Industrial Electronics*, 69(6):6441–6444, June 2022. ISSN 0278-0046. https://doi.org/10.1109/TIE.2021.3088332.

10 Bashir M. Ebrahimi and Jawad Faiz. Magnetic field and vibration monitoring in permanent magnet synchronous motors under eccentricity fault. *IET Electric Power Applications*, 6(1):35–45, January 2012. ISSN 1751-8660. https://doi.org/10.1049/iet-epa.2010.0159.

11 Yonghyun Park, Daniel Fernandez, Sang Bin Lee, Doosoo Hyun, Myung Jeong, Suneel Kumar Kommuri, Changhee Cho, David Diaz Reigosa, and Fernando Briz. Online detection of rotor eccentricity and demagnetization faults in PMSMs based on Hall-effect field sensor measurements. *IEEE Transactions on Industry Applications*, 55(3):2499–2509, May 2019. ISSN 0093-9994. https://doi.org/10.1109/TIA.2018.2886772.

12 Seyyed Mehdi Mirimani, Abolfazl Vahedi, Fabrizio Marignetti, and Roberto Di Stefano. An online method for static eccentricity fault detection in axial flux machines. *IEEE Transactions on Industrial Electronics*, 62(3):1931–1942, March 2015. ISSN 0278-0046. https://doi.org/10.1109/TIE.2014.2360070.

13 Taner Goktas, Mohsen Zafarani, Kun Wang Lee, Bilal Akin, and Terry Sculley. Comprehensive analysis of magnet defect fault monitoring through leakage flux. *IEEE Transactions on Magnetics*, 53(4):1–10, April 2017. ISSN 0018-9464. https://doi.org/10.1109/TMAG.2016.2617318.

14 Sergio M. A. Cruz and António J. M. Cardoso. Stator winding fault diagnosis in three-phase synchronous and asynchronous motors, by the extended Park's vector approach. In *Conference Record of the 2000 IEEE Industry Applications Conference*, volume 1, pages 395–401, Rome, Italy, October 2000. IEEE. https://doi.org/10.1109/IAS.2000.881141.

15 Mustafa Eker and Mehmet Akar. Eccentricity fault diagnosis in a permanent magnet synchronous motor under nonstationary speed conditions. *Turkish Journal of Electrical Engineering and Computer Sciences*, 25:1881–1893, 2017. ISSN 1300-0632. https://doi.org/10.3906/elk-1601-157.

16 Antonio Femia, José E. Ruiz-Sarrió, Giacomo Sala, José Antonino-Daviu, and Luca Zarri. Eccentricity fault diagnosis in permanent-magnet synchronous motors using the stray flux vector. In *2025 IEEE Workshop on Electrical Machines Design, Control and Diagnosis (WEMDCD)*, pages 1–6. IEEE, April 2025. https://doi.org/10.1109/WEMDCD61816.2025.11014202.

17 Zijian Liu, Jin Huang, and Shan He. Diagnosis of air-gap eccentricity and partial demagnetisation of an interior permanent magnet synchronous motor based on inverse transient complex inductance vector theory. *IET Electric Power Applications*, 12(8):1166–1175, May 2018. ISSN 1751-8660. https://doi.org/10.1049/iet-epa.2018.0003.

18 Qianming Shang, Tianyao Jin, and Mingsheng Chen. A new cross-domain motor fault diagnosis method based on bimodal inputs. *Journal of Marine Science and Engineering*, 12(8):1304, August 2024. ISSN 2077-1312. https://doi.org/10.3390/jmse12081304.

19 Jayarama Pradeep and Ramasamy Devanathan. Fault diagnosis of PMSM using artificial neural network. *International Review on Modelling and Simulations (IREMOS)*, 7(5):760, October 2014. ISSN 1974-9821. https://doi.org/10.15866/iremos.v7i5.3156.

20 Tribeni Prasad Banerjee and Swagatam Das. Multi-sensor data fusion using support vector machine for motor fault detection. *Information Sciences*, 217:96–107, December 2012. ISSN 0020-0255. https://doi.org/10.1016/j.ins.2012.06.016.

21 Hicham El Hadraoui, Oussama Laayati, Nasr Guennouni, Ahmed Chebak, and Mourad Zegrari. A data-driven model for fault diagnosis of induction motor for electric powertrain. In *2022 IEEE 21st Mediterranean Electrotechnical Conference (MELECON)*, pages 336–341. IEEE, June 2022. https://doi.org/10.1109/melecon53508.2022.9843046.

22 Insu Bae and Suan Lee. A multi-input convolutional neural network model for electric motor mechanical fault classification using multiple image transformation and merging methods. *Machines*, 12(2):105, February 2024. ISSN 2075-1702. https://doi.org/10.3390/machines12020105.

23 A. Abeena and Praveen N. Kumar. Investigation of dynamic eccentricity in interior permanent magnet synchronous motor through finite element method and machine learning. In *2024 Control Instrumentation System Conference (CISCON)*, pages 1–6. IEEE, August 2024. https://doi.org/10.1109/ciscon62171.2024.10696743.

24 M. Skowron and M. Krzysztofiak. Permanent magnet synchronous motor stator and rotor fault detection using transfer learning and field-circuit model. *IEEE Access*, 13:74555–74566, 2025. https://doi.org/10.1109/ACCESS.2025.3564631.

25 Zhiqiang Wang, Shangru Shi, Xin Gu, Zhezhun Xu, Huimin Wang, and Zhen Zhang. Fault diagnosis method of permanent magnet synchronous motor demagnetization and eccentricity based on branch current. *World Electric Vehicle Journal*, 16(4):223, April 2025. ISSN 2032-6653. https://doi.org/10.3390/wevj16040223.

26 Xueyan Wang. A fuzzy neural network-based automatic fault diagnosis method for permanent magnet synchronous generators. *Mathematical Biosciences and Engineering*, 20(5):8933–8953, 2023. ISSN 1551-0018. https://doi.org/10.3934/mbe.2023392.

27 Mojtaba Babaei, Jawad Faiz, Bashir Mahdi Ebrahimi, S. Amini, and Jalal Nazarzadeh. A detailed analytical model of a salient-pole synchronous generator under dynamic eccentricity fault. *IEEE Transactions on Magnetics*, 47(4):764–771, April 2011. ISSN 0018-9464. https://doi.org/10.1109/tmag.2011.2105498.

6

Short-circuit and Inter-turn Faults

6.1 Introduction

Short-circuit faults in permanent magnet machine stator windings represent critical failure mechanisms with significant economic impact and safety implications across industrial, transportation, and renewable energy applications. Statistical analysis of permanent magnet motor failure modes reveals that short-circuit and inter-turn faults account for approximately 21% of total faults, with stator winding-related issues constituting about 33% of all machine failures [1]. The severity and rapid progression potential of these faults, particularly when initial inter-turn shorts evolve into phase-to-phase or phase-to-ground failures, necessitates comprehensive understanding of underlying physical mechanisms, accurate predictive modeling capability, and robust diagnostic approaches enabling early detection before catastrophic damage occurs.

The fundamental characteristic distinguishing short-circuit faults from other PM machine failure modes lies in their progressive nature and the complex interplay of thermal, electrical, and mechanical stress factors driving insulation degradation. Unlike instantaneous bearing failures or sudden demagnetization events, short-circuit faults typically initiate as localized insulation weaknesses affecting small numbers of turns within a single coil. Without detection and intervention, these incipient faults propagate through adjacent turns and potentially extend across phase boundaries, driven by thermal feedback where fault current heating accelerates insulation breakdown in surrounding regions. This cascade progression creates an expanding fault zone that can transition from a detectable but manageable condition to complete motor failure within timescales ranging from hours to seconds depending on fault severity, loading conditions, and protection system response.

The economic consequences of undetected short-circuit faults extend beyond direct motor replacement costs to include unplanned production downtime, potential damage to driven equipment from vibration and torque pulsations, and safety incidents when ground faults energize normally grounded enclosures at dangerous voltages. In renewable energy applications such as wind turbine generators, inter-turn fault progression to phase-to-phase shorts can destroy multi-megawatt machines located in remote offshore installations where replacement requires specialized vessels and weather windows, creating outage durations measured in months with associated revenue losses exceeding millions of dollars. In electric vehicle traction drives, short-circuit faults create immediate safety concerns requiring rapid detection and graceful degradation strategies enabling continued operation at reduced capacity until controlled vehicle shutdown can occur safely.

Recent evolution in PM machine design and application environments has introduced new challenges for short-circuit fault mitigation. The transition to inverter-fed operation with pulse width modulation voltage waveforms subjects insulation to repetitive voltage surges and partial discharge stress uncharacterized by historical sinusoidal excitation test procedures. Higher power density designs achieving specific power exceeding 5 kW/kg employ thinner insulation systems operating at elevated temperatures approaching material thermal limits, reducing thermal margins and accelerating aging. Hairpin winding manufacturing processes introduce mechanical stress through tight-radius conductor bending that creates residual stress in insulation layers, potentially establishing weak points vulnerable to electrical breakdown during operation. These modern stressors compound traditional reliability challenges, motivating development of advanced diagnostic approaches capable of detecting incipient faults early in their development when preventive maintenance or operational derating can prevent catastrophic failure.

The scope of this chapter encompasses comprehensive treatment of short-circuit and inter-turn fault mechanisms, modeling methodologies, and diagnostic techniques appropriate for diverse PM machine topologies and application environments. Section 6.1 examines failure modes and physical mechanisms underlying insulation degradation and fault initiation, establishing the foundation for understanding how thermal, electrical, and mechanical stresses interact to produce short-circuit faults. Section 6.2 develops analytical and computational modeling approaches enabling prediction of electromagnetic behavior under asymmetric fault conditions, providing the theoretical basis for diagnostic indicator selection and threshold optimization. Section 6.3 presents comprehensive diagnostic techniques organized by signal processing methodology and reference frame transformation, including motor current signature analysis, DQO and $\alpha\beta 0$ frame processing, model-based parameter estimation, and experimental validation demonstrating robustness under challenging operating conditions. The synthesis of these complementary perspectives provides machine designers with criteria for insulation system selection and protection coordination, diagnostic algorithm developers with validated fault signatures and detection thresholds, and maintenance practitioners with condition monitoring strategies enabling transition from time-based to condition-based maintenance philosophies.

6.2 Failure Modes and Mechanisms

This section examines the physical processes, material properties, and stress factors that lead to short-circuit faults in permanent magnet machines, providing the foundation for modeling and diagnostic approaches discussed in subsequent sections.

6.2.1 Inter-turn Short Circuit Characteristics

Inter-turn short circuit faults originate from localized insulation breakdown between adjacent turns within a single coil. The fault begins when two or more turns in a coil experience electrical contact, creating a low-impedance current path that bypasses portions of the winding. This initial inter-turn fault represents the most common entry point for a cascade of increasingly severe winding failures [1].

The characteristic signature of inter-turn faults manifests through several observable phenomena in machine operation. When an inter-turn fault occurs in one phase, the impedance of that phase decreases due to the reduced number of effective turns. This impedance change causes unbalanced three-phase currents, with the faulty phase exhibiting altered current amplitude and phase angle relative to healthy phases [1]. The current in the short-circuited turns themselves increases dramatically, driven by the voltage induced across the faulted segment. This localized overcurrent generates intense heat within the affected coil, accelerating further insulation degradation in adjacent regions.

Surface-mounted permanent magnet motors exhibit higher short-circuit currents compared to interior permanent magnet configurations when subjected to inter-turn faults [2]. This distinction arises from the magnetic circuit characteristics and the direct exposure of surface magnets to armature reaction effects. The short-circuit current magnitude depends on the fault resistance, which varies from near-zero for complete conductor-to-conductor contact to higher values when carbonized insulation material provides a partially conductive path.

The electromagnetic torque of the machine experiences increased ripple under inter-turn fault conditions. For a typical permanent magnet synchronous motor operating with nominal torque ripple of 2 Nm in healthy condition, the ripple amplitude can increase to 4 Nm for moderate fault severity and 6 Nm for more severe faults [3]. These torque ripples propagate through the mechanical system, causing speed oscillations, increased vibration, and mechanical stress on bearings and coupling components. The torque ripple frequency typically corresponds to twice the electrical frequency for balanced fault configurations.

Air gap flux density distribution becomes asymmetric when inter-turn faults occur. The faulted phase produces altered magnetic field patterns that interact with the rotor permanent magnets, creating unbalanced magnetic forces. This asymmetry can lead to increased radial forces, contributing to bearing wear and potentially inducing dynamic eccentricity if the fault persists over extended operation. The flux density variations also induce additional frequency components in back electromotive force waveforms, providing diagnostic signatures that can be exploited for fault detection.

6.2.2 Insulation System Materials and Properties

The insulation system of permanent magnet machine windings serves as the critical barrier preventing electrical short circuits between turns, coils, and phases. Understanding the properties and capabilities of insulation materials provides essential context for analyzing failure mechanisms and predicting machine reliability under various operating conditions.

Enamel represents the most common turn insulation material used in electrical machines. This thin, uniform coating applied to copper or aluminum conductors provides the primary electrical isolation between adjacent turns. For medium and high voltage form-wound machines utilizing rectangular copper strands, more robust insulation systems become necessary. These applications employ Mica-glass turn tapes, Mica polyethylene terephthalate turn tapes, or Daglas insulation, which offer superior thermal and electrical performance compared to standard enamel coatings [4].

The dielectric strength of an insulation material quantifies its maximum electric field tolerance before breakdown occurs, typically expressed in kilovolts per millimeter. This property depends on multiple factors including electrode geometry, material thickness, rate of voltage change, and waveform characteristics. Table 6.1 presents dielectric strength values for various insulation materials commonly encountered in permanent magnet machine construction.

The wide range of dielectric strengths reflects the diverse requirements across different machine types and applications. High-performance mica-based materials offer exceptional dielectric strength exceeding 100 kV/mm, making them suitable for high voltage applications. Standard enamel coatings provide adequate protection for low and medium voltage machines while maintaining manufacturing simplicity and cost-effectiveness.

Table 6.1 Dielectric strength of different insulation materials.

Type	Material	Dielectric strength (kV/mm)
Gas	Air	3
Ceramics	Alumina (Al_2O_3)	13.4
	Boron nitride (BN)	37.4
	Porcelain	35–160
Glasses	Fused silica (SiO_2)	470–670
	Alkali-silicate glass	200
	Standard window glass	9.8–13.8
Micas	Muscovite, ruby, natural	118
	Phlogopite, amber, natural	118
	Fluorophlogopite, synthetic	118
	Glass-bonded mica	14.0–15.7
Enamel	Red enamel, fast cure	47.2–78.7
	Black enamel	47.2–70.9
Varnishes	Vacuum-pressure impregnated baking	68.9–70.9
	Baking type epoxy varnish	90.6–181.1
	Polyurethane coating	47.2–78.7
Thermoplastic polymers	Polybutylene terephthalate	15.7
	Polypropylene	23.6

Insulation strength degrades over time due to multiple stress factors. Excessive overheating accelerates chemical degradation processes within organic insulation materials. Mechanical shocks and vibration can cause physical cracking or delamination of insulation layers. Moisture ingress reduces dielectric strength by providing conductive paths and promoting electrochemical corrosion. Chemical contamination from environmental pollutants or degraded lubricants attacks organic insulation compounds. Inverter ripples from pulse width modulation drives subject insulation to repetitive voltage stress, causing cumulative damage through partial discharge and dielectric fatigue [4].

6.2.3 Insulation Temperature Classes

Permanent magnet machines employ insulation systems rated according to NEMA temperature classes, with Class F (155 °C) and Class H (180 °C) dominating modern designs. These higher-rated systems provide thermal margins necessary for inverter-fed applications experiencing harmonic content and thermal transients. The allowable winding temperature rise depends on insulation class rating minus ambient temperature and hot spot allowances.

6.2.4 Thermal Stress and the 10-degree Rule

Thermal stress represents one of the most significant factors in irreversible insulation degradation. The relationship between temperature and insulation life follows an exponential decay pattern captured by the widely referenced 10-degree rule. According to this principle, each additional 10 °C of operating temperature approximately halves the expected insulation service life [4]. This dramatic sensitivity to temperature elevation emphasizes the critical importance of thermal management in permanent magnet machine design and operation.

The aging process can be categorized into two distinct scenarios. Normal thermal aging follows the natural life cycle of the winding insulation system, occurring even when the machine operates within its designed thermal envelope. This gradual degradation results from slow chemical reactions, oxidation processes, and thermal cycling effects that accumulate over thousands of operating hours. Early thermal aging occurs when excessive heat accelerates degradation mechanisms, drastically reducing the expected service life. Such premature aging can result from various factors including voltage imbalances, cycling between high and low loads, continuous overloading, obstructed ventilation systems, or high ambient temperatures exceeding design assumptions [4].

Figure 6.1 illustrates thermal aging curves for the four NEMA insulation classes (A, B, F, and H), plotted using the Montsinger relationship $L(T) = L_{\text{rated}} \times 2^{(T_{\text{rated}}-T)/\Delta T}$ with a halving interval $\Delta T = 10\,°\text{C}$. Each curve intersects the rated design life of approximately 20,000 hours at its class-rated temperature, a benchmark established by IEC 60216, which defines the Temperature Index as the temperature at which insulation retains adequate properties after 20,000 hours of continuous thermal exposure [5]. This relationship, originally observed empirically by Montsinger in 1930 and later grounded in Arrhenius chemical rate theory by Dakin in 1948, provides the theoretical foundation for the 10-degree rule discussed above. As illustrated in the figure, a 10 °C increase on the Class H curve shifts the operating point from 20,000 hours down to 10,000 hours, exactly halving the expected service life. Operating below the rated temperature extends life exponentially: for example, a Class H system operating at 155 °C rather than its rated 180 °C achieves approximately 113,000 hours of service life, representing a more than fivefold improvement.

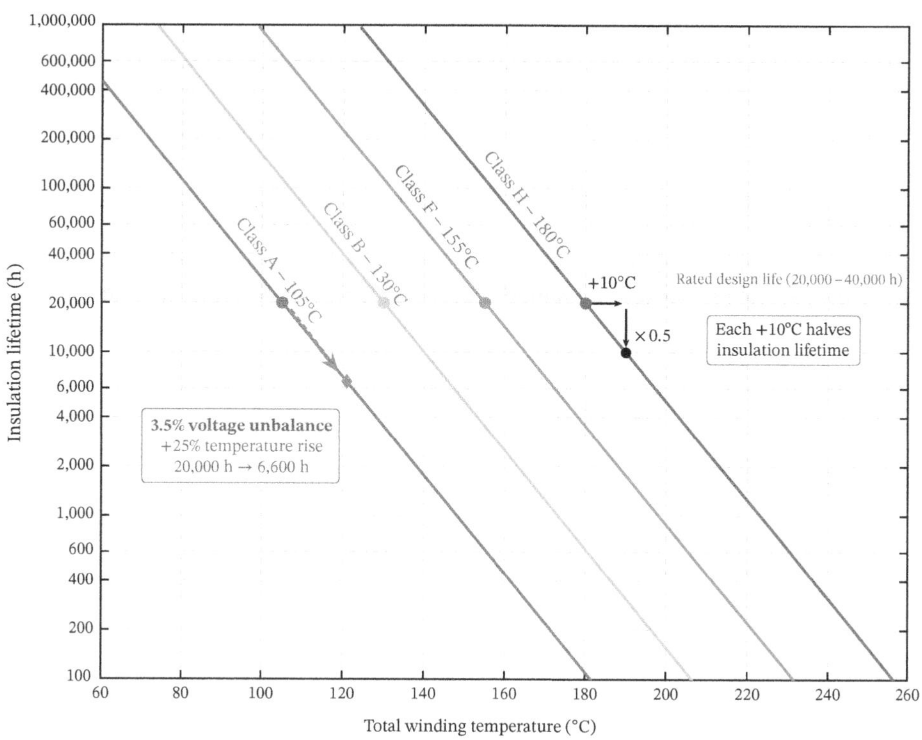

Figure 6.1 Thermal aging curves for NEMA insulation classes based on the Montsinger 10 ° rule, with each class intersecting the 20,000-hour rated design life at its rated temperature. The Class H annotation illustrates the halving of lifetime per 10 °C increase, while the Class A annotation shows how a 3.5% voltage unbalance reduces lifetime from 20,000 to approximately 6,600 hours. *Source:* Adapted from [4].

The practical consequences of thermal stress become particularly severe when combined with electrical imbalances. Research has quantified that a 3.5% voltage unbalance between phases increases the temperature of the phase winding carrying the highest current by approximately 25% [4]. As shown in Figure 6.1 for a Class A insulation system, this temperature rise shifts the effective operating point from 105 °C to approximately 121 °C, reducing the expected lifetime from 20,000 to roughly 6,600 hours, a reduction by a factor of three from a seemingly modest voltage imbalance. This localized temperature elevation creates uneven aging across the three phases, potentially leading to cascading failures as the weakest insulation region fails first, redistributing stress to the remaining phases.

Temperature control through effective cooling system design represents the primary defense against premature thermal aging. Liquid cooling systems provide superior heat extraction capability compared to air cooling, enabling higher power density while maintaining acceptable winding temperatures. Selection of an appropriate insulation class provides additional thermal margin, allowing the machine to withstand transient overtemperatures without immediately compromising reliability. During the design stage, the maximum continuous motor load must be carefully evaluated considering the application duty cycle and ambient conditions to ensure the selected insulation class provides adequate thermal capability.

6.2.5 Electrical Stress from Inverter Operation

The transition from sinusoidal utility power to pulse width modulated inverter drives has introduced new electrical stress mechanisms that significantly impact winding insulation degradation. Voltage source inverters generate output voltages through high frequency switching of power semiconductor devices, creating voltage waveforms containing rapid transitions and significant harmonic content. These characteristics pose particular challenges for motor winding insulation systems originally designed for sinusoidal excitation.

The aging process accelerates substantially for inverter-fed motors compared to line-fed machines. When long cables connect the inverter to the motor, voltage transmission reflections or traveling waves result from the interaction between pulse width modulation switching and impedance mismatch between machine and cable. These reflections generate voltage stress and overshoot on the stator winding terminals, potentially leading to inter-turn insulation breakdown [4]. The magnitude of voltage overshoot depends critically on cable length and pulse rise time, with slower rise times producing lower overshoot amplitudes.

The theoretical worst case occurs when machine impedance significantly exceeds cable impedance. Under these conditions, the voltage pulse experiences complete reflection at the motor terminals, generating an oscillating transient overvoltage with decaying magnitude. Peak voltages can reach twice the DC bus voltage under worst-case reflection conditions. Although cable and machine resistance provide damping that reduces peak magnitudes in practical systems, sustained exposure to these repetitive overvoltages degrades insulation through partial discharge and dielectric fatigue mechanisms.

Figure 6.2 illustrates voltage distribution over winding turns for three representative pulse rise times: 0.2 microseconds (typical of modern SiC MOSFET inverters), 1 microsecond, and 5 microseconds (typical of conventional IGBT inverters). If the voltage were distributed uniformly across all coils, each coil would carry an equal fraction of the total voltage, as indicated by the uniform reference line. In practice, shorter rise times produce highly nonlinear voltage distributions, with the first few turns near the terminal absorbing a disproportionate share of the applied voltage. At a rise time of 0.2 microseconds, the first coil can experience up to 80% of the total voltage drop, far

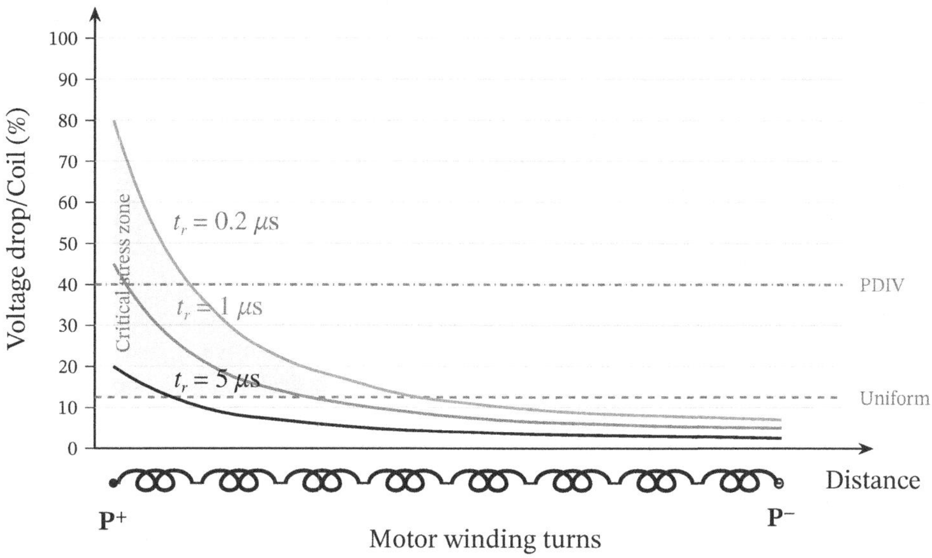

Figure 6.2 Voltage distribution over stator winding turns for different pulse rise times. *Source:* Adapted from [4].

exceeding the partial discharge inception voltage (PDIV) threshold and placing these turns firmly within the critical stress zone. Even at a rise time of 1 microsecond, the first-coil voltage reaches approximately 45%, still above the PDIV threshold. Only at the slowest rise time of 5 microseconds does the first-coil voltage remain below this critical level. These observations underscore that the first few turns of each phase winding are particularly vulnerable to partial discharge initiation and dielectric breakdown, especially in drives employing wide-bandgap semiconductor devices with fast switching transients.

Initially, a new motor with fresh insulation withstands high voltage surges without immediate damage. The insulation system design incorporates a margin above nominal operating voltage to accommodate transient overvoltages. However, as normal or early aging degrades the insulation dielectric strength over time, vulnerability to voltage surges increases. Each surge event that exceeds the instantaneous dielectric strength produces a localized arc, significantly accelerating the degradation process. This creates a positive feedback cycle where aging reduces dielectric strength, enabling more frequent partial discharge events, which further accelerates aging [4].

The cumulative effect of repetitive voltage stress continues until insulation strength decreases to levels corresponding to the nominal operating voltage. At this critical point, even normal operating conditions can trigger breakdown, initiating an inter-turn short circuit. The progression from initial partial discharge to complete insulation failure can occur rapidly once the critical degradation threshold is reached, providing limited warning time for preventive maintenance or controlled shutdown.

6.2.6 Fault Progression Mechanisms

Short-circuit faults in permanent magnet machine windings typically follow a characteristic progression sequence, beginning with localized inter-turn faults and potentially escalating to more severe failure modes. Understanding this progression helps establish appropriate diagnostic strategies and protection systems that can detect faults early in their development, preventing catastrophic damage.

The degradation and deterioration of insulation represents the primary cause of stator winding short-circuit faults. These faults originate from inter-turn failures where two or more turns in a coil experience an electrical short circuit. The initial inter-turn fault may involve relatively high fault resistance if degraded insulation provides a partially conductive path, or near-zero resistance if complete conductor-to-conductor contact occurs [1].

Once an inter-turn fault establishes a conductive path, the local current in the faulted segment increases dramatically. This overcurrent generates intense localized heating in the short-circuited turns. The thermal elevation caused by inter-turn fault current subsequently damages insulation in adjacent regions, enabling fault propagation. The progression typically follows the sequence: inter-turn fault to coil-to-coil fault to phase-to-phase short circuit or phase-to-ground fault. In some cases, thermal stress and mechanical forces can cause conductor fracture, resulting in open-circuit faults.

Phase-to-ground faults occur when the wall-ground insulation sustains damage, allowing electrical contact between the winding and the machine frame or magnetic core. This represents one of the most severe fault types due to the large current magnitudes typically involved and the potential safety hazards from exposed conductive surfaces at high voltage [1]. Ground fault protection systems generally act quickly to disconnect power when phase-to-ground faults occur, limiting damage but resulting in immediate machine shutdown.

Phase-to-phase short circuits develop when insulation between windings of different phases fails. This can result from progressive damage extending outward from an initial inter-turn fault, or from independent insulation failure in slot or end winding regions where phase conductors are in close proximity. Phase-to-phase faults produce severe electromagnetic unbalance and high circulating currents, typically triggering overcurrent protection devices rapidly.

Figure 6.3 provides photographic examples of various winding failure modes observed in permanent magnet machines. The images illustrate the physical damage associated with inter-turn faults, phase-to-phase shorts, phase-to-ground failures, and complete phase burnout resulting from sustained overheating. The visible carbonization, discolored insulation, and melted conductor material demonstrate the intense thermal conditions present during short-circuit fault progression. In the most severe case shown in Figure 6.3d, prolonged overcurrent and the resulting thermal runaway have completely destroyed the phase winding, leaving extensive charring and deformation of both conductors and insulation throughout the stator.

The temporal scale of fault progression varies widely depending on initial fault severity, operating conditions, and load level. An inter-turn fault under light load conditions may persist for extended periods, gradually degrading adjacent insulation before propagating to more severe failure modes. Conversely, under heavy load or during transient overload events, an inter-turn fault can escalate to complete phase failure within seconds or minutes. This variability in progression rate emphasizes the importance of continuous monitoring and rapid diagnostic response to maximize the window for protective intervention.

Undetected inter-turn faults can also cause permanent magnet demagnetization through armature reaction effects associated with the asymmetric magnetic fields [6]. The altered current distribution produces localized regions of high demagnetizing magnetic field strength that can partially or completely demagnetize magnets, particularly in surface-mounted permanent magnet configurations. This secondary damage mechanism can render the machine unusable even after winding repair, necessitating expensive rotor replacement.

(a) (b) (c) (d)

Figure 6.3 Different modes of winding failure. (a) Inter-turn fault, (b) phase-to-phase fault, (c) phase-to-ground fault, (d) complete phase burnout due to overheating. Courtesy of WEG S.A.

6.2.7 Turn-to-turn Versus Turn-to-ground Characteristics

While inter-turn and turn-to-ground faults both involve insulation breakdown, their mechanisms, signatures, and consequences differ significantly. Turn-to-turn short circuits, more commonly referenced as inter-turn faults, occur when the turn-to-turn insulation fails between adjacent turns within the same coil or between turns of different coils in the same phase. The fault current path completes through the winding conductors themselves, with current magnitude limited by the winding impedance and the induced voltage across the shorted segment.

Turn-to-ground faults involve breakdown of the ground wall insulation, creating a conductive path between winding conductors and the machine frame, stator core, or other grounded metallic components. The fault current magnitude and characteristics depend critically on the system grounding configuration. In grounded neutral systems, turn-to-ground faults produce large single-phase ground currents that typically trigger protective devices rapidly. In ungrounded systems, the first ground fault may not produce large currents but creates a voltage reference point that enables destructive line-to-line voltage stresses if a second ground fault occurs on a different phase.

The detection signatures for these fault types exhibit distinct characteristics. Inter-turn faults produce negative sequence current components, torque ripple increases, and altered harmonic content in phase currents and voltages. Turn-to-ground faults generate zero-sequence current components (in grounded systems), residual ground fault current, and common mode voltage signatures. These distinct fingerprints enable diagnostic systems to differentiate between fault types, supporting targeted maintenance responses.

Safety considerations differ substantially between fault types. Inter-turn faults primarily affect machine operation and reliability but generally do not create immediate personnel hazards unless fault progression leads to fire risk. Turn-to-ground faults, particularly in systems lacking proper grounding and ground fault protection, can create electrical shock hazards by placing exposed conductive surfaces at dangerous voltages. Industrial safety standards mandate ground fault protection for permanent magnet machines in most applications to mitigate these risks.

6.2.8 Phase-to-phase Short Circuit Mechanisms

Phase-to-phase short circuit faults represent an advanced stage of insulation failure where electrical contact occurs between conductors of two different phases. These faults can develop through multiple pathways, each associated with distinct precursor conditions and progression timelines.

The most common pathway involves progressive extension of an existing inter-turn fault. As local heating from inter-turn fault current damages adjacent insulation, the fault can propagate beyond the boundaries of a single-phase winding. When this propagation reaches conductors of an adjacent phase, either in the slot region or at the end windings, a phase-to-phase connection forms. This progression typically occurs over hours to days, depending on fault severity and loading conditions.

Alternative initiation mechanisms can produce phase-to-phase faults independent of prior inter-turn failures. Mechanical damage from vibration, thermal cycling, or foreign object intrusion can compromise phase-to-phase insulation directly. Contamination by conductive particles, moisture with dissolved ions, or tracking paths formed by sustained partial discharge can create phase-to-phase conductive paths. Manufacturing defects, including insufficient creepage distance, damaged insulation during assembly, or voids in impregnation compound, may not produce immediate failure but create weak points vulnerable to premature breakdown under normal operating stress.

The electromagnetic consequences of phase-to-phase faults significantly exceed those of inter-turn faults. The voltage difference between the two phases drives the fault current, producing magnitudes potentially orders of magnitude higher than inter-turn fault currents. This high current causes severe electromagnetic force unbalance, extreme torque pulsations, and rapid temperature rise in the fault region. The intense thermal conditions typically trigger protective devices within a few electrical cycles, limiting damage but requiring immediate shutdown.

Current signatures under phase-to-phase fault conditions exhibit distinctive characteristics useful for fault classification. Large negative and positive sequence current components appear, with magnitudes far exceeding those produced by inter-turn faults of comparable winding involvement. The phase currents become highly unbalanced, with the two affected phases carrying dramatically different currents than the unfaulted phase. Harmonic analysis reveals rich spectral content including fundamental frequency imbalance, even and odd harmonics, and sideband components.

The mechanical consequences extend beyond electromagnetic effects. The large electromagnetic forces generated during phase-to-phase faults can cause conductor motion, winding deformation, and mechanical shock to bearings and structural components. Repetitive phase-to-phase faults, such as those occurring during starting of motors with incipient faults, can cause cumulative mechanical damage that further compromises machine integrity.

Protection systems typically detect phase-to-phase faults rapidly through overcurrent relays, differential protection schemes, or negative sequence overcurrent elements. The challenge lies in detecting incipient phase-to-phase insulation degradation before breakdown occurs, enabling proactive maintenance rather than reactive repair after catastrophic failure. Partial discharge monitoring, insulation resistance trending, and advanced diagnostic algorithms processing multiple sensor signals offer promising approaches for early detection of phase-to-phase insulation deterioration.

6.2.9 Mechanical Stress Factors in Winding Insulation Degradation

Mechanical stress factors play critical roles alongside thermal and electrical mechanisms in accelerating insulation degradation and initiating short-circuit faults. These stresses arise from electromagnetic forces during operation and fault conditions, vibration from rotor dynamics and external sources, thermal cycling producing differential expansion, and manufacturing processes introducing residual stresses. The interaction between mechanical stress and other degradation mechanisms creates synergistic effects where combined stresses produce more rapid failure than individual factors operating independently.

Electromagnetic forces acting on stator windings result from interactions between permanent magnet fields, stator current distributions, and core structures. End winding regions experience the highest stress levels due to the absence of slot containment, with stress concentration at transition regions where conductors change direction. The cyclic nature of these forces at twice electrical frequency creates fatigue loading capable of initiating crack defects in insulation structures [7]. Force magnitudes scale with the square of winding current, meaning transient events including short circuits produce electromagnetic forces far exceeding steady-state design values.

Vibration from electromagnetic, mechanical, and external sources causes relative motion between insulation surfaces, progressively wearing material through abrasion. When structural resonances align with electromagnetic force frequencies, amplification factors of 10–50 can transform modest forces into large displacement oscillations. Variable frequency drive applications present particular challenges as they sweep through wide frequency ranges during acceleration and deceleration.

Thermal cycling subjects insulation to mechanical stress through differential expansion of materials with dissimilar coefficients. Copper conductors exhibit expansion coefficients of approximately 17 ppm/°C while organic insulation materials range from 50 to 200 ppm/°C. This mismatch creates interfacial shear stress during temperature excursions that accumulates damage through low-cycle fatigue. Applications with frequent duty cycles such as electric vehicle traction motors experience accelerated degradation from repeated thermal expansion and contraction [8].

Manufacturing processes introduce residual stresses through coil winding tension, forming operations, slot insertion, and impregnation resin cure shrinkage. These residual stresses superimpose on operational stresses, potentially creating combined stress states exceeding material capabilities even when individual components remain within design limits.

The most severe degradation scenarios involve synergistic interactions where mechanical, thermal, and electrical stresses combine. Mechanical defects locally concentrate electrical fields, initiating partial discharge that thermally degrades surrounding insulation. Multi-physics finite element models reveal combined stress peaks two to three times higher than single-physics predictions, corresponding to observed failure locations [7].

6.2.10 Phase-to-phase Short-circuit Fault Mechanisms

While inter-turn short-circuit faults represent the most common initial failure mode in permanent magnet machine stator windings, their undetected progression can lead to more catastrophic phase-to-phase short-circuit (PPSC) faults. Unlike inter-turn faults that primarily affect a localized region within a single-phase winding, phase-to-phase faults create direct electrical connection between separate phase windings, producing severe three-phase current imbalance, substantial torque ripple, and elevated thermal stresses that can rapidly escalate to complete motor failure [9]. Phase-to-phase short circuits represent one of the most severe damage scenarios in electrical motor installations, posing significant safety risks to connected equipment and potentially leading to fire hazards if not rapidly detected and isolated [10].

6.2.10.1 Fault Evolution: From Inter-turn to Phase-to-phase

The progression from inter-turn short-circuit to phase-to-phase short-circuit follows a predictable degradation path driven by thermal, mechanical, and electrical stresses. Understanding this evolution is critical for establishing appropriate detection thresholds and protection strategies.

6.2.10.2 Initiation Mechanisms

Phase-to-phase faults initiate through multiple pathways, most commonly as the endpoint of inter-turn fault progression. When an inter-turn fault remains undetected, the localized heating in the shorted turns accelerates insulation deterioration. The short-circuit current i_{sc} flowing through the fault loop can reach 2–5 times the rated phase current, producing localized hot spots that cause thermal runaway in adjacent insulation layers [11].

The thermal degradation follows an exponential relationship with temperature described by the Arrhenius equation:

$$L = L_0 \exp\left(\frac{E_a}{k_B T}\right) \tag{6.1}$$

where L is the insulation lifetime, L_0 is a pre-exponential factor, E_a is the activation energy (typically 0.9–1.1 eV for common winding insulations), k_B is Boltzmann's constant, and T is absolute temperature. A temperature increase of just 10 °C halves the insulation lifetime, meaning that the localized heating from an inter-turn fault can reduce the time to phase-to-phase fault from years to hours or even minutes during operation.

As the inter-turn fault severity increases (more turns short-circuited or lower fault resistance), the fault current grows according to:

$$i_{sc} = \frac{\mu N_{\text{total}} \omega_r \lambda_m \sin \theta_r}{R_f + \mu^2 R_s} \tag{6.2}$$

where μ is the fraction of shorted turns, N_{total} is the total turns per phase, R_f is the fault resistance, and R_s is the phase resistance. When R_f decreases due to continued insulation breakdown (solid insulation chars and becomes conductive at temperatures above 200–250 °C), i_{sc} increases dramatically, accelerating the degradation process [9].

6.2.10.3 Cascade Progression Sequence

The fault progression from initial inter-turn short-circuit to complete phase-to-phase fault typically follows this sequence:

Stage 1: Incipient Inter-turn Fault (Hours to Days) – Initial insulation breakdown affects 1–5% of turns in single phase – Fault resistance R_f is relatively high (0.5–2 Ω) – Short-circuit current 20–50% of rated phase current – Temperature rise localized to fault region (ΔT = 20–40 °C above nominal) – Fault detectable through parameter estimation or harmonic analysis.

Stage 2: Progressive Inter-turn Fault (Minutes to Hours) – Thermal stress expands fault zone to 5–15% of turns – Fault resistance decreases to 0.1–0.5 Ω as insulation chars – Short-circuit current reaches 50–150% of rated phase current – Temperature at fault exceeds 180–200 °C, approaching insulation thermal limit – Adjacent turn insulation begins accelerated aging – Detectable through current unbalance and elevated harmonic content.

Stage 3: Critical Inter-turn Fault (Seconds to Minutes) – Fault encompasses 15–30% of turns – Fault resistance drops below 0.1 Ω (near-metallic short) – Short-circuit current exceeds 150–300% of rated phase current – Insulation in fault region completely compromised, mechanical support degraded – Electromagnetic forces from fault current begin displacing conductors – Mechanical vibration increases due to rotor-stator force imbalance.

Stage 4: Phase-to-phase Fault Transition (Instantaneous to Seconds) – Degraded insulation between phase windings fails under combination of: – Thermal stress from adjacent overheated conductors – Mechanical stress from electromagnetic forces and vibration – Electrical stress from transient voltage spikes – Direct phase-to-phase connection established – Three-phase current severely unbalanced (negative sequence component 30–70% of positive) – Torque ripple increases to 40–100% of average torque – Protection systems must trip within milliseconds to prevent motor burnout.

The transition from Stage 3 to Stage 4 can occur very rapidly, especially if subjected to transient events like sudden load changes, voltage spikes from grid disturbances, or starting transients that temporarily increase electromagnetic forces and voltage stresses [10].

6.2.10.4 Accelerating Factors

Several operational conditions accelerate the inter-turn to phase-to-phase fault progression:

Thermal Cycling: Repeated thermal expansion and contraction during start-stop cycles create mechanical stress concentrations at conductor-insulation interfaces. The coefficient of thermal expansion mismatch between copper (17 ppm/°C) and typical organic insulations (50–200 ppm/°C) induces interfacial shear stresses that propagate cracks through insulation layers. Motors operating with frequent duty cycles experience 3–5× faster fault progression than continuously operating machines.

Vibration and Mechanical Resonance: When motor operating speeds align with mechanical resonances of the stator structure or winding support system, vibration amplitudes can increase by 10–50×. These vibrations cause conductor movement within slots, abrading insulation and creating conductive debris paths. In the presence of an existing inter-turn fault, vibration accelerates the spread to adjacent turns and phases [9].

Voltage Transients from PWM Switching: High-frequency voltage transients from inverter switching produce repetitive partial discharge activity at insulation weak points. The rise time of PWM voltage pulses (typically 100–500 ns for modern IGBTs) creates voltage stress distribution along windings that can be 3–5 × the DC voltage magnitude. This stress is highest at conductor interfaces near turn-to-turn and phase-to-phase boundaries, promoting progressive breakdown.

Contamination and Moisture: Conductive contamination (metal particles, carbon dust) or moisture ingress provides low-resistance pathways that bridge insulation gaps. In the presence of an inter-turn fault, the elevated local temperature causes moisture to condense in cooler adjacent regions, creating conductive films that facilitate phase-to-phase fault initiation. Industrial environments with high humidity or airborne contaminants reduce the time to phase-to-phase fault by 40–60%.

6.2.10.5 Phase-to-phase Fault Signatures

Phase-to-phase short-circuit faults produce distinctive signatures in electrical and mechanical quantities that differentiate them from inter-turn faults and enable rapid detection.

6.2.10.6 Current Unbalance Characteristics

The most prominent signature of phase-to-phase faults is severe three-phase current unbalance. While inter-turn faults create modest asymmetry (negative sequence component typically 5–15% of positive sequence), phase-to-phase faults produce dramatic imbalance with negative sequence components reaching 30–70% of positive sequence [10].

For a phase-to-phase fault between phases a and b, the fault current i_f divides between the two phases according to their respective impedances:

$$i_f = \frac{v_{ab}}{Z_f + Z_a \parallel Z_b} \tag{6.3}$$

where v_{ab} is the line-to-line voltage between phases a and b, Z_f is the fault impedance (typically 0.01–0.1 Ω for metallic faults), and Z_a, Z_b are the phase impedances. This fault current superposes on the normal three-phase currents, creating severe asymmetry.

Using symmetrical component analysis, the negative sequence current component is calculated as:

$$I_2 = \frac{1}{3}(I_a + a^2 I_b + a I_c) \tag{6.4}$$

where $a = e^{j2\pi/3}$ is the complex rotation operator and I_a, I_b, I_c are the phasor representations of phase currents. The unbalance factor is defined as:

$$\text{UF} = \frac{|I_2|}{|I_1|} \times 100\% \tag{6.5}$$

Experimental studies show that for phase-to-phase faults with fault resistance below 0.5 Ω, the unbalance factor exceeds 30% across all load conditions, providing clear discrimination from inter-turn faults (UF < 15%) and supply voltage unbalance (UF < 10%) [12].

6.2.10.7 Harmonic Content and Spectral Signatures

Phase-to-phase faults introduce specific harmonic patterns in stator currents that differ from inter-turn fault signatures. The primary distinguishing features are as follows:

Fundamental Frequency Asymmetry: The negative sequence component at fundamental frequency (f_e) becomes dominant. For inter-turn faults, the negative sequence primarily appears at twice the electrical frequency ($2f_e$) due to flux asymmetry, whereas phase-to-phase faults directly create negative sequence at f_e.

Odd Harmonics Amplification: Phase-to-phase faults significantly amplify odd harmonic components (3rd, 5th, 7th) due to the nonlinear fault impedance and the distorted spatial MMF distribution. The fifth harmonic amplitude can increase by 200–500% relative to a healthy baseline, compared to 20–50% increase for inter-turn faults [13].

Zero-sequence Components: Unlike inter-turn faults within a single phase, phase-to-phase faults can inject zero-sequence current if the motor neutral is accessible or if the fault creates an unintended path to ground. Zero-sequence current magnitude correlates with fault severity and provides additional diagnostic information.

The harmonic current spectrum for phase-to-phase fault between phases a and b can be approximated as:

$$i_a(t) = I_1 \cos(\omega_e t) + \sum_{h=3,5,7,\ldots} I_h \cos(h\omega_e t + \phi_h) + I_f \cos(\omega_e t + \phi_f) \tag{6.6}$$

where I_f is the fault current component and ϕ_f is its phase angle relative to the healthy current. The harmonic amplitudes I_h and phases ϕ_h depend on the fault location, fault impedance, and operating point.

6.2.10.8 Torque Ripple and Mechanical Signatures

Phase-to-phase faults create substantial torque pulsations due to the severe current unbalance. The instantaneous electromagnetic torque can be decomposed into average and ripple components:

$$T_e(t) = T_{\text{avg}} + T_{\text{ripple}}(t) \tag{6.7}$$

For phase-to-phase faults, the ripple component magnitude reaches 40–100% of average torque, compared to 5–20% for inter-turn faults [9]. This torque ripple manifests as:

Low-frequency Speed Oscillations: At twice the electrical frequency ($2f_e$), speed variations of ±5–15% of nominal speed occur, creating audible noise and mechanical stress on couplings and driven loads.

Vibration Signatures: Radial vibration at the motor frame increases by 300–800% relative to healthy operation, with peak amplitudes at fundamental mechanical frequency and its harmonics. Axial vibration also becomes significant (100–300% increase) due to asymmetric magnetic pull.

Bearing Stress: The unbalanced magnetic pull creates time-varying radial forces on bearings that can exceed 200% of the nominal bearing load. These forces accelerate bearing wear and can lead to secondary bearing failures within days of phase-to-phase fault occurrence.

The torque ripple frequency content serves as a diagnostic indicator, with the ratio of $2f_e$ ripple component to average torque providing fault severity quantification:

$$I_{\text{torque}} = \frac{T_{2f_e}}{T_{\text{avg}}} \times 100\% \tag{6.8}$$

Values exceeding 20% indicate phase-to-phase fault conditions requiring immediate protective action.

6.2.10.9 Protection Challenges and Detection Timing

Phase-to-phase faults impose stringent requirements on protection systems due to their rapid progression and severe consequences if unmitigated.

6.2.10.10 Detection Speed Requirements

The critical detection time for phase-to-phase faults is substantially shorter than for inter-turn faults. While inter-turn fault detection within 100–500 ms is generally acceptable (given the hours-to-days progression timeline), phase-to-phase faults require detection and tripping within 10–50 ms to prevent catastrophic motor damage and potential fire hazards [10].

This fast response requirement stems from the severe thermal stress imposed by phase-to-phase fault currents. The I^2R heating rate in conductors adjacent to the fault can reach:

$$\frac{dT}{dt} = \frac{i_f^2 R_{\text{conductor}}}{\rho c_p V_{\text{conductor}}} \tag{6.9}$$

where ρ is the conductor material density (8,960 kg/m^3 for copper), c_p is specific heat (385 J/(kg·K)), and $V_{\text{conductor}}$ is the conductor volume. For a phase-to-phase fault with $i_f = 500$ A in a typical motor winding, the heating rate exceeds 50 °C/s, reaching insulation thermal limits (200–220 °C) within 2–4 seconds from fault inception at nominal operating temperature (80–100 °C).

6.2.10.11 Sensitivity Versus Selectivity Trade-offs

Protection schemes for phase-to-phase faults must balance sensitivity (ability to detect low-severity faults) against selectivity (ability to distinguish faults from benign conditions). Several challenges complicate this balance:

Differentiation from Unbalanced Supply: Supply voltage unbalance up to 2–3% is common in industrial installations and creates current unbalance of 6–9% (unbalance amplification factor of 3–4 for typical motor impedances). This overlaps with incipient phase-to-phase fault signatures, requiring protection algorithms to incorporate additional discrimination criteria.

Modern protection schemes achieve discrimination by monitoring the harmonic content in addition to fundamental unbalance. Supply unbalance primarily affects the fundamental component with minimal harmonic distortion, while phase-to-phase faults significantly increase odd harmonics. A combined index is used:

$$I_{\text{combined}} = \alpha \cdot UF + \beta \cdot \frac{THD_i - THD_{i,\text{baseline}}}{THD_{i,\text{baseline}}} \tag{6.10}$$

where THD_i is the total harmonic distortion of phase current, and α, β are weighting factors (typically $\alpha = 0.6, \beta = 0.4$). Fault declaration occurs when I_{combined} exceeds a threshold (typically 0.25–0.35).

Load Transient Immunity: Sudden load changes create transient current asymmetries lasting 50–200 ms that can exceed steady-state phase-to-phase fault signatures. Protection systems must incorporate time delays or adaptive thresholds to prevent false trips during normal operation. Typical implementations use a two-stage threshold: - Instantaneous trip at $UF > 50\%$ (indicates severe fault) - Time-delayed trip (20–50 ms) at $UF > 25\%$ (allows transient settling).

CT Saturation Effects: Current transformer saturation during fault conditions can distort measured currents, potentially masking fault signatures or creating false indications. Flux-balance protection techniques or Rogowski coil current sensors avoid saturation issues but increase system cost.

6.2.10.12 Multiphase Fault Scenarios

In dual-redundancy or multiphase motor systems (five-phase, six-phase, nine-phase machines), phase-to-phase faults can occur within a single winding group or across winding groups, creating complex fault scenarios with distinct signatures [14].

Intra-group Phase Faults: Faults between phases within the same winding group (e.g., phases A-B in a nine-phase (1 × 9) configuration) primarily affect that group's contribution to torque production. The fault current is limited by the group's impedance and source impedance. Detection can leverage differential current measurements within each group.

Inter-group Phase Faults: Faults between phases in different winding groups create current circulation between groups that can bypass group isolation. These faults are particularly challenging to detect in separately-supplied configurations where each group has independent power electronics, as the fault current path includes machine internal paths not directly monitored.

For nine-phase machines, analysis shows that (9 × 1) separately-supplied configurations exhibit lower fault current peaks (180–250% rated) compared to (1 × 9) star-connected configurations (250–400% rated) for equivalent fault severity. However, the (1 × 9) configuration provides simpler fault detection due to current measurement at a common neutral point [14].

The fault detection complexity increases with phase number. For an n-phase machine, there are $\binom{n}{2} = n(n-1)/2$ possible phase-pair fault combinations. A three-phase machine has 3 possible phase-to-phase faults, while a nine-phase machine has 36 possible combinations, necessitating more comprehensive monitoring and diagnosis algorithms.

6.2.11 Turn-to-ground Fault Analysis

6.2.11.1 Characteristics and Mechanisms

Turn-to-ground faults, also referred to as phase-to-ground or ground faults, represent one of the most severe fault types in permanent magnet machine stator windings due to the large currents typically involved and significant safety implications. These faults occur when ground wall insulation between winding conductors and grounded machine components fails, creating a conductive path between an energized phase and the machine frame, stator core, or other grounded metallic structures. Ground faults can develop independently through direct insulation failure or evolve as the final stage of fault progression from initial inter-turn short circuits that propagate through thermal degradation of adjacent insulation layers.

The physical mechanisms initiating turn-to-ground faults involve breakdown of the ground wall insulation system designed to isolate phase conductors from grounded structural components. Unlike turn insulation providing electrical separation between adjacent conductor turns within a coil or between coils of the same phase, ground wall insulation must withstand the full phase-to-ground voltage potential while occupying space between conductors and slot walls in close proximity to grounded lamination stacks [15]. The demanding requirements placed on ground wall insulation combined with exposure to thermal, electrical, mechanical, and environmental stresses create vulnerability points where failure can occur.

Electrical stress represents a primary degradation mechanism for ground wall insulation. In inverter-fed permanent magnet machines, pulse width modulation voltage waveforms produce repetitive voltage surges that stress ground insulation through partial discharge mechanisms similar to those affecting turn insulation. However, the longer distance between phase conductors and ground creates higher voltage stress concentration compared to turn-to-turn spacing. Partial discharges within voids or at interfaces in ground insulation progressively erode dielectric material, reducing insulation thickness and electrical strength over thousands to millions of operating cycles.

Thermal stress compounds electrical degradation effects through accelerated aging of organic insulation materials. Ground wall insulation typically employs multilayer systems combining enamel conductor coatings, slot liners, tape wraps, and impregnating resins. Each material component exhibits characteristic temperature-dependent aging rates following exponential relationships similar to the ten-degree rule governing turn insulation life. Elevated temperatures from normal operation, localized hotspots from manufacturing defects, or thermal elevation resulting from existing inter-turn faults accelerate ground insulation degradation [16].

Mechanical stress contributes to ground fault initiation through several pathways. Electromagnetic forces during normal operation and transient fault conditions impose cyclic mechanical loading on conductors within slots, potentially causing abrasion between conductor surfaces and slot liners over extended operating periods. Thermal cycling creates differential expansion and contraction between copper conductors, insulation materials, and steel laminations with dissimilar thermal expansion coefficients. This differential movement generates mechanical stresses that can delaminate layered insulation structures or create cracks providing pathways for eventual electrical breakdown. Vibration from electromagnetic and mechanical sources transmits through the stator structure, causing conductor motion within slots that wears insulation surfaces over millions of operating cycles.

Environmental factors including moisture ingress, chemical contamination, and conducting particles can dramatically accelerate ground fault development. Moisture absorption reduces dielectric strength of many organic insulation materials while enabling electrochemical corrosion processes. Conducting particles from wear debris, manufacturing process residues, or external contamination can bridge insulation gaps, creating localized high field concentrations initiating partial discharge. Industrial atmospheres containing corrosive gases, oils, or chemical vapors may attack specific insulation compounds, leading to premature failure.

The fault progression from initial ground insulation weakness to complete phase-to-ground short circuit typically proceeds through several stages. Microscopic defects in ground insulation, whether introduced during manufacturing or developed through operational stress, create localized regions of enhanced electric field. These field enhancements initiate partial discharge activity when local fields exceed the inception voltage for the geometry and insulation material. Partial discharge erodes insulation material gradually, deepening channels through the insulation thickness. Eventually, the conducting channel penetrates completely through the ground insulation, establishing the phase-to-ground connection.

In some failure scenarios, inter-turn faults serve as precursors to ground faults through thermal mechanisms. The localized heating from short-circuited turn currents thermally degrades insulation in the immediate vicinity of the turn fault. If sufficient thermal elevation occurs, the temperature can exceed the thermal destruction threshold for ground insulation materials near the turn fault location. This thermal path enables rapid fault progression from turn-to-turn to turn-to-ground configurations within seconds to minutes under severe fault current conditions [17].

6.2.11.2 Grounding System Impact on Fault Characteristics

The system grounding configuration critically affects turn-to-ground fault current magnitudes and detection characteristics. Solidly grounded systems produce high fault currents approaching three-phase short circuit levels, enabling rapid overcurrent protection response but creating substantial electromagnetic forces and thermal stress. Resistance grounded configurations limit fault current to 25–400 A depending on design philosophy, reducing damage potential while maintaining detection capability. High resistance grounding further limits current to 5–10 A, requiring sensitive zero-sequence detection methods [18]. Ungrounded systems allow continued operation following first ground faults but create hazardous overvoltage conditions on healthy phases and risk catastrophic phase-to-phase faults if a second ground fault occurs [19]. The grounding method selection balances safety requirements, damage limitation, continuity of service, and protection system capabilities according to applicable industrial standards.

6.2.11.3 Detection Methods for Turn-to-ground Faults

Turn-to-ground fault detection employs techniques distinct from inter-turn fault monitoring due to the different electrical signatures and sensor accessibility. While inter-turn faults primarily affect phase current symmetry and harmonic content, ground faults produce characteristic zero-sequence currents and voltages, enabling specialized detection approaches.

6.2.11.4 Zero-sequence Current Detection

Zero-sequence current represents the common mode current component present when phase currents fail to sum to zero at the machine terminals. In a balanced three-phase operation without neutral connection, Kirchhoff's current law requires the instantaneous sum of the three-phase currents to equal zero. Any deviation from zero indicates either measurement asymmetry, current leakage to ground, or internal ground faults.

Ground fault detection using zero-sequence current typically employs a core balance current transformer encircling all three-phase conductors. In healthy operation, the net magnetic field from balanced three-phase currents cancels within the transformer core, producing zero output voltage. When ground fault current flows, it passes through only one phase conductor outbound and returns through ground rather than the other phase conductors, creating net flux in the core balance transformer and producing a measurable voltage proportional to ground fault current.

The sensitivity of zero-sequence current detection depends on the transformer turns ratio, burden resistance, and signal processing electronics. Modern ground fault relays achieve detection thresholds below 100 mA, enabling identification of incipient ground faults before they progress to destructive levels. However, capacitive charging current in machine windings and connecting cables can produce zero-sequence quantities even without faults, requiring careful threshold setting to avoid false trip conditions.

For multimachine installations, selective ground fault protection requires individual zero-sequence measurement for each machine rather than common bus monitoring. This selectivity ensures that only the faulted machine trips, allowing healthy machines to continue operating. Some advanced protection schemes compare the magnitude and phase of zero-sequence currents from multiple locations to identify the fault source when multiple machines connect to common buses.

6.2.11.5 Zero-sequence Voltage Monitoring

Zero-sequence voltage detection measures the common mode voltage relative to ground appearing at the machine terminals during ground faults. In grounded neutral systems, ground faults create neutral-to-ground voltage proportional to fault current and grounding impedance. Ungrounded or high impedance grounded systems develop substantial zero-sequence voltage during ground faults due to the voltage divider formed by the fault impedance and system impedance to ground.

Voltage transformers connected in broken delta configuration provide sensitive zero-sequence voltage detection. Three voltage transformers with primaries connected phase-to-neutral and secondaries connected in delta topology produce zero output voltage under balanced conditions. Ground fault events generate zero-sequence primary voltage producing a secondary winding voltage, enabling relay operation. This approach detects ground faults even when the fault current remains very small, making it particularly valuable for high resistance grounded or ungrounded systems where current-based detection becomes challenging.

Third harmonic voltage monitoring provides an alternative zero-sequence detection approach exploiting the natural third harmonic content in permanent magnet machines. During healthy operation, third harmonic voltages in each phase combine to produce a measurable neutral point voltage in star-connected windings without neutral ground connection. Ground faults disturb this harmonic voltage balance, creating detectable shifts in third harmonic magnitude or phase [18].

6.2.11.6 Insulation Resistance Monitoring

Periodic or continuous insulation resistance measurement provides early warning of ground insulation degradation before complete breakdown occurs. This preventive monitoring approach enables scheduled maintenance intervention, avoiding unplanned outages and potential safety incidents from unexpected ground faults.

Offline insulation resistance testing applies DC voltage between phase windings and ground with the machine de-energized, measuring leakage current through insulation. Standard test voltages range from 500 to 2,500 V depending on machine voltage rating. Measured resistance below specified thresholds, typically around 1 MΩ plus 1 kΩ per volt of rated voltage, indicates insulation degradation requiring further investigation or remedial action. Time-based trend analysis of insulation resistance measurements provides advanced indication of deteriorating insulation, enabling predictive maintenance scheduling [20].

Online insulation monitoring systems inject low frequency probing signals into the machine winding while operating, measuring the changing impedance to the ground as insulation degrades. These systems track insulation resistance trends during normal operation, avoiding the need for machine shutdown to perform testing. However, the probing signal amplitudes must remain carefully controlled to avoid interfering with normal operation or introducing additional insulation stress.

Partial discharge monitoring represents an advanced online technique detecting the high frequency electrical signals generated by partial discharge activity within degrading ground insulation. Specialized sensors capacitively couple to machine windings, capturing the microsecond-scale pulses produced when internal voids or surface tracking channels experience discharge events. Signal processing algorithms analyze the discharge magnitude, repetition rate, and phase relationship to applied voltage, characterizing insulation condition and estimating remaining life before ground fault occurrence [21].

6.2.11.7 Model-based Detection and Isolation

Advanced model-based fault detection algorithms employ observer structures comparing actual machine behavior against predictions from mathematical models. Discrepancies between measured and predicted quantities indicate fault presence, with the specific pattern of discrepancies enabling fault isolation and classification.

Sliding mode observers have demonstrated effectiveness for detecting turn-to-ground faults in permanent magnet machines by estimating winding parameters from terminal voltage and current measurements. The observer tracks resistance and inductance values for each phase winding independently. Ground faults alter the effective inductance and resistance seen from terminals, creating estimation errors that exceed healthy thresholds. This approach requires minimal additional sensing beyond voltage and current measurements already present for control purposes [17].

Extended Kalman filter approaches provide stochastic estimation frameworks incorporating measurement noise and modeling uncertainty. The filter predicts machine states based on dynamic equations, then corrects predictions using actual measurements weighted by estimated measurement reliability. Ground faults introduce unmodeled dynamics that manifest as increased innovation terms in the filter, enabling fault detection when innovation statistics exceed expected bounds. Parameter augmentation extends the state vector to include winding resistance parameters, enabling the filter to track resistance changes indicative of ground fault development.

The computational requirements of model-based approaches have historically limited deployment to research applications. However, increasing availability of inexpensive high-performance embedded processors now enables real-time implementation of these advanced algorithms in industrial motor drives and protection systems. The primary challenges involve robust parameterization of models to account for manufacturing variation, temperature dependencies, and aging effects without false fault indications from normal operating variations.

6.2.11.8 Differences from Inter-turn and Phase-to-phase Faults

Turn-to-ground faults exhibit distinct characteristics differentiating them from inter-turn short circuits and phase-to-phase faults in terms of current signatures, detection approaches, and protective response requirements.

The current magnitude in ground faults depends primarily on system grounding configuration rather than fault location within the winding. Inter-turn faults produce short circuit current constrained by the impedance of the turns involved in the fault, typically resulting in modest current increases except for severe faults involving many turns. Ground faults in solidly grounded systems can produce fault currents approaching three-phase short circuit levels regardless of turn count, creating more severe thermal and mechanical stress.

The electrical signature differences enable specialized detection strategies. Inter-turn faults create characteristic second harmonic components in synchronous reference frame currents and amplitude/phase asymmetry in stationary reference frame currents. Ground faults primarily manifest as zero-sequence current and voltage appearing only when conduction to ground occurs, with minimal second harmonic content in differential mode currents. This distinction allows protective relaying to differentiate fault types and apply appropriate response strategies.

The fault progression timescales vary significantly between fault types. Inter-turn faults can persist for extended periods if fault resistance remains high and thermal conditions allow continued operation without rapid escalation. Ground faults in grounded systems trigger immediate overcurrent protection due to large fault currents, preventing extended operation in faulted conditions. Ungrounded system ground faults may persist undetected for hours or days until a second ground fault occurs, creating catastrophic phase-to-phase conditions.

Safety implications distinguish ground faults from other winding fault types. Inter-turn and phase-to-phase faults primarily affect machine operation and reliability without necessarily creating personnel hazards. Ground faults can energize normally grounded machine frames at dangerous voltages if protective devices fail to operate, creating electric shock hazards. Electrical codes and industrial standards mandate ground fault protection for all medium and high voltage machines, while inter-turn fault monitoring remains optional depending on application criticality.

The repair complexity following faults varies with fault type. Minor inter-turn faults affecting a few turns in a single coil may be repairable by replacing the affected coil, if the machine design provides accessible coil connections. Ground faults typically cause more extensive damage due to the high current and energy involved, often requiring complete winding replacement rather than localized repairs. Phase-to-phase faults usually necessitate full rewinding due to the extensive insulation damage affecting multiple phases.

6.2.11.9 Safety Implications and Protection Requirements

Ground faults in permanent magnet machines create multiple safety hazards requiring comprehensive protective measures to protect personnel, prevent machine damage, and avoid secondary failures potentially affecting adjacent equipment.

The primary safety hazard from turn-to-ground faults involves energization of normally grounded metallic enclosures at dangerous voltages. When ground fault current flows, voltage appears across any impedance in the ground return path. If protective bonding conductors have insufficient cross-section or connections develop high resistance, substantial voltage can appear between different grounded components during fault conditions. Personnel contacting these surfaces while also touching other grounded conductors or standing on grounded floors may experience electric shock, with severity depending on voltage magnitude, current path through the body, and contact duration.

Ground fault protection systems must detect and clear faults rapidly enough to prevent dangerous touch voltages from persisting. Electrical codes specify maximum clearing times ranging from milliseconds to several seconds depending on fault current magnitude and system voltage level. Higher fault currents enable faster detection and shorter clearing times, while low current faults in high resistance grounded systems may require longer detection intervals, necessitating enhanced personnel protection through ground fault circuit interrupters or other protective devices.

Arc flash hazards represent another significant safety concern during ground fault conditions. When ground faults develop through arc channels rather than solid conductor-to-frame contact, the arc produces intense thermal radiation, pressure waves, and metal vapor release. Personnel performing maintenance or operation activities near the equipment during fault occurrence face severe burn injuries even without direct electrical contact. Arc flash hazard calculations must account for ground fault current magnitudes and protection system operating times to determine appropriate personal protective equipment and safety procedures.

Protective relaying for ground faults employs specialized schemes adapted to the grounding system configuration. Instantaneous ground overcurrent relays provide primary protection in solidly grounded systems, operating when ground fault current exceeds pickup settings typically set between 10% and 40% of machine rated current. Time-delayed ground overcurrent provides backup protection with intentional delays coordinating with upstream protective devices. Directional ground overcurrent protects systems with multiple sources, ensuring only devices between source and fault trip.

Sensitive ground fault protection using residual current detection enables fast response to low-magnitude faults in high-resistance grounded systems. These schemes measure the vector sum of phase currents through core balance transformers or electronic summation, detecting the zero-sequence component indicating ground fault presence. Settings typically range from 100 mA to several amperes depending on system charging current and desired sensitivity.

Differential protection schemes compare current entering and leaving protected zones, detecting faults through unbalanced current flow. Ground differential protection applies this principle specifically to ground fault detection, comparing neutral current against the sum of phase currents or employing multiple current transformer sets arranged to produce zero output during external faults but finite current for internal ground faults.

Regular testing and maintenance of ground fault protection systems ensures reliable operation when faults occur. Test procedures verify sensor accuracy, relay setting accuracy, circuit integrity from sensors through relay logic to trip circuits, and operating time measurements. Annual or more frequent testing intervals prevent undetected protection system failures that could leave equipment unprotected against ground faults.

6.3 Modeling and Analysis

Accurate modeling of short-circuit faults in permanent magnet machines requires analytical tools capable of predicting electromagnetic behavior under asymmetric winding conditions. Chapter 3 presents comprehensive coverage of modeling techniques for permanent magnet machines, including finite element methods, magnetic equivalent circuits, winding function theory, and field reconstruction methods. This section applies the magnetic equivalent circuit (MEC) methodology to inter-turn short circuit and phase-to-phase fault analysis, with particular emphasis on mathematical formulations in multiple reference frames that facilitate diagnostic indicator extraction.

Among the modeling approaches discussed in Chapter 3, the MEC method offers particular advantages for short-circuit fault analysis: physical flux path representation enabling intuitive interpretation of fault effects, straightforward incorporation of nonlinear magnetic characteristics, and computational efficiency suitable for parametric studies across varying fault severities [22, 23]. The MEC approach employed here incorporates refined air gap meshing using triangular cells to capture the asymmetric flux patterns characteristic of short-circuit conditions, achieving accuracy approaching finite element analysis while enabling comprehensive fault studies.

6.3.1 Reference Machine for Fault Analysis

Figure 6.4 illustrates the structure of a representative permanent magnet synchronous machine employed for model development and validation. The machine features outer rotor construction with surface-mounted permanent magnets, 40 poles arranged in 20 pole pairs, and 36 stator slots accommodating concentrated three-phase windings. Each phase comprises 12 coils with 80 turns per coil. Table 6.2 summarizes the geometric and electrical specifications of this machine, which serves as the reference design throughout the fault analysis development.

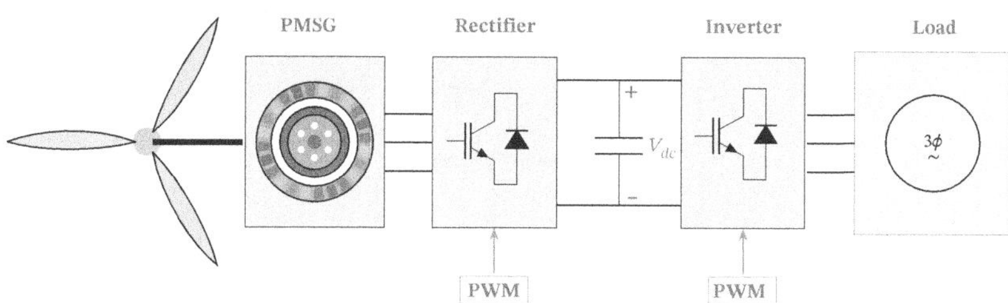

Figure 6.4 Geometrical structure and winding arrangement of the reference permanent magnet machine.

Table 6.2 Specifications of the reference permanent magnet machine.

Parameter	Value	Parameter	Value
Rated power (W)	1,500	Motor axial length (mm)	20
Rated RMS current (A)	3	Air gap length (mm)	1
Nominal speed (rpm)	600	PM thickness (mm)	2
Nominal frequency (Hz)	200	PM width (mm)	20
Number of stator pole pairs	20	PM flux density, B_{rem} (T)	1.23
Number of stator slots	36	Current density (A/mm^2)	4.5
Number of rotor pole pairs	20	Conductor size (AWG)	18

(Continued)

Table 6.2 (Continued)

Parameter	Value	Parameter	Value
Number of stator coil turns	80	Cogging torque (Nm)	0.25
Stator inner/outer diameter (mm)	160/250	PM material	NdFeB-N42
Rotor inner/outer diameter (mm)	252/270	Stator core material	M470-50A

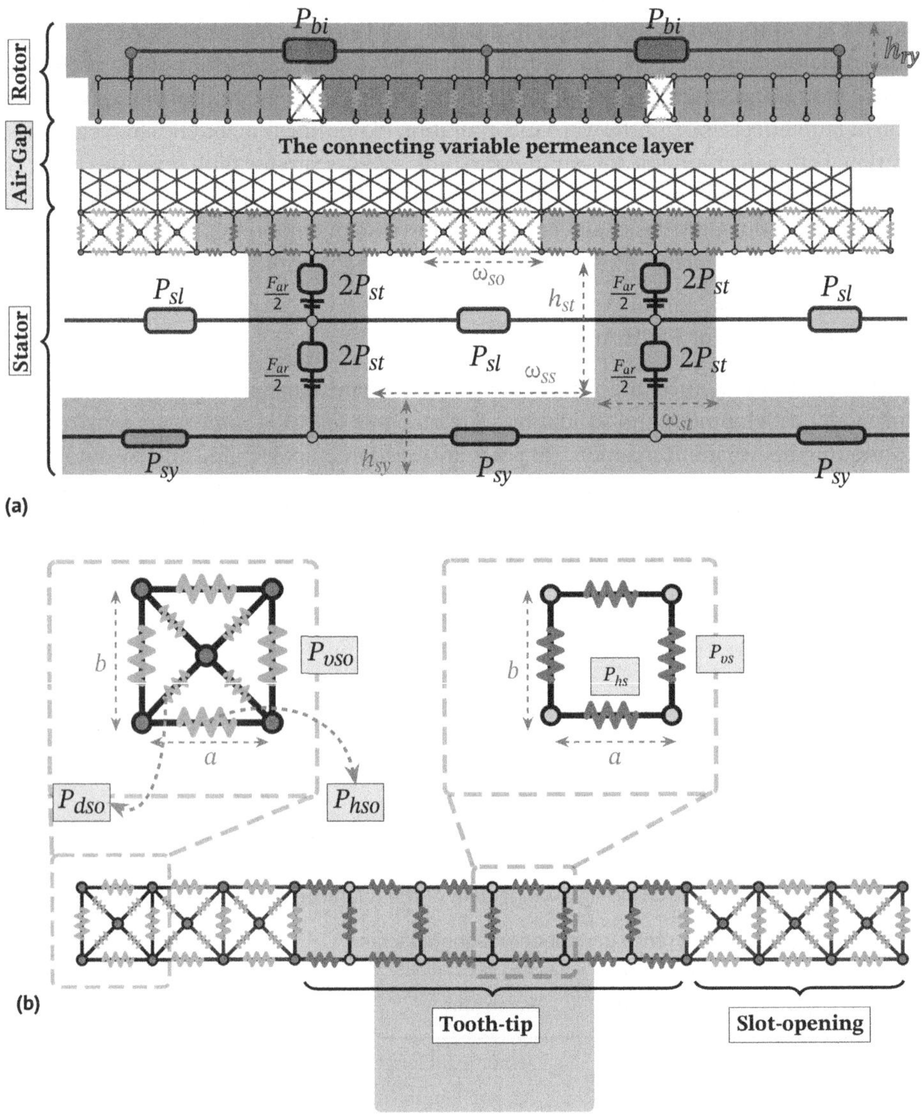

Figure 6.5 Magnetic permeance network modeling. (a) Stator yoke, teeth, and slot flux distribution, (b) detailed permeance network for tooth shoes and slot openings.

The MEC model follows the general methodology presented in Chapter 3, with the stator circuit divided into yoke, tooth, slot leakage, tooth shoe, and slot opening regions, and the rotor circuit comprising permanent magnets, rotor yoke segments, and inter-magnet air spaces. Figure 6.5

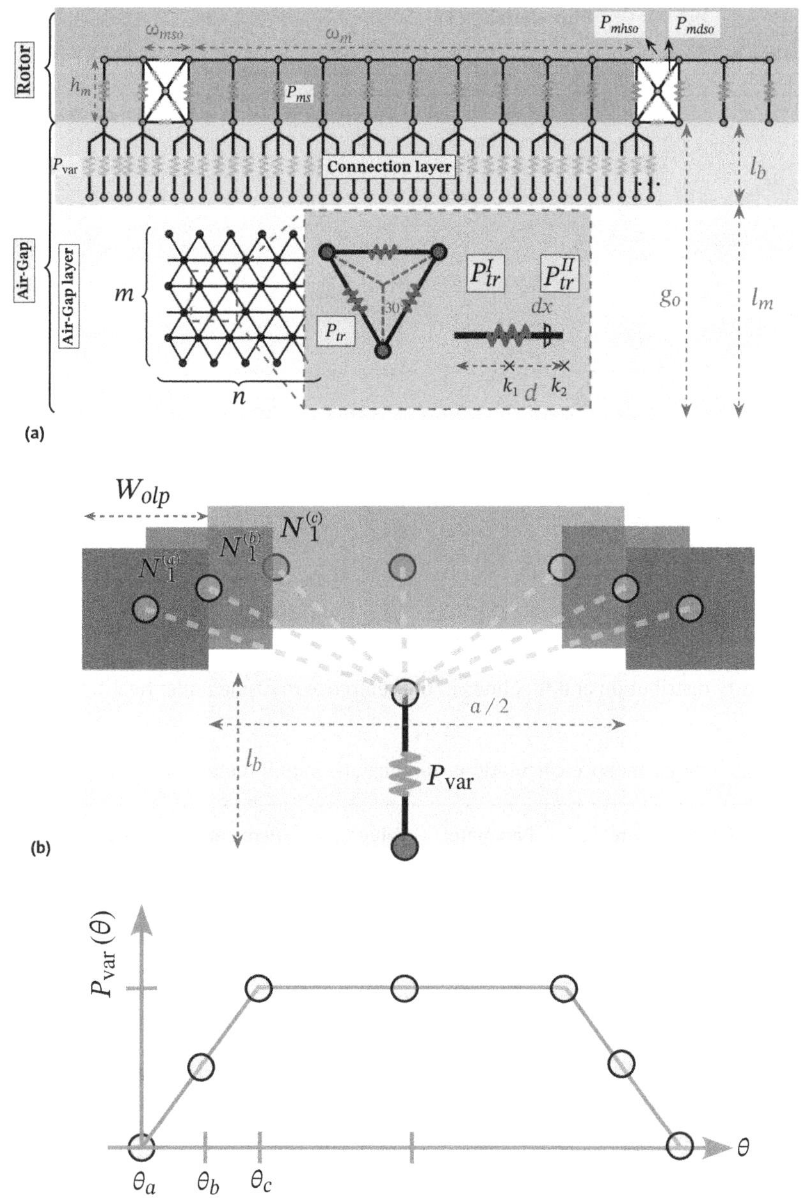

Figure 6.6 Rotor magnetic circuit representation. (a) Permeance network for permanent magnets and rotor yoke, (b) refined air gap meshing using triangular cells.

illustrates the permeance network topology employed for tooth shoes and slot openings, while Figure 6.6 shows the rotor circuit and refined air gap meshing.

Figure 6.7 presents the magnetic flux density distribution and field lines underrated load conditions. The flux pattern exhibits expected symmetry in healthy operation, with flux concentrating in tooth regions and spreading through the air gap to interact with rotor permanent magnets. Under fault conditions, this symmetry breaks down, producing diagnostic signatures exploited for fault detection.

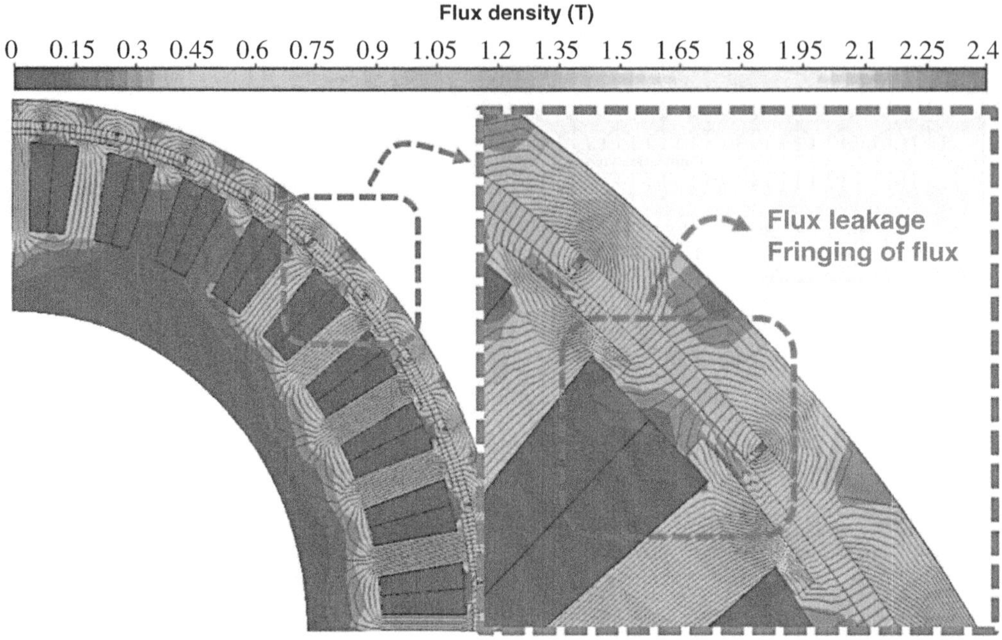

Figure 6.7 Magnetic flux density distribution and flux lines in the reference machine under healthy operation.

Table 6.3 Geometric parameters for permeance calculations (dimensions in millimeters).

Parameter	Value	Parameter	Value	Parameter	Value	Parameter	Value
R_{sy1}	80	W_{so}	8.1	h_m	2	m	5
R_{sy2}	125	W_{st}	8.5	W_m	20	n	1,600
R_{ry1}	126	W_{ss}	9.2	W_{mso}	0.4	a	2.7
R_{ry2}	135	h_{sy}	20.6	W_{olp}	0.05	b	2
l_{stk}	20	h_{ry}	6	W_{ps}	1	g_o	1
θ_{sy}	10 °	h_{so}	2	l_b	0.1	l_m	0.9
θ_{ry}	9 °	N_{km}	12	h_{st}	23	d	2

Table 6.3 summarizes the geometric parameters employed in permeance calculations for this machine.

6.3.2 Air Gap Modeling and Rotor-stator Coupling

The air gap region, where rotor and stator fields interact to produce electromagnetic torque, requires particular modeling attention. Conventional magnetic equivalent circuit approaches representing the air gap with single permeances between stator teeth and rotor surfaces cannot capture the detailed field patterns present under asymmetric conditions such as short-circuit faults. The enhanced approach employs triangular cell meshing, creating a dense permeance network throughout the air gap volume.

The air gap domain divides into m radial layers and n circumferential segments, creating $m \times n$ cells. Each cell further subdivides into two triangular elements, providing flexibility to represent arbitrary field patterns. The radial extent of each layer is

$$\Delta r = \frac{g_0 - l_b}{m} \tag{6.11}$$

and the circumferential span of each segment is

$$\Delta \theta = \frac{2\pi}{n} \tag{6.12}$$

where g_0 represents total air gap length and l_b denotes the final layer thickness adjacent to the rotor.

The permeances forming triangular cell edges require careful calculation to ensure conservation of flux and accurate representation of field patterns. Each triangular cell has three edges, and the permeance of each edge results from series combination of two permeances corresponding to the triangular regions on opposite sides of the edge. For an edge at radius r spanning angle $\Delta\theta$, the permeance is

$$P_{edge} = \mu_0 l_{stk} \int_{k_1}^{k_2} \frac{1}{r\Delta\theta} dr \tag{6.13}$$

where the integration limits k_1 and k_2 are selected to ensure all three edges of each triangle have equal permeance, creating a balanced network. This constraint leads to the result that edge permeances equal $\mu_0 l_{stk}$ independent of cell location within the mesh, greatly simplifying network construction while maintaining accuracy.

The air gap mesh maintains a fixed position relative to the stator while the rotor permanent magnets and yoke rotate at machine speed. Consequently, the permeance connections between the rotor magnetic circuit and the air gap mesh vary with rotor position. At each time step, the algorithm identifies which air gap mesh nodes overlap with rotor nodes and establishes appropriate permeance connections based on the overlap geometry.

The number of permeances connecting each rotor node to air gap nodes depends on the mesh resolution ratio. For the outer layer of air gap mesh nodes adjacent to the rotor, the number of connections to each rotor node is

$$N_{conn} = \left\lceil \frac{\theta_0}{\Delta\theta_{ag}} \right\rceil \tag{6.14}$$

where θ_0 represents the angular span of the overlap region and $\Delta\theta_{ag}$ denotes the angular width of air gap mesh cells. The overlap angle is

$$\theta_0 = \frac{N_{km}\theta_{ry}}{n} \tag{6.15}$$

where N_{km} is the number of rotor subdivisions and n is the number of circumferential air gap segments. This position-dependent coupling captures the rotating field interactions central to machine operation and enables prediction of torque ripple, cogging torque, and other phenomena depending on relative rotor-stator positioning. Under fault conditions, the asymmetric stator excitation creates circumferentially varying flux patterns that interact with rotor rotation to produce diagnostic signatures in time-domain signals.

6.3.3 Mathematical Representation of Short-circuit Faults

Short-circuit faults alter the winding inductance matrix, resistance matrix, and flux linkage relationships characterizing machine electrical behavior. Mathematical description of these alterations in multiple reference frames enables systematic development of fault detection algorithms exploiting specific signatures manifest in particular coordinate transformations.

6.3.3.1 Turn-to-turn Short Circuit in ABC Frame

Consider a turn-to-turn short circuit affecting phase A while phases B and C remain healthy. The fault involves a fraction ζ of the total phase A winding, where $\zeta = 1$ corresponds to a complete single-coil short circuit in a winding with multiple coils. The voltage equations for the three-phase machine in the ABC reference frame under fault conditions become

$$v^f_{abc} = R^f i^f_{abc} + \frac{d\lambda^f_{abc}}{dt} \tag{6.16}$$

where superscript f denotes fault conditions. The inductance and resistance matrices for the faulted machine are

$$L^f = \begin{bmatrix} (1-\zeta)^2 L & (1-\zeta)M & (1-\zeta)M \\ (1-\zeta)M & L & M \\ (1-\zeta)M & M & L \end{bmatrix} \tag{6.17}$$

$$R^f = \begin{bmatrix} (1-\zeta)^2 R & 0 & 0 \\ 0 & R & 0 \\ 0 & 0 & R \end{bmatrix} \tag{6.18}$$

where L represents healthy phase self-inductance, M denotes mutual inductance between phases, and R is phase resistance. These matrices reflect the reduced number of effective turns in the faulted phase, affecting both self-inductance (quadratic dependence on turn count) and mutual inductance (linear dependence).

The flux linkage of each phase includes contributions from permanent magnets and armature reaction. Under fault conditions, the permanent magnet flux linkage in phase A reduces proportionally to the fraction of remaining active turns

$$\lambda^f_{pm} = \begin{bmatrix} (1-\zeta)\lambda_m \cos(\theta_e) \\ \lambda_m \cos(\theta_e - 2\pi/3) \\ \lambda_m \cos(\theta_e + 2\pi/3) \end{bmatrix} \tag{6.19}$$

where λ_m represents the healthy phase permanent magnet flux linkage amplitude and θ_e is electrical angle.

The short-circuited turns themselves carry fault current determined by the voltage induced across the faulted segment and the fault impedance. The fault current is

$$i_f = \frac{\zeta \lambda_m \omega_s \sin(\theta_e) + \zeta(L - M)\frac{di_a}{dt}}{R_f + \zeta^2 R} \tag{6.20}$$

where ω_s represents electrical angular velocity and R_f denotes fault resistance. This current circulates through the faulted turns, producing localized heating and modifying the net magnetomotive force distribution.

6.3.3.2 Turn-to-turn Fault Signatures in DQO Reference Frame

Applying Park's transformation (see Chapter 3) to the faulted machine equations reveals distinctive fault signatures in the synchronous reference frame. Under turn-to-turn short circuit fault conditions, the DQO frame voltage equations become

$$v_d^f = R(1-\zeta)i_d^f + L_d\frac{di_d^f}{dt} - \omega_s L_q i_q^f - K_f\cos(2\omega_s t) \tag{6.21}$$

$$v_q^f = R(1-\zeta)i_q^f + L_q\frac{di_q^f}{dt} + \omega_s L_d i_d^f + \omega_s(1-\zeta)\lambda_m - K_f\sin(2\omega_s t) \tag{6.22}$$

$$v_0^f = Ri_0^f + L_0\frac{di_0^f}{dt} + K_f\sin(\omega_s t) \tag{6.23}$$

where K_f represents a fault-dependent coefficient incorporating fault severity and machine parameters. These equations reveal the diagnostic signatures of turn-to-turn faults in the DQO frame. The direct and quadrature axis currents develop second harmonic components at frequency $2\omega_s$, while the zero-sequence current develops a fundamental frequency component proportional to fault severity.

The DQ current solutions in fault conditions have the form

$$i_d^f = I_d^{dc} + I_d^{2\omega}\cos(2\omega_s t + \phi_d) \tag{6.24}$$

$$i_q^f = I_q^{dc} + I_q^{2\omega}\cos(2\omega_s t + \phi_q) \tag{6.25}$$

where the DC components represent steady-state torque-producing currents and the second harmonic components constitute the fault signature. The second harmonic amplitude increases with fault severity, providing a quantitative indicator for fault detection and progression monitoring.

6.3.3.3 Phase-to-phase Short Circuit Modeling

Phase-to-phase short circuits, where complete windings of two phases experience electrical contact, produce more severe electromagnetic imbalances than turn-to-turn faults. For a phase-to-phase fault between phases A and B, the inductance, resistance, and flux linkage matrices become

$$L^{ppsc} = \begin{bmatrix} L & L & M \\ L & L & M \\ M & M & L \end{bmatrix} \tag{6.26}$$

$$R^{ppsc} = \begin{bmatrix} R & R & 0 \\ R & R & 0 \\ 0 & 0 & R \end{bmatrix} \tag{6.27}$$

$$\lambda_{pm}^{ppsc} = \begin{bmatrix} \lambda_m \cos(\theta_e) \\ \lambda_m \cos(\theta_e - 2\pi/3) \\ \lambda_m \cos(\theta_e + 2\pi/3) \end{bmatrix} \tag{6.28}$$

where the identical first and second rows reflect the electrical paralleling of phases A and B. These matrices produce dramatically altered current distributions with very high amplitudes in the faulted phases, typically triggering protective devices rapidly.

In the DQO frame, phase-to-phase faults produce second harmonic components in d- and q-axis currents with substantially larger amplitudes than those from turn-to-turn faults of equivalent turn involvement. This amplitude distinction enables classification algorithms to differentiate fault types based on diagnostic indicator magnitudes.

6.3.3.4 Turn-to-turn Fault Signatures in $\alpha\beta 0$ Reference Frame

Clarke's transformation (see Chapter 3) provides an alternative representation in the stationary $\alpha\beta 0$ reference frame. In healthy balanced operation, the α and β current components have equal amplitudes and 90 ° phase difference, producing a circular trajectory in the $\alpha\beta$ plane. When a turn-to-turn fault affects phase A, this balance is disrupted and the $\alpha\beta 0$ currents become

$$i_\alpha^f = I_\alpha \cos(\omega_s t + \phi_\alpha) \tag{6.29}$$

$$i_\beta^f = I_\beta \cos(\omega_s t + \phi_\beta) \tag{6.30}$$

$$i_0^f = I_0 \cos(\omega_s t + \phi_0) \tag{6.31}$$

where the amplitude relationship $I_\alpha \neq I_\beta$ and phase difference $\phi_\beta - \phi_\alpha \neq 90\,°$ provide diagnostic signatures. The specific amplitude and phase relationships depend on which phase experiences the fault, enabling fault location identification from $\alpha\beta$ signal analysis.

For turn-to-turn faults in phases B and C, the current relationships become

$$i_\alpha^f = \frac{2}{3}\left(i_a - \frac{1}{2}i_b - \frac{1}{2}i_c\right) \tag{6.32}$$

$$i_\beta^f = \frac{2}{\sqrt{3}}(i_b - i_c) \tag{6.33}$$

leading to amplitude and phase signatures distinct from phase A faults. This phase-dependent signature variation enables diagnostic algorithms to identify not only fault presence and severity but also fault location among the three phases.

Phase-to-phase short circuits produce particularly distinctive signatures in the $\alpha\beta$ frame. For a fault between phases A and B, the currents become

$$i_\alpha^{ppsc} = \frac{4}{3}i_a - \frac{2}{3}i_c \tag{6.34}$$

$$i_\beta^{ppsc} = \frac{2}{\sqrt{3}}(i_a - i_c) \tag{6.35}$$

For phase-to-phase faults involving other phase combinations, analogous relationships apply. The phase difference between α and β signals under phase-to-phase fault conditions can reach 0 ° or 180 ° depending on fault location, providing unambiguous classification signatures distinct from turn-to-turn fault patterns where phase differences deviate from but remain near 90 °.

6.3.4 Model Validation and Parameter Studies

Validation of the magnetic equivalent circuit model requires comparison against finite element analysis, which provides benchmark accuracy through discretization of the continuous magnetic field problem. The finite element model employs the same geometric specifications as the magnetic equivalent circuit model to enable direct comparison.

Two-dimensional finite element analysis using ANSYS Maxwell software solves the magnetic vector potential formulation of Maxwell equations throughout the machine cross-section. The stator and rotor cores employ nonlinear B–H curves for M470-50A electrical steel with saturation characteristics matching the permeability functions used in the magnetic equivalent circuit model. The permanent magnets use demagnetization curves for NdFeB material with remanence of 1.23 T and relative recoil permeability of 1.05. Winding regions employ current density boundary conditions representing the actual conductor distribution.

The finite element mesh contains 329,486 elements with concentrated refinement in the air gap region and tooth shoe areas where field gradients concentrate. This fine discretization ensures solution convergence with numerical error below 1% for predicted torque and flux linkage. The transient solver uses 0.1 millisecond time steps to resolve fundamental and harmonic components in current and flux linkage waveforms. Twenty electrical cycles are simulated to reach periodic steady state and eliminate transient startup effects.

Figure 6.8 compares phase current waveforms predicted by magnetic equivalent circuit modeling, finite element simulation, and experimental measurement for both healthy operation and turn-to-turn short circuit conditions. The three methods show excellent agreement, with current amplitude predictions differing by less than 5% and phase relationships matching within 2 °. Under healthy conditions at rated load, the current amplitude is 4 A RMS with minor higher harmonic content from slot opening effects and magnetic saturation. The turn-to-turn fault reduces phase A current amplitude to approximately 3.7 A while slightly increasing phase B and C currents, consistent with the reduced effective turns in the faulted phase.

Phase-to-phase short circuit conditions produce dramatic current increases and severe waveform distortion. Figure 6.9 shows phase currents under phase-to-phase fault between phases A and B. The faulted phases carry very high circulating currents determined primarily by winding resistance and internal impedance of the power supply, while phase C current decreases. The magnetic equivalent circuit and finite element predictions again show close agreement, validating the model's capability to represent severe asymmetric fault conditions.

Additional validation examines other electromagnetic quantities including back electromotive force, air gap flux density distribution, and cogging torque. Figure 6.10 compares these predictions between magnetic equivalent circuit and finite element models.

The back electromotive force waveforms show sinusoidal shape with minor harmonic content from slot opening effects. The predicted fundamental amplitude differs by less than 2% between models. Air gap flux density distributions exhibit expected sinusoidal spatial patterns with local variations near slot openings. The cogging torque predictions match within 10%, which is acceptable considering the sensitivity of cogging torque to small geometric variations and mesh discretization effects.

Table 6.4 quantifies the computational requirements for finite element and magnetic equivalent circuit analyses. The magnetic equivalent circuit model achieves an approximately sixfold reduction in solution time compared to finite element analysis while maintaining accuracy within 5% for electromagnetic quantities of interest. This computational advantage enables comprehensive fault

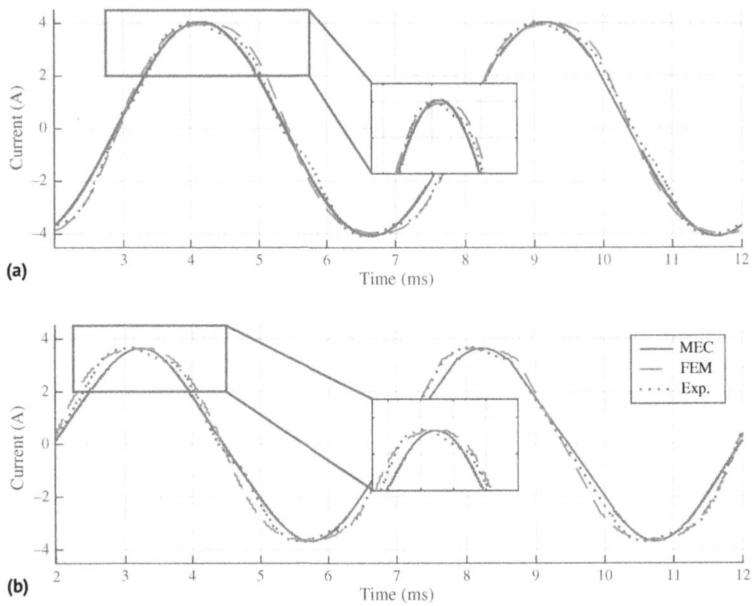

Figure 6.8 Current waveform comparison between magnetic equivalent circuit (MEC) modeling, finite element method (FEM) simulation, and experimental testing. (a) Healthy condition, (b) turn-to-turn short circuit condition.

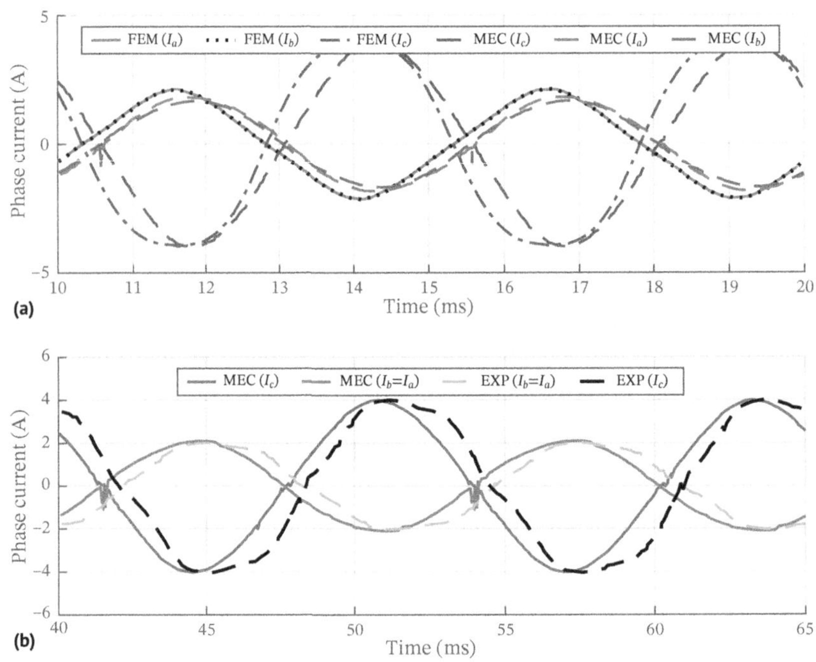

Figure 6.9 Phase current waveforms under phase-to-phase short circuit fault between phases A and B. (a) Magnetic equivalent circuit and finite element comparison, (b) magnetic equivalent circuit and experimental comparison.

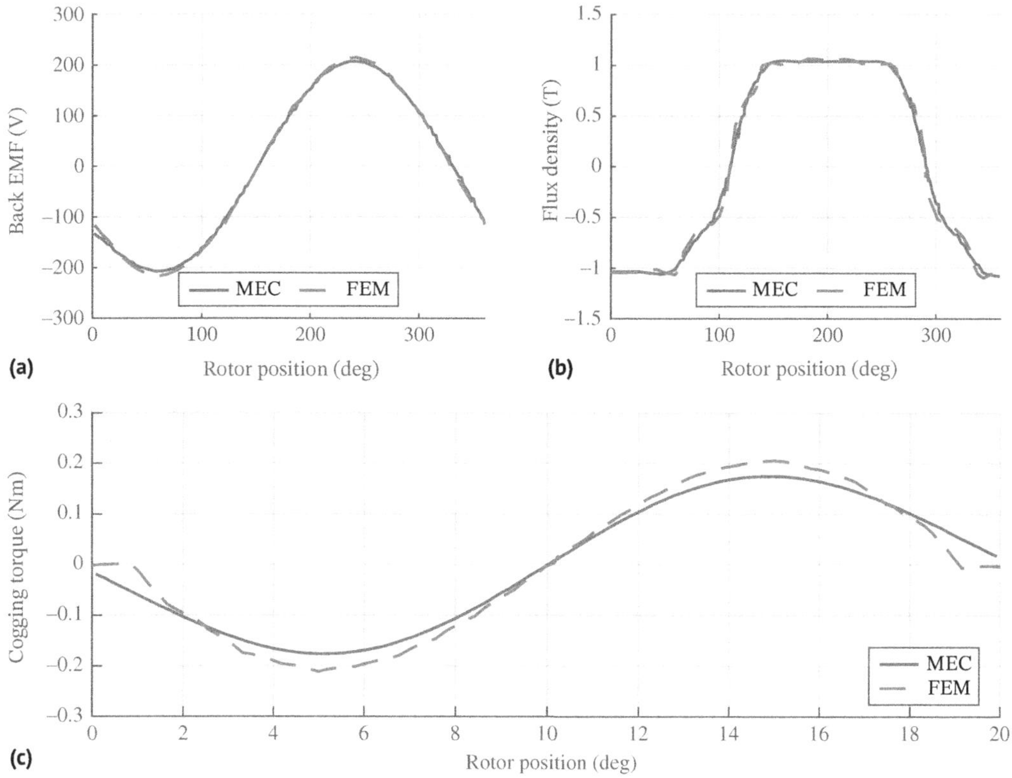

Figure 6.10 Comparison of magnetic equivalent circuit and finite element predictions for back electromotive force, air gap flux density, and cogging torque. (a) Back EMF waveforms, (b) air gap flux density distribution, and (c) cogging torque.

Table 6.4 Computational requirements comparison between finite element method and magnetic equivalent circuit modeling.

Parameter	FEM	MEC
Number of mesh elements	329,486	3,628
Number of circuit branches	—	7,124
Solution time under rated load (s)	22,520	3,620
Solution time under no load (s)	18,400	2,400
Cycles per mechanical rotation	20	20
Time per cycle at rated load (s)	1,000	175
Solution time step (ms)	0.1	0.1

studies evaluating numerous fault severities, locations, and operating conditions impractical with finite element methods.

The computation was performed on a system with Intel Core i7-4710HQ processor (2.5 GHz) and 16 GB RAM. The magnetic equivalent circuit solution employs Gauss elimination recursive methods for the nonlinear permeance network with iterative updates for saturation effects. Convergence typically requires 3–5 iterations per time step to achieve less than 1% change in flux values between successive iterations.

Understanding how fault detection indicators vary with fault severity informs threshold selection for diagnostic algorithms and guides sensor selection and signal processing requirements. The validated magnetic equivalent circuit model enables systematic parameter variation studies impractical with finite element or experimental approaches.

Turn-to-turn fault severity varies from incipient faults affecting small percentages of phase winding turns through complete single-coil short circuits. The fault severity parameter ζ ranges from 0.085 (corresponding to one coil shorted in a phase with 12 coils, representing 8.5% fault) to 1.0 (complete phase short circuit). Intermediate severities of 0.17, 0.34, 0.51, 0.68, and 0.85 represent progressive involvement of additional coils.

Figure 6.11 shows the q axis current as a function of fault severity. In healthy conditions ($\xi = 0\%$), the q axis current contains primarily a DC component near 4 A with minimal ripple. As fault severity increases, a second harmonic ripple develops in the q axis current, with amplitude proportional to fault severity. The DC component of q axis current decreases due to the reduced back electromotive force in the faulted phase. At 100% fault severity (complete single-phase short circuit), the second harmonic amplitude reaches approximately 1.5 A, clearly distinguishable from healthy operation.

The $\alpha\beta$ reference frame analysis reveals amplitude and phase relationships varying with fault severity and location. In healthy conditions, the α and β currents have equal amplitudes and $90\,°$ phase difference, producing a circular trajectory in the $\alpha\beta$ plane. As turn-to-turn fault severity increases, the trajectory becomes elliptical with the major axis oriented according to fault location, while phase-to-phase faults collapse the trajectory into a line whose orientation identifies the involved phases. These geometric signatures, analyzed in detail in the fault diagnosis section (Figure 6.16), provide robust indicators for both fault detection and classification.

These parameter variation results inform diagnostic algorithm development by establishing quantitative relationships between fault severity and observable signatures, guiding threshold selection to balance sensitivity to incipient faults against false alarm rates from measurement noise and normal operating variations.

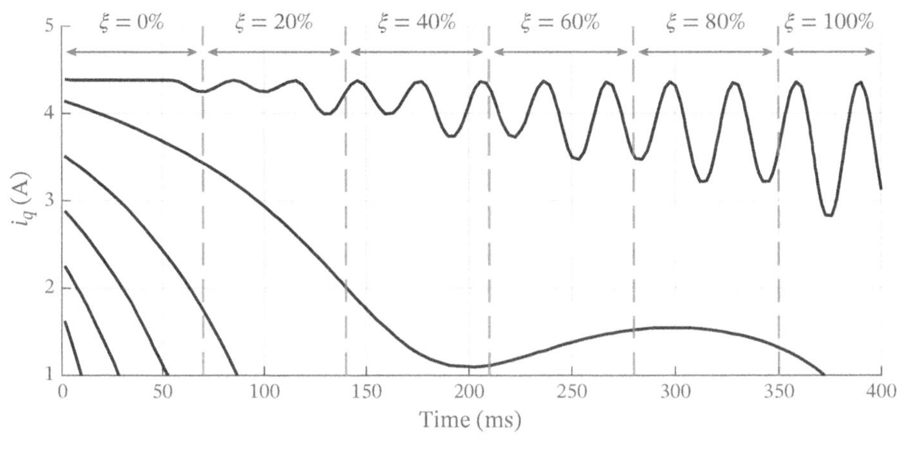

Figure 6.11 *Q* axis current for varying fault severity ξ.

The comprehensive modeling framework presented in this section provides the analytical foundation for fault diagnosis algorithm development detailed in Section 6.3. The magnetic equivalent circuit approach achieves computational efficiency, enabling extensive fault studies while maintaining accuracy within 5% of finite element benchmarks. The mathematical representations in ABC, DQO, and $\alpha\beta0$ reference frames reveal fault signatures exploitable for detection, classification, and localization. Experimental validation confirms model fidelity under both healthy and various short-circuit fault conditions, establishing confidence in predictions used to guide diagnostic indicator selection and threshold optimization.

6.4 Fault Diagnosis

The detection and classification of short-circuit faults in permanent magnet machines requires diagnostic algorithms capable of extracting fault signatures from available sensor signals while maintaining robustness against operating condition variations, measurement noise, and harmonic distortion from power electronic drives. This section presents comprehensive fault diagnosis approaches organized by signal type and processing methodology, progressing from fundamental motor current signature analysis through advanced reference frame transformations that enable speed-independent fault detection and discrimination between turn-to-turn and phase-to-phase fault types. Experimental validation with a 1.5 kW permanent magnet machine demonstrates the practical effectiveness of these techniques under diverse loading conditions, speed variations, and nonlinear loads representative of industrial applications.

6.4.1 Reference Frame Based Diagnostic Approaches

Motor current signature analysis fundamentals, including FFT-based spectral analysis, time-frequency methods, and general signal processing techniques, are presented in Chapter 4. This section focuses on the application of reference frame transformations specifically for inter-turn and short-circuit fault detection in permanent magnet machines, where these approaches offer distinct advantages over conventional frequency-domain analysis.

Under inter-turn short-circuit fault conditions, the rotating flux in the air gap becomes asymmetric, inducing second harmonic components in transformed currents beyond those present in healthy operation. These fault-related signatures arise from the interaction between the fundamental rotating field and the asymmetric impedance distribution caused by the faulted winding. While conventional FFT-based MCSA can detect these signatures, reference frame transformations provide speed-independent fault indicators that overcome the limitations of fixed-frequency analysis during variable-speed operation.

The coordinate transformations described in Chapter 4, specifically Park (DQO) and Clarke ($\alpha\beta0$) transformations, enable extraction of fault features that appear as distinct signatures more readily separated from normal operating characteristics. The DQO transformation converts fault-induced asymmetries into second harmonic ripple on otherwise DC quantities, while the $\alpha\beta0$ transformation reveals geometric trajectory deformations that indicate both fault presence and faulted phase location. The following subsections develop these reference frame-based diagnostic approaches for short-circuit fault detection.

6.4.1.1 DQO Reference Frame Diagnostic Applications

Applying the Park transformation (Chapter 4) to inter-turn fault conditions reveals characteristic second harmonic components in d and q axis currents. Under balanced steady-state healthy conditions, these quantities are DC values, but short-circuit faults introduce AC ripple at twice the electrical frequency with amplitude proportional to fault severity.

6.4.1.2 Signal Processing Algorithm for DQO Frame

The diagnostic methodology based on DQO reference frame processing exploits the characteristic second harmonic component that appears in d and q axis currents under short-circuit fault conditions. The processing algorithm incorporates short-time windowing to enable real-time fault detection and transient condition monitoring. Figure 6.12 illustrates the complete signal flow from three-phase current measurements through Park transformation, window-based processing, and fault index calculation.

The algorithm operates on the q axis current component, which contains the primary torque-producing current and exhibits clear fault signatures. In healthy operation at constant speed and load, the q axis current comprises primarily a DC component corresponding to steady-state torque production with minimal AC ripple. When turn-to-turn or phase-to-phase short circuits occur, second harmonic components develop with amplitudes proportional to fault severity.

The processing employs windowing with very short length, where each signal segment enters a window of predetermined duration. Within each window, the algorithm calculates the amplitude of AC ripple and the DC component. The window duration selection balances competing requirements: shorter windows enable faster fault detection and better transient response, while longer windows provide more accurate amplitude estimation and improved noise rejection. Typical window durations range from several fundamental periods to tens of periods depending on the required detection speed and signal quality.

After extracting amplitude and DC components within each window, the fault index calculation proceeds according to the mathematical formulation derived from the DQO frame fault equations. The q axis current in healthy and faulty conditions follows the form:

$$i_q = I_q^{dc}(1 + \eta_{sc}\cos(2\omega_s t)) \tag{6.36}$$

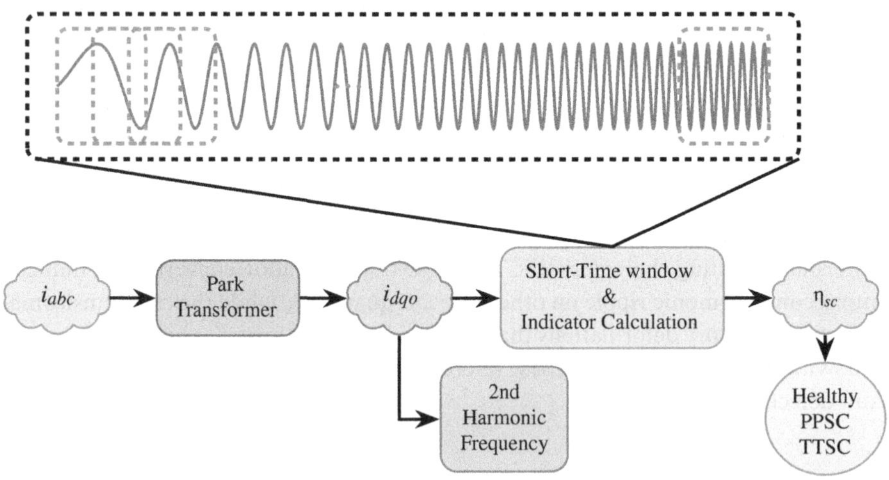

Figure 6.12 Flowchart of the DQO reference frame fault detection and classification algorithm.

where I_q^{dc} represents the DC component, η_{sc} denotes the normalized fault severity index, and ω_s indicates electrical angular frequency. This representation reveals that the fault signature manifests as amplitude modulation of the DC component at twice the electrical frequency.

6.4.1.3 Fault Index Calculation and Interpretation

The dimensionless fault index quantifying short-circuit severity is defined as:

$$\eta_{sc} = \frac{2 \times \text{Amplitude of AC ripple}}{\text{DC component}} \tag{6.37}$$

This formulation provides several advantageous properties. First, the index remains independent of machine speed because both numerator and denominator scale proportionally with speed variations, canceling speed dependence. Second, the dimensionless nature enables comparison across different machine sizes and ratings. Third, the index provides quantitative fault severity indication rather than binary fault detection.

For turn-to-turn short circuit faults, the theoretical value of the fault index depends on fault severity parameter ζ (representing the fraction of phase winding involved in the fault). When ζ equals 1.0, corresponding to complete short circuit of a single coil in a multicoil phase winding, the fault index approaches unity. For phase-to-phase short circuits, the fault index reaches approximately 2.0, providing clear discrimination between these fault types.

The speed-independent characteristic proves particularly valuable for applications involving variable speed operation. As machine speed varies, both the amplitude and frequency of q axis current change proportionally. However, because the fault index divides the AC amplitude by the DC component, and both quantities scale with the same speed dependence, the ratio remains constant regardless of speed. This enables application of fixed detection thresholds across the entire operating speed range without requiring speed-dependent threshold adaptation.

In healthy conditions, the fault index value remains negligible, limited primarily by measurement noise, minor manufacturing asymmetries, and numerical precision in the calculation. Typical healthy values remain below 0.05–0.10 depending on machine quality and measurement system characteristics. When phase-to-phase short circuits occur, the index rapidly increases to values near 2.0. For turn-to-turn faults, the index magnitude correlates with fault severity, increasing progressively as additional turns become involved in the fault. This quantitative relationship enables trending analysis to monitor fault progression over time and supports predictive maintenance strategies.

6.4.2 Experimental Validation of DQO Frame Methods

Experimental testing validates the DQO reference frame diagnostic approach using the 1.5 kW, 40-pole, 36-slot permanent magnet machine operating at 600 rpm. The test setup incorporates a power network analyzer with 8 kHz sampling rate to measure and record three-phase stator current waveforms. Turn-to-turn short circuit conditions are created by short-circuiting one coil of phase A through a variable resistor, enabling controlled fault severity variation from incipient faults through complete coil short circuits.

Figure 6.13 presents q-axis current waveforms comparing magnetic equivalent circuit modeling predictions with experimental measurements for healthy operation and turn-to-turn short circuit conditions. In healthy operation, the q axis current exhibits a 4 A DC component with minimal ripple, validating the expected behavior under balanced conditions. When turn-to-turn fault occurs,

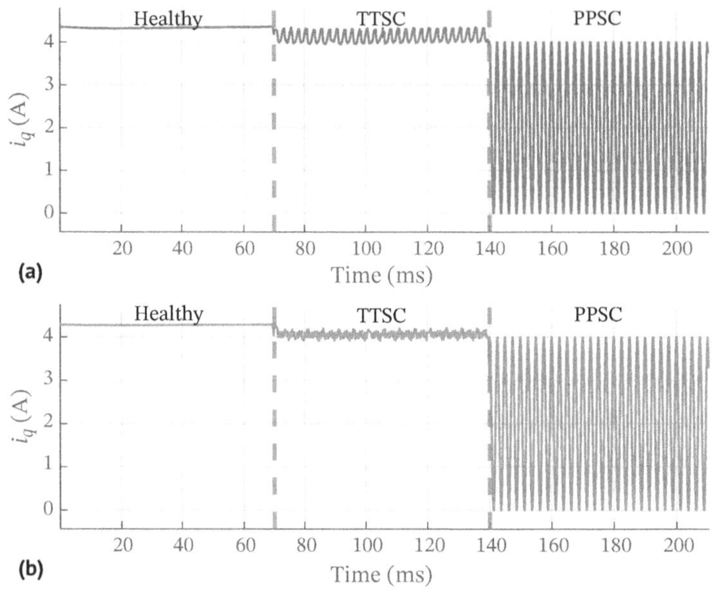

Figure 6.13 Comparison of q-axis current under healthy and turn-to-turn short circuit conditions showing increased second harmonic ripple indicating fault presence. (a) Magnetic equivalent circuit prediction, (b) experimental measurement.

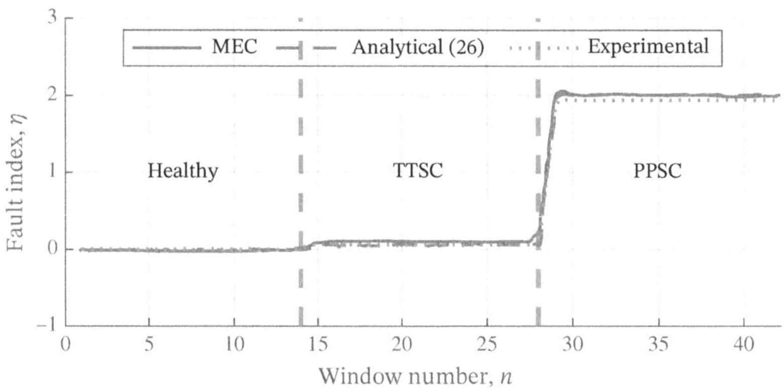

Figure 6.14 Fault index η comparison across fault severities: magnetic equivalent circuit modeling, analytical equation prediction, and experimental validation showing quantitative correlation with fault severity.

increased ripple amplitude becomes evident while the DC component experiences slight reduction consistent with decreased back electromotive force in the faulted phase.

Phase-to-phase short circuit conditions produce dramatically larger q axis current variations compared to turn-to-turn faults. The second harmonic amplitude increases substantially, providing clear discrimination between fault types based on the calculated fault index magnitude. Figure 6.14 demonstrates fault index values calculated from both magnetic equivalent circuit simulations and experimental data across the range from healthy operation through various fault severities to phase-to-phase short circuits.

The experimental results confirm the theoretical predictions with excellent agreement. For the 8.5% turn-to-turn fault (one coil shorted in a 12-coil phase), the measured fault index reaches approximately 0.15, matching the analytical prediction within experimental uncertainty. As fault severity increases to 100% (complete coil short circuit), the index approaches unity as expected. Phase-to-phase faults produce indices near 2.0, clearly distinguishable from even severe turn-to-turn faults and enabling unambiguous fault type classification.

6.4.2.1 Speed Variation Robustness Testing

A critical test of any diagnostic algorithm's practical utility involves performance under speed variations representative of actual operating conditions. Figure 6.15 presents experimental results demonstrating fault index calculation during speed transients. The machine operates initially at 600 rpm, then speed decreases to 300 rpm over a 0.2 second period, simulating a rapid load change or commanded speed reduction.

As speed decreases, both the amplitude and frequency of the three-phase currents reduce proportionally. The q axis current DC component and second harmonic amplitude both decrease with speed, but their ratio as quantified by the fault index η remains constant throughout the transient. This experimental confirmation validates the speed-independence property, demonstrating that

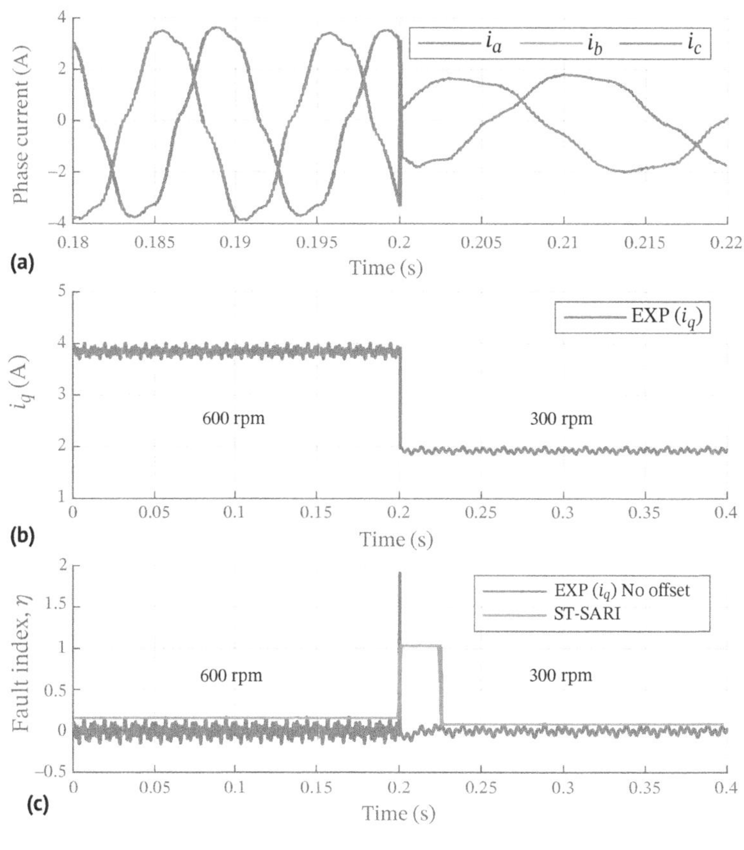

Figure 6.15 Experimental demonstration of speed-independent fault detection during transient speed reduction from 600 to 300 rpm. (a) Three-phase currents during speed reduction, (b) q-axis current showing amplitude and frequency reduction, (c) fault index η maintaining constant value despite speed change.

the diagnostic algorithm maintains reliable fault detection without requiring speed measurement or adaptive thresholding.

The ability to maintain constant fault index values during speed transients enables application in demanding industrial scenarios including traction drives, wind turbines, and variable speed industrial processes where speed varies continuously during normal operation. The speed-independent characteristic eliminates the need for precise speed measurement or estimation, simplifying implementation and improving reliability.

6.4.2.2 $\alpha\beta 0$ Reference Frame Diagnostic Applications

The Clarke transformation (Chapter 4) provides complementary diagnostic capability to the DQO approach, with the stationary $\alpha\beta 0$ frame revealing geometric trajectory signatures that enable both fault detection and faulted phase identification.

6.4.2.3 Amplitude and Phase Difference Based Detection

In healthy balanced three-phase operation, the α and β current components have equal amplitudes and maintain precisely 90 ° phase difference. Inter-turn short-circuit faults disrupt this symmetry, producing amplitude inequality and phase deviation in the $\alpha\beta$ plane that serve as the basis for fault detection and localization.

When turn-to-turn short circuit occurs in any phase, the affected phase current experiences altered amplitude and phase relative to healthy phases. This asymmetry propagates through the Clarke transformation, creating amplitude inequality and phase deviation in the $\alpha\beta$ plane. The specific pattern of amplitude and phase changes depends on which phase experiences the fault, enabling fault location identification. For phase A turn-to-turn faults, the β component amplitude becomes larger than α component, while the phase difference remains near but not exactly 90 °. Phase B faults produce different amplitude relationships with phase difference less than 90 °, and phase C faults generate distinct signatures.

Figure 6.16 presents $\alpha\beta$ current loci obtained from MEC modeling for turn-to-turn short-circuit faults applied to each phase at severities of 25%, 50%, 75%, and 100%. In healthy operation, the i_α-i_β trajectory forms a circular locus with uniform radius corresponding to the balanced phase current amplitude. As fault severity increases, the circular trajectory progressively deforms into an ellipse whose major axis shrinks and rotates depending on the faulted phase. For phase A faults, the ellipse narrows predominantly along the horizontal (i_α) direction, while phase B and phase C faults produce elliptical deformations oriented along distinct directions separated by approximately 120 °, consistent with the spatial displacement of the three-phase windings. This phase-specific directional sensitivity enables simultaneous identification of the faulted phase and estimation of fault severity from a single $\alpha\beta$ trajectory measurement. To improve the accuracy of trajectory-based detection in practice, the α and β current signals are sampled using a short-time window before processing, reducing the influence of transient disturbances and harmonic content.

Phase-to-phase short circuits produce particularly distinctive $\alpha\beta$ signatures compared to the turn-to-turn patterns shown in Figure 6.16. When two phases experience direct electrical connection, their currents become constrained to sum to zero in a star-connected winding without neutral conductor. This constraint collapses the two-dimensional elliptical trajectory into a one-dimensional lobe, providing unambiguous discrimination between turn-to-turn and phase-to-phase faults. The lobe orientation indicates which pair of phases experiences the short circuit.

6.4.2.4 Zero-crossing Based Phase Difference Extraction

The practical implementation of $\alpha\beta$ frame diagnostics requires efficient algorithms for extracting amplitude and phase difference from α and β signals in real-time. The proposed approach employs zero-crossing detection combined with short-time windowing to achieve computational efficiency suitable for real-time embedded implementation while maintaining accuracy even under harmonic distortion conditions.

The phase difference between α and β signals is calculated from their zero-crossing time measurements according to:

$$\phi_{\alpha\beta} = 360° \times f \times (t_\beta - t_\alpha) \tag{6.38}$$

where t_α and t_β represent zero-crossing times of α and β signals respectively, and f denotes the fundamental frequency within the analysis window. This approach offers significant computational advantages over alternative methods such as Fourier analysis or correlation techniques, requiring only simple sign detection and timestamp recording.

A critical implementation detail involves ensuring correct zero-crossing point selection. Within each analysis window, multiple zero-crossings may occur depending on window duration relative to signal period. Using incorrect zero-crossing pairs produces erroneous phase difference estimates. Figure 6.17 illustrates the correct and incorrect zero-crossing selections, demonstrating that the first zero-crossing of the β signal following window start may not represent the correct measurement point.

To ensure accurate phase difference calculation, the algorithm incorporates a simple conditional check verifying that the selected β zero-crossing occurs after the α zero-crossing within the same signal period. The robust phase difference calculation becomes:

$$\phi_{\alpha\beta} = 360° \times f \times (t_\beta - t_\alpha), \quad \text{where} \quad t_\beta > t_\alpha \tag{6.39}$$

and both times correspond to the same signal cycle. This condition effectively eliminates phase wraparound errors and ensures monotonic phase difference values, facilitating threshold-based fault detection.

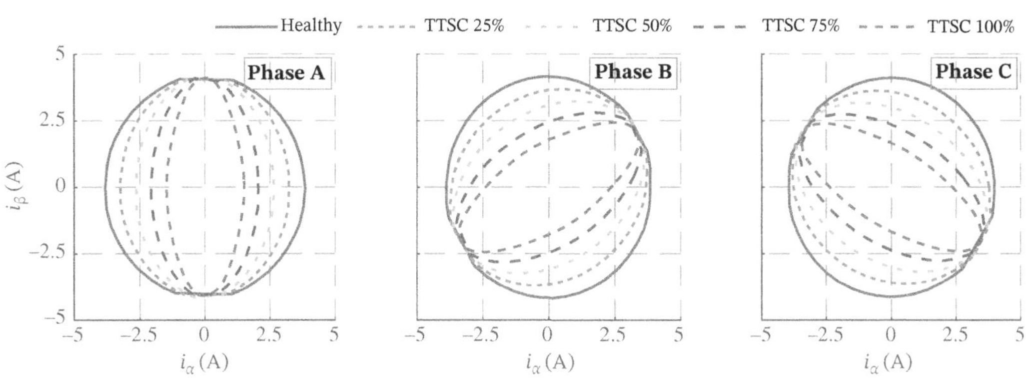

Figure 6.16 $\alpha\beta$ current loci from MEC modeling showing the effect of turn-to-turn short-circuit faults at varying severities for each phase. The healthy circular trajectory deforms into phase-specific elliptical patterns as fault severity increases.

The fundamental frequency f required for phase difference calculation can be determined directly from zero-crossing measurements within the window, eliminating the need for external frequency estimation or measurement. The period between successive same-direction zero-crossings provides instantaneous frequency information. This self-contained frequency estimation enables application in variable frequency drives without requiring additional sensors or communication with the drive controller.

6.4.2.5 Window-based Real-time Processing

The diagnostic algorithm employs sliding window processing where signal segments of fixed duration enter the analysis pipeline sequentially. Within each window, the algorithm extracts amplitude and phase difference, compares these parameters against healthy baseline values, and updates fault status indicators. Figure 6.18 illustrates the window-based processing concept showing how amplitude and phase difference parameters update as the analysis window advances through the signal time series.

The window duration selection involves similar tradeoffs as discussed for DQO frame processing. Shorter windows provide faster fault detection and better response to transient conditions, particularly valuable for phase-to-phase faults requiring rapid detection before catastrophic damage occurs. Longer windows improve parameter estimation accuracy and noise rejection but increase detection latency. Typical implementations employ windows spanning 2–5 fundamental periods, balancing these competing requirements based on application priorities.

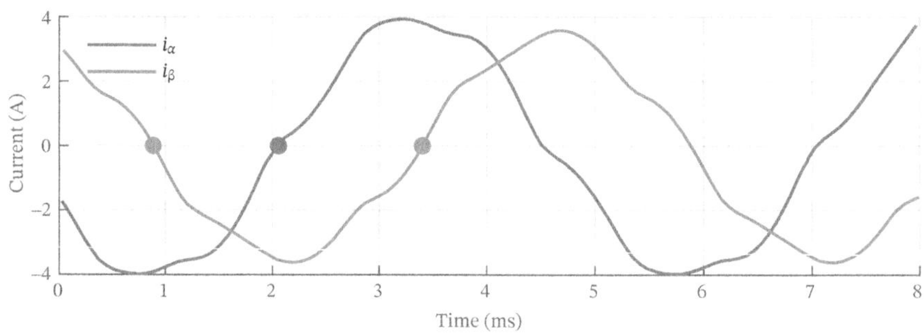

Figure 6.17 Window content showing α and β signals.

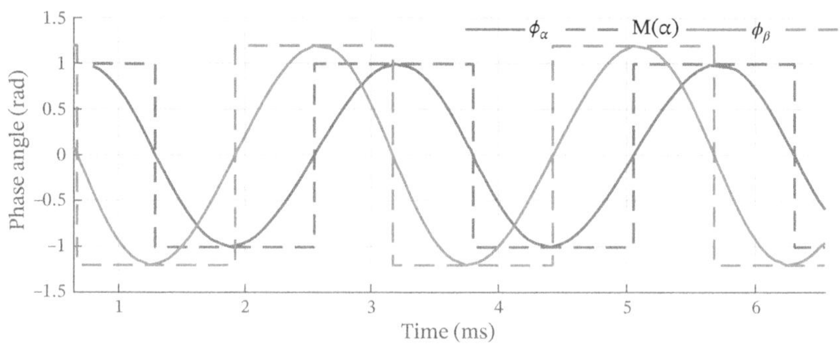

Figure 6.18 Phase angle difference between α and β signals.

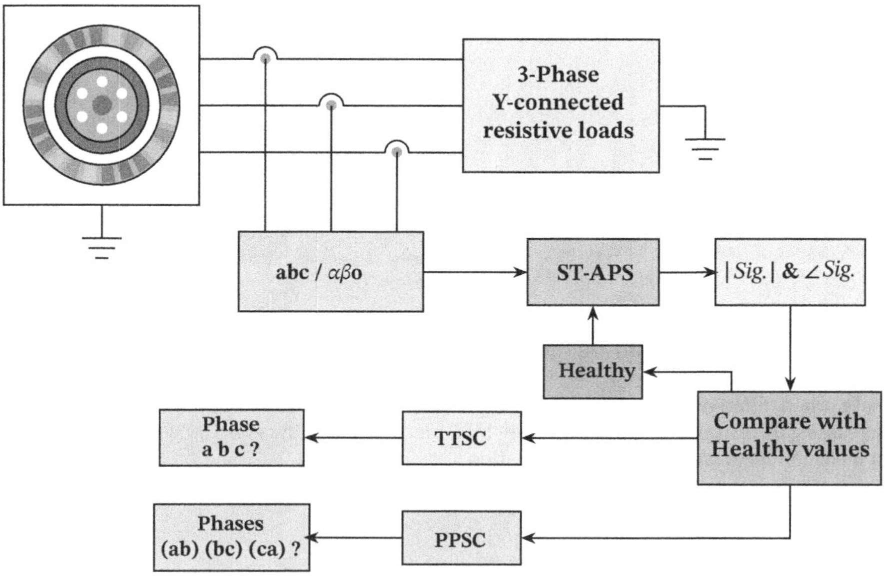

Figure 6.19 Fault detection and isolation algorithm flowchart for $\alpha\beta$ reference frame processing.

To extract signal amplitude within each window, the algorithm first removes the DC component through high-pass filtering or average subtraction. The peak value of the resulting AC signal provides amplitude estimation. The product of the DC-removed signal and its sign function (indicating signal polarity) enables efficient peak detection without requiring computationally intensive operations. This simple approach maintains computational efficiency while providing adequate accuracy for fault detection purposes.

Figure 6.19 presents the complete fault detection and isolation algorithm flowchart for $\alpha\beta$ reference frame processing, illustrating the sequential operations from three-phase current acquisition through Clarke transformation, window-based parameter extraction, and fault classification logic. The system acquires three-phase currents through current transformers, applies the Clarke transformation to obtain $\alpha\beta$ components, and extracts amplitude and phase information using short-time average power spectrum (ST-APS) processing. Comparison with healthy reference values enables classification into turn-to-turn short circuit (TTSC) or phase-to-phase short circuit (PPSC) faults, with subsequent identification of the affected phase or phase pair.

6.4.3 Experimental Validation and Phase Identification

The phase angle difference extracted from zero-crossing analysis provides complementary information for phase identification. Figure 6.20 shows phase angle evolution for healthy operation and turn-to-turn faults at two severity levels. The healthy phase angle maintains constant $90°$ separation between α and β signals. As fault severity increases from 8.5% to 100%, the phase angle deviation from $90°$ increases monotonically, providing quantitative fault severity indication analogous to the fault index in DQO frame but with additional phase location information.

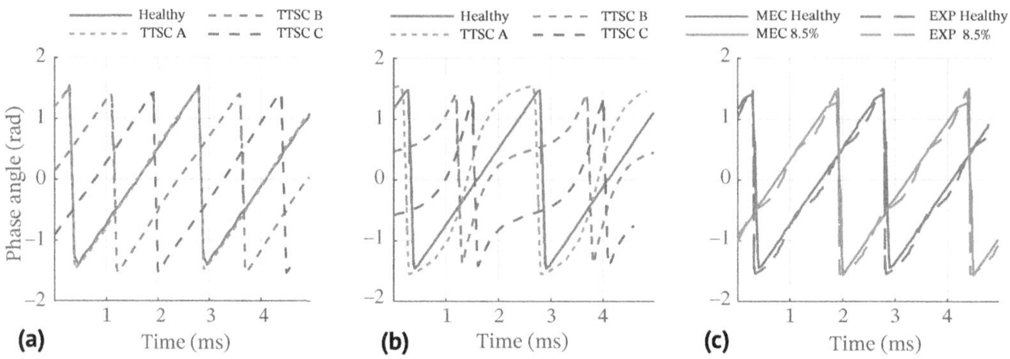

Figure 6.20 Phase angle difference between α and β signals showing progressive deviation from 90 ° as fault severity increases. (a) Healthy and 8.5% fault from MEC model, (b) healthy and 100% fault from MEC, (c) experimental measurements validating phase deviation.

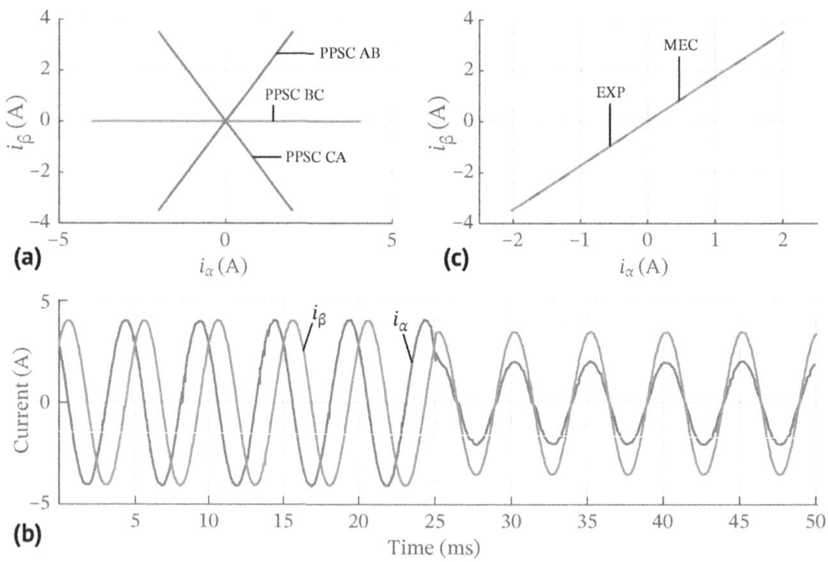

Figure 6.21 Phase-to-phase fault signatures in $\alpha\beta$ plane showing line trajectories enabling fault type classification and location identification. (a) Loci of all three phase-to-phase fault scenarios from MEC modeling, (b) experimental validation of fault signatures, (c) calculated currents in time using MEC.

6.4.3.1 Phase-to-phase Fault Discrimination

Phase-to-phase short circuits produce characteristic signatures in the $\alpha\beta$ frame distinct from turn-to-turn faults. The constraint that two directly connected phases must carry equal and opposite currents (in star connection) forces the $\alpha\beta$ trajectory to collapse from two-dimensional ellipse to one-dimensional line. This geometric distinction enables unambiguous classification between turn-to-turn and phase-to-phase fault types based solely on trajectory shape analysis.

Figure 6.21 presents $\alpha\beta$ current trajectories for phase-to-phase faults involving each possible phase pair combination. The three distinct line orientations correspond to the three possible phase-to-phase fault locations, enabling complete fault isolation to specific phase pairs. The comparison between magnetic equivalent circuit modeling and experimental measurements validates the diagnostic approach under these severe fault conditions.

The phase angle difference provides additional confirmation of phase-to-phase faults. When two phases become electrically connected, their phase relationship fundamentally changes from the balanced 120 ° separation to either 0 ° (same phase) or 180 ° (opposite phase) depending on fault location. This dramatic phase shift supplements the trajectory shape analysis, providing redundant information improving detection reliability and robustness against false alarms.

6.4.3.2 Robustness Under Challenging Operating Conditions

Practical diagnostic systems must maintain reliable performance under diverse operating conditions, including harmonic loads, nonlinear loads, speed variations, and transient events. Comprehensive experimental validation under these challenging conditions demonstrates the robustness of the developed diagnostic approaches and establishes confidence for industrial deployment.

6.4.3.3 Harmonic and Nonlinear Load Testing

Inverter-fed permanent magnet machines frequently supply harmonic or nonlinear loads including diode rectifiers, thyristor converters, and pulse width modulated inverter loads. These loads inject current harmonics that can potentially interfere with fault signatures, reducing detection sensitivity or causing false alarms. Systematic testing under four representative load types validates diagnostic performance under harmonic distortion conditions.

The test configuration connects the permanent magnet machine to four load types: resistive linear load (baseline), diode bridge rectifier with capacitive filter, thyristor controlled resistive load, and IGBT-based pulse width modulated inverter load. Each load type produces distinct harmonic spectra, with the linear load providing minimal harmonic content and the nonlinear loads generating significant harmonic distortion. Turn-to-turn short circuit at 8.5% severity in phase B is introduced at time $t = 0.4$ seconds, progressing to full 8.5% severity by $t = 0.5$ seconds, enabling evaluation of fault detection during the transient fault development period.

Figure 6.22 presents three-phase current waveforms before and after turn-to-turn fault occurrence for all four load types. The current waveform distortion increases progressively from linear load (nearly sinusoidal) through increasingly nonlinear loads (highly distorted with visible

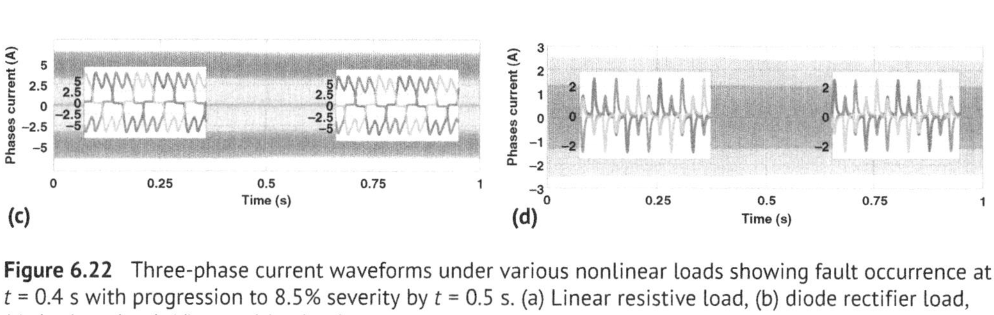

Figure 6.22 Three-phase current waveforms under various nonlinear loads showing fault occurrence at $t = 0.4$ s with progression to 8.5% severity by $t = 0.5$ s. (a) Linear resistive load, (b) diode rectifier load, (c) thyristor load, (d) capacitive load.

harmonic content). Despite this harmonic content, the fault-induced changes in phase B current remain difficult to identify through visual waveform inspection alone, emphasizing the need for sophisticated signal processing to extract fault signatures from harmonic-rich signals.

The $\alpha\beta$ reference frame processing algorithm successfully extracts fault signatures despite significant harmonic distortion. Figure 6.23 presents amplitude and phase difference parameters extracted from α and β signals for all four load types. In healthy operation before fault occurrence, the amplitude ratio between α and β signals remains near unity and phase difference stays at 90 ° regardless of load type. When turn-to-turn fault occurs at $t = 0.4$ seconds, clear transient responses

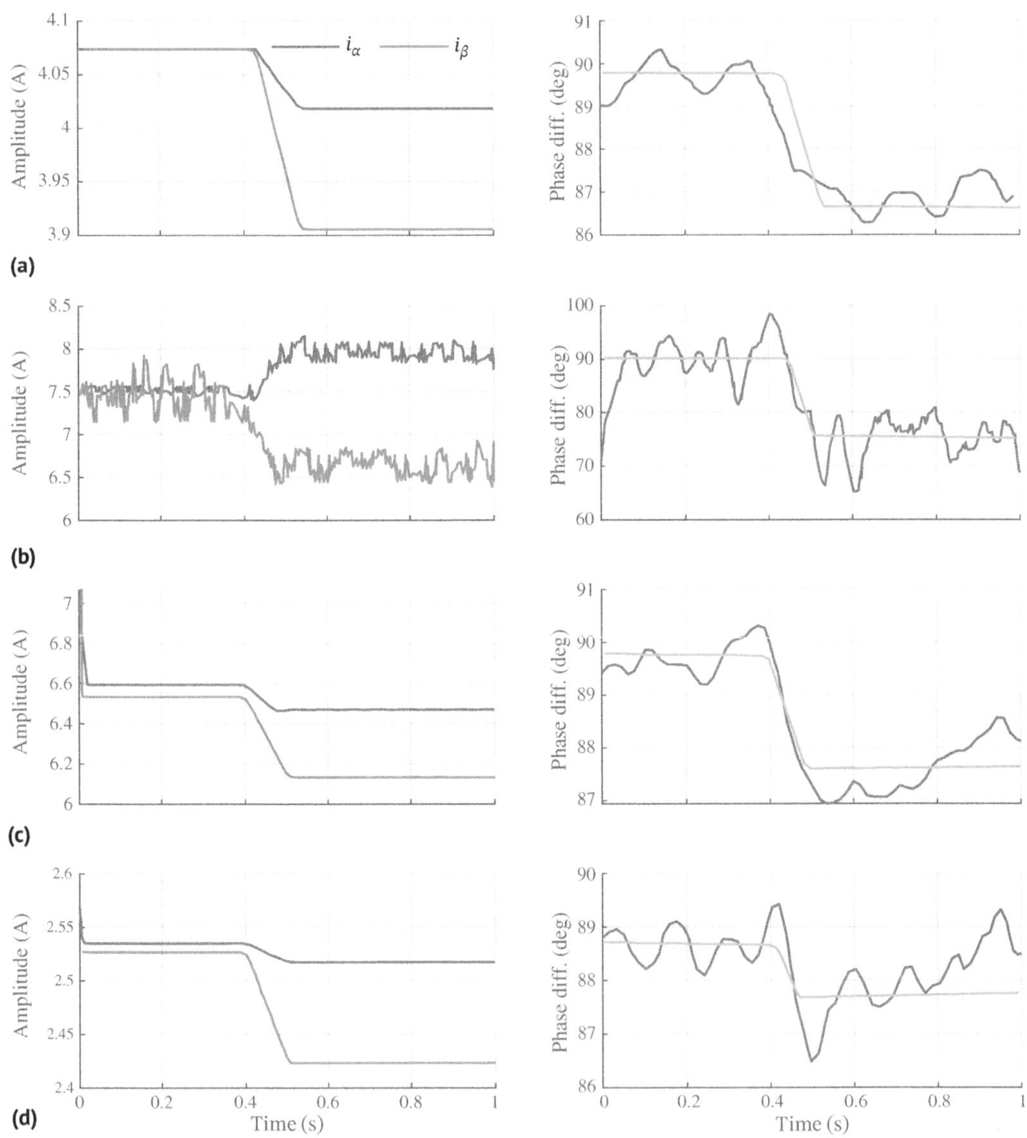

Figure 6.23 Amplitude and phase difference parameters extracted from $\alpha\beta$ signals under harmonic loads demonstrating robust fault detection despite current waveform distortion. (a) Linear load amplitude and phase difference, (b) diode rectifier load, (c) thyristor load, (d) capacitive load.

appear in both amplitude ratio and phase difference for all load types, followed by steady progression to final fault severity values by $t = 0.5$ seconds.

The successful fault detection across all load types validates the robustness of the zero-crossing-based phase difference extraction algorithm. Unlike peak detection methods that become unreliable under harmonic distortion due to multiple local maxima and altered peak locations, zero-crossing times remain stable and accurately represent fundamental frequency phase relationships even when substantial harmonic content exists. The zero-crossings of α and β signals maintain immunity to harmonic influence, enabling accurate phase difference extraction regardless of load type.

6.4.3.4 Concentrated Versus Distributed Winding Validation

Permanent magnet machines employ various winding configurations, including concentrated fractional-slot windings and distributed integral-slot windings. The winding type influences harmonic content, leakage inductance, and fault current paths. Validating diagnostic algorithms on both winding types ensures broad applicability across different machine designs.

The primary validation machine employs concentrated fractional-slot windings with 36 slots and 40 poles, producing a fractional slot per pole per phase ratio. To assess algorithm performance with distributed windings, a second machine with an identical power rating (1.25 kW) but 36 slots and 24 poles employing distributed windings provides comparative evaluation. Figure 6.24 illustrates the distinct winding patterns for concentrated and distributed configurations.

Turn-to-turn faults are introduced in each of the three phases with 8.5% severity (complete single coil short circuit). Figure 6.25 presents experimental results showing three-phase currents and extracted $\alpha\beta$ parameters for phase A, B, and C faults in the distributed winding machine. The algorithm successfully detects all three fault locations, with amplitude and phase difference parameters clearly indicating fault presence and enabling phase identification based on characteristic signature patterns.

The successful detection in both concentrated and distributed winding machines validates the fundamental basis of the diagnostic approach. The reference frame transformations and resulting fault signatures derive from electromagnetic principles applicable regardless of winding details. While specific harmonic content and minor waveform characteristics differ between winding types,

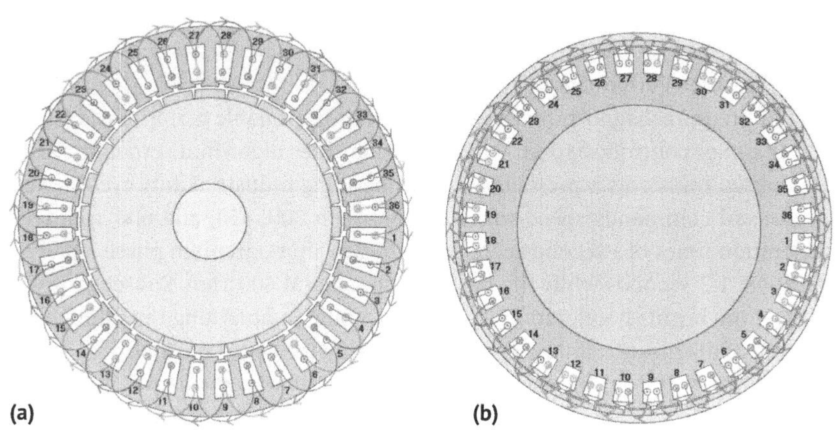

Figure 6.24 Comparison of winding configurations. (a) Concentrated fractional-slot winding, (b) distributed integral-slot winding.

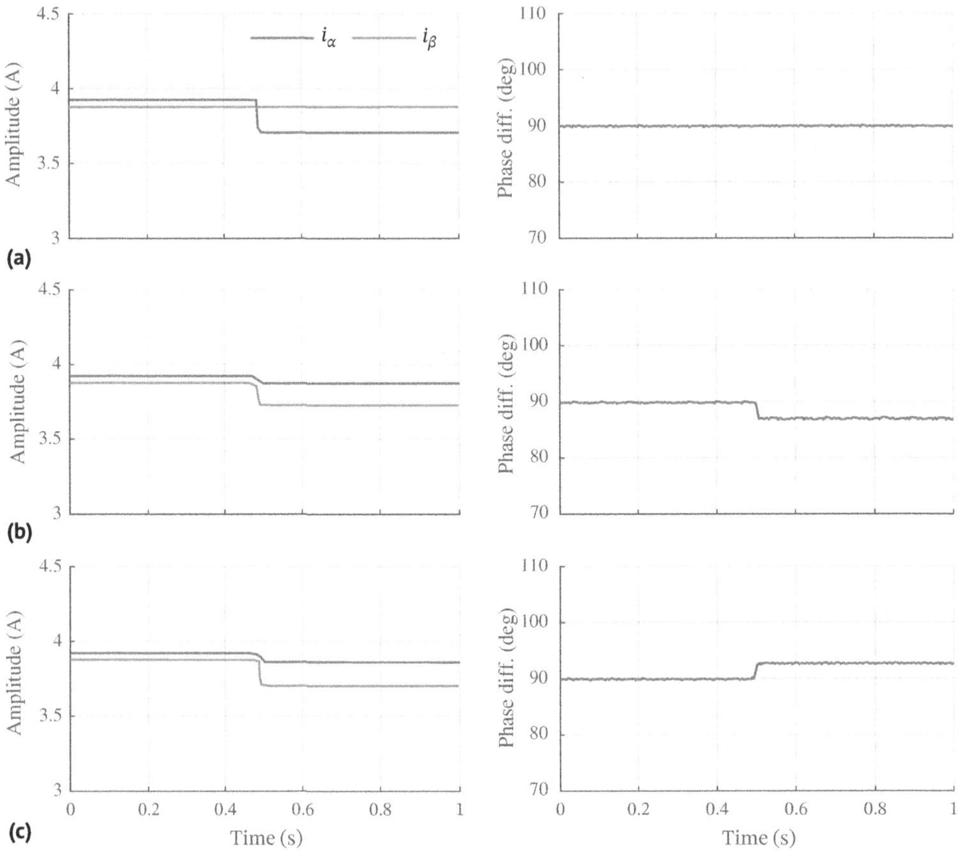

Figure 6.25 Turn-to-turn fault detection in distributed winding machine demonstrating algorithm effectiveness across different winding configurations. (a) Phase A turn-to-turn fault, (b) phase B turn-to-turn fault amplitude and phase difference, (c) phase C turn-to-turn fault.

the primary fault signatures based on amplitude asymmetry and phase deviation remain clearly identifiable in both cases.

6.4.3.5 Transient Speed Variation Performance

Variable speed applications require diagnostic algorithms maintaining reliable performance during speed transients. Testing under controlled speed variations quantifies algorithm response during acceleration and deceleration transients representative of demanding industrial duty cycles.

The experimental protocol commands speed variations between 300, 450, and 600 rpm with acceleration and deceleration times of 1 second each. Turn-to-turn short circuit in phase B at 8.5% severity is introduced at $t = 1.2$ seconds while the machine operates at 600 rpm. Subsequently, at $t = 3.0$ seconds, the machine begins deceleration to 300 rpm while maintaining the fault condition, enabling observation of diagnostic parameter behavior during the speed transient.

Figure 6.26 shows machine speed profile and three-phase current waveforms during the test sequence. As speed decreases, current amplitude and frequency reduce proportionally, but the fundamental current waveform shape remains similar. The fault signature in phase B current persists throughout the speed transient, though amplitude scaling makes visual identification challenging without signal processing.

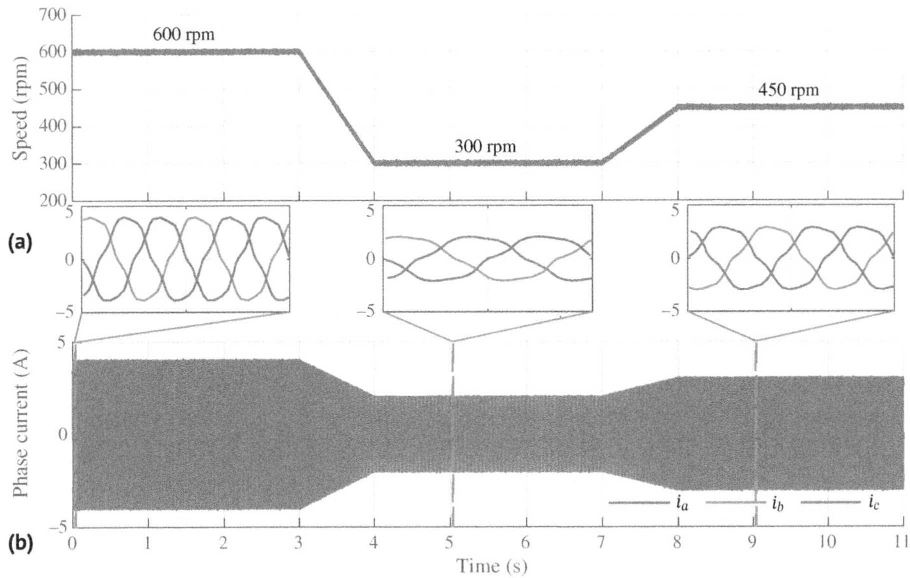

Figure 6.26 Speed variation test protocol demonstrating transient speed changes during maintained fault conditions. (a) Speed profile, (b) three-phase current waveforms during speed transient.

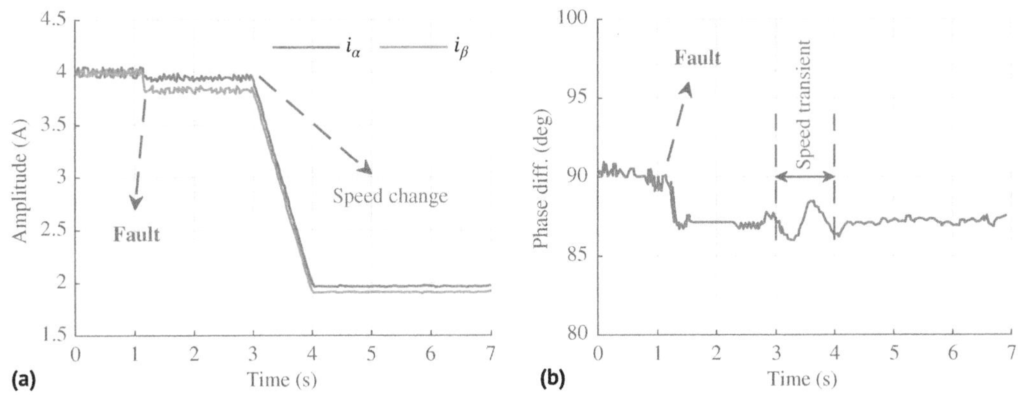

Figure 6.27 Amplitude and phase difference parameters during speed transient demonstrating maintained fault detection sensitivity despite speed variations. (a) α and β amplitude evolution, (b) phase difference evolution.

The $\alpha\beta$ diagnostic parameters successfully track the fault throughout the speed transient. Figure 6.27 presents α and β signal amplitude extraction and phase difference calculation during the test sequence. When the turn-to-turn fault initiates at $t = 1.2$ seconds, both amplitude ratio and phase difference immediately respond with step changes to values indicating fault presence. During the subsequent speed decrease from $t = 3.0$ to 4.0 seconds, the absolute amplitudes of both α and β signals decrease proportionally with speed, but their ratio remains constant. Similarly, the phase difference maintains stable values throughout the transient, confirming robust fault detection independent of speed variations.

This experimental validation confirms the robustness of $\alpha\beta$ frame diagnostics under speed transients. The algorithm's reliance on amplitude ratios and phase differences rather than absolute signal magnitudes provides inherent speed independence analogous to the DQO frame fault index. This characteristic enables application in variable speed drives, traction systems, and wind turbines where speed varies continuously without requiring speed measurement for diagnostic threshold adaptation.

6.4.3.6 Experimental Test Setup and Validation

Comprehensive experimental validation employs a carefully designed test platform enabling controlled fault introduction, accurate measurement, and systematic evaluation of diagnostic algorithm performance. This section describes the experimental hardware, measurement systems, and validation protocols employed throughout the investigation.

6.4.3.7 Test Machine and Instrumentation

The experimental test machine comprises a 1.5 kW, 40-pole, 36-slot permanent magnet synchronous machine with outer rotor construction and surface-mounted NdFeB magnets. The machine operates at a nominal speed of 600 rpm corresponding to 200 Hz electrical frequency. The stator employs concentrated winding configuration with 12 coils per phase, each coil containing 80 turns of AWG 18 copper conductor. The phase windings connect in star configuration without neutral conductor access.

Figure 6.28 presents the complete experimental test setup including the test machine, prime mover (electric motor or turbine simulator), torque transducer, power analyzer, and data acquisition system. A power network analyzer with 8 kHz sampling rate measures and records three-phase voltage and current waveforms with sufficient bandwidth to capture fundamental and harmonic content relevant for fault diagnosis.

6.4.3.8 Fault Introduction Methodology

Turn-to-turn short circuits are introduced through external connections to coil terminals with variable fault resistance enabling controlled severity variation. One coil terminal of phase A is accessible through a specially modified stator allowing connection of a variable resistor in parallel with

Figure 6.28 Experimental test setup with instrumentation for inter-turn short-circuit fault diagnosis validation.

the coil, creating a controlled short circuit path. The fault resistance varies from near-zero (hard fault) to moderate values (resistive fault) representing different fault development stages.

Phase-to-phase faults are created by directly connecting terminals of adjacent phase coils through low-impedance conductors. While phase-to-phase faults would typically evolve from existing turn-to-turn faults in field failures, the experimental implementation enables controlled introduction for diagnostic validation without risking permanent machine damage. All fault introduction mechanisms incorporate safety interlocks preventing accidental application of excessive fault severity that could cause irreversible winding or magnet damage.

6.4.3.9 Three-way Validation Methodology

The validation approach employs three-way comparison between magnetic equivalent circuit modeling predictions, finite element method simulation results, and experimental measurements. This comprehensive comparison strategy provides high confidence in both the modeling accuracy and diagnostic algorithm effectiveness. Agreement among all three methods confirms the fundamental understanding of fault physics, validates the simplified modeling approaches suitable for real-time implementation, and demonstrates practical feasibility.

The magnetic equivalent circuit model provides rapid prediction capability, enabling extensive parameter studies across fault severities and operating conditions impractical with finite element methods or experimental testing. The finite element simulations serve as a high-fidelity benchmark validating magnetic equivalent circuit simplifications and assumptions. Experimental measurements provide ultimate validation under real hardware conditions including manufacturing variations, measurement noise, and physical phenomena potentially neglected in modeling.

Results from all three validation methods for several key electromagnetic quantities, including phase currents, back electromotive force, and air gap flux density, show close agreement across methods, validating the entire diagnostic development chain from fundamental physics through modeling to practical implementation.

The comprehensive diagnostic approaches presented in this section provide robust methods for detecting, classifying, and localizing short-circuit faults in permanent magnet machines. The DQO reference frame methods enable speed-independent fault severity quantification and clear discrimination between turn-to-turn and phase-to-phase faults through the dimensionless fault index η. The $\alpha\beta 0$ reference frame techniques supplement this capability with phase identification, enabling fault location to specific phases. Extensive experimental validation demonstrates robustness under harmonic loads, different winding configurations, and speed transients, establishing confidence for industrial deployment in demanding applications. The combination of these complementary techniques provides a comprehensive diagnostic toolset adaptable to diverse application requirements balancing detection sensitivity, computational complexity, and implementation cost.

6.4.4 Model-based Parameter Estimation for Inter-turn and Short-circuit Fault Diagnosis

Model-based parameter estimation methods provide a framework for ITSC fault detection by continuously estimating machine electrical parameters and comparing them against healthy baseline values. The general algorithms for extended Kalman filters (EKF), unscented Kalman filters (UKF), sliding mode observers (SMO), and model reference adaptive systems (MRAS) are presented in Chapter 4. This section focuses on the ITSC-specific parameter variations, detection performance characteristics, and practical implementation considerations.

6.4.4.1 ITSC-specific Parameter Variations

For inter-turn short-circuit faults, the most sensitive parameters are phase resistance and self-inductance of the affected phase. An inter-turn fault in phase a with μ fraction of turns short-circuited produces characteristic parameter changes:

$$R_{a,\text{fault}} = R_{a,\text{healthy}} + \frac{R_f i_{sc}^2}{i_a^2} \tag{6.40}$$

$$L_{a,\text{fault}} = (1-\mu)^2 L_{a,\text{healthy}} \tag{6.41}$$

where R_f is the fault loop resistance, i_{sc} is the short-circuit current amplitude, and i_a is the healthy phase current [24]. Resistance estimation provides earlier and more reliable fault indicators than inductance-based approaches, with resistance increases of 5–15% reliably detectable for fault severities corresponding to 2–5% of total turns short-circuited.

Under ITSC conditions, the resistance estimate exhibits characteristic harmonic content at twice the electrical frequency ($2f_e$) due to the asymmetry introduced by the fault. Spectral analysis of parameter estimation residuals enables discrimination between transient load changes and genuine ITSC faults, with $A_2/R_{s,\text{DC}}$ serving as a normalized fault indicator robust to operating point variations [24].

6.4.4.2 Per-phase Estimation for Fault Localization

A key challenge in ITSC detection is distinguishing between balanced parameter drift (e.g., due to uniform temperature rise) and asymmetric fault conditions affecting individual phases. Per-phase resistance estimation in the natural abc frame enables fault localization by tracking individual phase resistances R_a, R_b, R_c separately [25]. Under healthy conditions, all estimates converge to similar values within 2–3%. When an ITSC fault occurs in phase a, the estimate $\hat{R}_a$ increases significantly while $\hat{R}_b$ and $\hat{R}_c$ remain near baseline. A fault indicator based on resistance asymmetry provides reliable fault phase identification for ITSC severities above 3% with false alarm rates below 1%.

6.4.4.3 Comparative Performance for ITSC Detection

Table 6.5 summarizes the ITSC detection performance of different observer techniques. Comparative studies reveal that UKF achieves 47% lower root-mean-square estimation error than EKF for 5% inter-turn faults, with 46% faster convergence (0.65 s vs. 1.2 s detection time) [26]. However, UKF requires approximately 2.8 times more floating-point operations. SMO exhibits superior robustness to model uncertainties (errors below 8% with ±30% inductance variation vs. 18% for EKF) and requires approximately 40% fewer operations than EKF [27]. MRAS offers the simplest implementation but slower fault response (60–150 ms vs. 30–65 ms for UKF) [25].

Detection latency of 20–150 ms is acceptable for ITSC fault protection since thermal time constants of winding insulation are on the order of seconds to minutes, providing adequate time for protective action before thermal damage propagation.

6.4.4.4 Discrimination Between ITSC and Other Parameter Variations

Different fault mechanisms can produce similar parameter variations, complicating diagnosis:

ITSC vs. Temperature Rise: Both increase apparent stator resistance. Discrimination requires monitoring resistance change rate (ITSC causes step change, thermal drift is gradual), checking phase asymmetry (ITSC affects single phase, temperature is balanced), or correlating with independent thermal indicators.

Table 6.5 Comparison of observer-based parameter estimation techniques.

Characteristic	EKF	UKF	SMO	MRAS
Computational cost (FLOPs/iteration)	800–1,200	2,200–2,800	400–600	300–500
RAM requirement	2–3 kB	3–4 kB	0.5–0.8 kB	0.3–0.5 kB
Detection latency (5% fault)	50–100 ms	30–65 ms	40–100 ms	60–150 ms
Model accuracy sensitivity	Moderate	Low	Low	Moderate
Robustness to noise	Moderate	High	High	Moderate
Minimum detectable severity	3%	2.5%	4%	5%
Tuning complexity	Moderate	High	Low	Low
Temperature compensation need	Essential	Essential	Essential	Essential
Suitable for sensorless drives	Yes	Yes	Yes	Excellent

ITSC vs. Demagnetization: Demagnetization reduces flux linkage, which can be misinterpreted as inductance or resistance change. Joint estimation of multiple parameters and cross-validation improves discrimination.

Parameter Drift vs. Fault: Manufacturing tolerances (±10%), aging, and environmental factors cause legitimate parameter variations requiring long-term trending and conservative thresholds.

6.4.4.5 Implementation Guidelines

Successful ITSC detection via parameter estimation requires attention to several practical factors. Inverter dead-time effects, IGBT voltage drops, and switching delays introduce voltage errors particularly significant at low speeds and light loads. Compensation improves resistance estimation accuracy by 35% at 10% rated speed [28]. Temperature compensation using thermal models or direct measurement is essential across all observer techniques. The recommended operating envelope spans 15–100% rated speed (below 15%, inverter nonlinearity dominates) and 20–120% rated torque (below 20%, insufficient excitation).

When properly implemented, model-based parameter estimation provides reliable ITSC fault diagnosis, complementing reference frame-based methods for comprehensive winding fault monitoring.

6.5 Summary

This chapter has presented a treatment of short-circuit and inter-turn faults in permanent magnet machines, encompassing their failure mechanisms, modeling approaches, diagnostic techniques, and experimental validation. Short-circuit faults represent a critical class of failures due to their progressive nature and the complex interplay of thermal, electrical, and mechanical stress factors driving insulation degradation. The severity and rapid progression potential of these faults, particularly when initial inter-turn shorts evolve into phase-to-phase or phase-to-ground failures, necessitates accurate predictive modeling and robust diagnostic approaches enabling early detection before catastrophic damage occurs.

The examination of failure modes established the fundamental taxonomy of short-circuit fault types and their progression pathways. Inter-turn short circuit faults originate from localized insulation breakdown between adjacent turns, with degradation driven by synergistic thermal, electrical, and mechanical stress factors. The Arrhenius relationship governs thermal aging such that a 10 °

temperature increase halves insulation lifetime, while inverter-fed operation subjects insulation to repetitive voltage surges and partial discharge. The fault progression typically follows a characteristic sequence from incipient inter-turn shorts through coil-to-coil faults to phase-to-phase short circuits or phase-to-ground failures. Phase-to-phase faults produce fault currents potentially orders of magnitude higher than inter-turn faults, while turn-to-ground faults create distinct zero-sequence signatures and present safety hazards. The temporal scale of progression varies widely from hours to seconds depending on fault severity and loading conditions, emphasizing the importance of continuous monitoring.

The modeling and analysis section applied the magnetic equivalent circuit methodology to inter-turn and phase-to-phase fault analysis in multiple reference frames. The MEC approach employs refined air gap meshing using triangular cells to capture asymmetric flux patterns, achieving accuracy approaching finite element analysis while providing computational efficiency suitable for parametric studies. Mathematical formulations in the ABC frame describe how fault severity parameter ζ modifies inductance and resistance matrices, while transformation to the DQO frame reveals second harmonic components in d and q axis currents proportional to fault severity. In the $\alpha\beta0$ frame, faults disrupt the balanced circular current trajectory into elliptical patterns, while phase-to-phase faults collapse the trajectory to a one-dimensional line whose orientation identifies the faulted phase pair. Three-way validation between MEC predictions, finite element simulations, and experimental measurements confirmed close agreement across methods.

The fault diagnosis section presented approaches organized by reference frame transformation, demonstrating speed-independent fault detection and discrimination between fault types. The DQO approach defines a dimensionless fault index η as the ratio of AC ripple amplitude to DC component in the q axis current, which remains speed-independent because both quantities scale proportionally with speed variations. Experimental validation confirmed that the index reaches approximately 0.15 for 8.5% turn-to-turn faults, approaches unity for complete coil short circuits, and reaches approximately 2.0 for phase-to-phase faults, enabling unambiguous fault type classification. The $\alpha\beta0$ technique provides complementary capability through amplitude and phase difference analysis, enabling both fault detection and faulted phase identification. Robustness testing validated diagnostic performance under nonlinear loads, across both concentrated and distributed winding configurations, and during transient speed variations. Model-based parameter estimation using Kalman filters, sliding mode observers, and Model Reference Adaptive Systems enables continuous monitoring within existing drive control systems, achieving detection latencies of 30–150 milliseconds for incipient faults. The integration of these complementary diagnostic approaches provides comprehensive fault monitoring capability, enabling protective intervention before thermal feedback mechanisms drive catastrophic fault progression.

References

1 Yuefeng Liao. Analysis of fault conditions in permanent magnet in-wheel motors. PhD thesis, KTH, Electrical Energy Conversion, 2011.

2 William T. Thomson and Mark Fenger. Current signature analysis to detect induction motor faults. *IEEE Industry Applications Magazine*, 7(4):26–34, July 2001. https://doi.org/10.1109/2943.930988.

3 Hans C. Roters. *Electromagnetic Devices*. 1970. https://books.google.com/books/about/Electromagnetic_Devices.html?hl=fr&id=vT3TXwAACAAJ.

4 Wei Tong. *Mechanical Design of Electric Motors*. CRC Press, 2014.

5 IEC. IEC 60216: Electrical insulating materials – thermal endurance properties. International Electrotechnical Commission, 2013.

6 David Faustner, Wolfgang Kemmetmüller, and Andreas Kugi. Magnetic equivalent circuit modeling of a saturated surface-mounted permanent magnet synchronous machine. *IFAC-PapersOnLine*, 48(1):360–365, 2015. https://doi.org/10.1016/j.ifacol.2015.05.033.

7 Haijun Zhang, Bangwei Zhang, Guowen Cao, and Shuhong Wang. Fatigue propagation characteristics of insulation crack failure in stator end winding for large traction motor based on magnetic–thermal–mechanical coupling model. *IEEE Transactions on Dielectrics and Electrical Insulation*, 31(4):2111–2119, 2024. https://doi.org/10.1109/TDEI.2024.3363119.

8 Zoltán Ádam Tamus and Richárd Cselkó. Effect of bending-induced mechanical stress on the dielectric response of rectangular PAI/PEEK insulated winding wire. In *2025 IEEE Electrical Insulation Conference (EIC)*, pages 1–4, 2025. https://doi.org/10.1109/EIC63069.2025.11123222.

9 Liang Liu, Kai Wang, Long L. Guo, and Jun Li. Analysis of inter-turn short circuit faults in dual three-phase PMSM for electromechanical actuator. *IEEE Transactions on Transportation Electrification*, 9(3):4059–4070, 2023. https://doi.org/10.1109/TTE.2023.3239632.

10 Woongkul Lee and Bulent Sarlioglu. Fault-tolerant six-phase motor drive topology and control technique for short-circuit failure. In *2023 IEEE Transportation Electrification Conference & Expo (ITEC)*, pages 1–6, 2023. https://doi.org/10.1109/ITEC55900.2023.10186950.

11 Jing Zhao, Xiaoqing Guan, Chenghai Li, Quansong Mou, and Zhen Chen. Comprehensive evaluation of inter-turn short circuit faults in PMSM used for electric vehicles. *IEEE Transactions on Intelligent Transportation Systems*, 22(1):611–621, 2021. https://doi.org/10.1109/TITS.2020.2987637.

12 Ujwala V. Dongare, Bhimrao S. Umre, and Makarand S. Ballal. Voltage–current locus-based stator winding inter-turn fault detection in induction motors. *International Journal of Circuit Theory and Applications*, 51(6):2889–2911, 2023. https://doi.org/10.1002/cta.3561.

13 Aymen Khlaief, Mohamed Boussak, and Mohamed Gossa. Open phase faults detection in PMSM drives based on current signature analysis. In *The XIX International Conference on Electrical Machines - ICEM 2010*, pages 1–6, 2010. https://doi.org/10.1109/ICELMACH.2010.5607977.

14 Carina Damhuis, Jörg Kammermann, and Hans-Georg Herzog. Analysis of internal short-circuit faults in the stator winding of a nine-phase permanent magnet synchronous machine using finite element analysis. In *2022 International Symposium on Power Electronics, Electrical Drives, Automation and Motion (SPEEDAM)*, pages 671–676, 2022. https://doi.org/10.1109/SPEEDAM53979.2022.9842057.

15 Ailiang Kang, Muqin Tian, Jiancheng Song, Lingyan Lin, and Zhipeng Lei. Slot partial discharge characteristics under electro-thermal aging of 10 kV motor stator windings. In *2018 12th International Conference on the Properties and Applications of Dielectric Materials (ICPADM)*, pages 517–520, 2018. https://doi.org/10.1109/ICPADM.2018.8401042.

16 Dimitrios Chronopoulos, Dimitrios Glykos, and Konstantinos N. Gyftakis. An investigation of the impedance spectroscopy suitability for traction motor windings insulation health assessment. In *2024 International Conference on Electrical Machines (ICEM)*, pages 1–6, 2024. https://doi.org/10.1109/ICEM60801.2024.10700204.

17 Maliheh Hashemi, Mohammad Ali Golkani, and Daniel Watzenig. A robust approach for inter-turn fault detection of PMSM used for autonomous vehicles. In *2022 International Conference on Connected Vehicle and Expo (ICCVE)*, pages 1–6, 2022. https://doi.org/10.1109/ICCVE52871.2022.9742916.

18 Digvijay Dabade, Sanjay Dambhare, and Murugesh Mudaliar. A case study on protection of pumped storage hydro-generator against stator earth fault. In *2024 3rd International Conference on Power, Control and Computing Technologies (ICPC2T)*, pages 733–738, 2024. https://doi.org/10.1109/ICPC2T60072.2024.10474944.

19 Xiaojun Zhao, Rui Wang, Xiaona Liu, and Lin Li. A dynamic hysteresis model for loss estimation of GO silicon steel under DC-biased magnetization. *IEEE Transactions on Industry Applications*, 57(1):409–416, 2021. https://doi.org/10.1109/TIA.2020.3038712.

20 Singaram Gopinath, Kumarasamy Sathiyasekar, Sanjeevikumar Padmanaban, and Bharatiraja Chokkalingam. Insulation condition assessment of high-voltage rotating machines using hybrid techniques. *IET Generation, Transmission & Distribution*, 13(2):171–180, 2018. https://doi.org/10.1049/iet-gtd.2018.6500.

21 Shihu Xiang, Guiheng Li, Feng Zhou, Shaopo Huang, Zijun Xu, Guanlong Jia, and Feng Niu. Life prediction of ground-wall insulation material in electric motors based on accelerated degradation test under different failure mechanisms. *IEEE Journal of Emerging and Selected Topics in Industrial Electronics*, 6(2):654–663, 2025. https://doi.org/10.1109/JESTIE.2024.3522001.

22 Manel Fitouri, Yemna Bensalem, and Mohamed Naceur Abdelkrim. Modeling and detection of the short-circuit fault in PMSM using finite element analysis. *IFAC-PapersOnLine*, 49(12):1418–1423, 2016. https://doi.org/10.1016/j.ifacol.2016.07.769.

23 Javier A. Rosero, Luis Romeral, José A. Ortega, and Ernesto Rosero. Short-circuit detection by means of empirical mode decomposition and Wigner–Ville distribution for PMSM running under dynamic condition. *IEEE Transactions on Industrial Electronics*, 56(11):4534–4547, November 2009. https://doi.org/10.1109/TIE.2008.2011580.

24 Yacine Maanani and Arezki Menacer. Modeling and diagnosis of the inter-turn short circuit fault for the sensorless input-output linearization control of the PMSM. *Periodica Polytechnica Electrical Engineering and Computer Science*, 63(3):159–168, 2019. https://doi.org/10.3311/PPee.13658.

25 Sofiane Bouslimani, Salah Meradi, Smail Drid, Lilia Chrifi-Alaoui, and Ali Bezziane. State estimation MRAS and identification of stator winding phase fault detection of the PMSG in wind energy based on the sliding mode control. *Power Electronics and Drives*, 8(1):109–127, 2023. https://doi.org/10.2478/pead-2023-0009.

26 Waseem El Sayed, Ahmed Aboelhassan, Amr Madi, Ahmed Hebala, and Michael Galea. Comparative analysis between unscented and extended Kalman filters for PMSG inter-turn fault detection. In *2021 IEEE Workshop on Electrical Machines Design, Control and Diagnosis (WEMDCD)*, pages 243–248, 2021. https://doi.org/10.1109/WEMDCD51469.2021.9425684.

27 Peien Luo, Zhonggang Yin, Zhe Zhang, Yanping Zhang, Pinjia Zhang, and Jing Liu. Diversified diagnosis strategy for PMSM inter-turn short-circuit fault via novel sliding mode observer. *IEEE Transactions on Power Electronics*, 39(4):4149–4159, 2024. https://doi.org/10.1109/TPEL.2024.3352077.

28 Ziqiang Ye, Tao Liu, Morris Fuller, and Gerald Griepentrog. Parameter identification of PMSM based on MRAS with considering nonlinearity of inverter. In *IECON 2019 - 45th Annual Conference of the IEEE Industrial Electronics Society*, 2019. https://doi.org/10.1109/IECON.2019.8927595.

7

Bearing Faults

7.1 Introduction

Rolling bearings provide essential mechanical support for rotating parts in permanent magnet (PM) motors and are extensively utilized in industrial machines. Despite rigorous design considerations, bearings in PM motors are subject to extremely demanding operating conditions that can lead to various failure modes. Bearing faults represent the most common mechanical failure mode in electrical machines, accounting for approximately 40–50% of all motor failures [1]. The performance and reliability of bearings under harsh environmental conditions, including inadequate lubrication, mechanical shocks, overloading, insufficient cooling, and thermal stresses, are critical factors determining the overall reliability of PM motor systems.

This chapter provides comprehensive coverage of bearing faults in PM motors, examining failure mechanisms, analytical modeling approaches, and diagnostic techniques specific to permanent magnet machines. The unique characteristics of PM motors, including higher radial loads due to permanent magnet forces, unbalanced magnetic pull (UMP) effects, compact designs with smaller air gaps, and high-speed operation capabilities, create distinct bearing fault signatures and challenges compared to conventional induction motors. Understanding these PM motor-specific considerations is essential for developing effective bearing fault diagnosis and condition monitoring strategies. Table 7.1 summarizes the key differences between PM motors and conventional induction motors with respect to bearing considerations.

7.2 Failure Modes and Mechanisms

7.2.1 Bearing Construction and Components

This chapter focuses on the single-row deep-groove radial ball bearing commonly employed in PM motors, consisting of four primary components illustrated in Figure 7.1: an inner ring (mounted on the motor shaft), an outer ring (fixed in the motor housing), rolling elements (balls) that make point contact with grooved raceways on both rings, and a cage (retainer) that maintains ball spacing [2]. Lubricants reduce metal-to-metal contact between rolling and sliding surfaces. Under radial loading, approximately one-third of the bearing circumference (the load zone) carries the applied load, making the condition of surfaces entering this zone critical to bearing performance and fault signature generation.

Table 7.1 Key bearing-related differences between induction motors and PM motors.

Aspect	Induction motors	PM motors
Radial forces and UMP	Moderate radial forces; UMP limited by larger air gaps (0.5–2 mm)	20–40% higher bearing loads from magnet attraction; strong UMP due to smaller air gaps (0.3–1 mm) creates positive feedback with bearing wear
Speed range	Narrow to moderate; limited by slip	Wide range with field weakening; high-speed designs (>20,000 rpm) introduce centrifugal and gyroscopic effects on bearings
Inverter-induced bearing currents	Lower dv/dt; less severe bearing current damage	High dv/dt from PWM inverters (4–20 kHz) causes EDM pitting, fluting, and accelerated raceway degradation
Torque pulsations	Relatively smooth torque	Cogging torque and PM-slot interaction create periodic bearing load modulation
Vibration damping	Rotor currents provide electromagnetic damping of vibrations	Lower electromagnetic damping; bearing fault signatures more pronounced in vibration spectra

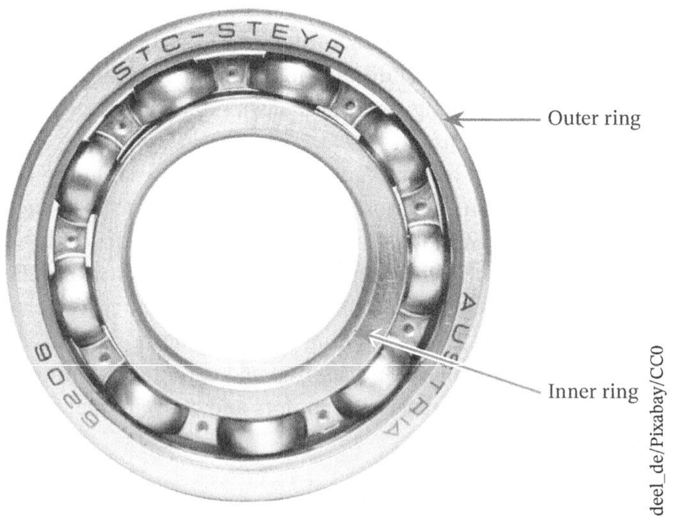

Figure 7.1 Ball bearing construction showing inner ring, outer ring, rolling elements (balls), and cage components.

7.2.2 Bearing Race Defects and Failure Mechanisms

Bearing race defects constitute one of the most common bearing failure modes and can occur on either the inner race or outer race surfaces. These defects primarily result from rolling contact fatigue, which is a statistical phenomenon influenced by the heterogeneous nature of the material microstructure. Even under normal operating conditions in a healthy machine, fatigue failures can occur due to the probability of subsurface-initiated fatigue spalling [3].

Flaking and Spalling: Flaking occurs when the bearing surface develops scales and peels off due to repeated contact load on the raceway and rolling surface. This localized fault is commonly observed at an early stage of bearing degradation. Pitting refers to the formation of small holes on

Figure 7.2 Bearing raceway defects: (a) flaking damage showing surface scales and peeling, (b) pitting damage showing small holes formed by rolling fatigue.

the raceway surface caused by rolling fatigue. The progression of surface fatigue in ball bearings typically begins with a V-shaped spall, which then grows in the rolling direction as material starts to flake away from the surface [4]. These localized faults often begin as single defects but can progressively spread over wider areas as damage accumulates. Figure 7.2 shows typical flaking and pitting damage patterns on bearing raceway surfaces.

Electric Pitting: In PM motors driven by power converters, electric pitting can result from partial melting of the bearing surface due to electric current passing through the contact point. Bearing discharge current, especially in inverter-fed machines, represents a significant cause of bearing failure that can manifest as either single point defects or generalized roughness faults. These failures are particularly relevant for PM motor applications where high-frequency inverter switching can induce shaft voltages and bearing currents [3].

PM Motor-specific Considerations: In permanent magnet motors, bearing race defects are exacerbated by higher radial loads resulting from strong permanent magnet fields, particularly in surface-mounted PM (SPM) configurations. The UMP phenomenon, which occurs due to air gap asymmetries and flux distortions, generates additional radial forces on bearings. When a bearing supports radial loads in the presence of UMP, the effective load on the bearing load zone increases significantly, accelerating race degradation and reducing bearing life. Larger shaft diameters in PM motors experience greater reduction in bearing life when UMP is present [3].

7.2.3 Ball/Rolling Element Defects

Ball and rolling element defects represent critical failure modes that can generate severe vibration signatures and accelerate bearing degradation. These defects primarily arise from subsurface stress concentrations, surface fatigue, and mechanical damage during operation.

Brinelling: Brinelling is characterized by indentations on the raceways where they meet the rolling elements, caused by plastic deformation. External factors contributing to brinelling include extremely heavy loads, vibrations, faulty mounts, or the presence of solid foreign particles. Electric current due to leakage can also be a potential cause. Brinelling damage typically initiates as a localized fault and gradually progresses into a distributed one. True brinelling occurs when static overload exceeds the material yield strength, while false brinelling results from fretting wear under vibration conditions [4].

Smearing: Smearing is a consequence of ball sliding or spinning, which often happens when the balls enter the load zone. This phenomenon creates surface damage through adhesive wear

Figure 7.3 Ball defect modes: (a) smearing damage from ball sliding/spinning, (b) brinelling indentations caused by plastic deformation at ball-raceway contact points.

and material transfer between contacting surfaces. In high-speed PM motors, where lightweight rotors enable rapid acceleration and deceleration, smearing can occur more readily during transient operating conditions. Figure 7.3 illustrates both smearing and brinelling damage mechanisms.

Subsurface-initiated Fatigue: Rolling contact fatigue of rolling element bearings is fundamentally a statistical phenomenon. Subsurface stress concentrations beneath the rolling contact area can initiate microcracks that propagate to the surface, resulting in spalling. The progression of fatigue failure follows a characteristic pattern: initial subsurface crack formation, crack propagation toward the surface, and eventual surface spalling. In continued operation with fatigue spalling, the dynamic loading resulting from contact with the rough spall surface becomes the main limiting factor determining the remaining useful life of the bearing [4].

High-speed PM Motor Considerations: PM motors designed for high-speed applications impose unique stresses on rolling elements. The compact bearing designs required for high-speed operation, combined with the centrifugal forces experienced by the balls, create elevated contact stresses. Additionally, the high-frequency vibrations characteristic of high-speed PM motors can accelerate subsurface fatigue processes. These factors make ball defect monitoring particularly critical in high-speed PM motor applications such as electric vehicle traction motors and aerospace actuators.

7.2.4 Cage Defects and Failure Modes

Cage defects, while less common than race or ball defects, can lead to catastrophic bearing failure if left undetected. Cage wear primarily results from friction between the cage and balls or ring surfaces, exacerbated by improper lubrication. Cage fracture occurs due to fatigue from cyclic loading, particularly under conditions of cage instability where irregular motion arises from inadequate ball guidance [3]. Cage instability is particularly relevant for PM motors with wide speed ranges: at low loads and speeds, the cage may not be properly guided, leading to erratic motion and potential fracture. This phenomenon is especially concerning in variable-speed PM motor applications where operation at partial load is common.

7.2.5 Lubrication-related Failures

Lubrication-related failures represent a major category of premature bearing failures. Bearing grease degrades chemically and physically when exposed to elevated temperatures, with oxidation, base oil evaporation, and thickener breakdown progressing to increased friction and accelerated wear [4]. Lubrication starvation from improper initial lubrication or grease migration, and contamination by foreign particles or moisture, further compromise bearing performance.

In PM motors, a thermal coupling between electromagnetic heating (stator losses) and mechanical heating (bearing friction) creates a positive feedback loop where increased bearing friction generates additional heat that further degrades the lubricant. Additionally, shaft voltages induced by inverter switching can cause electrical discharge machining (EDM) of bearing surfaces, creating pitting that disrupts lubricant film formation. These PM motor-specific mechanisms accelerate lubrication degradation beyond what is experienced in line-fed machines.

7.2.6 Contamination and Excessive Loading Effects

Contamination and overloading represent two major external factors that significantly shorten bearing life in PM motors.

Contamination Sources and Effects: Contamination by foreign matter represents a major cause of bearing wear and premature failure. Contaminants can enter the bearing through several pathways: inadequate sealing systems, degradation of seal materials, ingress during maintenance procedures, and particles generated from internal wear processes. Solid particles cause abrasive wear on bearing surfaces, creating additional surface roughness and stress concentrations that accelerate fatigue processes. Wear primarily results from sliding abrasion on various bearing components, including balls and raceways [4]. Figure 7.4 illustrates typical wear and fretting damage patterns observed in bearing components.

Corrosion: Corrosion refers to oxidation or dissolution occurring on the contacting surfaces of a bearing. Corrosion damage is primarily caused by sealing failure or decomposition of lubricant additives under high temperature conditions. The corrosion of a bearing can generate rust particles, which further exacerbate mechanical abrasion on balls and raceways. This type of defect typically manifests as distributed faults that affect large bearing surface areas. In PM motors operating in harsh environments (marine applications, outdoor installations), corrosion protection becomes particularly critical [4]. Figure 7.5 shows typical rust and corrosion damage patterns.

Fretting: Fretting is a phenomenon that occurs when slight sliding is repeatedly caused on the contact surface due to vibration loads. Initially, fretting wear tends to be localized but can quickly propagate to other components. In PM motors subjected to external vibrations or mounted on

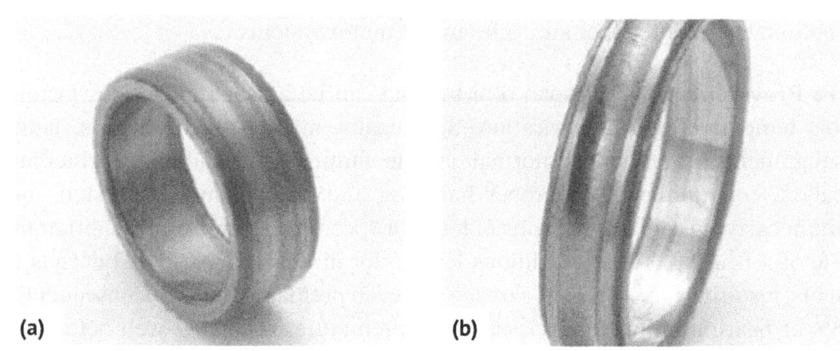

Figure 7.4 Bearing damage modes: (a) wear damage showing surface abrasion, (b) fretting damage caused by repeated slight sliding under vibration loads.

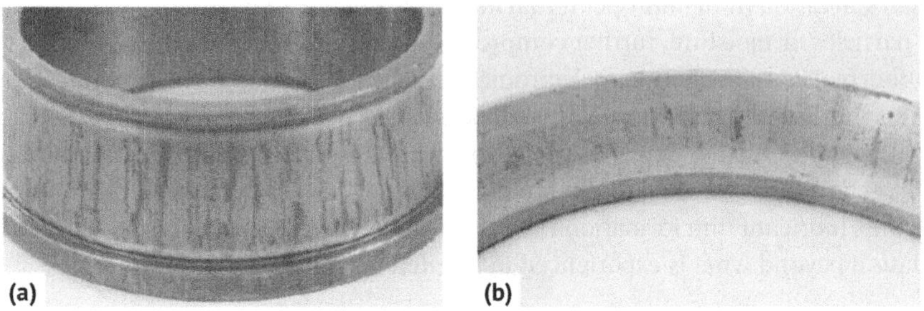

Figure 7.5 Bearing corrosion damage: (a) rust formation on bearing surfaces, (b) corrosion damage from lubricant decomposition under high temperature.

vibrating platforms, fretting damage can develop even when the motor is stationary if relative micromotion occurs between bearing components [4].

Excessive Loading and UMP: Overloading the bearing increases the likelihood of premature failure, occurring when contact pressure between the rolling element and bearing raceway exceeds the material's yield strength. In PM motors, the electromagnetic forces impose additional radial loads on bearings beyond the mechanical loads from the rotor and external forces. The UMP, which arises from air gap asymmetries, demagnetization faults, winding faults, or eccentricity conditions, induces substantial additional radial load on the bearing, leading to excessive wear [3].

The relationship between UMP and bearing degradation creates a progressive fault mechanism. Rotor eccentricity (which may initially be small) generates UMP, which increases bearing loads and accelerates bearing wear. As bearing clearances increase due to wear, rotor eccentricity worsens, further amplifying UMP in a positive feedback loop. Furthermore, rotor eccentricity causes additional mechanical stress, resulting in increased rubbing and fatigue on ball bearings. While 80% of mechanical faults result in an eccentric rotor and, ultimately, UMP in electrical machines, the severity of this effect is particularly pronounced in PM motors due to their smaller air gaps and stronger magnetic fields.

7.2.7 Statistical Analysis of Bearing Failure Rates

Understanding bearing failure statistics provides essential context for prioritizing condition monitoring efforts and optimizing maintenance strategies in PM motor systems.

Premature Failure Prevalence: The lifespan of a bearing can be affected by various factors, including loading, temperature, and lubrication. Specifically, manufacturing defects, faulty mountings, misalignments, overheating, normal fatigue failure, true and false brinelling, improper lubrication, contamination, improper handling and maintenance, corrosion, and exceeding maximum carrying capacity in terms of loads or speeds can significantly shorten the typical fatigue life of a bearing. These conditions lead to localized and distributed defects in bearing components, resulting in permanent damage and even premature failure. Consequently, an estimated 95% of bearing failures are classified as premature, occurring well before the calculated L_{10} bearing life.

Failure Mode Distribution: Bearing failure often occurs due to a combination of causes rather than a single factor. Bearing defects can be categorized as localized (cracks, pits, spalls) or distributed (surface roughness, waviness, misaligned races, off-size rolling elements). Statistical analysis of bearing failures shows that improper lubrication and contamination account for the majority of premature bearing failures, followed by installation errors and overloading conditions.

Bearing Life Calculations: Traditional bearing life calculations employ the Lundberg–Palmgren theory [5], which relates bearing life to the applied load through a power law relationship. The basic rating life L_{10} represents the number of revolutions that 90% of a bearing population will complete before the first evidence of fatigue develops. For ball bearings operating under radial loads, the relationship follows:

$$L_{10} = \left(\frac{C}{P}\right)^3 \times 10^6 \text{ revolutions} \tag{7.1}$$

where C represents the basic dynamic load rating and P denotes the equivalent dynamic bearing load. However, this basic calculation must be modified for PM motor applications to account for additional factors including temperature effects, contamination levels, and electromagnetic loading.

Impact of UMP on Bearing Life: The effect of UMP on bearing life in large PM machines demonstrates a significant reduction in bearing lifetime. Since bearing life follows an inverse cubic relationship with the applied radial load, the additional force introduced by UMP causes a nonlinear decrease in L_{10} life as eccentricity increases. As shown in Figure 7.6a, this effect is more pronounced for larger shaft diameters: machines with larger rotors generate proportionally

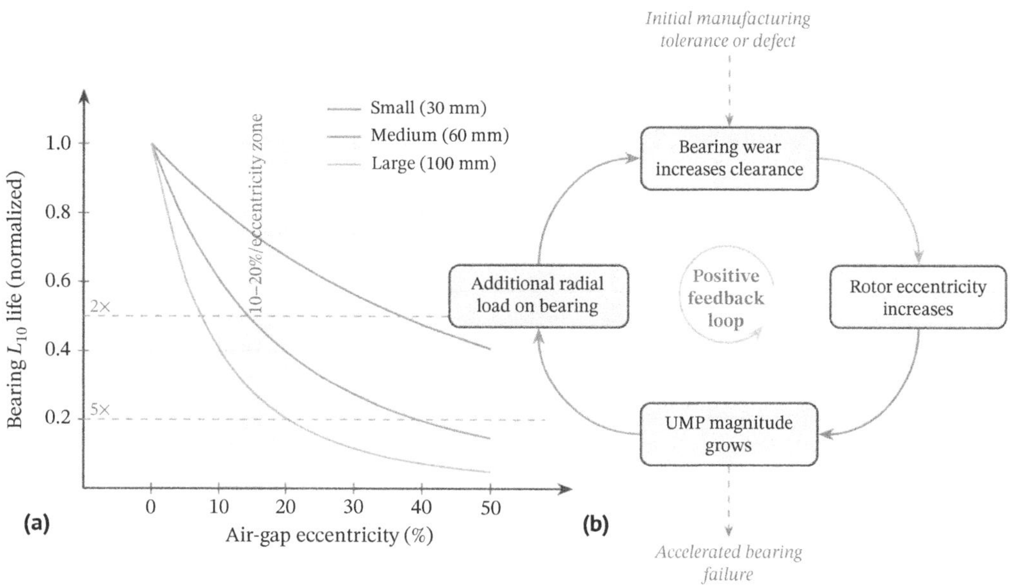

Figure 7.6 Effect of UMP on bearing life in PM motors. (a) Normalized bearing L_{10} life as a function of air-gap eccentricity for three representative shaft diameters, showing the nonlinear life reduction governed by the inverse cubic load-life relationship and the greater sensitivity of larger machines. Dashed lines indicate 2× and 5× life reduction levels. The shaded region highlights the 10–20% eccentricity zone where moderate faults produce significant life reduction, particularly in medium-to-large machines. (b) The UMP-bearing degradation positive feedback loop, where bearing wear, eccentricity, UMP growth, and increased radial loading form a self-reinforcing cycle that accelerates failure.

greater UMP for a given eccentricity level, resulting in steeper bearing life reduction. Even moderate eccentricity in the range of 10–20% of the air gap can reduce bearing life by factors of 2–5 in medium-to-large PM motors [3], highlighting the importance of eccentricity fault detection and UMP monitoring. Furthermore, as illustrated in Figure 7.6b, the UMP-bearing degradation interaction creates a positive feedback loop: initial bearing wear increases mechanical clearance, which introduces rotor eccentricity, which generates UMP that imposes additional radial load on the bearing, further accelerating degradation. This self-reinforcing mechanism emphasizes the critical importance of early bearing fault detection in PM motors, since without intervention the degradation rate increases progressively over time.

PM Motor-specific Life Factors: Several factors unique to PM motors influence bearing life calculations. The higher radial loads imposed by permanent magnet attraction forces increase the equivalent bearing load. High-speed operation reduces bearing life according to the relationship between rotational speed and the number of stress cycles. Thermal effects from localized motor heating can reduce lubricant effectiveness and material properties. Inverter-induced bearing currents introduce an additional degradation mechanism not captured in traditional bearing life models. The basic bearing life equation modified for electromagnetic effects becomes:

$$L_{10,PM} = \left(\frac{C}{P + F_{UMP,radial}}\right)^3 \times k_T \times k_{EC} \times 10^6 \text{ revolutions} \tag{7.2}$$

where k_T accounts for temperature effects on lubricant viscosity and material properties, and k_{EC} represents the life reduction factor due to electrical erosion from bearing currents. Accurate bearing life prediction for PM motors requires incorporating these factors into modified life calculation models.

The various bearing failure modes discussed in this section are systematically categorized in Table 7.2, which provides a comprehensive reference for understanding the characteristics, causes, and progression patterns of different bearing degradation mechanisms.

Table 7.2 Classification of bearing failure modes in PM motors. The type indicates whether the defect is typically localized (L), distributed (D), or progressive (P). PM motor-specific considerations highlight unique failure mechanisms and accelerating factors compared to conventional motors.

Failure mode	Type	Primary causes	Key indicators	PM motor specific considerations
Flaking	L	Rolling contact fatigue	Surface peeling; spalls	UMP increases contact stress
Pitting	L	Subsurface stress concentration	Raceway surface holes	Higher SPM radial loads
Electric pitting	L/D	Bearing currents; EDM	Frosting; fluting patterns	High dv/dt inverter switching
Brinelling (true)	L	Static overload	Raceway indentations	PM attraction radial loads
Brinelling (false)	D	Vibration under static load	Contact point wear	Transportation/storage vibration
Smearing	L	Ball sliding/spinning	Adhesive wear; material transfer	High-speed rapid transients
Subsurface fatigue	L	Cyclic subsurface stress	Subsurface cracks; spalling	Compact PM high contact stress

Failure mode	Type	Primary causes	Key indicators	PM motor specific considerations
Wear	D	Abrasive particles; poor lubrication	Increased clearance; roughness	UMP feedback loop
Fretting	D	Vibration-induced micromotion	Contact surface wear	Vibrating platform operation
Corrosion, rust	D	Moisture ingress; seal failure	Surface oxidation; rust particles	Harsh environments (marine, outdoor)
Grease degradation	D	Temperature; oxidation	Increased friction; temperature rise	Electromagnetic heating
Contamination	D	Seal degradation; maintenance ingress	Abrasive surface damage	Magnetic particle attraction
Cage wear	L/D	Friction; poor lubrication	Irregular cage motion	Low-load wide-speed-range operation
Overloading	P	Excessive loads; UMP	Plastic deformation; fatigue	UMP positive feedback loop

7.3 Modeling and Analysis

Having established the failure modes and mechanisms of bearing faults in Section 7.2, this section develops mathematical models and analytical frameworks for understanding bearing dynamics, defect-induced forces, and their electromagnetic coupling in PM motors. The analysis progresses from fundamental kinematic relationships through contact mechanics to comprehensive electromagnetic-mechanical interaction models. These models provide the foundation for fault diagnosis techniques presented in Section 7.4 and enable quantitative prediction of bearing fault signatures specific to PM motor applications.

7.3.1 Bearing Dynamics and Kinematic Modeling

7.3.1.1 Bearing Geometry and Fundamental Relationships

Key Dimensional Parameters: A deep-groove ball bearing consists of four main components: the inner race, outer race, rolling elements (balls), and cage that maintains the rolling elements' position. The key geometric parameters that govern bearing dynamics are illustrated in Figure 7.7 and defined as follows:

The bearing pitch diameter represents the mean of the inner and outer race contact diameters:

$$d_m = R_{\text{in}} + R_{\text{out}} \tag{7.3}$$

where R_{in} is the inner race contact radius and R_{out} is the outer race contact radius.

The diametral clearance, which allows for thermal expansion and manufacturing tolerances, is defined as:

$$c_d = 2(R_{\text{out}} - R_{\text{in}} - d) \tag{7.4}$$

where d is the ball diameter. This clearance parameter significantly influences bearing stiffness and vibration response, with larger clearances generally reducing bearing stiffness and increasing nonlinear dynamic behavior.

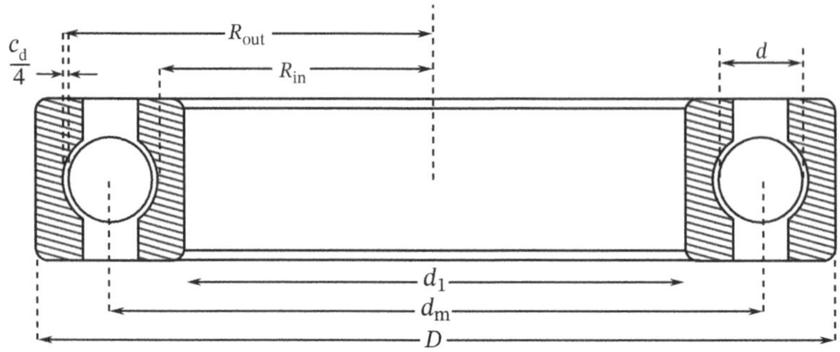

Figure 7.7 Axial cross-section of a deep-groove ball bearing showing key geometric parameters: outer diameter (D), bore diameter (d_1), pitch diameter (d_m), ball diameter (d), inner and outer raceway radii (R_{in}, R_{out}), and diametral clearance (c_d).

Race Conformity and Contact Geometry: Race conformity quantifies the geometrical fit between the ball and the race groove, defined as the ratio of the race groove radius to the ball diameter:

$$R_r = \frac{r}{d} \tag{7.5}$$

Perfect conformity corresponds to $R_r = 0.5$. In modern ball bearings, race conformity typically ranges between 0.51 and 0.54, with 0.52 being the most common value. This design compromise balances contact stress reduction (which favors higher conformity) against frictional heat generation (which favors lower conformity). For PM motors operating at high speeds, this design parameter becomes critical as excessive conformity can lead to thermal issues while insufficient conformity increases contact stress and accelerates fatigue damage [6].

Since the contact radius in the circumferential direction differs between inner and outer races ($R_{in} < R_{out}$), manufacturers typically design slightly larger race conformity ratios for the outer race. This ensures approximately equal curvature sums for inner and outer race contacts, resulting in similar maximum contact stresses in both races and a balanced fatigue life.

7.3.1.2 Bearing Kinematics and Characteristic Frequencies

Ball and Cage Motion: The kinematic relationships governing ball bearing motion form the foundation for understanding characteristic defect frequencies. When the inner race rotates at angular velocity ω_{in} while the outer race remains stationary, the angular velocity of the balls relative to the stationary outer race is:

$$\omega_b = \left(\frac{\omega_{in}}{2}\right)\left(\frac{d_m}{d}\right)\left[1 - \left(\frac{d\cos(\alpha)}{d_m}\right)^2\right] \tag{7.6}$$

The angular velocity of the cage with respect to the stationary outer race is:

$$\omega_c = \frac{\omega_{in}}{2}\left[1 - \frac{d\cos(\alpha)}{d_m}\right] \tag{7.7}$$

where α is the bearing contact angle [7]. For deep-groove ball bearings, the contact angle is typically close to zero degrees during radial loading, simplifying these relationships.

Characteristic Defect Frequencies: When a localized defect exists on any of the four bearing components (inner race, outer race, balls, or cage), specific vibration frequencies emerge as the defect interacts with mating surfaces. These characteristic frequencies are fundamental to bearing fault diagnosis and are derived from the kinematic relationships.

The Ball Pass Frequency of the Outer Race (BPFO), representing the frequency at which balls pass a fixed point on the outer race, is:

$$f_{\text{BPFO}} = \frac{N_b}{2} f_r \left[1 - \frac{d \cos(\alpha)}{d_m} \right] \tag{7.8}$$

The Ball Pass Frequency of the Inner Race (BPFI), representing the frequency at which balls pass a point on the rotating inner race, is:

$$f_{\text{BPFI}} = \frac{N_b}{2} f_r \left[1 + \frac{d \cos(\alpha)}{d_m} \right] \tag{7.9}$$

The Ball Spin Frequency (BSF), representing the rotation frequency of a ball about its own axis, is:

$$f_{\text{BSF}} = \frac{d_m}{2d} f_r \left[1 - \left(\frac{d \cos(\alpha)}{d_m} \right)^2 \right] \tag{7.10}$$

The Fundamental Train Frequency (FTF), representing the cage rotation frequency, is:

$$f_{\text{FTF}} = \frac{1}{2} f_r \left[1 - \frac{d \cos(\alpha)}{d_m} \right] \tag{7.11}$$

where N_b is the number of balls and f_r is the rotating frequency of the rotor [8].

These primary bearing frequencies form the foundation for fault detection and are summarized in Table 7.3 for quick reference. In PM motors, electromagnetic forces and mechanical vibrations interact, creating complex frequency spectra with coupling mechanisms detailed in Section 7.3.4.

Secondary Bearing Frequencies and Modulation Effects: Beyond primary characteristic frequencies, secondary frequencies emerge due to waviness, misalignment, and electromagnetic force variations. These manifest as sideband frequencies around primary defect frequencies, providing crucial diagnostic information.

Three categories of secondary frequencies are particularly significant [8]:

Modulation of f_{char} with f_r: When bearing component defects interact with shaft rotation, sidebands appear at:

$$f_{sb1} = f_{char} \pm f_r \tag{7.12}$$

Table 7.3 Bearing characteristic frequencies for fault diagnosis. These frequencies appear in vibration and current spectra when localized defects exist on the corresponding bearing components.

Frequency	Formula	Physical meaning
BPFO	$f_{\text{BPFO}} = \frac{N_b}{2} f_r \left[1 - \frac{d \cos(\alpha)}{d_m} \right]$	Frequency at which balls pass a fixed point on the stationary outer race; indicates outer race defects
BPFI	$f_{\text{BPFI}} = \frac{N_b}{2} f_r \left[1 + \frac{d \cos(\alpha)}{d_m} \right]$	Frequency at which balls pass a point on the rotating inner race; indicates inner race defects
BSF	$f_{\text{BSF}} = \frac{d_m}{2d} f_r \left[1 - \left(\frac{d \cos(\alpha)}{d_m} \right)^2 \right]$	Rotation frequency of a ball about its own axis; indicates ball/rolling element defects
FTF	$f_{\text{FTF}} = \frac{1}{2} f_r \left[1 - \frac{d \cos(\alpha)}{d_m} \right]$	Cage rotation frequency; indicates cage defects and modulates other frequencies

where f_{char} represents either BPFO or BPFI frequencies. This modulation indicates that defect severity is sufficient to cause measurable shaft displacement.

Modulation of two f_{char}: The interaction between two distinct bearing characteristic frequencies, most commonly FTF modulating BPFO or BPFI:

$$f_{sb2} = f_{char} \pm f_{FTF} \tag{7.13}$$

Modulation of f_r and two f_{char}: This signature emerges when faults distribute across bearing surfaces, characteristic of distributed bearing faults:

$$f_{sb3} = f_{char} \pm (f_r \pm f_{FTF}) \tag{7.14}$$

The presence of these modulation frequencies, particularly the third category, distinguishes distributed bearing faults from localized single-point defects. In PM motors, these sideband patterns become more complex due to electromagnetic force modulation. Table 7.4 summarizes these modulation patterns and their diagnostic significance.

7.3.1.3 Hertzian Contact Theory and Load Distribution

The contact mechanics between rolling elements and races govern both the magnitude of defect-induced force pulses and the long-term fatigue processes leading to bearing degradation. According to Hertzian contact theory, the ball-race contact forms an elliptical area whose geometry depends on the curvature properties of both contacting bodies. The detailed derivation of the contact geometry, including curvature sums, ellipticity parameters, and elliptic integral approximations, can be found in [5, 6].

The key result for bearing dynamics is the nonlinear force-deflection relationship. The contact stiffness coefficient K_c depends on the bearing geometry and material properties (elastic modulus E and Poisson's ratio ν of both ball and race). The deflection between ball and race under load F follows:

$$\delta_0 = \left(\frac{F}{K_c}\right)^{2/3} \tag{7.15}$$

Table 7.4 Secondary bearing frequencies and modulation patterns. These sideband frequencies provide crucial information about fault type, severity, and distribution, particularly valuable for distinguishing localized from distributed defects.

Modulation type	Frequency formula	Diagnostic significance	Fault type indication
Characteristic frequency modulated by rotation	$f_{sb1} = f_{char} \pm f_r$	Defect severity causes measurable shaft displacement; amplitude proportional to eccentricity	Localized defects on inner or outer race with sufficient severity to affect rotor position
Two characteristic frequencies	$f_{sb2} = f_{char} \pm f_{FTF}$	Cage modulation of bearing defect frequencies; indicates interaction between cage motion and race defects	Localized race defects with cage position effects; common in normal bearing operation with defects
Rotation and two characteristic frequencies	$f_{sb3} = f_{char} \pm (f_r \pm f_{FTF})$	Complex modulation indicating widespread surface degradation	Distributed bearing faults (wear, roughness, contamination) affecting multiple bearing surfaces

This 3/2 power-law relationship is fundamental to bearing dynamics and distinguishes rolling element bearings from linear spring systems. The total stiffness coefficient combining both inner and outer race contacts is:

$$K_c^{tot} = \left[\left(\frac{1}{K_c^{in}} \right)^{2/3} + \left(\frac{1}{K_c^{out}} \right)^{2/3} \right]^{3/2} \tag{7.16}$$

The contact force acting on ball i is then:

$$F_i = K_c^{tot} (\delta_i^{tot})^{1.5} \tag{7.17}$$

where δ_i^{tot} combines the elastic deformations at inner and outer race contacts. The resultant bearing forces are obtained by summing contributions from all loaded balls:

$$F_X = -\sum_{i=1}^{z} F_i \cos(\phi_i)\cos(\beta_i), \quad F_Y = -\sum_{i=1}^{z} F_i \cos(\phi_i)\sin(\beta_i) \tag{7.18}$$

where ϕ_i is the contact angle and β_i is the attitude angle of ball i, with the summation including only balls under compression ($\delta_i^{tot} > 0$) [6]. The time-varying attitude angle as the inner race rotates leads to fluctuations in bearing stiffness, known as the varying compliance effect, which generates vibrations even in geometrically perfect bearings.

Figure 7.8 illustrates these relationships for the SKF 6205-2RS bearing. The nonlinear stiffening characteristic in Figure 7.8a means that the effective bearing stiffness increases with load, which influences both natural frequencies and defect-induced force amplitudes. In Figure 7.8b, the polar load distribution shows that as bearing clearance increases due to wear, fewer balls carry load and the maximum ball force increases. This load zone narrowing concentrates stress on fewer rolling elements, accelerating fatigue and producing sharper defect-induced force pulses.

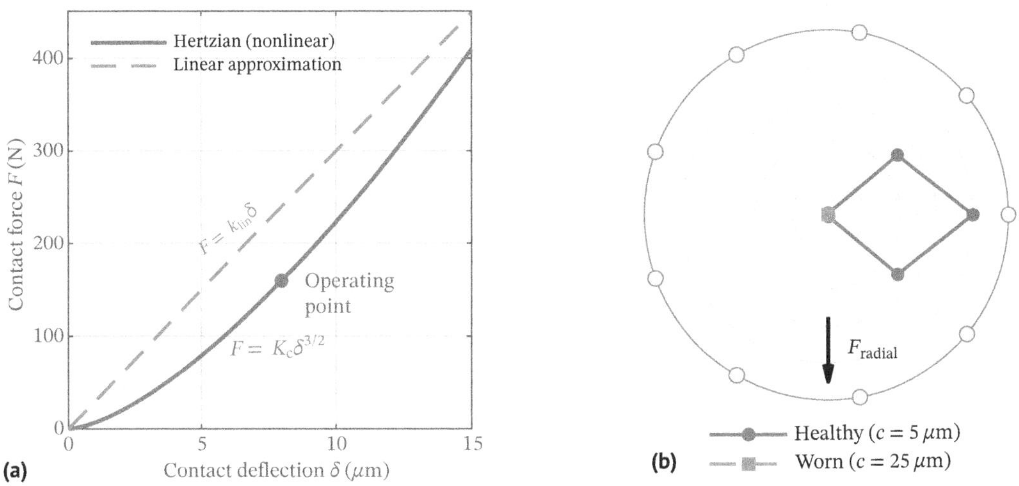

Figure 7.8 Hertzian contact mechanics and load zone distribution for SKF 6205-2RS bearing. (a) Nonlinear Hertzian force-deflection relationship ($F = K_c\delta^{3/2}$) compared with a linear spring approximation tangent at the operating point. (b) Polar plot of ball load distribution under radial load for healthy and worn bearings, showing load zone narrowing with increased clearance.

7.3.1.4 Speed-dependent Variations in PM Motor Applications

As bearing rotational speed increases, centrifugal forces push rolling elements outward, shifting the contact angle and altering load distribution. The centrifugal force on a ball is $F_{cf} = \frac{1}{2}m_b\omega_c^2 d_m$, where m_b is the ball mass and ω_c is the cage angular velocity [7]. At high speeds, gyroscopic moments further affect ball motion, potentially causing microslip at ball-race interfaces. Ball slip causes actual characteristic frequencies to differ from ideal kinematic predictions by typically 1–3%, and under variable speed operation this slip variation produces frequency smearing in vibration spectra that complicates fault diagnosis.

These speed-dependent effects have direct implications for PM motor bearing diagnostics across different operating regimes. In low-speed, high-torque applications (direct-drive wind turbines, robotics), electromagnetic radial forces and UMP dominate bearing loading, centrifugal effects are negligible, and characteristic defect frequencies closely match theoretical predictions. In the constant-power regime typical of traction motors, field weakening reduces electromagnetic forces while centrifugal and gyroscopic effects increase, and the lubrication film transitions from mixed to full elastohydrodynamic contact, changing vibration damping characteristics. In high-speed applications (spindle motors, aerospace generators above 6,000 rpm), centrifugal forces dominate, full-film lubrication establishes significant damping that can mask high-frequency defect signatures, and cage instability becomes a concern. Robust fault diagnosis algorithms for PM motors must account for these regime-dependent variations to avoid false alarms and missed detections across the full operating envelope.

7.3.1.5 Bearing Dynamics Under Faulty Conditions

Localized Defect-induced Force Pulses: When a rolling element passes over a localized defect such as a spall or pit (see Figure 7.2), an impulse force is generated. The shape and magnitude of this pulse depend on the defect geometry and operating conditions. The defect can be characterized by two parameters: length L_{defect} and depth h_{defect}.

The force pulse occurs when the attitude angle of ball i aligns with the defect position angle ϕ_{defect}:

$$\beta_i - \theta_{in} - \phi_{defect} = -n \cdot 2\pi, \quad n = 0, 1, 2, ... \tag{7.19}$$

This alignment condition triggers a reduction in contact force as the ball temporarily loses support when entering the defect zone, followed by an impact as the ball exits the defect. The duration of the pulse is:

$$\tau_{pulse} = \frac{L_{defect}}{R_{in}\omega_{rel}} \tag{7.20}$$

where ω_{rel} is the relative angular velocity between the defective component and the balls.

Distributed Defects and Surface Waviness: Distributed defects, including surface waviness, contamination-induced roughness, and progressive wear, modify the roundness profile of the bearing ring. The roundness profile can be expressed as a Fourier series:

$$R(\theta) = R_0 + \sum_{m=1}^{n} A_m \cos(m\theta + \phi_m) \tag{7.21}$$

where R_0 is the nominal radius, A_m is the amplitude of the m-th order waviness, and ϕ_m is the phase angle [6].

The effect of waviness on ball deformation is:

$$\Delta_i^{total} = \delta_i^{tot} + \sum_{m=1}^{n} A_m^{in} \cos[m(\beta_i - \theta_{in}) + \phi_m^{in}] + \sum_{m=1}^{n} A_m^{out} \cos[m(\beta_i - \theta_{out}) + \phi_m^{out}] \quad (7.22)$$

Waviness orders close to the number of balls ($m = N_b \pm 1$ and $m = N_b$) produce the most severe vibrations, generating frequencies at BPFI, BPFO, and their sidebands [8].

Nonlinear Dynamic Response: The nonlinear force-deflection relationship in ball bearings leads to complex dynamic behaviors, including:

- **Subharmonic resonances:** When bearing excitation frequencies are integer fractions of system natural frequencies, parametric resonances can occur, particularly with low-order waviness.
- **Superharmonic generation:** The Hertzian contact nonlinearity generates harmonics of the fundamental defect frequencies even for single-frequency excitation.
- **Jump phenomena:** In certain speed ranges, the bearing response can exhibit bistable behavior, suddenly transitioning between different vibration amplitudes.

For PM motors, these nonlinear effects interact with electromagnetic forces, potentially creating complex vibration patterns that challenge traditional linear fault diagnosis methods.

7.3.2 Defect-induced Force Modeling

7.3.2.1 Impulse Response Model for Localized Defects

Single-point Defect Force Pulse: When a ball encounters a localized defect such as a spall, pit, or crack on either race, a characteristic force pulse is generated. The magnitude and shape of this pulse depend on the defect geometry, entry and exit angles, and the operating speed. For a defect of length L_{defect} and depth h_{defect}, the force reduction when the ball enters the defect can be modeled as:

$$F_{defect}(t) = F_{normal} \cdot \left[1 - f_{defect}(t)\right] \quad (7.23)$$

where F_{normal} is the normal contact force and $f_{defect}(t)$ is a time-varying defect function. Several models have been proposed to represent this defect function, with the exponential model being commonly adopted:

$$f_{defect}(t) = \begin{cases} h_{defect}\left[1 - \cos\left(\dfrac{2\pi v_{rel} t}{L_{defect}}\right)\right], & 0 \le t \le \dfrac{L_{defect}}{v_{rel}} \\ 0, & \text{otherwise} \end{cases} \quad (7.24)$$

where v_{rel} is the relative velocity between the defect and the ball, given by:

$$v_{rel} = \omega_{rel} \cdot R_{race} \quad (7.25)$$

For an outer race defect, $\omega_{rel} = \omega_c$ (cage velocity), while for an inner race defect, $\omega_{rel} = \omega_{in} - \omega_c$.

Entry and Exit Impact Forces: The ball experiences two distinct impact events when traversing a defect: an entry impact as the ball drops into the defect, and an exit impact as the ball climbs out. These impacts generate broadband vibration excitation that excites bearing and housing resonances. The impact force magnitude can be approximated using:

$$F_{impact} = K_c\left[\delta_{impact}\right]^{1.5} \quad (7.26)$$

where the impact deformation δ_{impact} depends on the defect depth, ball velocity, and impact angle:

$$\delta_{impact} = h_{defect} + \frac{v_{entry}^2}{2k_{eff}} \tag{7.27}$$

The effective stiffness k_{eff} combines the Hertzian contact stiffness with material damping effects. The entry velocity v_{entry} can be estimated from the kinematics of the ball dropping distance equal to the defect depth.

Multi-point Defects and Fault Progression: As bearing degradation progresses, single-point defects often expand into multi-point or extended defects. The force signature transitions from a periodic pulse train to a more complex waveform. For n discrete defects distributed around a raceway, the total defect-induced force becomes:

$$F_{total}(t) = \sum_{j=1}^{n} F_{defect,j}(t - t_j) \tag{7.28}$$

where t_j represents the time delay between successive defect encounters. When defects are irregularly spaced, this produces amplitude modulation of the characteristic defect frequencies.

7.3.2.2 Distributed Defect Force Models

Surface Roughness and Waviness: Distributed defects resulting from wear, contamination, or corrosion manifest as increased surface roughness or waviness. Unlike localized defects that produce discrete impulses, distributed defects generate continuous vibration excitation. The additional deformation induced by surface irregularities can be modeled using a random process or a deterministic waviness function.

For surface roughness characterized by an amplitude distribution $A_{rough}(\theta)$, the contact force variation at ball i becomes:

$$F_i = K_c^{tot} \left[\delta_i^{tot} + A_{rough}(\beta_i)\right]^{1.5} \tag{7.29}$$

Statistical characterization of A_{rough} typically employs metrics such as root-mean-square (RMS) roughness (R_q) or arithmetic average roughness (R_a). The frequency content of roughness-induced vibration is broadband, with energy concentrated at the ball passage frequencies and their harmonics.

Wear-induced Clearance Increase: Progressive bearing wear increases internal clearance, fundamentally altering the load distribution and reducing bearing stiffness. The modified diametral clearance evolves as:

$$c_d(t) = c_{d,0} + \Delta c_{wear}(t) \tag{7.30}$$

where Δc_{wear} increases with cumulative operating hours and load cycles. Increased clearance causes:

- reduced radial stiffness by a factor that can reach 50% or more,
- load zone narrowing, concentrating loads on fewer balls,
- increased ball slip and microslip, and
- greater susceptibility to cage instability.

The force distribution in a worn bearing with clearance c_d shows that fewer balls carry the load, with maximum ball load increasing according to:

$$F_{max,worn} = F_{max,new}\left(\frac{c_{d,worn}}{c_{d,new}}\right)^{1.5} \tag{7.31}$$

This nonlinear relationship accelerates the degradation process as clearance increases.

7.3.2.3 PM Motor-specific Force Modulation

Electromagnetic Force Coupling: In PM motors, the total radial force on a bearing combines mechanical and electromagnetic contributions:

$$F_{radial,total} = F_{mechanical} + F_{UMP}(\theta_r, i_a, i_b, i_c, e) \tag{7.32}$$

where F_{UMP} represents the unbalanced magnetic pull depending on rotor position, phase currents, and eccentricity. When bearing faults induce rotor vibration, the resulting eccentricity variations modulate F_{UMP}, creating diagnostic sidebands at $f_{sideband} = 2f_e \pm f_{defect}$ where f_e is the electrical frequency. This electromagnetic-mechanical coupling distinguishes PM motor bearing faults from purely mechanical systems.

Additionally, torque ripple effects from cogging torque ($f_{cog} = N_s \cdot p \cdot f_r$) create periodic tangential forces that interact with bearing defects, producing combination frequencies $f_{combined} = f_{defect} \pm n \cdot f_{cog}$ that provide diagnostic signatures unique to PM motors. Section 7.3.4 provides a comprehensive analysis of these coupling mechanisms.

7.3.2.4 Time-domain Force Simulation

Numerical Integration Approach: Complete dynamic simulation of bearing forces under faulty conditions requires time-domain numerical integration. The equations of motion for the rotor supported by bearings are:

$$m\ddot{x} + c_x\dot{x} = F_{x,bearing} + F_{x,EM} + F_{x,external} \tag{7.33}$$

$$m\ddot{y} + c_y\dot{y} = F_{y,bearing} + F_{y,EM} + F_{y,external} \tag{7.34}$$

where m is the rotor mass, c_x and c_y are damping coefficients, and the force terms include bearing forces, electromagnetic forces, and external loads. The bearing forces $F_{x,bearing}$ and $F_{y,bearing}$ are calculated at each time step using:

$$F_{x,bearing} = \sum_{i=1}^{N_b} F_i \cos(\beta_i) \tag{7.35}$$

$$F_{y,bearing} = \sum_{i=1}^{N_b} F_i \sin(\beta_i) \tag{7.36}$$

where each ball force F_i incorporates defect effects through the modified deformation:

$$F_i = \begin{cases} K_c^{tot}\left[\delta_i + \delta_{defect,i}(t)\right]^{1.5}, & \delta_i + \delta_{defect,i} > 0 \\ 0, & \text{otherwise} \end{cases} \tag{7.37}$$

Resonance Excitation: The impulse forces generated by bearing defects excite structural resonances of the bearing housing, motor frame, and mounting structure. Each resonance frequency $f_{res,k}$ has an associated mode shape and damping ratio ζ_k. When defect pulse repetition rates approach these resonances, the vibration amplitude amplifies significantly.

The frequency response function (FRF) between defect force input and vibration acceleration output exhibits peaks at each resonance:

$$H(f) = \sum_k \frac{A_k}{1 + 2j\zeta_k(f/f_{res,k}) - (f/f_{res,k})^2} \tag{7.38}$$

where A_k is the amplitude coefficient for mode k. This FRF must be convolved with the defect force time history to predict the measured vibration signal.

Simulation Validation: Time-domain simulations have been validated against experimental measurements showing excellent agreement for:

- pulse repetition rates matching characteristic defect frequencies,
- amplitude modulation patterns,
- harmonic and sideband structure, and
- resonance excitation levels.

These validated models enable parametric studies of defect severity, operating conditions, and motor design parameters, providing insights not readily available from experiments alone. They also form the basis for model-based fault diagnosis algorithms and digital twin implementations for prognostics.

7.3.3 Vibration Transmission Path Analysis

The measured vibration signal from a bearing defect is shaped by the transmission path between the defect location and the sensor. The measured signal can be expressed as a convolution of the defect force and the structural impulse response:

$$x_{measured}(t) = \int_{-\infty}^{\infty} F_{defect}(\tau) h_{path}(t - \tau) d\tau + n(t) \tag{7.39}$$

where $h_{path}(t)$ captures the dynamic characteristics of the bearing housing, endshields, and motor frame, and $n(t)$ represents noise and other vibration sources. The transmission path acts as a filter: bearing characteristic frequencies (10–500 Hz for typical industrial motors) transmit efficiently through the rigid motor structure, while bearing defect impulses excite structural resonances in the 1–10 kHz range with quality factors of 20–100, producing significant amplification. This resonance amplification is the physical basis for envelope analysis, which filters around these structural resonances to enhance defect signatures. At frequencies above 10 kHz, attenuation increases with distance, favoring sensor placement close to bearings for acoustic emission monitoring.

A key challenge for PM motor bearing diagnostics is separating bearing vibrations from electromagnetic vibration sources that propagate through the same structural paths. PM motors generate vibration at multiples of the electrical frequency ($f_{EM} = kpf_r$), slot harmonic frequencies ($N_s f_r$), and their interactions. When electromagnetic frequencies coincide with bearing characteristic frequencies, careful spectral analysis and modulation pattern recognition are required to distinguish bearing faults from normal electromagnetic excitation. In inverter-driven PM motors, PWM switching harmonics (typically 2–20 kHz) create additional high-frequency vibration that can intermodulate with bearing defect signatures. Temperature variations shift structural resonance frequencies by 0.5–2% per 100 °C, which must be accounted for in long-term condition monitoring.

7.3.4 Bearing-motor Coupling Effects

This section analyzes the bidirectional electromagnetic-mechanical coupling unique to PM motors, building upon the PM motor-specific considerations introduced throughout Sections 7.2 and 7.3. While previous sections noted that electromagnetic forces affect bearing loading and bearing defects modulate electromagnetic behavior, this section provides comprehensive quantitative analysis of these interactions, their impact on diagnostic signatures, and their implications for bearing life prediction.

7.3.4.1 Electromagnetic-mechanical Interaction Fundamentals

Bidirectional Coupling Mechanism: In PM motors, bearing faults and electromagnetic phenomena interact bidirectionally through two primary mechanisms: bearing defects induce rotor vibration and eccentricity that modulate electromagnetic forces and flux distribution, while electromagnetic forces (particularly UMP) affect bearing loading and accelerate degradation. The coupled system can be represented by:

$$\begin{cases} F_{bearing} = f_{mech}(\delta, \dot{\delta}, \ddot{\delta}, defect) \\ F_{EM} = f_{EM}(\delta, i_a, i_b, i_c, \theta_r) \\ m\ddot{\delta} = F_{bearing} + F_{EM} + F_{external} \end{cases} \tag{7.40}$$

where δ represents rotor displacement, $F_{bearing}$ represents bearing reaction forces, F_{EM} represents electromagnetic forces, and the third equation expresses Newton's second law. This nonlinear coupled system exhibits complex dynamics requiring simultaneous electromagnetic and mechanical analysis.

Air Gap Variation and Flux Modulation: Rotor eccentricity caused by bearing defects or wear modulates the air gap length around the rotor periphery. As introduced in Chapter 3 for the general eccentricity modeling framework, for a rotor displaced by distance e from the geometric center, the air gap at angular position θ becomes:

$$g(\theta) = g_0 - e\cos(\theta - \phi_e) \tag{7.41}$$

where g_0 is the nominal air gap, e is the eccentricity magnitude, and ϕ_e is the eccentricity direction angle. In bearing fault contexts, e is not a static design parameter but a time-varying quantity driven by bearing defect dynamics and wear progression, as described in the kinematic models of the preceding sections. This air gap variation directly affects the magnetic permeance, causing flux density modulation:

$$B(\theta) = B_0 \frac{g_0}{g(\theta)} \approx B_0\left[1 + \frac{e}{g_0}\cos(\theta - \phi_e)\right] \tag{7.42}$$

This flux modulation generates characteristic frequency components in stator current and electromagnetic force spectra that correlate with bearing-induced vibration frequencies.

Figure 7.9 summarizes this bidirectional coupling as a signal-flow block diagram. In the forward mechanical path, bearing defects generate periodic force pulses that excite rotor vibration and produce time-varying eccentricity. The electromagnetic path then converts this eccentricity into air gap modulation, flux perturbation, and ultimately measurable current sidebands. The UMP feedback loop closes the system: the asymmetric air gap produces an attractive radial force that

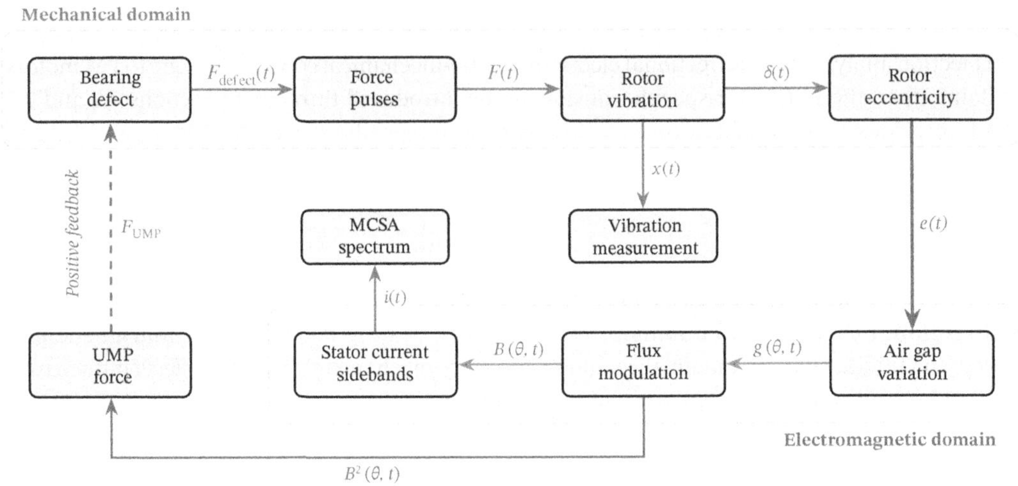

Figure 7.9 Signal-flow block diagram of the electromagnetic-mechanical coupling in PM motors with bearing faults. The mechanical path (top) maps bearing defects to rotor eccentricity, which drives air gap and flux modulation in the electromagnetic path (bottom). The dashed UMP feedback loop closes the system, accelerating bearing degradation.

increases bearing loading, further accelerating wear. This closed-loop structure explains why bearing degradation in PM motors can progress rapidly once a threshold eccentricity is reached, and highlights the two complementary diagnostic access points, stator current and vibration, that are analyzed in the remaining subsections.

7.3.4.2 Current Signature Modulation by Bearing Faults

MCSA Sideband Generation: Bearing-induced rotor vibration creates eccentricity variations that modulate the back-EMF and inductance of the stator windings. This modulation appears as sidebands around the fundamental supply frequency in the stator current spectrum. For a bearing characteristic frequency f_{char} (BPFI, BPFO, BSF, or FTF), sidebands appear at:

$$f_{bearing-MCSA} = f_s \pm k f_{char} \tag{7.43}$$

where f_s is the supply frequency and k is an integer (typically $k = 1, 2$). These sidebands arise from the amplitude and phase modulation of the supply current by the time-varying air gap permeance.

The sideband amplitude is proportional to the eccentricity magnitude and the bearing defect severity:

$$I_{sideband} \propto \frac{e_{defect}}{g_0} I_{fundamental} \tag{7.44}$$

where e_{defect} represents the vibration amplitude induced by the bearing defect. This relationship enables quantitative bearing condition assessment from current measurements.

Mixed Eccentricity Signature: Bearing clearance increase due to wear produces mixed eccentricity (combination of static and dynamic eccentricity). The characteristic MCSA signature for mixed eccentricity includes frequency components at:

$$f_{mix-ecc} = f_s \pm kf_r \quad (7.45)$$

where f_r is the rotational frequency. When bearing faults coexist with eccentricity, the current spectrum contains both bearing-related sidebands and eccentricity-related sidebands, creating a complex but information-rich signature. Distinguishing these components requires careful analysis of sideband spacing and modulation patterns.

Load-dependent Signature Variations: The amplitude of bearing-fault-induced current sidebands varies with motor load. At higher loads, the bearing load zone concentrates, making defects produce more pronounced force variations. Additionally, electromagnetic forces increase with load current, amplifying the coupling effect. Consequently, bearing fault signatures in MCSA become more prominent under loaded conditions:

$$\frac{I_{sideband}}{I_{fundamental}} \propto T_{load}^{\alpha} \quad (7.46)$$

where T_{load} is the load torque and α is an exponent typically ranging from 0.5 to 1.0 depending on motor design and bearing configuration.

Figure 7.10 illustrates the sideband structure predicted by these equations for a 12-pole PMSM with an SKF 6205-2RS bearing operating at 1,000 rpm. In Figure 7.10a, the outer race and inner race fault sidebands appear at distinct offsets from the supply frequency, while the mixed eccentricity components at $f_s \pm f_r$ occupy a separate spectral region closer to the fundamental. The relative spacing of these sideband families allows each fault type to be identified independently. Figure 7.10b shows that sideband amplitude grows linearly with the eccentricity ratio e/g_0, but remains buried in the noise floor for small eccentricities. This establishes a practical detection threshold below which MCSA alone cannot reliably detect bearing degradation, motivating the complementary use of vibration-based and data-driven methods discussed in later sections.

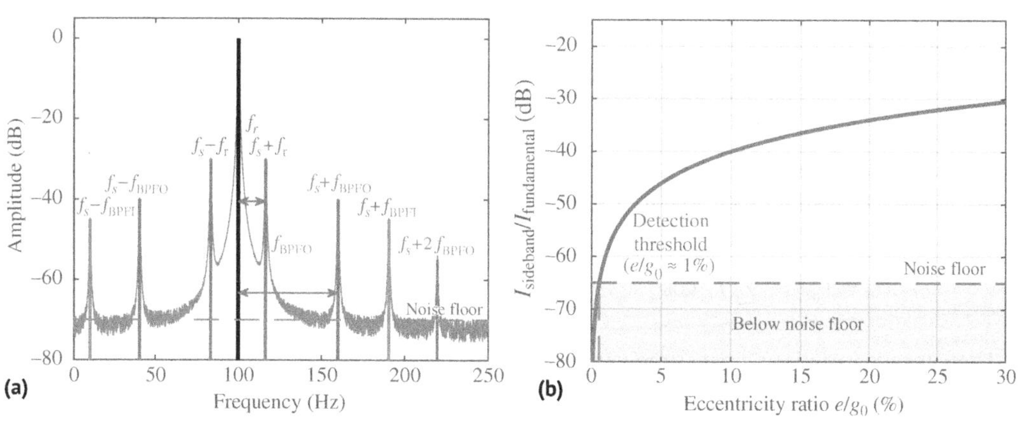

Figure 7.10 Bearing fault signatures in stator current spectrum of a 12-pole PMSM. (a) Idealized current spectrum showing sidebands induced by bearing characteristic frequencies around the supply frequency f_s: outer race fault sidebands at $f_s \pm kf_{BPFO}$, inner race fault sidebands at $f_s \pm kf_{BPFI}$, and mixed eccentricity components at $f_s \pm kf_r$. (b) Bearing-fault sideband amplitude relative to the fundamental as a function of eccentricity ratio e/g_0, showing the detection threshold where sidebands emerge above the noise floor.

7.3.4.3 Unbalanced Magnetic Pull and Bearing Life

UMP Force Characteristics: Unbalanced magnetic pull (UMP) arises from air gap asymmetry and creates an attractive radial force pulling the rotor toward the stator region with the smallest air gap. For small eccentricity $e << g_0$, the UMP can be approximated as:

$$F_{\text{UMP}} = k_{\text{UMP}} \frac{B_0^2 A_{\text{rotor}}}{2\mu_0} \frac{e}{g_0} \tag{7.47}$$

where B_0 is the nominal air gap flux density, A_{rotor} is the rotor surface area, μ_0 is the permeability of free space, and k_{UMP} is a geometry-dependent coefficient. For PM motors with strong permanent magnet fields, B_0 can exceed 1 T, producing substantial UMP forces even for small eccentricities.

The UMP force acts in the direction of eccentricity, creating a destabilizing effect that can amplify rotor vibration. This force is approximately proportional to eccentricity for small displacements but becomes highly nonlinear as eccentricity approaches the air gap length.

Bearing Life Reduction and Feedback Mechanism: As discussed in Section 7.2.7, UMP significantly reduces bearing life by increasing the equivalent bearing load according to a cubic power law relationship. The positive feedback degradation mechanism where bearing wear increases eccentricity, which amplifies UMP, which further accelerates wear, creates a concerning runaway effect that can rapidly progress from minor clearance increase to catastrophic failure. This feedback loop makes early detection and intervention critical in PM motors.

7.3.4.4 Torque Ripple and Bearing Load Modulation

Cogging Torque Effects: Cogging torque, resulting from the interaction between permanent magnets and stator slot openings, creates periodic tangential force variations. These variations modulate the bearing load distribution as the rotor rotates. The cogging torque can be expressed as a Fourier series:

$$T_{cog}(\theta_r) = \sum_{n=1}^{\infty} T_{cog,n} \sin(nN_{cog}\theta_r + \phi_n) \tag{7.48}$$

where $N_{cog} = LCM(N_s, 2p)$ is the least common multiple of slot number and pole number. This periodic torque creates tangential forces on the bearing that can excite torsional vibration modes and modulate radial bearing forces through gyroscopic coupling in high-speed applications.

Torque Ripple Under Load: Under loaded conditions, PM motors exhibit torque ripple arising from current harmonics, back-EMF harmonics, and reluctance torque variations. The total electromagnetic torque contains components at various frequencies:

$$T_{EM}(t) = T_{avg} + \sum_{k} T_k \cos(k\omega_e t + \phi_k) \tag{7.49}$$

These torque variations create time-varying tangential forces that couple into the bearing system. In high-performance servo applications where torque ripple is minimized through advanced control, bearing loading becomes more uniform. Conversely, in cost-optimized designs with significant torque ripple, bearing loads exhibit greater cyclic variation, potentially affecting fatigue life.

Interaction with Bearing Defects: When bearing defects produce periodic force pulses, these pulses can synchronize with torque ripple frequencies if f_{char} is close to an electromagnetic frequency. This synchronization can either amplify or attenuate the measured vibration depending on the phase relationship, creating intermittent fault signatures that challenge diagnosis algorithms.

7.3.4.5 High-speed Effects and Gyroscopic Coupling

Critical Speed Considerations: PM motors designed for high-speed operation must avoid critical speeds where rotor natural frequencies coincide with operating frequencies. Bearing stiffness significantly influences the rotor critical speeds. For a simple rotor-bearing system, the first critical speed is:

$$\omega_{critical} = \sqrt{\frac{k_{bearing}}{m_{rotor}}} \tag{7.50}$$

where $k_{bearing}$ is the effective bearing stiffness (which decreases with bearing wear) and m_{rotor} is the rotor mass. As bearing clearance increases due to wear, bearing stiffness reduces, potentially bringing a critical speed into the operating range and causing severe vibration amplification.

Gyroscopic Effects: In high-speed PM motors with high rotor inertia, gyroscopic moments couple the radial and tilting motions of the rotor. The gyroscopic moment is:

$$M_{gyro} = I_p \Omega \dot{\beta} \tag{7.51}$$

where I_p is the polar moment of inertia, Ω is the rotational speed, and $\dot{\beta}$ is the angular velocity of the rotor axis precession. Bearing defects that induce rotor tilt generate gyroscopic moments that redistribute forces between bearing locations, potentially causing unexpected vibration patterns.

Electromagnetic Damping and Stiffness: The stator windings provide electromagnetic damping and stiffness effects that interact with mechanical bearing dynamics. Short-circuited stator windings (or resistive loads in generator applications) generate electromagnetic damping through eddy current effects. This damping coefficient can be estimated as:

$$c_{EM} = \frac{3p^2 B_0^2 l_{stack} r_{rotor}^2}{\omega R_{phase}} \tag{7.52}$$

where l_{stack} is the stack length, r_{rotor} is the rotor radius, and R_{phase} is the phase resistance. This electromagnetic damping attenuates rotor vibration, partially mitigating bearing-induced oscillations. However, it also reduces the measured vibration amplitude, potentially masking incipient bearing faults.

7.3.4.6 Inverter Effects on Bearing Performance

Bearing Current Mechanisms: In inverter-driven PM motors, high dv/dt of PWM voltage pulses induce common-mode voltages that drive currents through bearing films. The bearing current magnitude depends on the rate of voltage rise:

$$i_{bearing} \propto C_{bearing} \frac{dv_{CM}}{dt} \tag{7.53}$$

where $C_{bearing}$ is the bearing capacitance and v_{CM} is the common-mode voltage. These currents cause electrical discharge machining (EDM) of bearing surfaces, creating frosting (generalized surface roughness) or fluting (corrugation patterns) damage [9].

Fluting Pattern and Characteristic Frequencies: Bearing current-induced fluting creates axial grooves on the bearing race at regular intervals [10]. The fluting wavelength corresponds to:

$$\lambda_{fluting} = \frac{2\pi r}{N_{slots} \times p} \tag{7.54}$$

This distributed defect generates vibration at characteristic fluting frequencies that combine with traditional bearing defect frequencies, creating a complex signature unique to inverter-driven PM motors [9]. The fluting-related vibration frequencies are:

$$f_{fluting} = f_r N_{slots} p \pm f_{char} \tag{7.55}$$

Mitigation Impact on Vibration Signatures: Bearing current mitigation techniques (insulated bearings, shaft grounding, common-mode filters) eliminate fluting damage but may affect vibration transmission paths. Insulated bearings introduce a dielectric layer that can slightly alter bearing stiffness and damping. Shaft grounding brushes add a mechanical interface that may introduce additional vibration sources. Condition monitoring systems for PM motors must account for these mitigation measures when establishing baseline vibration signatures.

7.3.4.7 Integrated Modeling Approaches

Co-simulation Frameworks: Accurate prediction of bearing-motor coupling effects requires co-simulation of electromagnetic and mechanical domains. The electromagnetic modeling techniques introduced in Chapter 3, including FEM, MEC, and machine learning-based models, serve as the electromagnetic component within these coupled frameworks. Modern approaches employ:

- **FEA-MBS Coupling:** Finite element electromagnetic analysis coupled with multibody dynamics simulation, where the FEA component follows the methodology described in Chapter 3 while the multibody dynamics captures bearing kinematics from the preceding sections.
- **Reduced-order Models:** Computationally efficient models such as MEC or lookup-table approaches (see Chapter 3) that approximate full FEA results for real-time simulation of bearing-motor interaction.
- **Digital Twins:** Physics-based models updated with measured data for prognostics, extending the digital twin concept discussed in Chapter 3 to incorporate bearing degradation dynamics.

The co-simulation time step must be sufficiently small to capture bearing contact events (typically 1–10 microseconds), making computational efficiency critical for practical implementation.

Parameter Identification: Coupled electromechanical models require accurate parameters for both domains. Bearing parameters (stiffness, damping, clearance) can be identified from vibration measurements using inverse methods. Electromagnetic parameters are typically obtained from FEA calibrated to experimental back-EMF and torque measurements. The coupling parameters, particularly the UMP coefficients, require joint electromagnetic-mechanical measurements during controlled eccentricity tests.

7.3.5 FEM Analysis of Bearing-induced Eccentricity

Chapter 3 presented FEM as a general-purpose electromagnetic modeling tool for PM machines, including its application to eccentricity modeling. This section focuses on the aspects specific to bearing-induced eccentricity, where the rotor displacement is a dynamic quantity driven by bearing defect mechanics rather than a fixed geometric offset. The key distinctions are the time-varying nature of the eccentricity, the need for mesh handling strategies that accommodate continuous rotor motion, and the coupling between electromagnetic forces and bearing dynamics.

For bearing fault simulations, time-stepping FEA captures the dynamic evolution of electromagnetic fields as the rotor rotates with time-varying eccentricity. The mesh must be updated at each step through sliding surface, remeshing, or mortar element methods to accommodate the changing geometry. Since bearing fault simulations may require thousands of rotor revolutions, full FEA is computationally prohibitive, making reduced-order models essential for practical applications.

Parametric FEA studies reveal that UMP increases approximately linearly with eccentricity ratio $\varepsilon = e/g_0$ for small eccentricities ($\varepsilon < 0.2$) and quadratically with air gap flux density ($F_{UMP} \propto B_g^2$), explaining why PM motors with strong magnets experience more severe UMP than wound-field machines. Bearing wear typically produces mixed eccentricity (static plus dynamic components), generating complex UMP patterns with combination frequencies. Studies on PM motors show that 2D FEA predictions of UMP agree with experiments within 10–20%, while 3D FEA achieves 5–10% accuracy.

The electromagnetic consequences of bearing-induced eccentricity include unbalanced back-EMF waveforms, with amplitude modulation appearing as sidebands at $f_s \pm f_r$ in the current spectrum, providing an MCSA signature for bearing-induced eccentricity. Even-order harmonics that are absent in concentric machines appear under eccentricity, with the second harmonic amplitude proportional to $(e/g_0)^2$, offering early warning of developing bearing wear. For IPM motors, eccentricity creates additional spatial saliency that interacts with the inherent d-q saliency, generating reluctance torque components at f_r.

Multi-physics coupling is significant: bearing-induced eccentricity creates localized temperature increases in stator teeth adjacent to the minimum air gap, while temperature rise affects UMP through reduced magnet remanence (from −0.1% to −0.12% per °C for NdFeB) and altered air gap clearance from thermal expansion. For typical operating temperatures (100–150 °C), UMP may decrease by 15–25% compared to cold conditions. UMP also causes stator bore ovalization, creating a positive feedback mechanism.

For real-time applications, reduced-order models bridge the gap between detailed FEA and practical requirements. Lookup table approaches store FEA results for grids of operating points, enabling interpolation during coupled simulations. Surrogate models based on neural networks or Gaussian process regression evaluate in microseconds compared to minutes for full FEA. MEC models, as described in Chapter 3, offer 1,000 times faster computation than FEA with 10–20% error in force prediction. These reduced-order approaches enable digital twin implementations for online bearing condition monitoring and remaining useful life prediction.

The modeling and analysis presented in this section, from bearing kinematics and contact mechanics through defect-induced forces to electromagnetic-mechanical coupling, provides the quantitative foundation for the fault diagnosis techniques presented next. The characteristic defect frequencies derived from kinematic models identify what to look for in diagnostic signals, the force models predict the amplitude and modulation patterns of fault signatures, and the coupling analysis explains how bearing faults manifest in electrical measurements such as stator current and flux.

7.4 Fault Diagnosis

7.4.1 Vibration Monitoring Techniques for Bearing Faults

Vibration analysis represents the most established and widely utilized diagnostic approach for bearing fault detection in permanent magnet machines. Externally mounted accelerometers detect vibration signatures associated with rolling bearing faults, providing a noninvasive means of condition assessment. The historical development of vibration-based bearing diagnostics began in the 1980s with FFT-based spectral analysis and has progressed through increasingly sophisticated time-frequency domain methods enabled by advances in accelerometer design and data acquisition systems.

Vibration signal analysis for bearing fault diagnosis encompasses three primary analytical domains: time domain, frequency domain, and time-frequency domain. Each domain offers distinct advantages for extracting bearing fault information from vibration measurements. Time-domain analysis relies on statistical measures such as RMS, standard deviation (STD), kurtosis, and skewness to characterize fault patterns and assess bearing condition. Among these metrics, kurtosis and crest factors have proven particularly sensitive to bearing degradation, with their diagnostic effectiveness further enhanced through advanced algorithms such as step-varying vibrational resonance (SVVR) [11]. Experimental investigations have revealed axis-dependent sensitivity to bearing faults, wherein the X-axis typically exhibits the most pronounced fluctuations following bearing failure, thereby serving as an effective defect indicator [11]. While time-domain features provide computationally efficient fault indicators suitable for continuous monitoring, they offer limited diagnostic specificity regarding the nature and location of bearing defects. Figure 7.11 presents raw vibration waveforms from the CWRU bearing dataset for four conditions recorded at 1 HP motor load (1,772 rpm). The healthy bearing signal (a) exhibits low-amplitude background vibration, while the faulty cases display distinct impulsive patterns whose repetition rates correspond to their respective characteristic frequencies. The outer race fault (b) produces impulses repeating at the BPFO period, the inner race fault (c) at the BPFI period, and the ball fault (d) at the BSF period. Despite the clear visual differences in amplitude and impulsivity, these time-domain signatures alone cannot reliably distinguish fault type without supplementary spectral analysis.

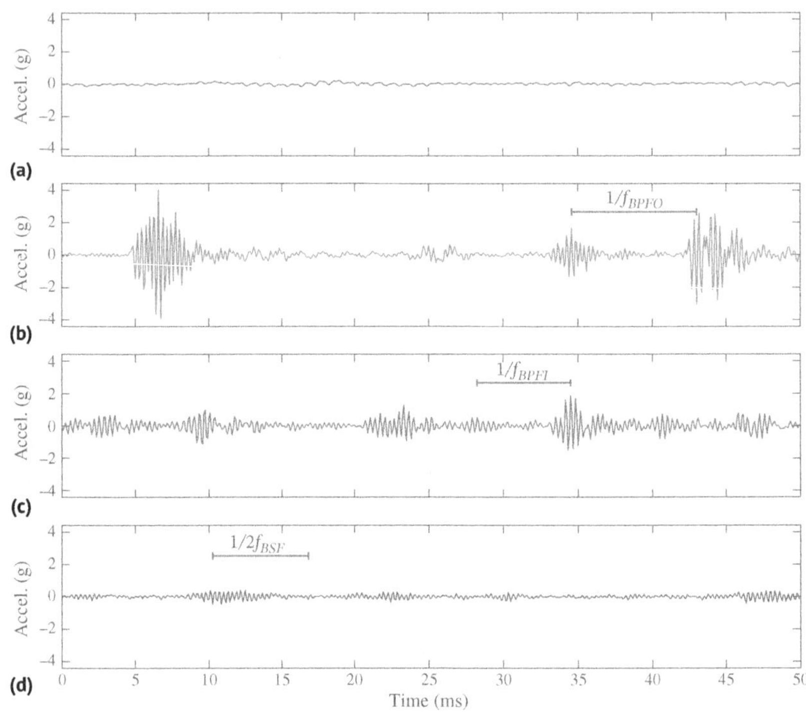

Figure 7.11 Raw vibration time-domain waveforms from the CWRU bearing dataset (SKF 6205-2RS, 1 HP load, 1,772 rpm) for four bearing conditions: (a) healthy bearing; (b) outer race fault (0.021 in.); (c) inner race fault (0.021 in.); and (d) ball fault (0.021 in.). Annotations indicate the fault repetition periods corresponding to BPFO, BPFI, and BSF, respectively.

Frequency domain analysis constitutes the most widely used technique for bearing diagnostics, leveraging the fact that bearing defects generate periodic impacts at predictable characteristic frequencies. FFT serves as the fundamental tool for transforming time-domain vibration signals into the frequency domain, enabling identification of these characteristic frequencies. The theoretical foundation for frequency-based bearing fault detection rests on the generation of bearing characteristic frequencies that appear in the vibration spectrum when localized defects are present. These characteristic frequencies include the BPFO, BPFI, BSF, and FTF, each corresponding to specific bearing defect locations. The ability to identify which characteristic frequency dominates the spectrum enables determination of the specific bearing component (inner race, outer race, or rolling element) experiencing degradation. Figure 7.12 presents the normalized FFT amplitude spectra of the same CWRU fault signals shown in Figure 7.11, with the theoretical characteristic frequencies marked as dashed vertical lines. The outer race fault spectrum (a) exhibits peaks near BPFO and its harmonics, though the harmonic amplitudes do not decay monotonically because the system's structural transfer function amplifies certain harmonics over others. The inner race fault spectrum (b) shows components near BPFI harmonics, while the ball fault spectrum (c) displays activity near even multiples of BSF ($2f_{BSF}$, $4f_{BSF}$), consistent with the ball defect contacting both the inner and outer races per revolution. Notably, the characteristic frequency peaks in all three cases coexist with shaft harmonics and other spectral components, illustrating that fault identification from raw spectra alone requires careful interpretation and is not always straightforward.

Despite its widespread adoption, practical experience has revealed several important limitations when analyzing bearing characteristic frequencies directly in raw vibration spectra. First, the low

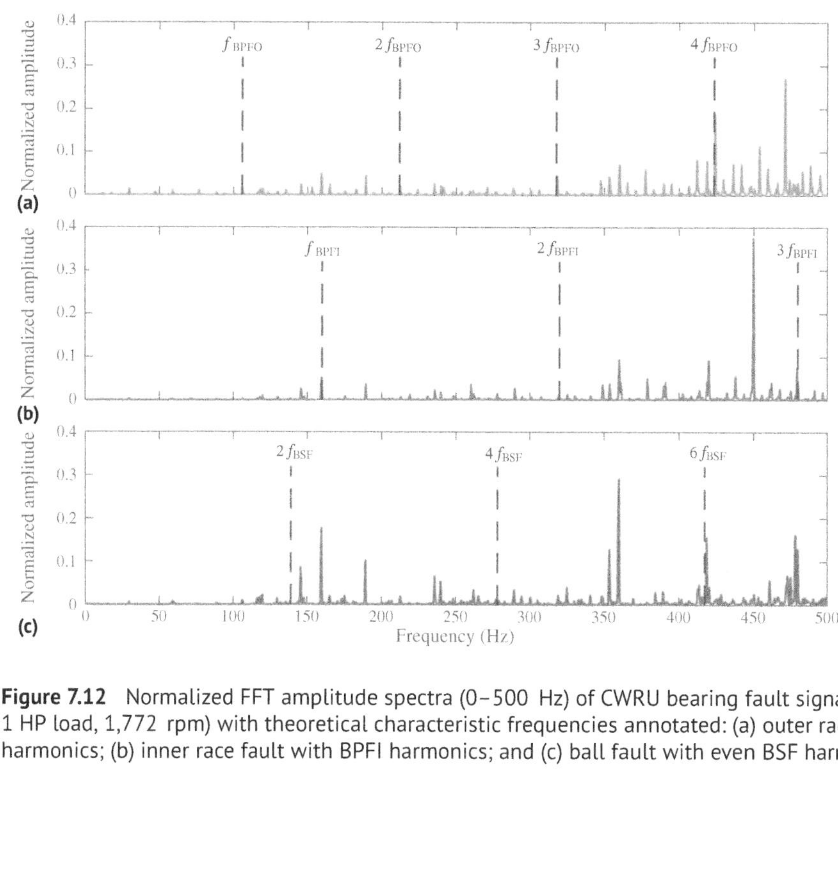

Figure 7.12 Normalized FFT amplitude spectra (0–500 Hz) of CWRU bearing fault signals (SKF 6205-2RS, 1 HP load, 1,772 rpm) with theoretical characteristic frequencies annotated: (a) outer race fault with BPFO harmonics; (b) inner race fault with BPFI harmonics; and (c) ball fault with even BSF harmonics.

harmonics of bearing characteristic frequencies typically exhibit very low magnitude and are easily masked by other machine components, particularly when the structural resonance frequency is two orders of magnitude higher than the repetition frequency of bearing impacts, with measurable acceleration levels appearing at tens of kHz. This high-frequency nature of bearing fault signatures necessitates appropriate sensor selection and sampling rates to capture the relevant diagnostic information. Second, even small amounts of slip (as little as 0.75%) cause the higher harmonics of characteristic frequencies to smear over one another in the frequency spectrum, complicating their identification. Third, bearing fault signals are inherently impulsive, but this impulsivity cannot be properly recognized unless the analyzed frequency range encompasses at least ten harmonics of the characteristic frequency; otherwise, the signal appears sinusoidal with no detectable impulsive character. Figure 7.13 illustrates these phenomena using synthetic bearing fault signals based on the SKF 6205-2RS bearing parameters from the CWRU dataset. The time-domain signals (a, b) show impulsive responses excited at the structural resonance frequency, appearing nearly identical with and without slip. However, the raw frequency spectra (c) reveal a striking difference: without slip, individual harmonics of the BPFO are clearly resolved, whereas even 0.75% random timing fluctuation causes these harmonics to smear into a smooth broadband envelope, rendering them unidentifiable. In contrast, the envelope spectra (d) demonstrate that amplitude demodulation successfully recovers clear BPFO harmonic peaks regardless of slip, confirming the superiority of envelope analysis for bearing fault detection.

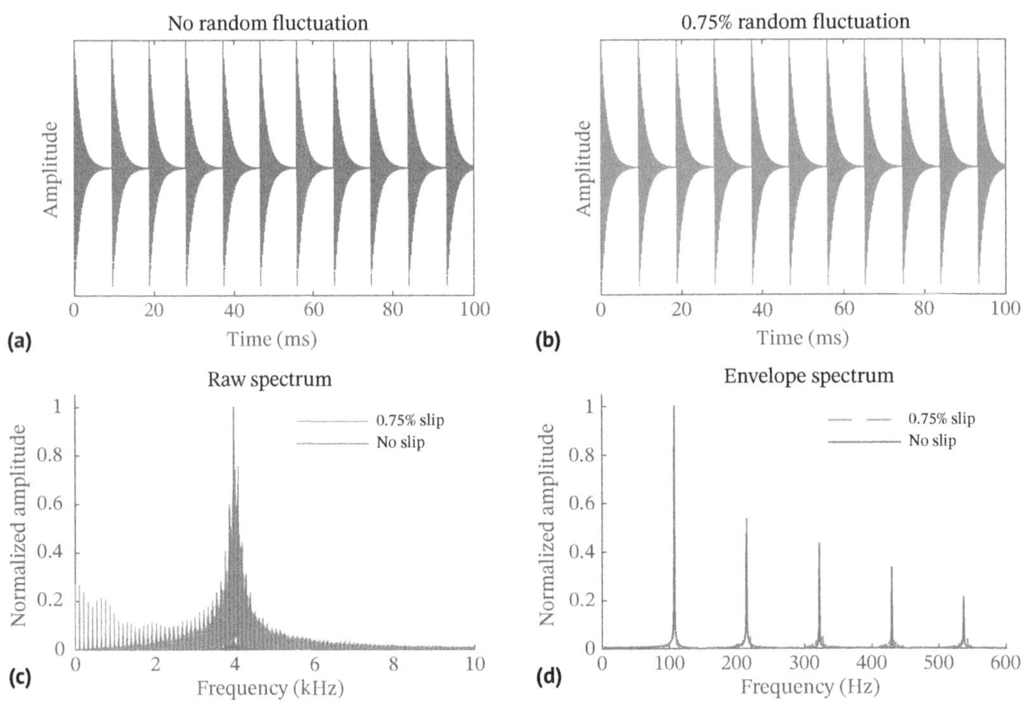

Figure 7.13 Synthetic bearing fault signals based on CWRU bearing parameters (SKF 6205-2RS, BPFO at 107.4 Hz): (a) time signal without slip; (b) time signal with 0.75% random timing fluctuation; (c) raw spectra comparison showing harmonic smearing due to slip; and (d) envelope spectra showing clear characteristic frequency peaks in both cases.

These practical challenges have motivated the development of advanced signal processing techniques that combine the strengths of multiple analytical domains. Envelope analysis, also known as amplitude demodulation, addresses the limitations of direct spectral analysis by focusing on the impulsive nature of bearing signals. This technique extracts the envelope of high-frequency vibration signals where bearing impacts excite structural resonances, then analyzes the spectrum of this envelope to identify bearing characteristic frequencies with improved signal-to-noise ratio. Envelope analysis proves more effective than direct spectral analysis because it separates the low-frequency bearing characteristic frequencies from the high-frequency structural resonances they modulate, rather than attempting to identify low-amplitude characteristic frequencies in heavily contaminated raw spectra. Time-frequency domain methods extend this concept further by providing simultaneous representation in both time and frequency, enabling analysis of nonstationary signals encountered during variable speed operation or transient events. These advanced techniques form the foundation for modern bearing diagnostic systems that must operate reliably across diverse operating conditions.

Beyond signal processing advances, advanced vibration monitoring implementations employ multisensor configurations to enhance diagnostic reliability and provide comprehensive fault coverage. Recent studies have demonstrated that one-dimensional convolutional neural networks (1D CNN) can identify and diagnose six distinct types of electric motor faults with high accuracy using data from two accelerometers measuring along different axes [11]. This multidirectional measurement strategy recognizes that bearing fault signatures manifest differently along various spatial axes depending on the defect location and the position of the bearing load zone. The integration of multiple vibration measurement points provides more robust fault detection capabilities while reducing the likelihood of missed detections that can result from unfavorable sensor placement.

Contemporary developments in vibration-based bearing diagnostics increasingly focus on integrating multiple analytical approaches to address practical challenges including strong masking signals from other machine components (particularly gears), variable speed operation, and early fault detection requirements. Modern diagnostic frameworks combine traditional time-domain and frequency-domain features with advanced statistical indicators and pattern recognition algorithms. This multidomain integration enables reliable bearing health assessment across the diverse operating conditions encountered in permanent magnet machine applications, improving diagnostic accuracy and robustness compared to single-domain analysis approaches. Table 7.5 provides a comprehensive comparison of the primary vibration monitoring techniques discussed in this section.

7.4.2 MCSA Application for Bearing Defect Detection

The fundamental principle underlying MCSA for bearing diagnostics is that bearing faults introduce mechanical vibrations that modulate the air gap permeance, subsequently introducing additional frequency components into the stator current signals and contributing to oscillations. Consequently, the motor current signals undergo modulation that can be analyzed to detect and characterize bearing faults.

The application of MCSA to bearing fault diagnosis has been demonstrated through various implementations. Through induction motor stator current analysis, researchers have successfully achieved diagnosis of variable-speed direct-drive wind turbines based on single-phase stator current monitoring [12]. Advanced demodulation techniques, such as the Teager–Kaiser energy operator, have been employed to demodulate the stator current and reveal bearing fault signatures that would otherwise be obscured by the fundamental frequency component and other dominant

Table 7.5 Comparison of vibration monitoring techniques for bearing fault diagnosis. The choice of technique depends on operating conditions (steady vs. variable speed), fault stage (incipient vs. advanced), computational resources, and required diagnostic specificity. Modern systems often combine multiple techniques for robust diagnosis.

Technique	Dom.	Advantages	Limitations	Cost
TDS (RMS, Kurt., CF)	T	Simple; low cost; real-time capable; effective for trending	Low diagnostic specificity; no fault localization; transient-sensitive	Very Low
FFT	F	Direct fault frequency identification; established theory; mature technology	Harmonics masked by noise; slip causes frequency smearing	Low
EA / Demod.	F (env.)	Enhanced SNR; separates defect from resonance frequencies; robust to slip	Requires optimal band selection; sensitive to filter design	Medium
T-F (STFT, WT, HHT)	T-F	Handles nonstationary signals; variable speed capable; transient localization	High computational cost; parameter selection critical; complex interpretation	High
SK / Kurtogram	F	Automatic demodulation band selection; objective; enhances impulsive content	Computationally intensive; may select noise bands	High
Cepstrum	Q	Identifies periodic spectral spacing; effective for sidebands and harmonic families	Requires log operation; complex interpretation; limited for aperiodic faults	Medium
OT	O	Eliminates speed variations; direct cross-speed comparison; clear fault signatures	Requires tachometer or encoder; resampling complexity	Med.-High

TDS = Time-domain Statistical, RMS = Root Mean Square, Kurt. = Kurtosis, CF = Crest Factor, FFT = Fast Fourier Transform, EA = Envelope Analysis, Demod. = Demodulation, SNR = Signal-to-noise Ratio, T-F = Time-frequency, STFT = Short-time Fourier Transform, WT = Wavelet Transform, HHT = Hilbert–Huang Transform, SK = Spectral Kurtosis, OT = Order Tracking. Dom.: T = Time, F = Frequency, Q = Quefrency, O = Order.

harmonics. Since the current signal experiences modulation in both phase angle and amplitude due to bearing-induced air gap variations, specialized signal processing techniques are required to extract the bearing fault information embedded in these modulations.

Recent developments in MCSA have focused on extracting fault features from multiphase current signals. A time-domain fault feature extraction method based on the mutual information between two-phase current signals has been proposed to enhance bearing fault detection sensitivity [12]. This approach leverages the fact that bearing faults affect multiple phases in a correlated manner, and the mutual information metric can capture these inter-phase relationships more effectively than single-phase analysis. Additionally, research has established the usefulness of current monitoring by comparing the correlation between current frequencies and vibration signatures, demonstrating that bearing characteristic frequencies manifest in both domains [13]. The development of the stator current square envelope spectrum has provided an alternative representation that enhances the visibility of bearing fault indicators in the current signal.

The integration of MCSA with modern machine learning techniques has significantly enhanced bearing fault classification performance. Studies have combined Motor Current Signature Analysis with two-dimensional convolutional neural networks (2D CNN), employing the Gramian Angular Field (GAF) transformation to convert time-domain current data into two-dimensional images suitable for image-based classification algorithms [11]. This innovative approach circumvents the conventional signal processing steps typically required in MCSA, allowing the deep learning architecture to automatically learn discriminative features directly from the transformed current representations. Experimental results have demonstrated that this combined 2D CNN and MCSA methodology achieves high classification accuracy exceeding 99% while requiring shorter computation time compared to traditional feature-based approaches [11].

The practical advantages of MCSA for bearing fault diagnosis include the availability of current measurements in most motor drive systems without additional sensor installation, the potential for continuous online monitoring, and the noninvasive nature of the measurement. However, challenges remain in MCSA-based bearing diagnostics, particularly regarding the relatively weak coupling between bearing mechanical faults and electrical signatures compared to the strong relationship between bearing faults and vibration signals. The bearing fault-related components in the current spectrum are typically small in magnitude and can be masked by harmonics introduced by power electronic converters, supply voltage unbalance, and electromagnetic asymmetries. Nevertheless, ongoing research continues to develop more sensitive current-based indicators and advanced signal processing techniques to overcome these limitations, establishing MCSA as a valuable complementary diagnostic modality for comprehensive bearing health monitoring in permanent magnet machine systems.

7.4.3 Envelope Analysis and Demodulation Methods

Envelope analysis has been established over many years as the benchmark method for bearing diagnostics, serving as the primary recommended technique for extracting bearing fault signatures from vibration signals [14]. The fundamental principle of envelope analysis is that the analysis of envelope signals provides more diagnostic information than analysis of raw vibration signals, particularly when bearing faults excite high-frequency structural resonances that are modulated by the repetitive impacts occurring at bearing characteristic frequencies. The envelope spectrum reveals bearing fault frequencies more clearly than conventional FFT analysis of raw signals, as it demodulates the high-frequency resonances and focuses on the modulation patterns characteristic of bearing defects.

The theoretical foundation of envelope analysis rests on the concept of spectral correlation and its relationship to the envelope spectrum. When a bearing defect generates repetitive impulses, these impacts excite structural resonances at frequencies typically much higher than the bearing characteristic frequencies themselves. The assumption of high resonance frequency relative to bearing fault frequencies is quite common in practical applications, with resonances often occurring at tens of kHz while bearing characteristic frequencies remain below a few kHz [14]. The envelope analysis procedure extracts the amplitude modulation of these high-frequency resonances through bandpass filtering followed by demodulation, revealing the underlying bearing fault periodicity in the envelope spectrum where the BPFI and BPFO components appear distinctly. Figure 7.14 demonstrates this complete pipeline using the CWRU outer race fault signal (OR021, 1 HP). The raw frequency spectrum (a) reveals a broad resonance band in the 2–5 kHz region excited by fault impacts, which is isolated by a fourth-order Butterworth bandpass filter. Panel (b) superimposes the raw and filtered signals, showing how the filter retains only the high-frequency burst patterns associated with each impact event while removing the low-frequency shaft harmonics and background content. The Hilbert transform extracts the envelope of the filtered signal (c), whose periodic peaks repeat at the BPFO interval. The envelope spectrum (d) confirms this periodicity, displaying clear BPFO harmonics that are far more distinct than those observable in the raw spectrum (Figure 7.12).

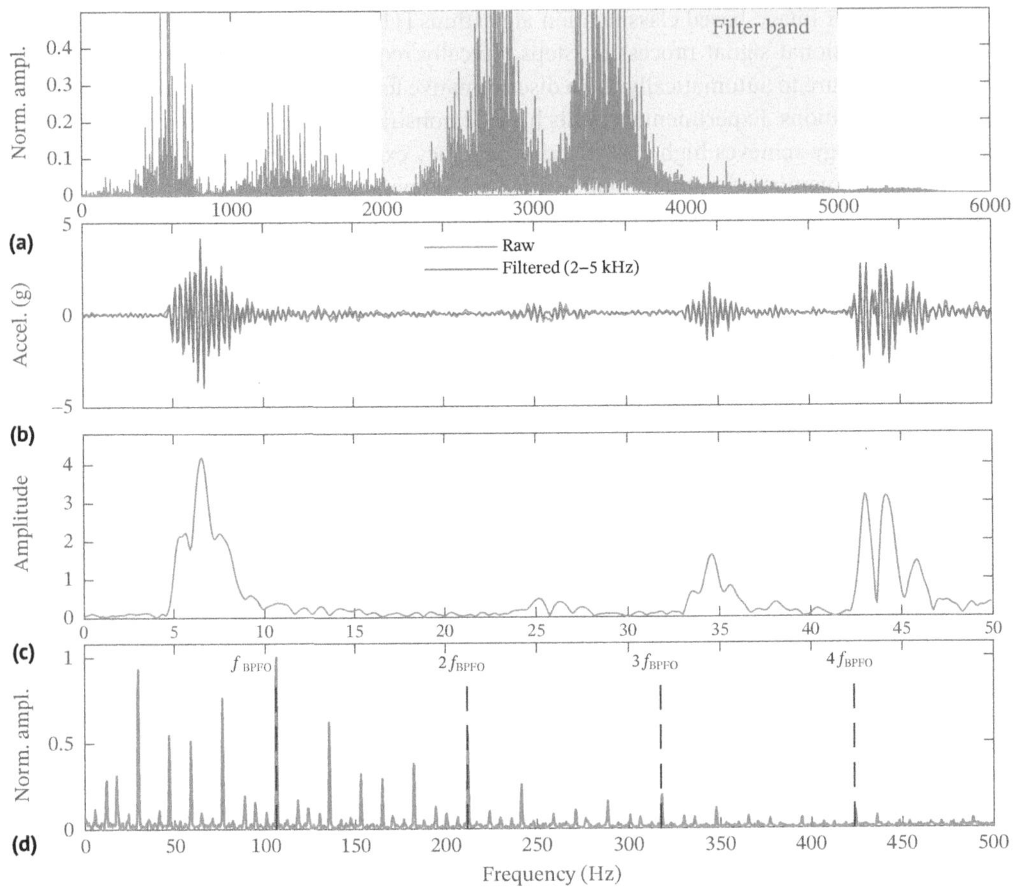

Figure 7.14 Envelope analysis pipeline applied to the CWRU outer race fault signal (SKF 6205-2RS, OR021, 1 HP load, 1,772 rpm): (a) frequency spectrum of the raw signal with the 2–5 kHz bandpass filter region highlighted; (b) raw vibration signal and bandpass-filtered signal superimposed; (c) envelope extracted via the Hilbert transform; and (d) envelope spectrum with BPFO harmonics annotated.

Notably, the absence of shaft-speed sidebands around the BPFO peaks is consistent with an outer race defect, since the outer race is stationary and the defect position does not rotate through the load zone.

Recent research has focused on enhancing envelope analysis performance through optimization of the demodulation frequency band. A genetic algorithm-based demodulation band optimization approach has been proposed specifically for rolling element bearing fault diagnosis, recognizing that the selection of the optimal frequency band for envelope analysis critically affects diagnostic sensitivity [15]. This demodulation band optimization determines the frequency range where bearing fault-excited resonances exhibit the strongest response, maximizing the signal-to-noise ratio of bearing fault indicators in the resulting envelope spectrum. The development of the FERgram (Fast Envelope Resonance gram) [16] provides a visual representation of envelope spectrum characteristics across different frequency bands, facilitating the identification of optimal demodulation parameters.

Various enhanced envelope analysis techniques have been developed to address challenges in bearing fault detection under adverse conditions. The Enhanced Envelope Spectrum method improves the visibility of weak bearing fault signatures in the presence of strong background noise and masking signals from other machine components. Envelope Spectrum L-kurtosis leverages higher-order statistical measures to identify the optimal frequency bands for envelope analysis while being robust to outliers and non-Gaussian noise distributions. The Squared Envelope Spectrum applies nonlinear processing to the envelope signal before spectral analysis, enhancing the contrast between fault-related peaks and background noise. Wavelet Packet Envelope combines wavelet packet decomposition with envelope analysis to provide adaptive frequency band selection and improved time-frequency resolution.

Demodulation transformation methods extend the envelope analysis framework to provide more sophisticated signal processing capabilities. The demodulation transformation signal analysis method has been specifically developed to perform envelope detection analysis of vibration signals with low signal-to-noise ratios, which commonly occur in early-stage bearing fault conditions. Signal optimization-based generalized demodulation transform provides a unified framework for demodulation-based bearing fault detection that encompasses various envelope analysis variants as special cases. These generalized approaches allow adaptive selection of demodulation parameters based on signal characteristics and fault severity.

The integration of envelope analysis with other signal processing techniques has yielded hybrid diagnostic methods with superior performance. A frequency shift-based hybrid signal processing strategy combines the Hilbert Transform (HT) and Discrete Wavelet Transform (DWT) to enhance envelope analysis effectiveness. The Hilbert transform provides one mathematical approach to envelope extraction by constructing the analytic signal, and its application to motor stator current has successfully identified bearing fault-related frequency components [12]. Spectral kurtosis-dependent algorithms, including the wavelet kurtogram and fast kurtogram [17], have been integrated with envelope analysis to automatically identify the optimal frequency band for demodulation based on kurtosis maximization criteria. A recalculation method of the impact envelope signal has been proposed to improve the accuracy of envelope extraction under conditions where multiple resonances are simultaneously excited. Figure 7.15 demonstrates the kurtogram applied to a CWRU inner race fault signal (IR014, 1 HP). The kurtogram (a) computes the spectral kurtosis across a grid of frequency bands and bandwidths using the STFT at multiple window lengths, producing a 2D map in which bright regions indicate high impulsiveness. The algorithm automatically identifies the optimal demodulation band centered near 2.6 kHz (marked by the white rectangle and star) without requiring any prior knowledge of the structural resonance frequencies. The resulting envelope spectrum (b), obtained by bandpass filtering at the selected band followed by Hilbert transform demodulation, reveals clear BPFI harmonics accompanied by shaft-speed sidebands at $\pm f_r$. These sidebands, absent in the outer race case (Figure 7.14), arise because the inner race defect rotates with the shaft, causing the impact amplitude to vary as the defect passes through and out of the load zone once per revolution.

Complementary signal processing techniques enhance envelope analysis performance in specific scenarios. Cepstrum analysis, which involves the spectrum of the logarithm of the spectrum, provides an alternative domain for identifying periodic components in bearing vibration signatures, with cepstrum pre-whitening and high-pass filtering demonstrating effectiveness for bearing fault diagnosis. The Hilbert–Huang Transform (HHT) combines Empirical Mode Decomposition (EMD) with the Hilbert transform to provide adaptive time-frequency analysis suitable for nonstationary bearing vibration signals. For current-based bearing diagnostics, envelope analysis of the current residue signal exposes the bearing characteristic fault frequencies that are embedded in the motor current modulation [13]. Vibration signal demodulation using envelope analysis has been

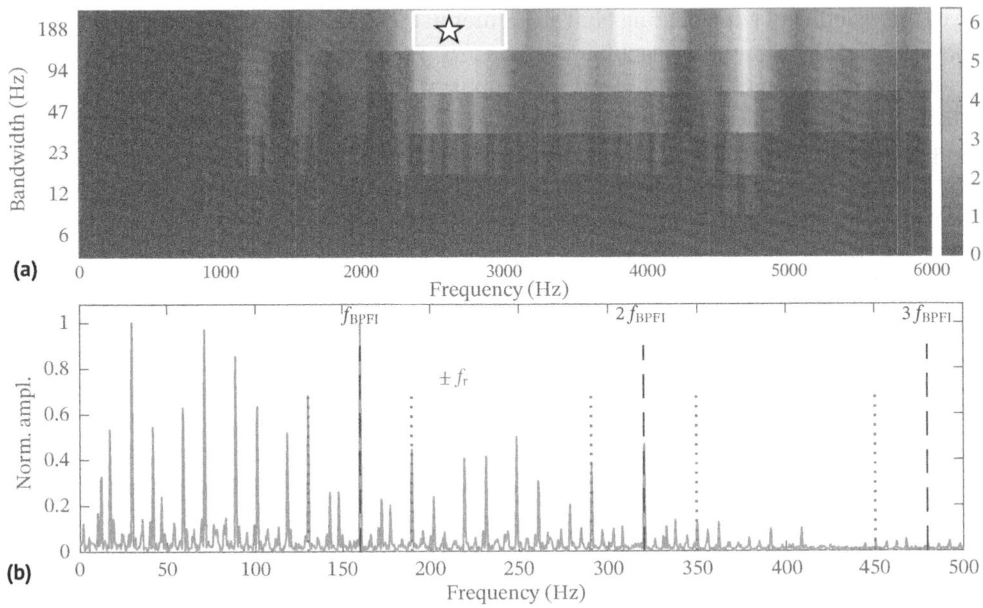

Figure 7.15 Kurtogram-based automatic demodulation band selection applied to the CWRU inner race fault signal (SKF 6205-2RS, IR014, 1 HP load, 1,772 rpm): (a) kurtogram showing spectral kurtosis as a function of frequency and bandwidth, with the optimal band highlighted; and (b) envelope spectrum from the optimal band, with BPFI harmonics and shaft-speed sidebands ($\pm f_r$) annotated.

compared with the kurtogram approach, with both methods showing complementary strengths depending on the specific fault characteristics and operating conditions.

The success of envelope analysis in bearing diagnostics derives from several key advantages. Unlike methods requiring separation of individual harmonics, envelope analysis performs equally well or better without requiring harmonic separation, making it more robust to closely spaced frequency components. The method inherently addresses the challenge of high-frequency resonance excitation by leveraging these resonances as carriers of bearing fault information rather than treating them as interference. The envelope spectrum provides clear visualization of bearing fault frequencies and their harmonics, facilitating straightforward fault identification and severity assessment. These characteristics establish envelope analysis and related demodulation methods as essential tools in the bearing diagnostics portfolio for permanent magnet machine condition monitoring systems.

7.4.4 Data-driven and Machine Learning Approaches

Data-driven techniques have become central to bearing fault diagnosis, driven by the availability of large public benchmark datasets (CWRU, Paderborn, NASA IMS) and the complexity of bearing fault signatures that benefit from automated pattern recognition. The fundamental advantage of deep learning over traditional machine learning (SVM, KNN, ANN) is the elimination of manual feature engineering: deep networks automatically learn discriminative representations from raw or minimally processed sensor data, consistently achieving higher diagnostic accuracy [18].

Three categories of deep learning architectures address distinct aspects of bearing diagnostics. Convolutional neural networks (CNNs) are the most widely adopted, with 1D CNNs processing raw vibration time series and 2D CNNs operating on time-frequency representations such as spectrograms or Gramian Angular Field (GAF) images [19]. Multihead and multiscale CNN variants extract features at multiple frequency scales or from multiple sensor channels simultaneously, achieving classification accuracies of 95–99% on standard benchmarks. Residual connections (ResNet) enable training of deeper networks for complex fault patterns. Recurrent architectures, particularly LSTM and GRU networks, capture temporal degradation patterns essential for prognostics, and CNN-LSTM hybrids combine spatial feature extraction with temporal modeling of bearing degradation evolution.

Several practical challenges in bearing diagnostics have motivated specialized architectural innovations. Data scarcity and class imbalance, where healthy bearing data vastly outnumbers fault data, are addressed by Generative Adversarial Networks (GANs) that synthesize realistic fault samples [12, 20], and by few-shot learning approaches such as Siamese networks that achieve over 97% accuracy with as few as three training samples per fault class [21]. Transfer learning enables knowledge reuse across different machines and operating conditions with minimal target domain data, which is particularly valuable for deploying diagnostics on new PM motor installations without extensive fault injection testing [22]. Physics-informed CNNs incorporate bearing fault physics (kurtosis, envelope characteristics) directly into network structure or loss functions, improving both accuracy and interpretability compared to purely empirical approaches [23]. Semi-supervised learning leverages abundant unlabeled operational data alongside limited labeled fault annotations, achieving near-supervised performance with a fraction of the labeling effort [24].

For PM motor applications, the primary challenge is the weak coupling between mechanical bearing faults and electrical signals when current-based approaches are used, requiring more sophisticated feature extraction compared to vibration-based implementations. Real-time deployment has been demonstrated on embedded and FPGA platforms, enabling edge computing for continuous bearing monitoring without centralized cloud processing [25]. Table 7.6 summarizes the principal deep learning architectures employed for bearing fault diagnosis. Most architectures achieve 95–99% accuracy on standard benchmark datasets (CWRU, Paderborn), though actual performance varies with dataset quality, preprocessing, and operating conditions. Data requirements range from very low (3–10 samples per class for few-shot methods) to high (tens of thousands of samples for deep residual networks), making architecture selection dependent on the available training data for the target application.

Table 7.6 Specialized deep learning architectures addressing key bearing diagnostic challenges. Standard architectures (CNN, LSTM, ResNet, CNN-LSTM hybrids) are discussed in Chapter 4; the architectures below target specific challenges in bearing fault diagnosis such as data scarcity, domain shift, and interpretability.

Architecture	Input type	Key advantage	Bearing diagnostic application
Autoencoder / GAN	Any sensor data	Unsupervised feature learning; anomaly detection; synthetic data generation addresses class imbalance	Novelty detection; limited labels; imbalanced datasets; data augmentation
Siamese Network	Paired samples	Few-shot learning; similarity learning with minimal labels	Limited fault data; new bearing types; rapid deployment
Physics-informed CNN	Raw vibration + physical constraints	Incorporates domain knowledge; better generalization; interpretable features	Small datasets; physical understanding required
Capsule Network	Time-series or images	Spatial relationships; viewpoint invariance; routing by agreement	Complex spatial patterns; orientation-invariant features

7.4.5 Wavelet Transform and Time-frequency Methods

The transient, impulsive nature of bearing fault signatures makes wavelet analysis particularly well-suited to bearing diagnostics, as the variable time-frequency resolution naturally matches the multiscale characteristics of defect-induced impacts: short windows capture high-frequency transient impacts while long windows resolve low-frequency modulation patterns.

Figure 7.16 compares two time-frequency representations applied to the same inner race fault signal (IR021, 1 HP) to illustrate this resolution trade-off. The STFT spectrogram (a), computed with a fixed 128-sample Hann window (10.7 ms at 12 kHz), applies uniform time-frequency resolution across all frequencies: the window duration and spectral bandwidth remain constant regardless of the analysis frequency. The CWT scalogram (b), using the Morlet wavelet ($\omega_0 = 6$), adapts its effective analysis window at each frequency, with shorter wavelets at high frequencies providing finer time resolution and longer wavelets at low frequencies yielding finer frequency resolution. In both representations, the periodic broadband impulses generated by the inner race defect at BPFI intervals (approximately 6.2 ms) appear as vertical stripes spanning a wide frequency range. However, the CWT achieves noticeably sharper temporal localization of these impulses at higher frequencies, where the effective wavelet duration is much shorter than the fixed STFT window. This adaptive resolution property makes wavelets particularly effective for bearing diagnostics, where transient fault impacts contain both high-frequency broadband content and lower-frequency modulation patterns that benefit from different resolution trade-offs.

The continuous wavelet transform (CWT) has been successfully applied to extract fault features from stator currents in motor drive systems [12], while DWT provides efficient multiscale decomposition suitable for real-time monitoring. Wavelet packet decomposition (WPD) extends the DWT

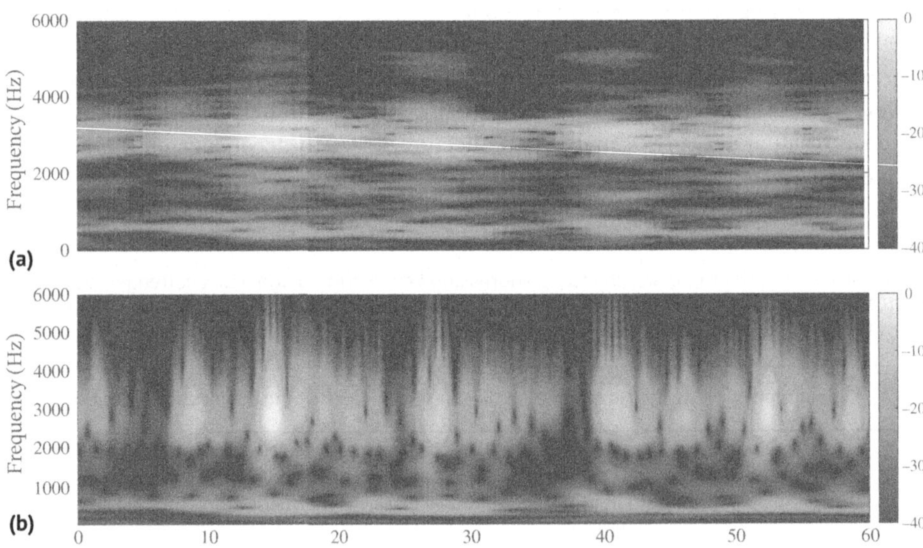

Figure 7.16 Time-frequency analysis of a CWRU inner race fault signal (SKF 6205-2RS, IR021, 1 HP load, 1,772 rpm, 12 kHz sampling): (a) STFT spectrogram with a 128-sample Hann window (10.7 ms, frequency resolution 93.75 Hz); and (b) CWT scalogram using the Morlet wavelet ($\omega_0 = 6$). Periodic broadband impulses at the BPFI interval (approximately 6.2 ms) appear as vertical stripes in both representations; the CWT provides improved time localization at higher frequencies, illustrating the adaptive resolution property of wavelet analysis.

by offering finer frequency resolution in high-frequency bands where bearing fault signatures often reside, and has been combined with deep CNNs for automated feature extraction and classification. The dual-tree complex wavelet transform (DTCWT) [26] addresses shift-sensitivity limitations of conventional DWT, providing nearly shift-invariant decomposition important for consistent bearing fault detection.

A critical practical challenge is selecting the optimal frequency band for analysis. The wavelet kurtogram, based on complex Morlet wavelets, addresses this by quantifying impulsiveness across different scales and time locations, automatically identifying the band where bearing fault signatures are most visible [14]. The optimized Morlet wavelet is particularly suitable for bearing diagnostics due to its similarity to the transient impacts generated by bearing defects.

Adaptive decomposition methods have been developed for the nonstationary, multicomponent bearing fault signals. Complete ensemble empirical mode decomposition with adaptive noise (CEEMDAN) decomposes vibration signals into mono-component intrinsic mode functions that isolate bearing fault signatures from background noise [27]. Improved variational mode decomposition (IVMD) offers superior noise robustness for detecting early-stage bearing degradation where fault-related transients are subtle [28]. These decomposition methods are frequently combined with machine learning classifiers: DWT-LSTM hybrids leverage wavelet multiresolution decomposition for feature extraction and recurrent network sequence modeling for temporal pattern recognition, while wavelet features serve as inputs to CNNs and SVMs for automated fault classification.

Wavelet denoising also serves as an important preprocessing step for early-stage fault diagnosis with low signal-to-noise ratios. The principle relies on the observation that bearing fault impulses concentrate energy in specific wavelet coefficient scales while noise distributes energy across all scales, enabling separation through thresholding strategies [14].

7.4.6 Multisignal Fusion for Bearing Diagnosis

Bearing faults manifest through multiple physical phenomena simultaneously, including mechanical vibrations, thermal effects from increased friction, and electromagnetic disturbances from air gap variations. Multisignal fusion leverages these complementary information sources to achieve diagnostic performance superior to any single modality.

Fusion strategies for bearing diagnostics operate at three levels. Feature-level fusion creates comprehensive feature vectors combining vibration envelope statistics, current spectral amplitudes at bearing characteristic frequencies, and thermal indicators, processed through dimensionality reduction techniques such as PCA to derive hybrid health indices. Decision-level fusion enables independent classification from each sensor modality, with bearing-specific strategies that weight vibration-based classifiers more heavily given their superior sensitivity to mechanical faults. Signal-level fusion combines raw or preprocessed measurements before feature extraction, with vibro-acoustic fusion (vibration plus acoustic emission) proving particularly effective for early-stage fault detection, where acoustic emissions reveal incipient defects before they manifest as strong vibration signatures [12].

For PM motors specifically, the combination of vibration monitoring with MCSA provides a particularly powerful fusion strategy. Vibration analysis offers high sensitivity to mechanical bearing defects, while MCSA detects the electromagnetic consequences of bearing-induced eccentricity through current sidebands at $f_s \pm kf_{char}$. The two modalities respond differently to operating conditions: vibration sensitivity improves under heavy load (concentrated load zone), while MCSA sensitivity increases with electromagnetic coupling. Combining both provides a robust diagnosis across the full PM motor operating envelope.

Deep learning architectures designed for multisensor bearing diagnostics include multihead 1D CNNs that process signals from multiple accelerometer axes through parallel branches, and hybrid LSTM-CNN models that capture both spatial features and temporal degradation evolution across sensor channels [11]. Practical implementation must address sensor synchronization, data alignment across different sampling rates, and graceful degradation when individual sensors provide unreliable data.

7.4.7 Remaining Useful Life Estimation

Remaining useful life (RUL) estimation extends beyond fault detection to predict the time remaining before bearing failure, enabling the transition from reactive to predictive maintenance. Bearing degradation progresses through distinct stages, from initial surface wear and fatigue crack initiation through accelerated spalling and increased contact stresses, with each stage exhibiting different vibration signature characteristics [15, 18].

Two complementary approaches address RUL estimation. Physics-based models incorporate bearing failure mechanics (fatigue life calculations, wear models, lubrication degradation) and are particularly relevant for PM motors where the UMP-bearing wear feedback loop described in the modeling section accelerates degradation beyond standard bearing life predictions. The modified bearing life equation incorporating UMP and electrical erosion factors provides a physics-based framework for predicting how PM motor-specific loading conditions affect remaining life. Data-driven approaches leverage run-to-failure datasets (notably NASA IMS) to learn degradation patterns through health indices that aggregate multiple condition indicators into scalar metrics tracking progression toward failure [18]. Hybrid approaches combining physics-informed constraints with data-driven learning have shown improved accuracy and generalization compared to either method alone.

For PM motors, RUL estimation must account for the coupling between bearing degradation and electromagnetic performance. As bearing clearance increases, the resulting eccentricity growth and UMP amplification accelerate degradation nonlinearly, making early-stage prognostics particularly critical. Robust prognostic systems incorporate uncertainty through probabilistic frameworks, providing RUL estimates with confidence intervals that account for variability in manufacturing quality, operating conditions, and measurement noise [13].

7.4.8 Emerging Sensor Technologies and Noninvasive Methods

Recent advances in sensor technology have introduced novel noninvasive diagnostic approaches that complement traditional vibration monitoring, offering advantages in terms of installation flexibility, cost reduction, and applicability to challenging environments where conventional accelerometer placement is impractical.

Thermal infrared imaging has emerged as a promising contactless bearing fault diagnosis method that detects temperature anomalies associated with bearing degradation without requiring physical sensor installation. This approach leverages the principle that bearing faults generate localized heat due to increased friction, inadequate lubrication, or mechanical damage. Recent implementations combine thermal imaging with DWT to extract discriminative features from two-dimensional thermal images, enabling classification of multiple bearing conditions including healthy state, outer race faults, inner race faults, and lubrication deficiency [29]. The noninvasive nature of thermal imaging proves particularly valuable for rotating machinery where sensor installation poses safety concerns or where existing vibration sensors cannot be easily retrofitted. Advanced implementations employ convolutional neural networks to automatically learn fault signatures from thermal image sequences, achieving high diagnostic accuracy while eliminating the need for contact-based sensors [29].

Fiber bragg grating (FBG) sensors represent an advanced fiber-optic sensing technology offering distinct advantages over conventional piezoelectric accelerometers for bearing fault detection. FBG sensors detect mechanical strain through wavelength shifts in reflected light, providing immunity to electromagnetic interference, high sensitivity, and the capability for multiplexed sensing with multiple gratings on a single fiber [30]. Recent studies have demonstrated FBG-based bearing fault detection in three-phase induction motors with the capability to diagnose outer race faults across 47 different characteristic frequencies. The integration of FBG measurements with support vector machines and principal component analysis enables effective dimensionality reduction and classification of diverse bearing fault types. The lightweight, compact nature of FBG sensors, combined with their electromagnetic immunity, makes them particularly suitable for harsh industrial environments and applications involving high electromagnetic noise levels where conventional sensors exhibit degraded performance.

Electrostatic sensing technology provides an alternative approach for bearing condition monitoring through detection of charge variations associated with debris generation and tribological phenomena in bearing contacts. Electrostatic sensors combined with spectral regression techniques enable diagnosis of roller bearing performance degradation and prediction of RUL. These sensors extract features from both time and frequency domains of electrostatic signals, with Gaussian models employed to predict bearing RUL based on degradation trends. The noncontact nature of electrostatic sensing allows monitoring without mechanical coupling to the rotating components, reducing sensor installation complexity and eliminating sensor-induced loading effects.

Acoustic emission sensing has gained increasing attention for early-stage bearing fault detection due to its superior sensitivity to incipient defects compared to conventional vibration analysis. Acoustic emission signals originate from transient elastic waves generated by sudden stress redistribution events such as crack propagation, plastic deformation, and surface asperity interactions occurring during bearing operation. The high-frequency nature of acoustic emission signals enables detection of fault-related events before they manifest as measurable vibration signatures. Recent implementations employ defective signature wavelet images (DWSI) combined with deep convolutional networks to process acoustic emission signals for bearing fault diagnosis. Advanced systems utilize FPGA-based multicore architectures for real-time bearing fault signature extraction from ultra-high sampling rate acoustic emission measurements, enabling online monitoring with minimal latency. The combination of acoustic emission sensing with spectrograms and EfficientNet optimizers has demonstrated effective bearing fault classification and degradation level assessment, with the acoustic signature analysis providing complementary information to vibration-based diagnostics.

Stray flux measurement represents a noninvasive electromagnetic approach that monitors the magnetic flux leakage around the machine housing to detect bearing faults through their influence on air gap permeance variations. Unlike Motor Current Signature Analysis, which requires electrical connection to the power supply, stray flux measurements employ external flux probes positioned near the machine without any electrical or mechanical contact. Power Spectral Density analysis of stray flux signals reveals distinct spectral signatures for different bearing fault types, with experimental studies demonstrating that stray flux measurements exhibit greater sensitivity to certain bearing faults compared to conventional vibration analysis. The technique proves particularly valuable for sealed or enclosed machines where vibration sensor installation is challenging and for applications requiring monitoring without machine modification. Recent developments have established the correlation between stray flux signatures and bearing characteristic frequencies, validating the physical mechanism underlying flux-based bearing diagnostics and demonstrating applicability across diverse machine types and bearing configurations.

The integration of multiple emerging sensor modalities enables a comprehensive bearing health assessment that leverages the complementary strengths of different physical sensing principles. Multisensor fusion approaches combining thermal imaging, acoustic emission, fiber-optic strain sensing, and electromagnetic flux measurements provide robustness against individual sensor failures while enhancing diagnostic reliability through cross-validation of fault indicators derived from independent physical phenomena. The trend toward wireless sensor networks incorporating these diverse sensing technologies enables distributed bearing monitoring across large machines with multiple bearing locations, facilitating comprehensive condition assessment without extensive wiring infrastructure.

Table 7.7 provides a comprehensive summary of bearing fault diagnostic signatures across the various measurement modalities discussed throughout this section, serving as a quick reference for practitioners selecting appropriate diagnostic techniques for specific PM motor applications.

7.4.9 Experimental Validation and Case Studies

Experimental validation provides empirical evidence for diagnostic algorithm effectiveness and establishes performance benchmarks. Several publicly available bearing fault datasets have become standard resources for algorithm development and comparative evaluation [18]. The CWRU dataset, the most widely utilized benchmark, encompasses vibration data from a 2-horsepower motor with EDM-induced bearing faults at multiple severities [31, 32]. The Paderborn dataset [33] complements CWRU by including both artificially induced and naturally developed faults under diverse operating conditions. For prognostics research, the NASA-IMS dataset provides 17 run-to-failure tests capturing complete bearing life cycles [34], while the XJTU-SY dataset offers multiple bearings per operating condition for statistical analysis.

Cross-dataset validation, where algorithms trained on one dataset are tested on another, assesses generalization capability and robustness to differences in test configurations. This is particularly important because most public datasets use induction motors rather than PM motors, and the electromagnetic-mechanical coupling effects unique to PM machines (UMP, inverter-induced bearing currents, cogging torque modulation) are not captured. Experimental validation on PM motor test rigs under both constant and variable speed settings remains an important area for bridging this gap [11].

An important consideration for PM motor bearing diagnostics is the complementary role of model-based and experiment-based techniques. While data-driven methods dominate current research, model-based approaches offer the ability to predict fault signatures for conditions not yet tested experimentally and to account for PM motor-specific effects such as UMP feedback and electromagnetic force modulation [13]. Hybrid approaches that combine physics-based models with experimental validation provide the most reliable diagnostic frameworks for PM motor applications.

7.4.9.1 Case Study: Few-shot Learning for Bearing Fault Diagnosis

A representative case study demonstrates the practical application of deep learning methods for bearing fault diagnosis under limited data conditions. The study addressed key industrial challenges including insufficient fault data, varying operating conditions, and noise sensitivity using a

Table 7.7 Bearing fault diagnostic signatures across measurement modalities.

Fault type	Vibration	MCSA	Thermal	Acoustic emission	Stray flux	Best method
Outer race	BPFO harmonics; FTF-modulated	$f_s \pm k \cdot f_{BPFO}$; weaker than inner race	Stationary localized hot spot	HF bursts at BPFO rate; good for early detection	BPFO peaks; strong permeance effect	Envelope analysis; AE for early faults
Inner race	BPFI harmonics; shaft-speed modulated	$f_s \pm k \cdot f_{BPFI}$; stronger than outer race	Rotating pattern in load zone	HF transients at BPFI rate	BPFI peaks; rotor-position modulated	Envelope analysis; MCSA if inaccessible
Ball/roller	BSF and 2×BSF; FTF-modulated	$f_s \pm k \cdot f_{BSF}$; weak	Intermittent; less pronounced	Irregular HF events; random modulation	Weak; often masked	Time-frequency analysis; AE
Cage	FTF harmonics; subharmonics	Very weak; FTF sidebands near f_s	Overall increase; not localized	LF impacts; rattle signatures	FTF flux modulation; very weak	Time-domain analysis; LF accelerometers
Distributed wear	Broadband noise; elevated RMS; reduced kurtosis	Elevated noise floor; no discrete peaks	Gradual uniform increase	Elevated continuous level	Broadband variation; reduced SNR	RMS trending; thermal; multisensor fusion
Lubrication	Friction noise; HF content	Minimal direct signature	Rapid rise; most sensitive indicator	Continuous asperity emission	Indirect via eccentricity	Thermal (primary); vibration (secondary)
Electric pitting	Roughness; slot frequency harmonics	Strong if bearing currents present	Moderate increase	Continuous HF emission	Strong electromagnetic interaction	Vibration + inverter frequency analysis
Generalized roughness	Broadband increase; nonimpulsive	Weak; broadband noise	Moderate distributed increase	Elevated background; no transients	Reduced modulation depth	RMS/kurtosis trending; baseline comparison

few-shot learning approach based on Siamese neural networks. Experiments utilized the CWRU dataset with SKF bearings subjected to electro-discharge machining to create controlled fault conditions at three damage severities (0.007, 0.014, and 0.021 inches) across inner race, outer race, and rolling element locations, yielding ten distinct classes: healthy bearing plus three severity levels for each of the three fault types.

The Siamese network architecture employs twin wide deep convolutional neural networks (WDCNN) with shared weights, each consisting of five convolutional layers followed by a fully connected layer producing a 100-dimensional feature embedding. For a given pair of input samples, the network computes the Euclidean distance between their embeddings and passes this distance through an additional fully connected layer and sigmoid activation to produce a similarity probability. Training uses regularized cross-entropy loss with the Adam optimizer, presenting the network with pairs of same-class and different-class samples so that it learns to map similar fault conditions close together and dissimilar conditions far apart in the embedding space. During testing, an unknown sample is paired with reference samples from each class, and the class yielding the highest similarity probability is assigned. In the one-shot setting, a single reference sample represents each class; in the N-shot setting, N reference samples per class are used, and their similarity scores are aggregated.

A key contribution of the study is a preprocessing pipeline applied to each 2048-point vibration window before it enters the network. The raw signal is first converted to a mel-frequency spectrogram, which is then transformed into a grayscale image. Histogram equalization is applied to enhance contrast across the spectrogram, followed by gamma correction with a value of 2.5 to further improve the representation of fault-related spectral features. This preprocessing chain produces an input representation that is substantially more informative than the raw time-domain signal.

The importance of input representation quality can be illustrated independently of the deep learning model. Figure 7.17 visualizes the CWRU dataset using t-distributed stochastic neighbor embedding (t-SNE), which projects high-dimensional feature vectors into two dimensions for visualization. When only basic time-domain statistical features are used, including mean, standard deviation, root-mean-square, kurtosis, and skewness (Figure 7.17a), the resulting clusters exhibit substantial overlap, particularly among fault conditions affecting the same bearing component at different severities. Inner race faults at 0.007 and 0.014 inch severity, for instance, occupy largely overlapping regions, and ball defect classes intermingle with both inner race and outer race clusters. This overlap reflects the limited discriminative capacity of simple statistical descriptors, which capture amplitude distribution characteristics but discard the spectral and temporal structure that distinguishes bearing fault types. In contrast, when comprehensive time-frequency features are extracted (Figure 7.17b), incorporating spectral band energies, spectral entropy, and envelope analysis features derived from the Hilbert transform, the t-SNE projection reveals substantially improved class separation. This comparison underscores that the quality of the input representation fundamentally constrains the achievable classification performance, motivating the mel spectrogram preprocessing pipeline adopted in the few-shot learning approach.

The experimental protocol used the 12 kHz drive-side sensor data, divided equally between training and testing sets. Training employed sliding windows of 2,048 sample points with 80-point increments, while testing used nonoverlapping 2,048-point windows. Four distinct experimental scenarios evaluated diagnostic performance under realistic industrial constraints. In the limited data experiment, using only three training samples per fault class, the proposed method achieved 97.73% five-shot classification accuracy, compared to 82.6% for a conventional WDCNN Siamese

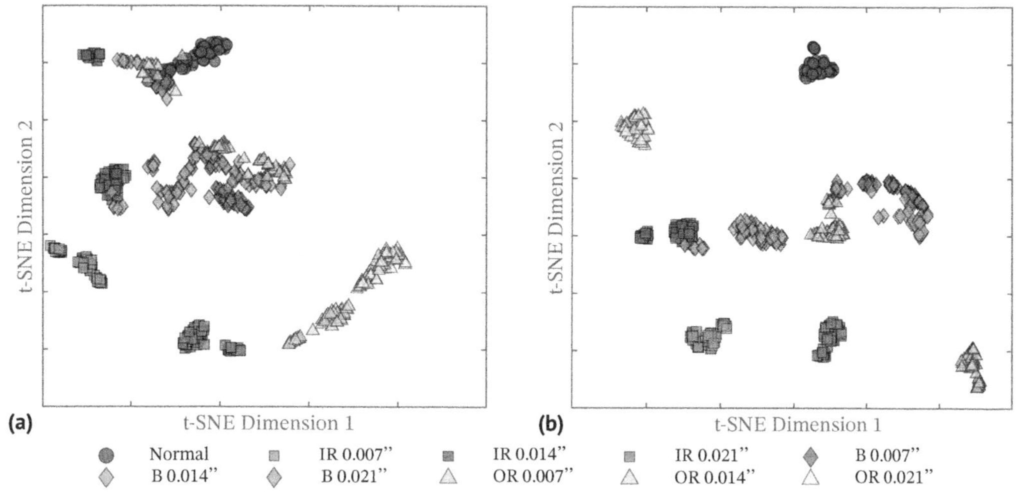

Figure 7.17 Visualization of CWRU bearing fault data using t-SNE for 10 fault conditions across four motor loads: (a) basic time-domain statistical features showing significant class overlap due to limited discriminative capacity, and (b) comprehensive time-frequency features including spectral and envelope analysis demonstrating substantially improved class separation.

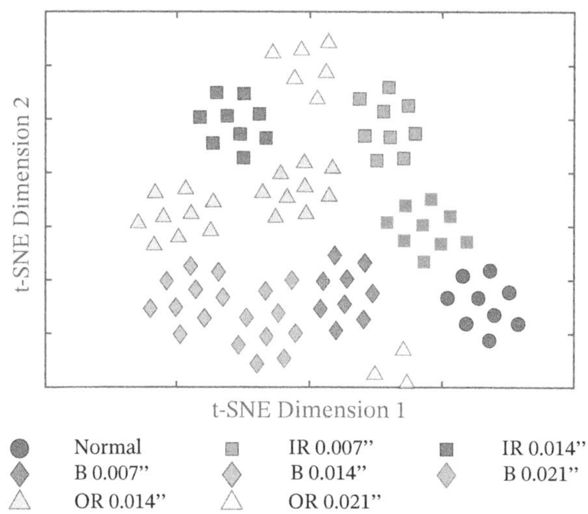

Figure 7.18 Visualization of extracted features using t-SNE for bearing fault classification with the proposed few-shot learning approach. The Siamese network produces well-separated clusters for each of the ten bearing fault conditions, demonstrating effective feature discrimination learned through contrastive training on mel spectrogram inputs with histogram equalization and gamma correction preprocessing.

baseline without the preprocessing pipeline. With 36 training samples per class and all three motor loads combined, accuracy reached 99.85%.

Figure 7.18 illustrates the quality of feature extraction achieved by the few-shot learning method using t-SNE visualization of the 100-dimensional embeddings produced by the Siamese network. The learned feature space exhibits well-defined clusters for each of the ten bearing fault conditions, with clear separation between fault types and severity levels. Comparing this learned representation with the hand-crafted features in Figure 7.17 reveals the advantage of end-to-end deep learning: while even comprehensive time-frequency features (Figure 7.17b) retain some inter-class proximity, the Siamese network's contrastive training objective explicitly optimizes for inter-class separation and intra-class compactness, producing tighter and more distinct clusters. This result is particularly notable given the few-shot training regime, demonstrating that the mel spectrogram preprocessing pipeline combined with metric learning can discover highly discriminative feature mappings from limited labeled data.

The confusion matrices presented in Figure 7.19 quantify the classification performance improvement, where correct predictions along the diagonal and misclassifications appear in the off-diagonal elements. The conventional WDCNN Siamese network without the preprocessing pipeline (Figure 7.19a) demonstrates difficulty distinguishing between certain fault classes, with notable off-diagonal entries particularly for ball defect and inner-race fault categories at similar severity levels. The proposed method with mel spectrogram preprocessing (Figure 7.19b) achieves a near-perfect diagonal confusion matrix, indicating accurate classification across all ten bearing fault conditions. This improvement is particularly significant for distinguishing between faults of different severities affecting the same bearing component, a critical capability for prognostics applications requiring assessment of fault progression.

The noise robustness experiment evaluated performance across varying signal-to-noise ratios, revealing that the preprocessing approach maintained superior accuracy in noisy industrial environments. The generalization capability experiment assessed detection of previously unseen fault classes, achieving 89.71% accuracy when 30% of fault types were excluded from training data, demonstrating the method's ability to identify novel bearing defects without model retraining. The variable operating condition experiment trained the network on data from two rotational speeds (from the set 1,772, 1,750, 1,730 rpm) and tested on the third speed, achieving accuracy ranging from 89.32% to 99.80% depending on the speed combination.

These results illustrate the practical viability of advanced machine learning methods for bearing diagnostics in permanent magnet machines operating under realistic conditions. The case study demonstrates that modern diagnostic approaches can overcome traditional limitations of extensive data requirements and achieve robust fault detection across varying operating conditions and noise levels representative of industrial applications.

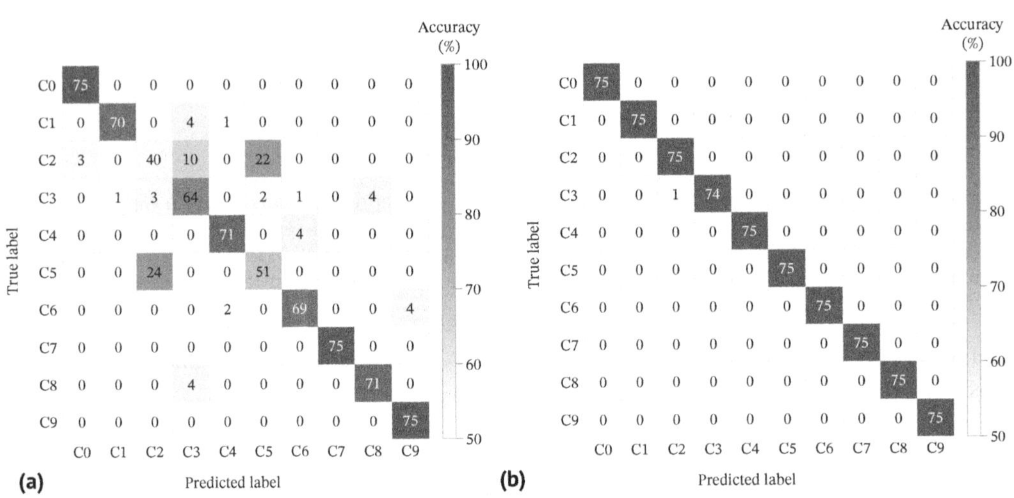

Figure 7.19 Confusion matrices comparing classification performance: (a) baseline deep learning method with notable misclassification errors in the off-diagonal elements, and (b) proposed few-shot learning method achieving near-perfect diagonal matrix indicating high classification accuracy across all bearing fault conditions.

7.5 Summary

Bearing faults account for 40–50% of electrical machine failures, and permanent magnet motors present unique challenges that distinguish their bearing diagnostics from conventional machines. The strong permanent magnet fields create higher radial loads and unbalanced magnetic pull (UMP) that accelerate bearing degradation through a positive feedback mechanism: bearing wear increases eccentricity, which amplifies UMP, which further accelerates wear. Smaller air gaps in PM motors make this coupling more severe, while inverter-driven operation introduces bearing currents that cause electrical erosion damage absent in line-fed machines.

The modeling framework developed in this chapter connects bearing kinematics and contact mechanics to electromagnetic coupling effects, establishing the quantitative relationships between bearing defect characteristics and their observable signatures in both vibration and electrical measurements. The characteristic defect frequencies (BPFO, BPFI, BSF, FTF) derived from kinematic analysis identify what to detect, while the electromagnetic-mechanical coupling analysis explains how bearing faults manifest as current sidebands at $f_s \pm kf_{char}$ and flux modulation patterns specific to PM motors.

For fault diagnosis, vibration-based envelope analysis remains the most reliable primary technique for bearing fault detection, with its effectiveness rooted in the physical mechanism of structural resonance excitation by defect impulses. MCSA provides a valuable complementary modality that requires no additional sensors, though the weaker coupling between mechanical bearing faults and electrical signatures demands more sophisticated signal processing. Data-driven approaches, particularly deep learning architectures, have demonstrated high classification accuracy on benchmark datasets and offer automated fault detection suitable for continuous monitoring. The combination of vibration monitoring with MCSA through multisignal fusion provides the most robust diagnostic strategy across the full PM motor operating envelope. For prognostics, the UMP-bearing degradation coupling unique to PM motors must be incorporated into remaining useful life models, as standard bearing life predictions significantly overestimate bearing life when electromagnetic loading effects are neglected.

References

1 Peter Tavner, Li Ran, Jim Penman, and Howard Sedding. *Condition Monitoring of Rotating Electrical Machines*, volume 56. Institution of Engineering and Technology, 2008. ISBN 9780863417412. https://doi.org/10.1049/pbpo056e.

2 STC-STEYR. Ball bearing components. Online, 2020. URL https://stc-steyr.com/home/.

3 Wei Tong. *Mechanical Design of Electric Motors*. CRC Press, April 2014. ISBN 9780429145025. https://doi.org/10.1201/b16863.

4 SKF. Bearing damage and failure analysis. https://www.skf.com/binaries/pub12/Images/0901d196 8064c148-Bearing-failures---14219_2-EN&_tcm_12-297619.pdf.

5 Theodore A. Harris and Wei Kuei Yu. Lundberg-palmgren fatigue theory: Considerations of failure stress and stressed volume. *Journal of Tribology*, 121(1):85–89, January 1999. https://doi.org/10.1115/1.2833815.

6 Juhani Sopanen and Aki Mikkola. Dynamic model of a deep-groove ball bearing including localized and distributed defects. Part 1: Theory. *Proceedings of the Institution of Mechanical Engineers, Part K: Journal of Multi-body Dynamics*, 217(3):201–211, September 2003. https://doi.org/10.1243/14644190360713551.

7 Abdelkader Rezig, Abdoulaye N'Diaye, Mohamed R. Mekideche, and Abdelaziz Djerdir. Modelling and detection of bearing faults in permanent magnet synchronous motors. In *2012 XXth International Conference on Electrical Machines*, pages 1778–1782, September 2012. https://doi.org/10.1109/icelmach.2012.6350122.

8 Mojtaba Afshar, Mehrdad Heydarzadeh, and Bilal Akin. A comprehensive investigation of fault signatures and spectrum analysis of vibration signals in distributed bearing faults. *IEEE Transactions on Industry Applications*, 61(1):515–526, January 2025. https://doi.org/10.1109/tia.2024.3462921.

9 Hugh E. Boyanton and Gary Hodges. Bearing fluting: The results of a long-term investigation into bearing fluting on AC motors, DC motors, and rolls on paper machines. *IEEE Industry Applications Magazine*, 8(5):53–57, September 2002. https://doi.org/10.1109/mia.2002.1028391.

10 Chenfei Song, Yulong Ren, Qirui Huang, Yuhang Li, Jianhua Cheng, Xianjuan Pang, and Yongzhen Zhang. The evolution process of fluting damage on bearing outer races during rolling with a shaft current. *Wear*, 556–557:205502, November 2024. https://doi.org/10.1016/j.wear.2024.205502.

11 Meiyanto Eko Sulistyo, Didik Djoko Susilo, Muhammad Nizam, and Ubaidillah. A literature review: Bearing fault in BLDC motor based on vibration and thermal signals. *Journal of Electrical, Electronic, Information, and Communication Technology (JEEICT)*, 7(1):10–15, April 2025. https://doi.org/10.20961/jeeict.7.1.100165.

12 Guoguo Wu, Tanyi Yan, Guolai Yang, Hongqiang Chai, and Chuanchuan Cao. A review on rolling bearing fault signal detection methods based on different sensors. *Sensors*, 22(21):8330, October 2022. https://doi.org/10.3390/s22218330.

13 Arnab Das and Susanta Ray. A review on diagnostic techniques of bearing fault and its modeling in induction motor. In *2020 IEEE Calcutta Conference (CALCON)*. IEEE, 2020. ISBN 978-1-7281-4283-8.

14 Robert B. Randall and Jérôme Antoni. Rolling element bearing diagnostics—A tutorial. *Mechanical Systems and Signal Processing*, 25(2):485–520, 2011. https://doi.org/10.1016/j.ymssp.2010.07.017.

15 Ruonan Liu, Boyuan Yang, Enrico Zio, and Xuefeng Chen. Artificial intelligence for fault diagnosis of rotating machinery: A review. *Mechanical Systems and Signal Processing*, 108: 33–47, 2018. https://doi.org/10.1016/j.ymssp.2018.02.016.

16 Yao Cheng, Shengbo Wang, Bingyan Chen, Guiming Mei, Weihua Zhang, Han Peng, and Guangrong Tian. An improved envelope spectrum via candidate fault frequency optimization-gram for bearing fault diagnosis. *Journal of Sound and Vibration*, 523:116746, April 2022. https://doi.org/10.1016/j.jsv.2022.116746.

17 Yongxiang Zhang and Robert B. Randall. Rolling element bearing fault diagnosis based on the combination of genetic algorithms and fast kurtogram. *Mechanical Systems and Signal Processing*, 23(5):1509–1517, July 2009. https://doi.org/10.1016/j.ymssp.2009.02.003.

18 Krish K. Raj, Shahil Kumar, and Rahul R. Kumar. Systematic review of bearing component failure: Strategies for diagnosis and prognosis in rotating machinery. *Arabian Journal for Science and Engineering*, 50:5353–5375, 2025. https://doi.org/10.1007/s13369-024-09866-x.

19 Bin Han, Hui Zhang, Ming Sun, and Fengtong Wu. A new bearing fault diagnosis method based on capsule network and Markov transition field/Gramian angular field. *Sensors*, 21(22):7762, November 2021. https://doi.org/10.3390/s21227762.

20 Ke Zhao, Hongkai Jiang, Chaoqiang Liu, Yanfeng Wang, and Ke Zhu. A new data generation approach with modified Wasserstein auto-encoder for rotating machinery fault diagnosis with limited fault data. *Knowledge-Based Systems*, 238:107892, February 2022. https://doi.org/10.1016/j.knosys.2021.107892.

21 Daehwan Lee and Jongpil Jeong. Few-shot learning-based light-weight WDCNN model for bearing fault diagnosis in Siamese network. *Sensors*, 23(14):6587, July 2023. https://doi.org/10.3390/s23146587.

22 Peng Chen, Yu Li, Kesheng Wang, and Ming J Zuo. A novel knowledge transfer network with fluctuating operational condition adaptation for bearing fault pattern recognition. *Measurement*, 158:107739, 2020.

23 Mohammadkazem Sadoughi and Chao Hu. Physics-based convolutional neural network for fault diagnosis of rolling element bearings. *IEEE Sensors Journal*, 19(11):4181–4192, June 2019. https://doi.org/10.1109/JSEN.2019.2898634.

24 Weiwei Zhang, Deji Chen, Yang Xiao, and Haibing Yin. Semi-supervised contrast learning based on multiscale attention and multitarget contrast learning for bearing fault diagnosis. *IEEE Transactions on Industrial Informatics*, 19(10):10056–10068, October 2023. https://doi.org/10.1109/TII.2023.3233960.

25 Minh Tuan Pham, Jong-Myon Kim, and Cheol Hong Kim. Deep learning-based bearing fault diagnosis method for embedded systems. *Sensors*, 20(23):6886, December 2020. https://doi.org/10.3390/s20236886.

26 Ivan W. Selesnick, Richard G. Baraniuk, and Nick G. Kingsbury. The dual-tree complex wavelet transform. *IEEE Signal Processing Magazine*, 22(6):123–151, November 2005. https://doi.org/10.1109/MSP.2005.1550194.

27 Aida Kabla and Zahir Asradj. Bearing fault diagnosis using combined complete empirical mode decomposition and TKEO methods. *Studies in Engineering and Exact Sciences*, 5(2):e11837, December 2024. https://doi.org/10.54021/seesv5n2-698.

28 Qing Li, Xia Ji, and Steven Y. Liang. Incipient fault feature extraction for rotating machinery based on improved AR-minimum entropy deconvolution combined with variational mode decomposition approach. *Entropy*, 19(7):317, June 2017. https://doi.org/10.3390/e19070317.

29 Zhiyi He, Haidong Shao, Xiang Zhong, Yu Yang, and Junsheng Cheng. An intelligent fault diagnosis method for rotor-bearing system using small labeled infrared thermal images and enhanced CNN transferred from CAE. *Advanced Engineering Informatics*, 46:101150, October 2020. https://doi.org/10.1016/j.aei.2020.101150.

30 Xiaofeng Wang, Yongxing Guo, and Li Xiong. Hybrid fiber Bragg grating sensor for vibration and temperature monitoring of a train bearing. *Chinese Optics Letters*, 16(7):070604, 2018. https://doi.org/10.3788/col201816.070604.

31 Case Western Reserve University. Bearing data center, 2024. https://engineering.case.edu/bearingdatacenter.

32 Jacob Hendriks, Patrick Dumond, and David A. Knox. Towards better benchmarking using the CWRU bearing fault dataset. *Mechanical Systems and Signal Processing*, 169:108732, 2022. https://doi.org/10.1016/j.ymssp.2021.108732.

33 Christian Lessmeier, James Kuria Kimotho, Detmar Zimmer, and Walter Sextro. Condition monitoring of bearing damage in electromechanical drive systems by using motor current signals of electric motors: A benchmark data set for data-driven classification. In *PHM Society European Conference*, volume 3, 2016. https://papers.phmsociety.org/index.php/phme/article/view/1577.

34 Christopher Teubert. IMS bearing dataset. NASA Prognostics Data Repository, NASA Ames Research Center, 2007. https://data.nasa.gov/dataset/ims-bearings.

8

Demagnetization Faults

8.1 Introduction

Permanent magnets provide the magnetic excitation that is fundamental to the operation of PM synchronous machines, and their degradation through demagnetization represents one of the most consequential fault modes in these drives. Unlike electrical or mechanical faults that often produce abrupt changes in machine behavior, demagnetization typically develops progressively as thermal stress, electrical fault events, or material degradation gradually reduce the remanent flux density of the magnets. This progressive nature makes demagnetization particularly insidious: by the time performance degradation becomes operationally apparent, the magnet damage is irreversible, and the machine may require costly rotor replacement or complete remanufacturing.

The diagnostic challenge posed by demagnetization is compounded by the distinction between its two spatial patterns. Local demagnetization, where specific poles or pole regions lose flux, breaks the magnetic symmetry of the rotor and generates characteristic harmonic signatures in currents, voltages, and vibration signals that can be exploited for detection. Uniform demagnetization, where all magnets degrade similarly due to sustained thermal exposure, preserves magnetic symmetry while reducing overall flux magnitude. This symmetry preservation renders uniform demagnetization essentially invisible to many conventional diagnostic methods that rely on asymmetry-induced harmonics, creating a significant detection gap that has motivated the development of specialized techniques including signal injection, model-based flux linkage estimation, and multisignal fusion approaches.

From an economic perspective, demagnetization faults carry disproportionate repair costs relative to other fault types. Rare-earth permanent magnet materials, particularly NdFeB grades with high-energy products suitable for high-performance machines, represent a substantial fraction of the motor bill of materials. The repair process typically requires complete rotor disassembly, magnet replacement, and re-magnetization under controlled conditions, with associated downtime costs that can exceed the material costs in critical applications such as electric vehicle propulsion, wind energy generation, and aerospace actuation systems. Furthermore, demagnetization can initiate cascading failure mechanisms: reduced magnet flux requires higher stator current to maintain torque, increasing both thermal loading and demagnetizing field exposure, which accelerates further magnet degradation in a positive feedback loop.

This chapter provides a comprehensive treatment of demagnetization faults organized into three parts. The first section examines failure modes and mechanisms, covering thermal stress-induced demagnetization, electrical fault-induced demagnetization, manufacturing defect-related causes, and the distinction between partial and complete demagnetization patterns, including the progressive feedback mechanisms that can lead to cascading failure. The second section addresses modeling and analysis techniques, presenting finite element methods, analytical approaches based on magnetic equivalent circuits and vector potential formulations, lumped-parameter models for computationally efficient simulation, and frequency domain analysis of demagnetization-induced harmonic signatures. The third section covers fault diagnosis methods, including back-EMF analysis, motor current signature analysis, flux monitoring techniques, signal injection methods, model-based parameter estimation, multisignal fusion, and acoustic and vibration signatures, concluding with experimental validation on laboratory PM motors. The general principles of the diagnostic techniques applied in this chapter are presented in Chapter 4; the treatment here focuses on their specific adaptation and application to demagnetization fault detection.

8.2 Failure Modes and Mechanisms

Permanent magnets constitute critical components in permanent magnet synchronous machines (PMSMs), providing magnetic excitation without the need for external power supply. The magnetic properties of PMs, however, are susceptible to degradation through various physical mechanisms that can compromise machine performance and reliability. Understanding these demagnetization failure modes is essential for developing effective diagnostic strategies and implementing appropriate mitigation measures in machine design and operation.

Permanent magnets, also known as hard magnetic materials, are characterized by their ability to maintain magnetization and generate persistent magnetic fields. The magnetic behavior of PMs is described by their hysteresis loops, with the demagnetization curve in the second quadrant of the B–H characteristic being particularly significant for understanding demagnetization phenomena [1, 2].

Figure 8.1 illustrates the hysteresis characteristics of a permanent magnet in the first and second quadrants. The virgin curve (dotted) shows the initial magnetization process: starting from the unmagnetized state at the origin, the flux density rises steeply as magnetic domains align with the applied field, then gradually saturates at B_{sat}. Once the external magnetizing field is removed, the material does not return to zero but retains a residual flux density B_r (remanence), which is the starting point of the demagnetization curve in the second quadrant. Two distinct curves characterize the second-quadrant behavior. The normal demagnetization curve (B–H) represents the total flux density in the magnet, which includes both the intrinsic magnetization of the material and the contribution of the applied field ($B = J + \mu_0 H$). This curve crosses zero at the coercive field H_c. The intrinsic demagnetization curve (J–H) isolates the magnet material's own polarization by removing the $\mu_0 H$ contribution; it remains nearly flat over a wide range of demagnetizing field before dropping steeply at the intrinsic coercivity H_{ci}. Since $H_{ci} > H_c$, the magnet material retains its magnetization even after the normal flux density has reached zero, it is the intrinsic curve that determines the true onset of irreversible demagnetization.

Modern PM materials such as neodymium-iron-boron (NdFeB) and samarium-cobalt (SmCo) exhibit a linear demagnetization characteristic until reaching a critical point known as the knee point, after which the flux density drops sharply. The key parameters characterizing PM performance include saturated magnetic flux density (B_{sat}), residual magnetic flux density (B_r), coercive magnetic field (H_c), and inherent coercive magnetic field (H_{ci}). The residual flux density

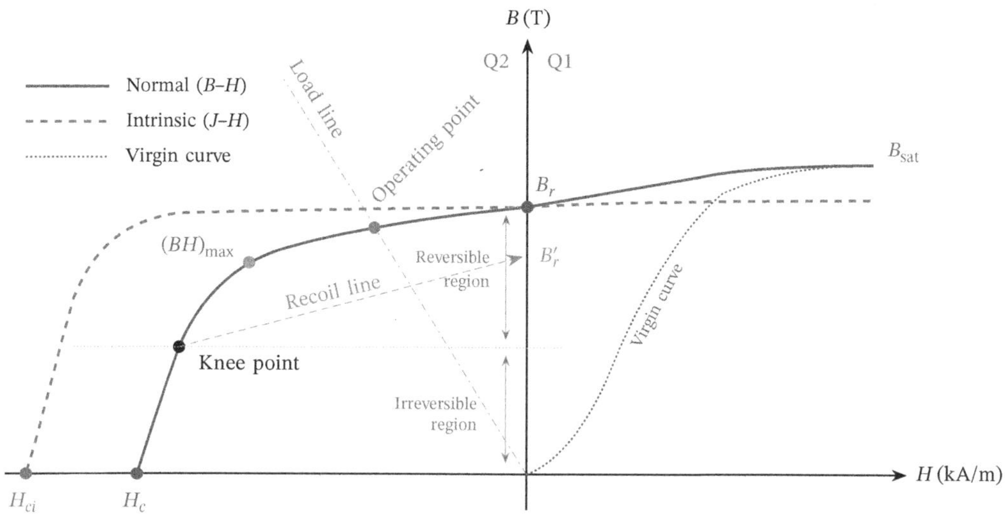

Figure 8.1 Hysteresis characteristics of a permanent magnet in the first and second quadrants, showing the virgin curve, normal (B–H) and intrinsic (J–H) demagnetization curves, and key magnetic parameters.

determines the air gap flux magnitude, while the inherent coercivity indicates resistance to irreversible demagnetization. The maximum energy product $(BH)_{max}$, located at the point on the normal demagnetization curve where the product of B and H is maximized, quantifies the energy storage capability of the magnet and serves as a key design parameter for selecting magnet grades and sizing magnetic circuits.

The operating point of a PM in an electrical machine is determined by the intersection of the PM demagnetization curve (in the second quadrant of the magnetization characteristic) and the load line, as shown in Figure 8.1. The load line is defined by the magnetic circuit geometry and the armature current magnitude. Under normal operating conditions, the operating point remains within the linear region of the demagnetization characteristic (above the knee point), allowing reversible movement along the curve as load conditions vary. When operating within this reversible region, the PM can return to its original magnetic state after the demagnetizing field is removed.

8.2.1 Field-induced Demagnetization

Irreversible demagnetization occurs when external factors force the operating point beyond the knee of the demagnetization curve into the nonreversible region. In this regime, the magnetic field strength is sufficient to reorient magnetic dipoles within the PM material. Upon removal of the demagnetizing field, the PM follows a recoil line rather than returning along the original demagnetization curve, resulting in a permanently reduced residual flux density B_r' as indicated in Figure 8.1. This fundamental mechanism underlies all forms of demagnetization faults in PM machines.

Figure 8.2 illustrates the effect of armature current on the PM operating point. The load line slope is determined by the magnetic circuit geometry (permeance coefficient) and remains constant for a given machine; armature current shifts the load line along the negative H-axis by an amount proportional to the demagnetizing magnetomotive force. Under no-load conditions, the load line

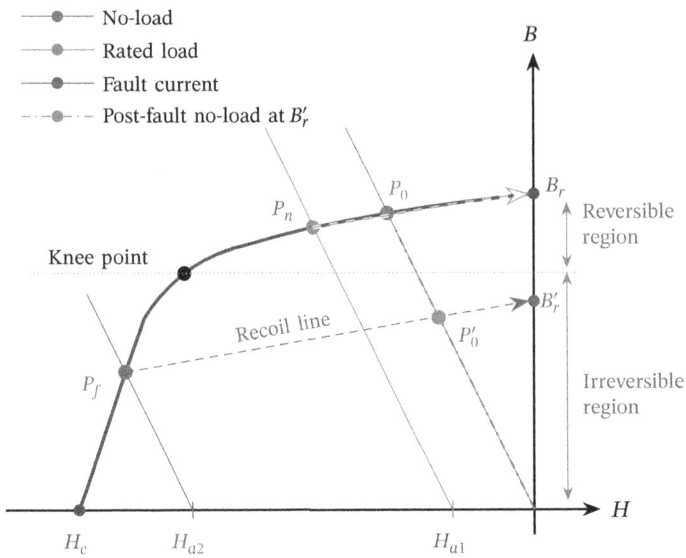

Figure 8.2 Effect of armature current on PM operating point, showing parallel load lines shifted by demagnetizing field H_a for no-load, rated load, and fault conditions.

passes through the origin and the PM operates at point P_0, well within the reversible region. At rated load, the load line shifts left by H_{a1}, moving the operating point to P_n, which remains above the knee; upon load removal, the PM recovers to B_r along the demagnetization curve. During a fault condition with excessive current, the load line shifts further left by H_{a2}, driving the operating point to P_f below the knee. Upon removal of the fault current, the PM follows a recoil line to a permanently reduced residual flux density B_r'. If the same no-load condition is subsequently applied, the PM now operates at point P_0' on the recoil line, at significantly reduced flux compared to the original operating point P_0.

8.2.2 Thermal Stress-induced Demagnetization

Temperature represents one of the most significant factors influencing PM demagnetization behavior. Thermal effects on demagnetization manifest through multiple mechanisms, including temperature-dependent changes in magnetic properties, metallurgical structure modifications, and thermal degradation at elevated temperatures exceeding material limits.

The magnetic properties of PM materials exhibit strong temperature dependence, with both residual flux density and coercivity varying as operating temperature changes. The temperature coefficient of magnetic properties differs significantly among PM material types. For NdFeB and SmCo materials, the residual flux density decreases with increasing temperature, while the coercive force exhibits more complex behavior. In ceramic PMs, coercivity increases with temperature rise, whereas SmCo and NdFeB materials show decreasing coercivity at elevated temperatures. Table 8.1 summarizes the temperature-dependent characteristics of common permanent magnet materials used in electrical machine applications.

This temperature sensitivity creates conditions where demagnetization can occur under normal load conditions if the operating temperature increases sufficiently. Consider an NdFeB magnet operating at temperature T_1 with its rated-load operating point P_n in the safe linear region, just above the knee of the T_1 demagnetization curve. If the temperature rises to T_2, the entire

Table 8.1 Comparison of temperature-dependent magnetic properties for common permanent magnet materials.

Material	Curie temp. (°C)	B_r temp. coeff.	H_c temp. coeff.	H_c vs. temperature	Demagnetization risk
NdFeB	310–400	Negative	Negative	Decreases	High temperature, fault currents
SmCo	700–850	Negative	Negative	Decreases	Extreme thermal conditions
Ceramic/ferrite	450–460	Negative	Positive	Increases	Low temperature, cold start

*Negative coefficient indicates the parameter decreases with increasing temperature; positive coefficient indicates the parameter increases with increasing temperature.

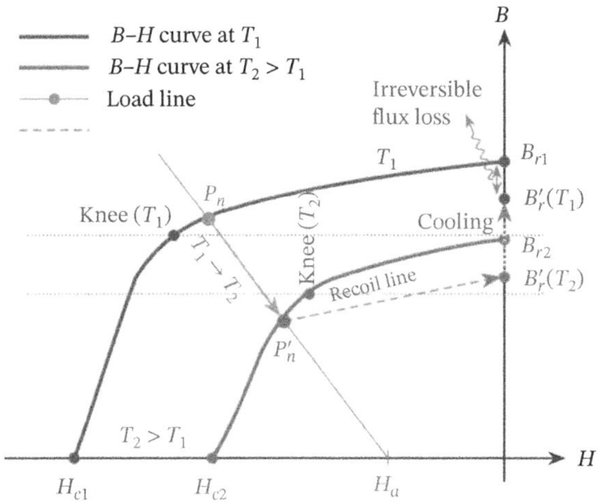

Figure 8.3 Temperature-dependent demagnetization mechanism for an NdFeB permanent magnet. At T_1, the rated-load operating point P_n is above the knee (safe region). Temperature rise to T_2 shifts the curve, moving the operating point to P'_n below the knee (irreversible region). The recoil line reaches $B'_r(T_2)$, and subsequent cooling to T_1 recovers the reversible thermal component to $B'_r(T_1)$, with the remaining gap to B_{r1} representing irreversible flux loss.

demagnetization curve shifts downward and to the right: both the residual flux density B_{r2} and the coercivity H_{c2} decrease, while the knee point moves toward higher flux density values. Under the same load conditions, the operating point shifts to P'_n on the T_2 curve, which now falls below the elevated knee. When the operating point crosses into the irreversible region, the PM suffers permanent magnetization loss and follows a new recoil line upon field removal, reaching a reduced residual flux density $B'_r(T_2)$ at the elevated temperature.

Subsequent cooling back to temperature T_1 recovers the reversible thermal component of the flux density, raising the residual flux density to $B'_r(T_1)$. However, $B'_r(T_1)$ remains below the original B_{r1}, and the difference represents the irreversible flux loss caused by the demagnetization event. The PM now operates on a parallel recoil line with permanently reduced magnetization.

Figure 8.3 illustrates this temperature-dependent demagnetization mechanism for an NdFeB magnet. At temperature T_1, the rated-load operating point P_n lies safely above the knee. When the temperature rises to T_2, the same load produces operating point P'_n, which falls below the T_2 knee in the irreversible region. The recoil line from P_n' reaches $B_r'(T_2)$ on the B-axis. Upon cooling back to T_1, the reversible thermal recovery raises the residual flux density to $B_r'(T_1)$, but the gap between B_{r1} and $B_r'(T_1)$ represents the permanent irreversible flux loss.

The combined effect of temperature variations and demagnetizing magnetomotive force (MMF) from armature currents significantly increases demagnetization risk in NdFeB and SmCo materials. Conversely, ceramic PMs face higher demagnetization risk at low temperatures where the coercivity decreases and the knee point shifts toward the origin.

Multiple heat sources contribute to temperature rise in PM machines, potentially leading to thermal demagnetization. Stator winding losses constitute a primary heat source, particularly under fault conditions when excessive currents flow through the windings. Short circuit faults, whether symmetrical or asymmetrical, generate substantial heat through I^2R losses in the windings. Turn-to-turn short circuits create localized hot spots that can elevate PM temperature in specific regions.

Eddy current losses in the PM material represent another significant heat source. The armature magnetic field contains high-order time and spatial harmonics that induce eddy currents in the electrically conductive PM material. These eddy currents generate losses proportional to the square of the frequency and flux density, contributing to PM temperature rise. The effect is particularly pronounced in surface-mounted PM (SPM) configurations where the PMs are directly exposed to the armature field harmonics.

Cooling system failures exacerbate thermal demagnetization risk by allowing PM temperature to rise above design limits. Blocked cooling passages, coolant pump failures, or inadequate heat sink design can result in excessive PM temperatures during normal or fault operation. Environmental conditions also influence thermal behavior, with high ambient temperatures reducing the thermal margin available before demagnetization occurs. At the extreme, if PM temperature exceeds the Curie temperature (approximately 310–400 °C for NdFeB, 700–850 °C for SmCo), magnetization vanishes entirely. Prolonged exposure to elevated temperatures, even below the Curie point, can cause irreversible metallurgical changes in rare-earth PM materials that permanently reduce magnetic performance.

Thermal demagnetization can manifest in either uniform or local spatial patterns depending on the heat distribution within the machine. Uniform thermal demagnetization occurs when the entire machine temperature rises uniformly, such as during sustained overload operation. In this case, all PMs experience similar thermal conditions and demagnetize with the same pattern. The magnetic flux distribution retains its symmetry but with reduced amplitude.

More commonly, thermal demagnetization exhibits local patterns where specific regions of PMs experience elevated temperatures. Localized heating can result from several mechanisms. Design-related thermal concentration occurs when machine geometry creates regions of high magnetic field strength and correspondingly high eddy current losses. PM pole tips frequently experience higher demagnetizing fields from armature reaction and may be designed with reduced thickness for flux shaping purposes, making them particularly vulnerable to thermal demagnetization under high current conditions [3]. In many machine designs, demagnetization of pole tips can occur even when the pole center remains properly magnetized, creating a characteristic partial demagnetization pattern. This vulnerability exists even in well-designed machines where pole shaping is intentionally used to achieve sinusoidal air gap flux distribution and minimize cogging torque.

Fault-induced local heating creates asymmetric thermal conditions. Turn-to-turn short circuits in specific stator coils generate heat in localized regions, elevating PM temperature in the adjacent magnetic poles. Unbalanced supply conditions or asymmetric inverter faults can result in nonuniform current distribution and corresponding spatial variation in PM temperature. When localized thermal demagnetization occurs, the resulting asymmetric flux distribution generates unbalanced magnetic forces and electromagnetic effects that facilitate fault detection.

High armature currents generated during electrical faults represent another critical demagnetization mechanism. Short circuit faults in PM machine windings generate transient currents that can reach 10–20 times the rated current magnitude, depending on machine design and fault impedance. These excessive currents produce demagnetizing fields that can irreversibly demagnetize PMs if the resulting operating point falls into the nonreversible region. The demagnetization risk from short circuit faults varies significantly with fault type, with asymmetrical faults generally creating higher demagnetization risk than symmetrical faults [1, 4].

Detailed studies of various electrical fault conditions have revealed significant differences in demagnetization risk among fault types [5]. Asymmetrical faults generally pose a higher demagnetization risk than symmetrical faults because they concentrate demagnetizing MMF on specific poles. Among all fault scenarios, single-switch unsymmetrical short circuit faults present the highest threat, followed by phase-to-phase short circuits (8–15 times rated current) and symmetrical short circuits (10–20 times rated current, but balanced across all poles). Phase-leg open-circuit faults and uncontrolled generator mode present lower risk. Turn-to-turn short circuits create highly localized demagnetizing fields adjacent to the shorted coil. Surface-mounted PM motors are generally more vulnerable than interior PM motors, where the rotor core provides additional magnetic shielding for the embedded magnets [5].

8.2.3 Manufacturing Defect-related Demagnetization

Manufacturing processes and quality control play crucial roles in determining PM susceptibility to demagnetization. Defects introduced during motor fabrication can create localized weaknesses that lead to premature demagnetization under operating conditions that would not affect properly manufactured machines.

PM motors are typically magnetized after the complete assembly of the rotor structure. This post-assembly magnetization process involves applying a strong pulsed magnetic field to orient the magnetic domains in the PM material. Incomplete magnetization can occur due to insufficient magnetizing field strength, improper positioning of magnetizing fixtures, or shielding effects from surrounding magnetic circuit components. Incompletely magnetized PMs exhibit lower residual flux density than their material specification would indicate, creating a smaller operating margin before the knee of the demagnetization curve and increasing susceptibility to demagnetization under normal operating transients.

Quality control procedures for verifying complete and uniform magnetization are essential but sometimes inadequate. Back-EMF testing at controlled speed can detect gross magnetization defects but may not identify subtle regional variations. More sophisticated magnetization verification requires dedicated flux measurement equipment or detailed magnetic field mapping, which may not be economically feasible for all production environments.

In surface-mounted PM motor designs, improper adhesive bonding of magnets to the rotor surface can alter the effective air gap geometry, reduce flux density, and impede heat dissipation between the PM and rotor core, all of which reduce demagnetization margins. Variations in PM material properties within a production lot also create vulnerabilities, as magnets at the lower end of the acceptable coercivity range have reduced resistance to demagnetization. In mixed-grade pole designs that place high-coercivity materials at vulnerable pole tips, manufacturing errors in grade placement can create severe demagnetization vulnerabilities.

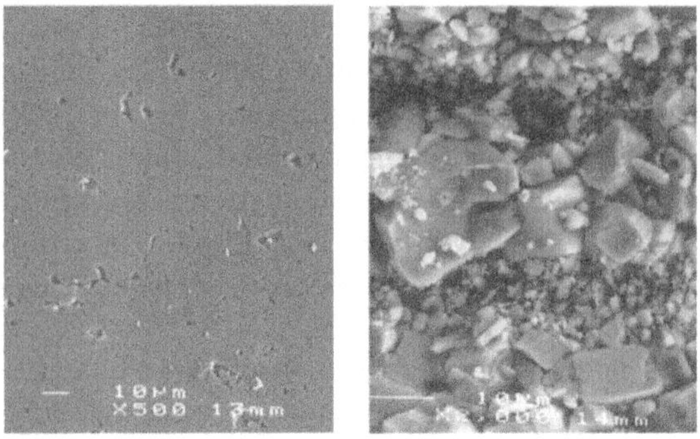

(a) (b)

Figure 8.4 Corrosion of NdFeB permanent magnet surface under microscope showing oxidation and structural degradation resulting from exposure to humid and chloride-rich environments. *Source:* [6]/with permission of Elsevier.

Physical damage to permanent magnets during manufacturing, assembly, or operation also compromises magnetic performance. Permanent magnet materials, particularly rare-earth materials like NdFeB, are mechanically brittle and susceptible to fracture under mechanical stress. Breakage can occur during handling, machining operations, press-fitting into rotor structures, or due to mechanical shock during operation. Mechanical damage creates multiple demagnetization-related problems, including disruption of the continuous magnetic domain structure and flux concentration at fracture sites. A particularly serious consequence of PM tip breakage is the potential for fractured pieces to detach from the rotor and fall into the air gap during operation. These loose PM fragments can become trapped between the rotor and stator, causing immediate mechanical damage to the stator winding insulation. The resulting insulation damage can lead to catastrophic turn-to-turn or phase-to-ground short circuits, transforming an initially localized mechanical defect into a severe electrical fault that further accelerates demagnetization through the mechanisms discussed previously.

Environmental degradation through corrosion and oxidation is particularly relevant for NdFeB materials, which are susceptible due to their electrochemically active rare-earth elements [6]. Surface oxidation creates a nonmagnetic oxide layer that reduces active PM volume, while the oxidized regions become significantly more brittle and susceptible to fracture under mechanical stress. Figure 8.4 shows corrosion damage on an NdFeB magnet surface. Protective coatings (nickel plating, epoxy, zinc phosphate) are commonly applied, but coating defects can expose the underlying material to corrosive attack.

8.2.4 Partial Versus Complete Demagnetization Patterns

Demagnetization faults are classified based on their spatial distribution into uniform (complete) and local (partial) patterns. This classification has significant implications for fault detectability and machine operational characteristics. Table 8.2 provides a comprehensive comparison of these two demagnetization patterns, and Figure 8.5 illustrates their distinct effects on the back-EMF waveform using a linear (unrolled rotor) representation.

It should be noted that the specific pattern of back-EMF reduction due to demagnetization is highly dependent on the stator winding configuration, as the flux linkage of each phase coil is determined by the spatial distribution of coils relative to the demagnetized poles. This interaction

Table 8.2 Comparison of uniform and local demagnetization characteristics and diagnostic implications.

Characteristic	Uniform demagnetization	Local demagnetization
Spatial distribution	All permanent magnets affected equally	Limited number of PMs or specific PM regions affected
Magnetic symmetry	Maintained (balanced across all poles)	Broken (asymmetric flux distribution)
Back-EMF pattern	Reduced magnitude, symmetric waveform	Rotational variation in magnitude, harmonic components generated
Current balance	Three-phase currents remain balanced	Harmonic components appear, potential phase imbalance
Torque impact	Reduced torque production capability uniformly	Variable torque ripple, pulsating torque components
UMP	No UMP generated (symmetry maintained)	UMP forces generated causing vibration and bearing stress
Diagnostic detectability	Challenging (no electrical signal asymmetries)	Easier (harmonic signatures, UMP, vibration)
Primary causes	Sustained overload, cooling system failure, uniform thermal events	Turn-to-turn shorts, localized hot spots, mechanical damage, manufacturing defects
Observable symptoms	Increased current draw at constant torque, reduced maximum torque	Characteristic harmonics in voltage/current, mechanical vibration, acoustic noise
Fault progression	Typically gradual across all poles	Can be sudden in affected poles, progressive expansion possible

between winding topology and demagnetization signatures is discussed in detail in the fault diagnosis section of this chapter.

Uniform demagnetization occurs when all permanent magnets in the machine experience the same demagnetizing conditions, resulting in similar flux reduction across all magnetic poles. This condition typically arises from global thermal events affecting the entire rotor, such as sustained overload operation or cooling system failure that elevates the temperature of the complete rotor assembly. Uniform demagnetization maintains magnetic symmetry in the machine, with the air gap flux distribution retaining its spatial pattern but with reduced magnitude. Since the three-phase windings continue to experience balanced flux linkages despite the reduced magnitude, the resulting back-EMF and current waveforms remain balanced. This symmetry makes uniform demagnetization challenging to detect using conventional fault diagnosis techniques that rely on identifying asymmetries in electrical signals.

The primary symptom of uniform demagnetization is reduced torque production capability. To develop the same torque with reduced PM flux, higher stator current is required. At constant torque operating points, uniform demagnetization manifests as increased current draw. For current-controlled drives, uniform demagnetization results in reduced maximum torque capability when current reaches its limit.

Local demagnetization affects a limited number of PMs or specific regions within individual PMs. This localized degradation creates asymmetry in the magnetic flux distribution, distinguishing it from uniform demagnetization. Local demagnetization typically results from asymmetric fault conditions such as turn-to-turn shorts affecting specific coils, localized thermal hot spots, mechanical damage to individual PMs, or manufacturing defects in particular magnets.

The asymmetric flux distribution created by local demagnetization generates several observable effects that facilitate fault detection. The reduced flux from demagnetized poles creates rotational

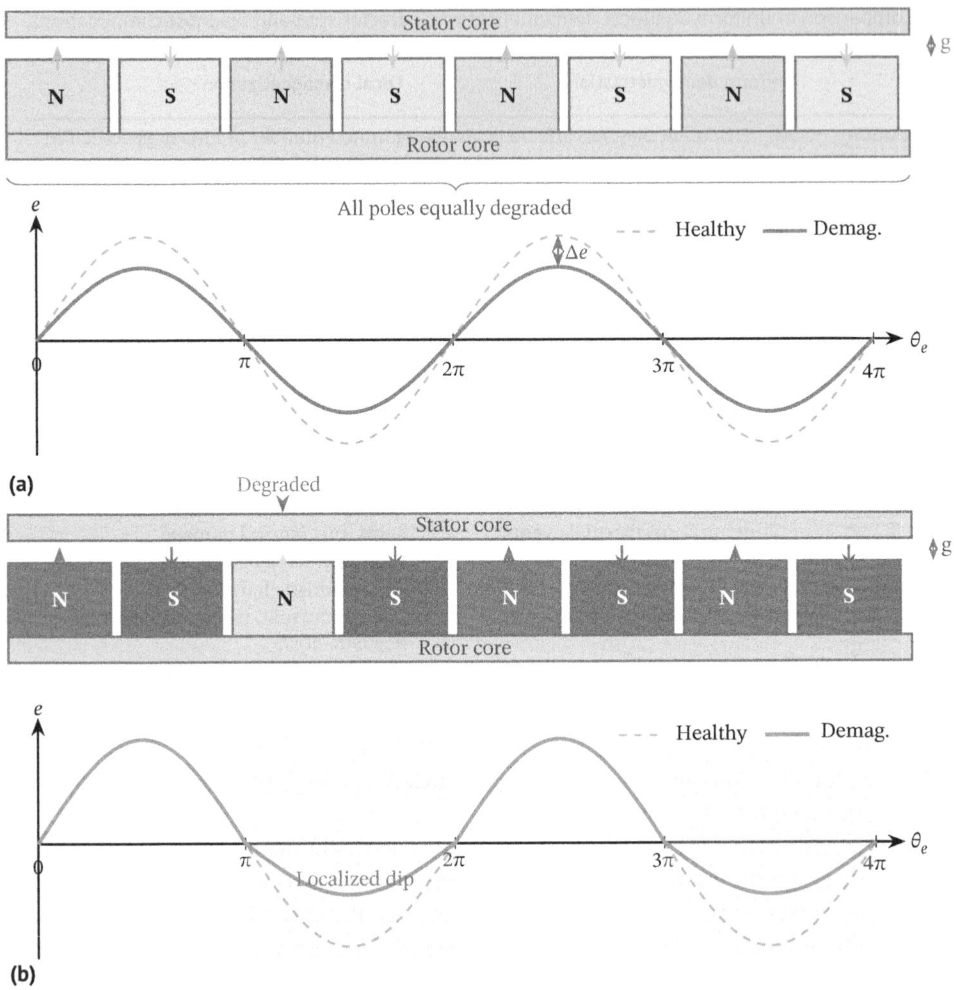

Figure 8.5 Uniform and local demagnetization patterns shown on an unrolled rotor. (a) Uniform: all poles equally degraded, symmetric back-EMF reduction. (b) Local: single degraded pole, asymmetric back-EMF with localized dips.

variation in back-EMF magnitude as the rotor rotates. This time-varying flux linkage generates characteristic harmonic components in the stator voltages and currents that serve as fault indicators. The spatial distribution of magnetic flux becomes nonuniform around the air gap, creating unbalanced magnetic pull (UMP) forces on the rotor that can cause mechanical vibration and bearing stress.

A particularly important category of local demagnetization involves partial demagnetization of individual PM poles rather than complete demagnetization of entire magnets. This pattern commonly occurs in machines where PM pole tips experience concentrated demagnetizing fields from armature reaction. The pole tips may fall into the irreversible demagnetization region while the pole centers remain properly magnetized. Partial pole demagnetization creates complex flux distribution patterns where the spatial harmonic content of the air gap flux field changes as the pole shape effectively becomes altered by the demagnetized regions. From a design perspective, partial pole demagnetization risk can be addressed through mixed-grade pole construction where high-coercivity PM materials protect vulnerable regions.

8.2.5 Progressive Demagnetization Mechanisms

Demagnetization in PM machines often exhibits progressive characteristics where initial degradation creates conditions that accelerate subsequent demagnetization, leading to cascading failure if not detected and addressed.

The relationship between demagnetization and machine operating conditions creates a positive feedback mechanism that can lead to rapid deterioration. When PM flux decreases due to partial demagnetization, the machine must draw higher stator current to produce the required torque at a given operating point. This increased current has two detrimental effects that accelerate further demagnetization. First, higher stator current increases the demagnetizing MMF opposing the PM flux, pushing the operating point of the remaining magnetized regions closer to the knee of the demagnetization curve. Second, the elevated current increases resistive losses in the stator windings, raising the thermal loading on the machine. The increased temperature shifts the PM demagnetization curve and reduces thermal margins, making thermal demagnetization more likely.

This coupled electromagnetic-thermal feedback creates a positive feedback loop, as illustrated in Figure 8.6. Partial demagnetization reduces the PM flux Φ_m, and since torque is proportional to the product $T = k_t\Phi_m I_s$, the machine must draw higher stator current I_s to maintain the required torque. The increased current drives further demagnetization through two parallel paths. Along the electromagnetic path, the higher current increases the demagnetizing MMF, pushing the operating point below the knee of the demagnetization curve and causing additional irreversible flux loss. Along the thermal path, elevated winding losses I_s^2R raise the PM temperature, shifting the B–H demagnetization curve and reducing the coercivity, which also promotes further demagnetization. Both paths feed back into cumulative damage that compounds with each cycle. If uninterrupted, this cycle can progress to complete loss of magnetization and machine failure. The time

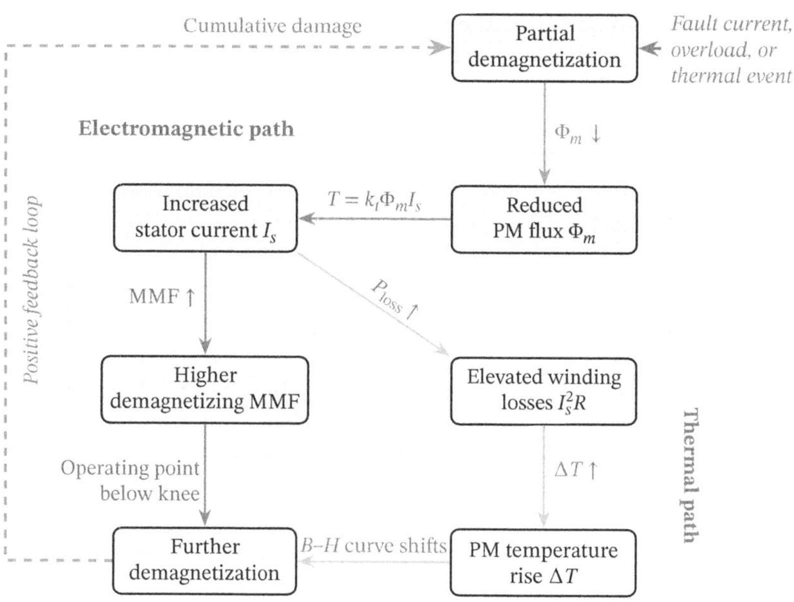

Figure 8.6 Progressive demagnetization cycle in constant torque PM machines showing the electromagnetic and thermal feedback paths that drive cumulative degradation.

constant of progressive demagnetization depends on several factors, including the severity of the initial demagnetization event, the machine's demagnetization margins, the thermal time constants of the machine, and the torque demands placed on the machine.

Repeated thermal cycling during normal machine operation can interact with partial demagnetization to create progressive degradation. During high-power operation, all regions of the machine heat up, but the increased stator current required to compensate for reduced total flux creates higher winding losses. During cooling phases when the machine is de-energized or lightly loaded, temperature gradients within the rotor structure can create thermal stress. Differential thermal contraction between PM materials and rotor core materials can induce mechanical stress at PM bonding interfaces. Over many thermal cycles, the combined effects of mechanical stress, gradual corrosion at exposed surfaces, and magnetic stress from operation can progressively expand the demagnetized region.

Machines operating in electrically harsh environments may experience repeated fault events such as voltage sags, momentary short circuits, or frequent high-torque transients. Each individual fault event may not be severe enough to cause significant irreversible demagnetization. However, cumulative effects from repeated exposure can progressively degrade PM properties. Each time the operating point is driven toward or slightly into the irreversible region, a small fraction of magnetic domains may reverse orientation. While the flux loss from a single event might be negligible, repeated events create accumulated degradation. This cumulative damage mechanism is particularly relevant for machines in automotive traction applications where frequent regenerative braking, acceleration transients, and environmental thermal cycling create repeated stress conditions.

8.2.6 Magnetic Flux Reduction and Electromagnetic Impact

Demagnetization fundamentally alters the electromagnetic behavior of PM machines through reduction of air gap magnetic flux and the creation of spatial asymmetries in the magnetic field distribution. Understanding these electromagnetic consequences is essential for both fault detection and assessment of demagnetization impact on machine performance.

In a healthy PM machine, the air gap magnetic flux distribution exhibits spatial periodicity corresponding to the pole pitch. Demagnetization disrupts this spatial distribution. For uniform demagnetization, the spatial pattern of flux distribution is preserved, but the magnitude is proportionally reduced across all pole positions. Local demagnetization creates more complex modifications, with flux density in the affected pole regions decreasing while remaining relatively constant in unaffected regions. This spatial variation introduces asymmetry into the air gap field. The specific harmonic components introduced depend on the spatial location and extent of the demagnetized region. Demagnetization affecting a single pole pair in a machine with p pole pairs introduces harmonics at frequencies that are noninteger multiples of the fundamental electrical frequency.

The stator winding flux linkage is determined by integration of the air gap flux over the coil spans of the distributed winding. Demagnetization-induced reduction in PM flux directly reduces the magnitude of flux linkage in the affected phase windings. For uniform demagnetization affecting all poles equally, the flux linkage in all three phases reduces proportionally, and the back-EMF magnitude decreases by the same factor. Local demagnetization creates time-varying flux linkage as the rotor rotates, generating additional frequency components in the back-EMF spectrum. The reduction in flux linkage has important implications for machine control, potentially degrading control performance in field-oriented control systems.

Electromagnetic torque in PM machines is produced by interaction between the PM-generated air gap flux and the MMF produced by stator currents. Demagnetization reduces the available PM flux, directly reducing the torque production capability for a given stator current level. For uniform demagnetization, if the PM flux is reduced to a fraction k of its original value, the torque production is similarly reduced to k times the original torque for the same current magnitude. Local demagnetization creates additional torque complications beyond simple magnitude reduction, with the asymmetric flux distribution causing the torque to vary as a function of rotor position even under steady current conditions.

Local demagnetization creates radial asymmetry in the magnetic field distribution, which generates UMP forces on the rotor. Significant UMP can deflect the rotor shaft, creating dynamic eccentricity where the rotor center of rotation does not coincide with the stator bore center. Unbalanced radial forces increase bearing loading beyond the design values, accelerating bearing wear and potentially leading to premature bearing failure. The time-varying nature of UMP under rotating conditions creates alternating stress on bearing components, promoting fatigue failure mechanisms. The detrimental effect of UMP on bearing service life, including the significant reduction in bearing lifetime resulting from unbalanced magnetic forces, is discussed in detail in Chapter 7.

The distorted air gap flux distribution created by demagnetization induces harmonic components in the stator voltages and currents. The specific harmonic frequencies present in the stator currents depend on the spatial distribution of the demagnetization. Analysis based on winding functions and spatial harmonic decomposition reveals that demagnetization of a single pole in a machine with p pole pairs generates current harmonics at frequencies $f_h = f_s(1 \pm k/p)$ where f_s is the fundamental stator frequency and k is an integer. These characteristic frequencies serve as diagnostic indicators for demagnetization detection.

The electromagnetic forces between stator and rotor create vibration in the stator structure. Demagnetization introduces additional force harmonics at frequencies corresponding to the flux distribution harmonics, exciting mechanical vibration in the stator core and frame structure. When electromagnetic force frequencies coincide with mechanical resonances of the stator structure, vibration amplitude can be significantly amplified. The acoustic signature of a demagnetized machine differs from that of a healthy machine, providing another diagnostic modality.

Demagnetization degrades machine efficiency through multiple mechanisms. The increased current required to develop a given torque with reduced flux increases stator copper losses. For a uniform demagnetization that reduces flux to k times the original value, copper losses increase by a factor of $1/k^2$ at constant torque. Core losses are also affected by demagnetization, with the altered flux distribution potentially increasing flux density in certain regions while decreasing it in others. Power factor degradation accompanies demagnetization, as the reduction in PM flux requires higher current magnitude to produce the required torque. Poor power factor has implications for inverter design and sizing, as the inverter must be rated for the increased current even though active power remains constant.

8.3 Modeling and Analysis

Accurate modeling of demagnetization faults is essential for understanding fault characteristics, developing diagnostic methods, and evaluating machine performance under faulty conditions. Simulation using models can resolve the need for high-cost experimental tests during initial investigation of fault diagnosis methods. Furthermore, in some diagnostic techniques, the model itself forms part of the fault diagnosis method, known as model-based techniques [7]. This section examines how the modeling techniques introduced in Chapter 3 are specifically adapted for demagnetization fault analysis, and presents demagnetization-specific formulations including analytical field solutions and lumped-parameter approaches.

8.3.1 Finite Element Method for Demagnetization Analysis

Finite element method (FEM), as discussed in Chapter 3, provides the most comprehensive approach for modeling demagnetization fault conditions in PM machines [8, 9]. For demagnetization analysis, time-stepped FEM is preferred over fixed-mesh approaches because demagnetized machines lose their magnetic symmetry, requiring simulation of the full machine geometry rather than exploiting periodicity to reduce computational cost.

For surface-mounted PM synchronous machines with skewed rotor magnets, specialized two-dimensional FEM approaches have been developed [10]. These methods obtain the magnetic field distribution from skewed PMs with accuracy very close to practical cases. The attractive feature of such approaches is their easy implementation in any two-dimensional FE electromagnetic software package. These methods offer high accuracy, simplicity, short computational time, and quick simulation compared with multilayer two- and three-dimensional FEA. However, when dealing with faulty motors that lose their symmetry due to demagnetization, it becomes impossible to exploit machine symmetry, and simulation of only a partial section of the motor is not feasible, resulting in increased computational burden.

To simulate a faulty BLDC motor under demagnetization conditions, two-dimensional time-stepping FEA can be applied [8]. The objectives of such a simulation are first to validate results with measured data from the healthy motor and second, to verify the model under fault conditions. PM breakage has been considered a major fault mechanism, and its impact upon stator back-EMF and current of PMSMs has been studied using fast Fourier transform analysis [11]. The effect of PM gap size and shape on fault characteristics has been investigated by simulating three types of gaps with identical area but different shapes, including vertical, horizontal, and zigzag configurations.

Back-EMF in PMSMs is the induced voltage in the stator phase due to variations in rotor PM flux linkage with the stator winding. For an open-circuit stator winding, the back-EMF caused by flux linkage variation with the rotating rotor at constant speed is expressed as:

$$e_a = \frac{d\lambda_a}{dt} = \frac{d\theta_r}{dt}\frac{d\lambda_a}{d\theta_r} = \omega_r \frac{d\lambda_a}{d\theta_r} \tag{8.1}$$

where e_a is the induced voltage in the stator winding (back-EMF), λ_a is the stator flux linkage, ω_r is the rotor electrical speed, and θ_r is the rotor angular position in electrical degrees. Any fault in the rotor affects the air gap flux density, which reflects directly on the back-EMF. This makes back-EMF a very suitable variable for analyzing both rotor and stator faults.

Comparison of simulation results between healthy and faulty motors with vertically cracked PMs shows that demagnetization-induced fault signals exhibit the following frequency pattern:

$$f_{BM,k} = \left(\frac{k}{P/2}\right) f_s = k f_m \quad k = 1, 2, 3, \ldots \tag{8.2}$$

where $f_{BM,k}$ is the k-th frequency component in the spectrum, f_s is the fundamental frequency of the back-EMF (electrical frequency), f_m is the rotation frequency, and P is the number of poles. This equation reveals that fault signals appear at integer multiples of the mechanical frequency. In healthy motors, some inherent fault signals are observable at these frequencies due to manufacturing imperfections and asymmetric components in the motor. However, when demagnetization faults occur in the rotor, the flux density and back-EMF are considerably influenced. For many motor configurations, the 0.25-th and 0.5-th harmonics exhibit the highest sensitivity to demagnetization faults.

8.3.2 Analytical Models for Demagnetization

Analytical models based on magnetic equivalent circuits (MEC), winding function methods, and field-based formulations offer valuable alternatives to FEM for demagnetization analysis, providing reduced computational burden while maintaining acceptable accuracy. Several approaches have been specifically adapted for demagnetization fault modeling in PM machines.

Local demagnetization risk in tubular PM machines can be evaluated analytically [12]. Open-circuit fields and armature reaction are obtained analytically in a cylindrical coordinate system, and for local demagnetization risk evaluation, the fields in PM regions are superimposed. This method has been applied to tubular PM machines with quasi-Halbach magnetization and modular stator windings, with results confirmed by FEM.

Permeance network methods have been applied to model demagnetization in outer-rotor PM machines [13]. Using the permeance method and superposition principle, computational tools can estimate major electromagnetic parameters, including electromagnetic torque and induced voltage, with acceptable accuracy and significantly reduced simulation time compared with FEM. The permeance model may use FEM to initialize flux path determination, which are then converted to permeances calculated from geometrical and material characteristics of the machine.

An analytical model based on back-EMF has been extended for PM fault diagnosis [14]. The rationale for choosing EMF as the calculated parameter is that it provides a direct representation of flux generated by PMs. The objective is to develop a fault diagnosis tool that not only detects faults but also identifies fault location. This analytical model is based on formulating vector potential using Maxwell equations. PMs are divided into smaller elements, equations are written for each element, and a new form of air gap vector potential equation is developed as a function of residual magnetic flux density B_r of all PM elements.

In applying governing equations, each PM element is assumed to have a uniform property distribution. The Maxwell governing equations are:

$$\nabla \times \vec{H} = 0 \tag{8.3}$$

$$\nabla \cdot \vec{B} = 0 \tag{8.4}$$

$$\vec{B}_I = \mu_0 \vec{H}_I \quad \text{in the air gap} \tag{8.5}$$

$$\vec{B}_{II} = \mu_0 \mu_r \vec{H}_{II} + \mu_0 \vec{M} \quad \text{in the PM elements} \tag{8.6}$$

where $\vec{H}$ is magnetic field intensity, $\vec{B}$ is magnetic flux density, $\vec{M}$ is the magnetization vector, μ_0 is vacuum permeability, and μ_r is relative permeability. For radially magnetized PMs, the magnetization vector is:

$$\vec{M} = M_r \vec{r} \tag{8.7}$$

For a magnetic pole divided into multiple elements, the magnetization vector for each element is presented by Fourier series depending on element position. Figure 8.7 illustrates the discretization of a magnetic pole into 25 elements for analytical modeling. This subdivision allows detailed representation of local demagnetization patterns where only certain PM elements may have reduced magnetization.

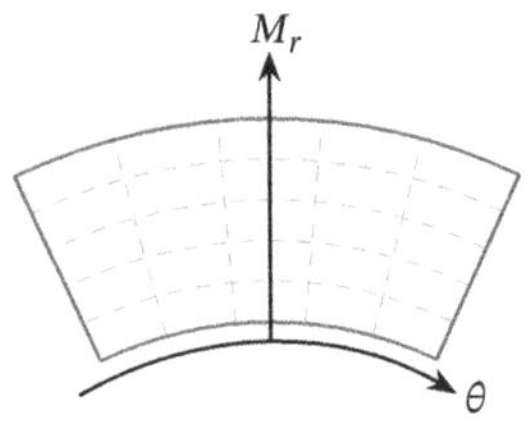

Figure 8.7 Discretization of a magnetic pole into 25 elements for analytical modeling, allowing representation of local demagnetization patterns in permanent magnet machines.

By solving field equations in both air gap and PM regions with appropriate boundary conditions, the vector potential and subsequently the back-EMF can be evaluated. Parameter estimation methods use regression processes to minimize differences between observed and calculated results. The objective function is defined in the least squares form to estimate residual flux density of PM elements by comparing calculated and measured back-EMF.

Reluctance network models have also been applied to surface-PM motors for demagnetization analysis [15], accounting for nonlinear characteristics, slot effects, saturation, and rotor motion while providing a good compromise between accuracy and computation time.

8.3.3 Lumped-parameter Model for Demagnetization

Simplified lumped-parameter models (LPMs) have been developed for PM machines to enable fast computation while maintaining sufficient accuracy for fault diagnosis studies. These models incorporate winding and slot position effects directly in back-EMF calculations through Fourier series representations, enabling implementation of different control techniques and fault-tolerant control strategies.

The healthy back-EMF of phase a is expressed through Fourier series as:

$$E_a(\theta_r) = \omega_r \sum_{n=-\infty}^{+\infty} K_{wn} E_n e^{jn\omega_0\theta_r P/2} \tag{8.8}$$

where ω_r is rotor speed, K_{wn} is the winding factor accounting for coil angular positions and current directions, E_n are back-EMF Fourier coefficients, P is the number of poles, and $\omega_0 = 1$ for a 2π period. Since PM permeability is approximately equal to that of air, demagnetization does not change the machine inductance profile, and the fault effect can be modeled entirely through back-EMF modification. The key modeling concept represents the flux reduction from a demagnetized pole as an inverted pulse waveform multiplied with the healthy coil back-EMF. The Fourier series of this pulse can be expressed as:

$$\text{pls}(\theta_r) = \sum_{l=-\infty}^{+\infty} P_l e^{j\omega_0 l\theta_r} \tag{8.9}$$

where P_l is the complex Fourier coefficients of the inverse pulse, F is a real number between 0 and 1 as the fault severity of one pole pair, θ_τ/T is the pulse duty cycle which determines the area of the demagnetized pole which is less than or equal to T/P, the pulse period $T = 2\pi$ for one mechanical cycle, and $\text{sinc}(x) = \sin(\pi x)/\pi x$. The DC component is:

$$P_0 = 1 - \frac{2F\theta_\tau}{T} \tag{8.10}$$

and the harmonic coefficients for $l \neq 0$ are:

$$P_l = -\frac{F\theta_\tau}{T}\,\text{sinc}\left(\frac{Fl\theta_\tau}{T}\right)\left(e^{-j\omega_0 l\theta_\tau/2} + e^{-j\omega_0 l(\theta_\tau/2+T/P)}\right) \tag{8.11}$$

In this expression, each phase shift term corresponds to either entrance or exit of the defected magnet in a specific coil region. Depending on the location and span of the fault on a magnet pole, the phase shift terms and θ_τ should be adjusted. In the case of uniform demagnetization, θ_τ is set to T/P (the whole magnet region), while this value is less than T/P for local demagnetization depending on the fault span.

The faulty back-EMF coefficients are obtained through discrete convolution of the healthy coefficients E_n with the pulse coefficients P_l down-sampled by factor $P/2$:

$$E_{mf} = \sum_{n=-\infty}^{+\infty} E_n P_{m-nP/2} \tag{8.12}$$

The resultant phase back-EMF under demagnetization is:

$$E_{af}(\theta_r) = \omega_r \sum_{m=-\infty}^{+\infty} K_{wm} E_{mf} e^{jm\omega_0\theta_r} \tag{8.13}$$

For modeling multiple demagnetized poles, the inverted pulse is applied with appropriate phase shifts (multiples of $2\pi/(P/2)$) corresponding to each faulty pole and its severity. The calculated back-EMF is substituted into the stator voltage equation for a star-connected winding to complete the electrical subsystem. For the linear model where inductances depend only on rotor position, the voltage equation takes the form:

$$v_{abc} = r_s i_{abc} + \frac{dL_s}{d\theta_r}\omega_r i_{abc} + L_s \frac{di_{abc}}{dt} + e_{abc} \tag{8.14}$$

where v_{abc}, i_{abc}, and e_{abc} are vectors of three-phase stator voltages, currents, and back-EMFs, r_s is the per-phase stator resistance, and L_s is the 3×3 inductance matrix containing self and mutual inductances. The second term accounts for the variation of inductance with rotor position, which contributes to reluctance torque. The electromagnetic torque is obtained from the summation of torque produced by the interaction of currents and back-EMFs and the reluctance torque due to inductance variation:

$$T_e = \frac{P}{2}\left(\frac{i_{abc}^T e_{abc}}{\omega_e} + \frac{1}{2} i_{abc}^T \frac{dL_s}{d\theta_e} i_{abc}\right) \tag{8.15}$$

where ω_e and θ_e are the electrical speed and angle, respectively. The mechanical dynamics governing the rotor motion are:

$$T_e - T_L = J\frac{d\omega_r}{dt} + B\omega_r \tag{8.16}$$

where T_L is the load torque, J is the moment of inertia, and B is the viscous damping coefficient. Figure 8.8 presents the block diagram of this LPM showing how the electrical, magnetic, and mechanical subsystems are interconnected for simulation under demagnetization faults.

The derivation presented above assumes a linear magnetic circuit where inductances are independent of current and rotor position. This assumption is reasonable for surface-mounted PM motors where the effective air gap seen by the armature field is large due to the low-permeability

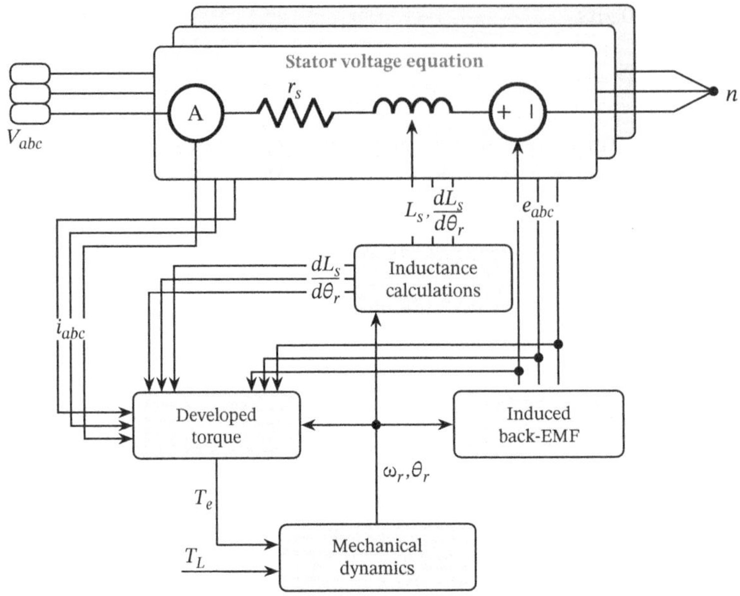

Figure 8.8 Block diagram of the LPM for BLDC motor showing the integration of electrical equations, back-EMF calculation with demagnetization modeling, electromagnetic torque computation, and mechanical dynamics.

magnets, and neglecting harmonics of inductances introduces minimal error. However, when magnetic saturation effects become significant, the linear model formulation requires extension. Taking flux as the state variable rather than current, the voltage equation can be rewritten in inductance-based form as:

$$v_{abc} = r_s i_{abc} + \frac{d\phi_{abc}}{d\theta_r}\omega_r + \frac{d\phi_{abc}}{di_{abc}}\frac{di_{abc}}{dt} \tag{8.17}$$

where the flux-linkage terms depend on both current and rotor position and are restored from look-up tables (LUTs) obtained through FEA. In this formulation, the second term represents a generalized BEMF that accounts for the position-dependent flux variation under actual loading conditions, while the third term is modeled using an inductance block with an incremental inductance matrix as its input. The block diagram structure of the nonlinear model remains the same as the linear case, with the difference that LUTs replace the analytical Fourier series expressions for electromagnetic torque, incremental inductance matrix, and generalized BEMF as functions of phase currents and rotor position. In comparison with LUT approaches, the Fourier series formulation of the linear model has the advantage that fault severity can be adjusted simply by changing the corresponding parameter, without requiring large multidimensional LUTs as a function of fault severity and rotor position. The results presented in the following sections are for validation of the linear model.

8.3.4 FEM Simulation Case Studies

To validate the LPM, case studies of four BLDC motors with concentrated windings and different rotor and stator configurations were conducted: Motor A (12s/10p, double-layer all-teeth winding), Motor B (12s/10p, single-layer alternative-teeth winding), Motor C (15s/10p, double-layer all-teeth winding), and Motor D (9s/8p, all-teeth winding) [16]. Motors A, B, and C are rated at 230 W with 36 V supply and 400 rpm nominal speed, while Motor D is rated at 133 V supply and 3000 rpm nominal speed. Figure 8.9 shows the finite element models of these configurations. Motors A and B share the same 12-slot, 10-pole geometry but differ in winding configuration: Motor A uses all-teeth winding where coils are wound on every tooth, while Motor B uses alternate-teeth winding where coils are placed on every other tooth with doubled turns per coil, resulting in the same total series turns per phase and identical back-EMF amplitude at nominal speed. Motor D serves as the experimental validation case, enabling a complete three-way comparison between the LPM, finite element simulations, and physical measurements. Demagnetization is realized in FEM simulations by eliminating a portion of PM volume from the rotor, with both uniform and local demagnetization patterns considered. For uniform demagnetization, the PM volume is reduced over the entire magnet region, while for local demagnetization, parts of a specific magnet are removed.

LPM parameters (h, θ_w, θ_z, and the number of Fourier harmonics of the BEMF waveform) are identified by minimizing the error between FEM (or measured) back-EMF waveforms and the model output for the healthy motor. This can be formulated as a constrained optimization problem where the objective function is the sum of squared errors (SSE) between the calculated and desired back-EMF at each rotor position:

$$\min_x J = \sum_{k=1}^{n} (e_c(x, \theta_k) - e_d(\theta_k))^2 \tag{8.18}$$

$$\text{s.t.} \quad \max(e_c(x, \theta_k)) - \max(e_d(\theta_k)) \leq 0 \quad \forall \quad k = 1, 2, \ldots, n \tag{8.19}$$

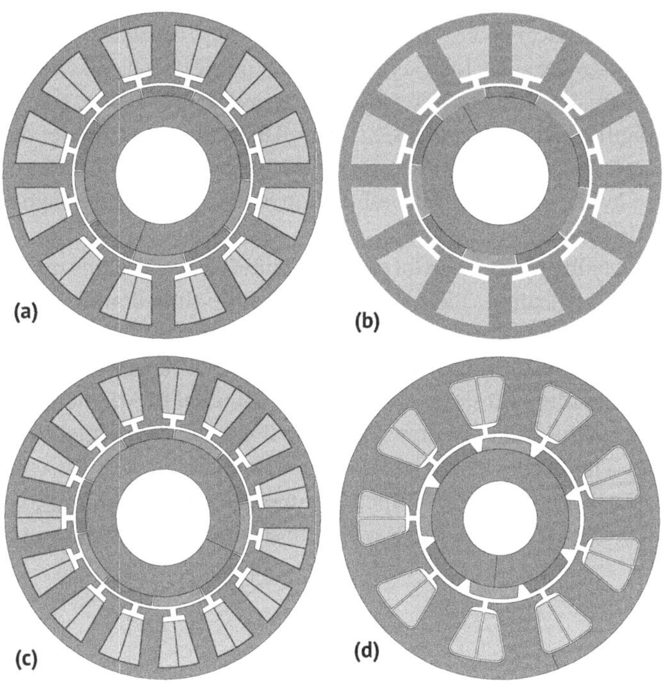

Figure 8.9 Cross-sectional views of the four BLDC motor topologies used for demagnetization fault validation. (a) Motor A: 12-slot, 10-pole, double-layer all-teeth winding. (b) Motor B: 12-slot, 10-pole, single-layer alternate-teeth winding. (c) Motor C: 15-slot, 10-pole, double-layer all-teeth winding. (d) Motor D: 9-slot, 8-pole, all-teeth winding.

where $e_c(x, \theta_k)$ is the calculated value of BEMF using the proposed model at the k-th rotor position, $e_d(\theta_k)$ is the desired value of BEMF at that position provided by FEA or experimental results, and x is the vector of parameters to be optimized. The constraint ensures that the model peak value does not exceed the reference peak, yielding physically meaningful parameters. An evolutionary algorithm such as a genetic algorithm is suitable for this optimization.

A notable advantage of this model is that it requires only the number of rotor poles and stator slots together with a reference BEMF waveform (from FEA or measurement) to identify parameters. No detailed geometric or winding information is needed. This property extends to automatic identification of the winding layout when the winding arrangement of the machine is not known a priori. Since PM machines with side-by-side windings (coil step of one) can be configured as either single-layer (alternate-teeth) or double-layer (all-teeth), and each layer type admits multiple valid winding dispositions defined by coil current directions and spatial positions, the model can be optimized for every valid winding configuration. The true winding layout is then the one yielding the minimum SSE. For example, optimization across all valid all-teeth and alternate-teeth configurations for Motor A identifies the winding disposition $AA'A'A$ with spatial offsets [0 30 180 210] as the true layout with the lowest SSE. Similarly, for Motor B, the alternate-teeth layout AA' with spatial offset [0 180] is identified as the true configuration. This capability is particularly useful in fault diagnosis applications where winding details may be unavailable or proprietary.

With 50% demagnetization of one rotor PM (5% of the total machine), the back-EMF reduction pattern depends critically on winding configuration. When the defective magnet enters or exits the region of a stator coil, it reduces the BEMF of that coil for one electrical period. The net phase BEMF reduction pattern differs between winding types because the winding configuration determines which coils are affected simultaneously. In double-layer Motor A, the BEMF exhibits 10% reduction in half-cycles corresponding to adjacent coils. In single-layer Motor B, because the number of turns per coil is doubled (to maintain the same total series turns per phase), each coil spans a larger portion of the back-EMF waveform, resulting in 10% reduction across all half-cycles. Due to the symmetry in rotor and stator configurations of Motor C (15s/10p), all half electrical cycles of the net BEMF are reduced by a uniform 5%. For Motor D (9s/8p), the reduction pattern is nonuniform because of the asymmetry introduced by the 9-slot/8-pole combination, where partially demagnetized regions align differently with different coils.

The BEMF and current waveforms of faulty Motor D are of particular interest because this motor is subsequently used for experimental verification. Figure 8.10 presents the time-domain comparison of FEA and LPM results for Motor D with 50% demagnetization of one pole. The BEMF waveform shows the characteristic nonuniform reduction pattern, where the defected magnet causes localized flux reduction as it passes different stator coils. This reduction in BEMF increases the peak values of the stator current, as the machine draws additional current to maintain the required torque under vector control. The proposed model results are compatible with FEA for both BEMF and current waveforms, confirming that the lumped-parameter approach captures the essential electromagnetic behavior of the faulty machine with sufficient accuracy for fault diagnosis applications.

8.4 Fault Diagnosis

8.4.1 Back-EMF Analysis for Demagnetization Detection

Back-electromotive force (back-EMF) represents one of the most direct indicators of permanent magnet health, as it reflects the fundamental relationship between magnetic flux and machine operation. Demagnetization-induced reduction in PM flux manifests directly in the back-EMF

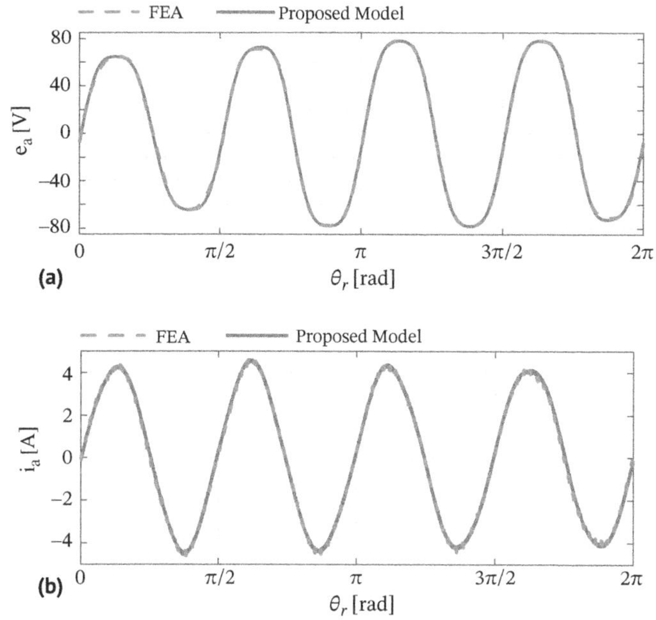

Figure 8.10 Comparison of FEA and lumped-parameter model results for Motor D with 50% demagnetization of one pole at nominal speed. (a) Phase-a back-EMF waveform showing nonuniform reduction pattern due to the defected magnet. (b) Phase-a stator current waveform.

waveform, making back-EMF analysis a fundamental approach for demagnetization fault detection. The voltage induced in stator windings due to the rotating magnetic field of the permanent magnets provides information about both the magnitude and spatial distribution of the air gap flux. Since demagnetization fundamentally alters this flux distribution, back-EMF monitoring offers a physically intuitive diagnostic pathway that can be implemented through either direct measurement or model-based estimation techniques.

The relationship between flux reduction and back-EMF characteristics forms the theoretical foundation for this diagnostic approach. In a healthy permanent magnet machine, the back-EMF magnitude is proportional to the rotor speed and the flux linkage produced by the permanent magnets. When demagnetization occurs, the reduced flux linkage directly decreases the back-EMF amplitude. For uniform demagnetization affecting all poles equally, the back-EMF maintains its waveform shape but exhibits proportional amplitude reduction across all phases. Local demagnetization affecting specific poles creates more complex signatures, introducing distortion in the back-EMF waveform and generating characteristic harmonic components that serve as fault indicators. The distinction between these two patterns provides diagnostic information about not only the presence of demagnetization but also its spatial distribution within the machine [17].

Demagnetization faults induce characteristic fractional harmonics in back-EMF that serve as the origin of stator current harmonics and torque ripple in BLDC motors. These fractional harmonics can be expressed as:

$$f_d = \left(1 \pm \frac{k}{P/2}\right) f_e \tag{8.20}$$

where f_d are demagnetization fault frequencies, f_e is the electrical frequency, k is an integer, and P is the number of poles. Table 8.3 provides the characteristic harmonic frequencies for different motor configurations.

Table 8.3 Characteristic harmonic frequencies for demagnetization faults in different motor configurations.

Configuration (slot/pole)	Harmonic relationship	Fractional orders in back-EMF	Filtered from current*	Special characteristics
12s/10p (double-layer)	$(1 \pm k/5)f_e$	$k/5, k = 1, 2, 4, 6, 7, 9, 11, 13$	$k/5, k = 3, 9$	Fractional odd harmonics appear; 11/5 and 13/5 most prominent in current
12s/10p (single-layer)	$(1 \pm k/5)f_e$	$k/5, k = 1, 2, 4, 6, 7, 9, 11, 13$	$k/5, k = 3, 9$	Larger harmonic amplitudes than double-layer due to higher winding factors
15s/10p (double-layer)	$(1 \pm k/5)f_e$	None (symmetry suppression)	All harmonics filtered	Slot/pole ratio of 3/2; symmetrical configuration suppresses all fractional harmonics
9s/8p	$(1 \pm k/4)f_e$	$k/4, k = 1, 2, 3, 5, 6, 7$	$k/4, k = 3, 6$	Both odd and even fractional harmonics; experimental validation available
General *p*-pole	$(1 \pm k/(P/2))f_e$	Depends on *P*	Harmonics with 0° phase shift	Linear scaling with fault severity and speed

*Filtered harmonics are neutralized in star connection. Visible current harmonics are the back-EMF orders minus the filtered set.

Phase back-EMF power spectra for different motor configurations reveal distinct patterns. For both 12-slot/10-pole motors (A and B), fractional odd harmonics appear in power spectra. The impact of winding layer on the fault harmonics can be studied by comparing Motor A (all-teeth, double-layer) with Motor B (alternate-teeth, single-layer): harmonic amplitudes are higher for the alternate-teeth winding because the winding factors for single-layer configurations are larger than for double-layer windings. However, for motors with slot-to-pole ratios that are multiples of 3/2 (such as the 15-slot, 10-pole Motor C), no fractional components appear in the phase back-EMF power spectrum. This result, which leads to identical and symmetrical back EMF reduction due to stator and rotor configuration, is fully justifiable. Machines with slot per pole ratio of multiples of 3/2 belong to this category.

The 9-slot/8-pole Motor D exhibits a distinctly different harmonic pattern from the 12s/10p motors. Since the harmonic relationship for this configuration is $(1 \pm k/4)f_e$, both odd and even fractional harmonics are engendered in the BEMF power spectrum. The harmonic orders 1/5, 11/5, and 13/5 are the most affected components in the 12s/10p motors considering winding factors, while the most significant fault harmonics in Motor D appear at orders 1/4, 2/4, 5/4, and 7/4. Figure 8.11 illustrates the back-EMF power spectra for the four motor configurations under healthy and 50% demagnetization conditions.

The effect of local vs. uniform demagnetization on the frequency spectrum can be studied by comparing the spectra of Motor D under both fault patterns. For uniform demagnetization (where the PM volume is reduced uniformly over the entire pole), the fractional harmonics follow the same frequency pattern as local demagnetization, but their amplitudes are generally lower. For local demagnetization, where the same amount of PM material is removed from only a portion of the pole, the amplitude of fractional harmonics is higher, especially for the sidebands of

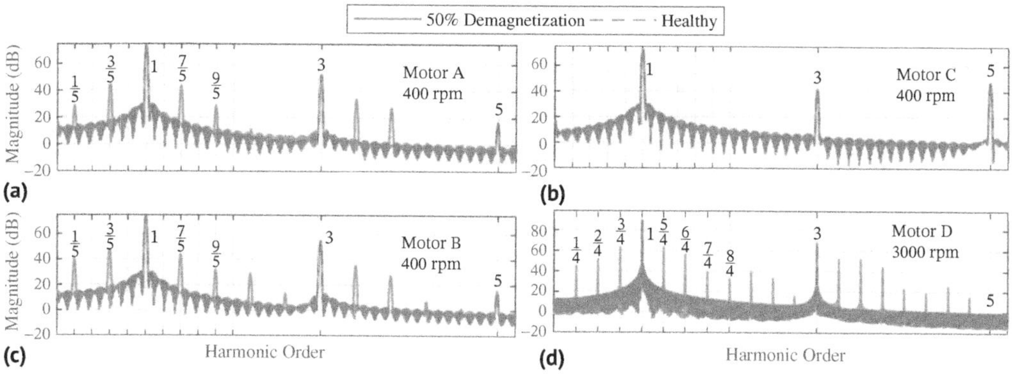

Figure 8.11 Back-EMF power spectra at nominal speed for four motor configurations comparing healthy (baseline) and 50% demagnetization conditions: (a) Motor A (12s/10p, all-teeth), (b) Motor B (12s/10p, alternate-teeth), (c) Motor C (15s/10p), and (d) Motor D (9s/8p) under uniform and local demagnetization, respectively, showing characteristic fractional harmonics in Motors A, B, and D but not in Motor C due to its symmetrical configuration.

harmonic 3. In addition, harmonic 5 appears in the power spectrum of the local demagnetization case but is absent in the uniform case. This distinction arises because local demagnetization creates sharper spatial flux density transitions in the air gap compared to uniform faults, generating stronger high-frequency harmonic content and introducing additional spectral components that provide diagnostic information about the spatial extent of the fault.

The impact of fault severity on back-EMF harmonic components has been investigated by progressively simulating different severities of demagnetization including 25%, 50%, 75%, and 100% removal of one pole for Motor D using both the proposed model and FEA. The amplitude of the BEMF fault harmonics 1/4 and 2/4 increases linearly with the change in fault severity. Furthermore, the behavior of these harmonics with speed variations shows that, at a certain fault severity, changes in amplitude of the BEMF sideband harmonics with speed are also linear. This linear dependence on both fault severity and rotor speed confirms that the fractional harmonics arise from fundamental electromagnetic induction phenomena rather than nonlinear effects such as saturation or eddy currents. For the 12s/10p motors, different fault severities (20%, 40%, 60%, 80%, and 100% of one pole) also produce linearly increasing harmonic components. Comparison at different speeds (200 and 400 rpm) indicates that harmonic components are directly proportional to both speed and fault severity, with the rate of variation differing among harmonic orders and between motor configurations. Some harmonics (such as the 7/5 order) exhibit similar variation rates for different motor types at the same speed, while others show significant differences between configurations.

Dynamic behavior of demagnetization fault signals vs. speed has been examined for various fault severities across a wide speed range. All fault signals vary approximately linearly with speed. Harmonics closer to the fundamental frequency (high-frequency harmonics on the left side and low-frequency harmonics on the right side of the fundamental) have higher values, and these relative magnitudes remain constant as speed increases. Comparison between single- and double-layer winding motors indicates that although fault harmonics in stator back-EMF change linearly with speed for both configurations, the amplitude range for single-layer winding motors is larger than for double-layer winding motors.

Direct measurement of back-EMF represents the most straightforward diagnostic approach but requires specific operating conditions. In generator applications or during coasting periods when the machine operates in open-circuit mode, the terminal voltage directly reflects the back-EMF without the influence of resistive voltage drops or inductive voltage components. This direct measurement approach enables offline fault diagnosis where the back-EMF can be measured accurately during controlled testing conditions [18, 19]. The open-circuit voltage waveform provides a clear indication of demagnetization through both amplitude reduction and waveform distortion. However, the requirement for open-circuit operation limits the applicability of direct measurement for continuous online monitoring in motor applications where the machine must remain under load during normal operation.

For online fault diagnosis in motor drives, indirect measurement techniques estimate the back-EMF from terminal voltage and current measurements using analytical models of the machine. These model-based approaches reconstruct the back-EMF by subtracting the resistive and inductive voltage components from the measured terminal voltage. The spatial harmonic method decomposes the estimated back-EMF into its harmonic components and analyzes the harmonic spectrum for fault-related frequencies [20, 21]. Demagnetization introduces fractional harmonics at frequencies determined by the pole number and the spatial distribution of the fault. The inverse transform method provides an alternative analytical framework for back-EMF estimation under faulty conditions [22]. These indirect estimation techniques enable continuous online monitoring without requiring dedicated test procedures, but they introduce dependence on accurate knowledge of machine parameters such as stator resistance and inductance. Temperature variations that modify resistance values and saturation effects that alter inductance can compromise the accuracy of back-EMF estimation, potentially reducing diagnostic reliability and limiting the detection of low-severity faults.

A novel analytical method for partial demagnetization detection in air-cored permanent magnet axial flux generators derives mathematical equations for back-EMF under demagnetization conditions to extract fault-related frequency components [23]. This approach uses magnetic flux sensors to estimate fault severity through the normalized peak-to-peak voltage induced in an auxiliary winding, which depends solely on the demagnetization level and provides a direct fault indicator. The auxiliary winding voltage monitoring technique offers advantages in terms of fault sensitivity while requiring additional sensing hardware. Signal processing methods can detect faults using frequency amplitude components under both constant and variable speed operating conditions. An important limitation of this technique is that uniform demagnetization produces zero total magnetic flux through the auxiliary winding, restricting fault detection capability to partial demagnetization patterns. The method applies to machines with implementable auxiliary windings or those with an even number of pole steps. For machines with odd pole step numbers, the auxiliary winding signal remains non-zero even under healthy conditions, necessitating initial calibration to establish baseline characteristics. Three-dimensional finite element modeling validates the proposed method under both constant and variable speed conditions [23].

Advanced signal processing techniques enhance back-EMF analysis for demagnetization detection under dynamic operating conditions. The S-transform applied to back-EMF signals provides superior capability in expressing signal characteristics compared to conventional Fourier analysis, particularly for transient and time-varying fault signatures [24]. An analytical back-EMF model for permanent magnet synchronous linear motors (PMSLM) under demagnetization enables extraction of frequency-time parameters that establish characteristic vectors for fault identification. The S-transform decomposes the back-EMF signal into time-frequency representations where demagnetization features manifest more clearly than in time-domain or pure frequency-domain analysis. Characteristic parameters extracted from the frequency–time curves serve as inputs to classification algorithms such as particle swarm optimization combined with the least squares support vector machines (PSO-LSSVM) for automated fault diagnosis. The hierarchical diagnosis structure improves classification accuracy by leveraging the enhanced feature clarity provided by time-frequency analysis. Advantages of this approach include superior frequency-time resolution that clarifies fault features, dimension reduction of characteristic vectors through standard deviation and similarity evaluation, and improved diagnosis accuracy through the hierarchical PSO-LSSVM model [24].

Zero-sequence voltage analysis provides an alternative back-EMF based diagnostic approach that exploits specific harmonic components rather than fundamental amplitude. The air gap spatial harmonic components containing all multiple three odd harmonics generate zero-sequence voltage in stator phase voltages [25]. Although the fundamental back-EMF is not directly measurable during online motor operation, the third harmonic component can be extracted through zero-sequence voltage measurement when the neutral point of the motor is accessible. Demagnetization causes reduction in the zero-sequence voltage magnitude, which can be measured through a resistive network connected to the machine terminals. The magnitude of this reduction correlates with demagnetization severity, enabling quantitative fault assessment. This diagnostic approach requires physical access to the neutral point, which may not be available in all motor configurations. The method provides sensitivity to demagnetization without requiring sophisticated signal processing or parameter-dependent models, but it cannot distinguish between uniform and local demagnetization patterns based solely on zero-sequence voltage magnitude.

For brushless DC machines with trapezoidal back-EMF waveforms, zero-crossing point (ZCP) detection offers a sensitive fault diagnosis method with practical implementation advantages. The zero-crossing instants of the back-EMF waveform shift when demagnetization alters the flux distribution, creating a more tangible diagnostic effect than simple amplitude reduction. Most BLDC motors employ sensorless control drives that detect zero-crossing points for commutation timing, making the ZCP information readily available without additional sensing hardware. The ZCP-based diagnostic method exhibits low sensitivity to machine parameter variations, temperature effects, and other faults such as static eccentricity, providing robust fault detection under varying operating conditions. The method can detect demagnetization during normal motor operation without requiring special test procedures or dedicated measurement equipment beyond the sensing already present for motor control.

Model-based approaches for back-EMF estimation under demagnetization employ observer structures to reconstruct flux linkage and identify fault conditions. An improved sliding mode observer based on equivalent input disturbance (EID) estimation reconstructs demagnetization faults without being affected by motor speed variations, enabling fault-tolerant control system implementation [26]. This approach models both healthy and demagnetized permanent magnet synchronous machine conditions, then uses the observer to estimate and compensate for demagnetization effects in real time. The overall system stability is analyzed using the globally uniformly ultimately bounded (GUUB) concept, providing theoretical guarantees for observer convergence. Hardware-in-the-loop simulation demonstrates the validity of the method under realistic operating conditions. A comprehensive analytical model estimates partially lost magnet flux during demagnetization caused by overcurrent events for six-phase surface-mounted PMSM based on back-EMF waveforms [27]. The online diagnostic algorithm accounts for effects of magnetic orientation (radial or parallel magnetization) in the analysis. The technique includes correction of the control system encoder angular position to achieve maximum torque under post-fault conditions, integrating fault diagnosis with fault-tolerant control strategies.

Analytical models based on Maxwell's equations provide detailed frameworks for back-EMF calculation under demagnetization. These approaches solve the magnetic field distribution in both the air gap and permanent magnet regions, deriving the back-EMF as a function of rotor position and PM magnetization state [14, 28]. For local demagnetization fault diagnosis in permanent magnet AC machines, the back-EMF function vs. rotor position is evaluated using finite element methods and entered into gradient-based optimization algorithms. The quick and precise convergence characteristics make these models useful for both offline fault analysis and online diagnostic implementations. For axial flux permanent magnet synchronous machines, scalar potential equations are solved to determine three-phase terminal voltages given three-phase input currents and the electromagnetic properties of the machine. Magnetization coefficients representing the magnetization state of each permanent magnet are determined through the modeling process, and deviations from nominal values indicate demagnetization. The obtained results demonstrate the effectiveness of analytical approaches for monitoring demagnetization fault status, with the potential for integration into control system implementations [28].

The primary advantages of back-EMF based diagnostic methods include their direct physical relationship to the fault mechanism, relatively simple implementation for machines with accessible back-EMF signals, and independence from some fault types that complicate current-based diagnostics. However, limitations must be recognized. Indirect estimation methods depend on accurate machine parameters that vary with temperature and saturation. Uniform demagnetization detection remains challenging due to the maintained symmetry in back-EMF waveforms. Low-severity faults may produce back-EMF changes within the noise floor of measurement systems. For symmetrical concentrated winding configurations, fractional harmonics in back-EMF may not appear, similar to the behavior observed in current-based methods [29]. Despite these limitations, back-EMF analysis continues to serve as a fundamental diagnostic approach, particularly when combined with advanced signal processing techniques and integrated with other diagnostic modalities in multisignal fusion frameworks. Table 8.4 provides a comprehensive comparison of the principal back-EMF analysis techniques.

8.4.2 MCSA Application for Demagnetization Signatures

The general principles, signal requirements, and implementation framework for motor current signature analysis are presented in Chapter 4. This section focuses on the specific application of MCSA to demagnetization fault detection, including the characteristic harmonic signatures, detection limitations, and challenges unique to demagnetization monitoring in permanent magnet machines.

When local demagnetization affects specific permanent magnets, the resulting asymmetry in the air gap magnetic field generates rotational variations in flux linkage, inducing characteristic harmonic components in the back-EMF that subsequently appear as current harmonics under load [30, 31]. For permanent magnet machines experiencing demagnetization, a sequence of fractional harmonics appears in the stator current spectrum. The frequencies of these fault-related harmonics can be expressed as:

$$f_{demag} = f_e\left(1 \pm \frac{k}{p}\right) \tag{8.21}$$

where f_{demag} represents the demagnetization fault harmonic frequencies, f_e is the fundamental electrical frequency, k is a positive integer, and p is the number of pole pairs. This relationship has been confirmed through both experimental measurements and finite element simulations [32, 33].

Table 8.4 Back-EMF diagnostic methods comparison for demagnetization detection.

Method	Measurement type	Operating condition	Detects uniform	Detects local	Parameter dependency	Implement. complexity	Primary advantages	Primary limitations
Direct open-circuit measurement	Direct voltage	Open circuit	Yes	Yes	None	Low	Direct measurement; no parameter dependency	Requires open-circuit; not for continuous monitoring
Model-based indirect estimation	Voltage and current estimation	Under load	No	Yes	High (R, L)	Medium	Online; no additional sensors	Parameter temperature/saturation dependent
Inverse transform method	Analytical reconstruction	Under load	Limited	Yes	High (R, L)	Medium-High	Analytical framework; online capable	Complex; parameter dependent
S-transform with PSO-LSSVM	Advanced signal processing	Variable speed	Limited	Yes	Medium	High	Superior time-frequency resolution; automated classification	Computational complexity; requires training
Zero-sequence voltage	Third harmonic monitoring	Under load	Yes	Yes	Low	Low	Simple; robust; neutral point access	Requires neutral point; cannot distinguish patterns
ZCP detection (BLDC)	Zero-crossing timing	Under load	Limited	Yes	Very Low	Low	Available in sensorless drives; no additional hardware	BLDC only; limited uniform sensitivity
Sliding mode observer	Observer-based	Under load	Yes	Yes	Medium	High	Speed-independent; fault-tolerant integration	Complex design; computational requirements
Analytical model	Field calculation	Any	Yes	Yes	Medium	High	Detailed spatial information; fault localization	Computational intensity; requires geometry data

The amplitude of each harmonic component provides information about fault severity and can be determined as:

$$A_{demag} = \frac{V_{slot} K_{demag}}{k\pi} \sin\left(\frac{k\pi}{2p}\right) \tag{8.22}$$

where V_{slot} represents the slot volume, K_{demag} is a constant related to demagnetization severity and machine geometry, k is the harmonic order, and p is the number of pole pairs. This equation reveals that harmonic amplitudes depend on both the geometric configuration of the machine and the severity of the demagnetization fault, with different harmonic orders exhibiting varying sensitivity to the fault condition.

Practical implementation of MCSA for demagnetization diagnosis requires systematic current harmonic analysis across different operating conditions. Diagnostic methods based on current harmonic monitoring have been successfully applied to both permanent magnet synchronous machines and brushless DC motors [11, 32, 33]. The diagnostic procedure involves monitoring the amplitude of generated current harmonics at frequencies predicted by the theoretical relationship, conducting measurements at different load levels and rotational speeds to distinguish demagnetization signatures from other operational effects. The variation of harmonic amplitudes with operating conditions provides additional diagnostic information that enhances fault discrimination capability.

An important distinction exists between the detectability of local vs. uniform demagnetization using current signature analysis. Local demagnetization affecting a limited number of permanent magnets creates magnetic asymmetry in the machine, generating the characteristic harmonic components described above. In normal operating conditions, local demagnetization can be readily diagnosed using these harmonic signatures. However, uniform demagnetization where all magnets experience similar flux reduction does not create asymmetry in the magnetic field distribution. Consequently, uniform demagnetization produces no distinctive harmonic components in the current spectrum, making it essentially invisible to conventional MCSA techniques [11]. This fundamental limitation represents a significant constraint on the applicability of current signature analysis for comprehensive demagnetization monitoring.

The specific harmonic content generated by demagnetization depends critically on the number of affected permanent magnets, their relative positions on the rotor, and the stator winding configuration. Analytical investigations of local and uniform demagnetization patterns reveal that full demagnetization of a permanent magnet generates new even and fractional harmonics in the current spectrum. These harmonic sets become particularly prominent when the spectrum already contains even harmonics and when the relative position between totally demagnetized permanent magnets equals an integer multiple of the stator pole pitch. The relationship between fault location, machine geometry, and harmonic generation creates complex dependencies that must be understood for reliable fault diagnosis.

Winding configuration exerts a substantial influence on the current spectrum under demagnetization conditions. Comprehensive analysis using magnetic equivalent circuit methods has examined the impact of permanent magnet demagnetization on stator variables, revealing that winding configuration determines which fault-related harmonics appear in the measurable current signals [11]. Different winding configurations produce distinct current spectra for identical demagnetization patterns. The number, location, and orientation of defective permanent magnets interact with winding arrangement to determine fault signal characteristics. Tests and simulations across different speeds and loads demonstrate that fault signal amplitude exhibits relatively weak dependency on torque but strong dependency on rotational speed, a relationship that must be carefully considered in variable speed control applications.

The harmonic components of back-EMF generated by demagnetization are reflected in stator current waveforms. However, some harmonics neutralize each other due to identical phase shifts in three-phase star-connected winding currents and do not appear in phase current power spectra. For the 12-slot/10-pole motors, harmonics 3/5 and 9/5 are eliminated from the current power spectrum of both Motor A and Motor B, while harmonics 11/5 and 13/5 become larger in the current spectrum instead. For Motor C with its symmetrical 15s/10p configuration, no fractional harmonics appear in either the BEMF or current power spectrum. For Motor D (9s/8p), harmonics 3/4 and 6/4 are eliminated from the sidebands of the fundamental harmonic in the stator current power spectrum. As with the BEMF spectra, the fractional harmonics of local demagnetization are larger than those of uniform demagnetization in the current spectra. Figure 8.12 presents the stator current power spectra for the four motor configurations under healthy and 50% demagnetization conditions. Comparing these spectra with the back-EMF results in Figure 8.11 confirms the selective harmonic cancellation imposed by the star-connected winding topology, while the remaining fault-induced harmonics retain sufficient amplitude for diagnostic purposes.

A critical challenge in applying MCSA for demagnetization diagnosis arises from the fact that other machine conditions can generate harmonics at frequencies identical to those produced by demagnetization. Load torque oscillations and static eccentricity both generate harmonic components that can create frequency signatures similar to local demagnetization [34]. Static eccentricity, in particular, produces a frequency spectrum pattern closely resembling that of local demagnetization, creating potential for misdiagnosis. However, important differences exist in the dynamic behavior of these components under varying operating conditions. In the presence of static eccentricity, stator inductances differ from those in a machine with permanent magnet defects due to the nonuniform air gap created by rotor displacement. This inductance difference enables the appearance of phase current components that would be absent under pure demagnetization conditions, generating a distinctive pattern that can be exploited to differentiate between these fault types [34].

The practical implementation of fault discrimination between demagnetization and eccentricity faces complications due to manufacturing imperfections. In practice, intrinsic manufacturing defects create some level of these harmonic components in machine signals even under healthy conditions, complicating the establishment of diagnostic thresholds and reducing reliability. Additionally, fault components at odd multiples of three times the mechanical frequency may disappear

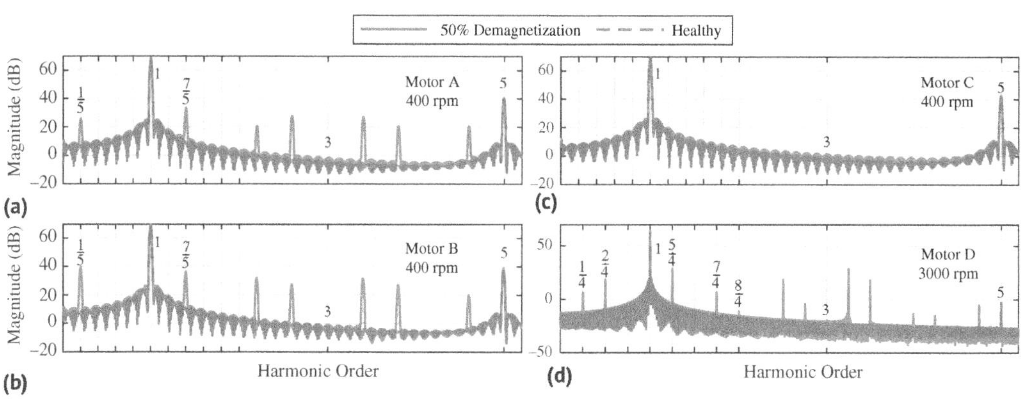

Figure 8.12 Stator current power spectra at nominal speed for four motor configurations comparing healthy and 50% demagnetization conditions: (a) Motor A (12s/10p, all-teeth), (b) Motor B (12s/10p, alternate-teeth), (c) Motor C (15s/10p), and (d) Motor D (9s/8p), showing selective elimination of certain fractional harmonics due to star-connected winding phase cancellation.

depending on the connection type of the machine winding. For example, in star-connected three-phase windings, triplen harmonics that are balanced across phases sum to zero in the line currents and become undetectable. This potential loss of diagnostic information must be carefully considered when designing MCSA-based fault detection systems [11].

Winding topology introduces another layer of complexity in current signature interpretation. In permanent magnet synchronous machines with symmetrical concentrated windings, harmonic components of the induced voltage appear in each series winding section [29]. However, these harmonics neutralize each other when summed across the complete winding due to the symmetrical configuration. Consequently, fractional harmonics do not appear in the overall back-EMF or current of symmetrically wound machines, similar to the signature of uniform demagnetization [29]. This phenomenon creates a diagnostic blind spot where certain machine configurations cannot reveal local demagnetization through standard harmonic analysis, necessitating alternative diagnostic approaches.

Beyond frequency domain analysis, examination of stator currents in the dq reference frame provides complementary diagnostic information. The relationship between d-axis and q-axis currents can be visualized by plotting i_d vs. i_q, which produces an elliptical trajectory under healthy operating conditions. Demagnetization alters the characteristics of this ellipse, changing both its radius and shape [17]. The increasing current magnitude required to maintain constant torque under demagnetized conditions manifests as expansion of the elliptical trajectory. Analysis of these geometric changes provides an alternative fault indicator that can complement frequency-based diagnostics. However, this method exhibits dependency on the control system configuration, with different control strategies producing varying ellipse characteristics [35]. Furthermore, other rotor faults affect the dq current trajectory in ways that can be difficult to distinguish from demagnetization effects.

Despite widespread commercial adoption of motor current signature analysis, significant limitations constrain its reliability for comprehensive demagnetization monitoring in permanent magnet machines. Table 8.5 summarizes the principal challenges affecting MCSA-based demagnetization diagnosis and potential mitigation strategies.

Recent analytical investigations have revealed that the second harmonic of stator current may be fundamentally related to demagnetization phenomena, presenting a new diagnostic indicator

Table 8.5 MCSA limitations and challenges for demagnetization diagnosis.

Challenge	Impact on diagnosis	Mitigation strategies
Uniform demagnetization	No distinct harmonics; fault undetectable	Flux monitoring; signal injection; fundamental amplitude trending
Eccentricity harmonic overlap	Identical fault frequencies; misdiagnosis risk	Inductance analysis; multiload testing; vibration correlation; dq-axis trajectories
Winding configuration filtering	Fractional harmonics canceled in symmetrical windings	Neutral point voltage monitoring; asymmetric sensing
Manufacturing tolerances	Intrinsic harmonics obscure fault thresholds	Baseline characterization; adaptive thresholds; trend analysis
Load dependency	Harmonic amplitudes vary with torque	Multiload protocols; normalized fault indices; load-independent indicators
Speed dependency	Signal variation with speed; frequency overlap	Normalized ratios; order tracking; RLS extraction; VKF filtering

but also introducing challenges in fault discrimination. Detailed analytical examination of partial demagnetization harmonics in the stator current spectrum, considering the effects of stator winding configuration, pole number, and coils per phase, demonstrates that three-phase wye connection leads to additional cancellation of certain fault signatures. Since various rotor faults in permanent magnet machines manifest as functions of mechanical frequency, including eccentricity, misalignment, and bearing faults, identifying the origin of observed harmonics becomes crucial for planning appropriate corrective actions before catastrophic failure occurs. The second harmonic relationship to demagnetization creates potential ambiguity with stator faults, requiring careful analysis to distinguish between demagnetization-induced current signatures and those arising from other fault mechanisms.

Advanced signal processing techniques have been developed to enhance the extraction of demagnetization-related harmonic components from stator current measurements. An algorithm based on the recursive least squares method has been developed for detecting partial demagnetization under both stationary and nonstationary operating conditions [36]. This approach extracts fault-related harmonic components adaptively from the current signal and employs the ratio of these harmonics to the fundamental component as a diagnostic index for partial demagnetization. The recursive nature of the algorithm enables real-time implementation suitable for online condition monitoring applications. The consideration of harmonic ratios rather than absolute amplitudes provides improved robustness against variations in operating conditions that affect overall current magnitude.

Despite the widespread commercial adoption of MCSA and its advantages of noninvasiveness and integration with existing motor control systems, significant limitations must be recognized when applying this technique to demagnetization fault diagnosis in permanent magnet machines. The numerous challenges associated with MCSA for demagnetization detection, including insensitivity to uniform demagnetization, harmonic signature overlap with other fault types, cancellation effects in certain winding configurations, and sensitivity to manufacturing tolerances, collectively indicate that this method lacks the desirable reliability for comprehensive permanent magnet machine condition monitoring. These limitations do not entirely preclude the use of MCSA for demagnetization diagnosis, but they necessitate careful consideration of machine configuration, operating conditions, and the potential for complementary diagnostic methods. Further investigation combining MCSA with other diagnostic modalities represents a more promising path toward reliable demagnetization detection than reliance on current signature analysis alone.

8.4.3 Flux Monitoring Techniques

Demagnetization faults fundamentally manifest as reduction and distortion of magnetic flux in permanent magnet machines, making flux monitoring a direct and physically intuitive diagnostic approach. The general principles of motor flux signature analysis, including sensor technologies (Hall-effect sensors, search coils, fluxgate magnetometers, magneto-resistors, magnetostrictive sensors), installation positions (internal and external), and multisensory configurations are detailed in Chapter 4. This section focuses on the specific application of flux monitoring methods to demagnetization detection, emphasizing the trade-offs between measurement accuracy and implementation complexity that are particularly relevant for characterizing magnet degradation.

8.4.3.1 Direct Flux Measurement Methods

Table 8.6 compares principal methodologies for magnetic flux assessment in demagnetization diagnosis.

Table 8.6 Flux monitoring techniques comparison for demagnetization detection.

Method	Invasiveness	Detects uniform	Detects local	Accuracy	Online capable	Cost	Advantages	Limitations
Hall-effect sensors	High (air gap)	Yes	Yes	Very high	No (typically)	Medium	Excellent accuracy; quantitative measurement; spatial resolution	Requires disassembly; not for IPM; offline only
Gauss-meters	High (PM surface)	Yes	Yes	Very high	No	Medium	Precision; PM defect and crack detection	Requires rotor disassembly; offline testing only
Search coils	Medium-high	Yes	Yes	High	Yes	Low-medium	Direct flux; on-load capable; no advanced algorithms	Requires winding installation/retrofit
Magnetostrictive FBG sensors	Medium (slot wedge)	Yes	Yes	High	Yes	High	Continuous monitoring; minimal modification; quantitative severity	Retrofit complexity; specialized sensor; material cost
Fluxgate (stray flux)	None (external)	Yes	Yes	Medium	Yes	Low-medium	Noninvasive; external mounting; no disassembly	Small signal; environmental interference; reduced sensitivity
Model-based estimation	None	Yes	Yes	Medium	Yes	Very low	Noninvasive; existing sensors; real-time; no hardware	Parameter dependency; temperature sensitivity; limited low-severity
MEC reconstruction	None	Yes	Yes	Medium-high	Yes	Low	Computationally efficient vs. FEM; spatial localization	Requires initial FEM; reduced accuracy vs. full FEM
High-frequency injection	None	Yes	Yes	Medium-high	Yes	Low	Parameter/temperature estimation; inverter-integrated	Modeling complexity; operational disturbance
Stray flux measurement	None (external)	Yes	Yes	Medium	Yes	Medium	Noninvasive; external sensors; multidomain	Signal strength; positioning sensitivity; noise

Direct measurement using Gauss-meters and Hall-effect sensors positioned in the air gap or on PM surfaces provides the most accurate assessment of permanent magnet health, enabling detection of both local and uniform demagnetization as well as PM defects and cracks [2, 37]. These instruments can identify subtle flux distribution variations indicating incipient demagnetization. However, accessing the air gap or PM surfaces requires rotor disassembly, precluding online diagnosis during normal operation. Furthermore, these methods are inapplicable to interior permanent magnet designs where magnets are embedded within the rotor core.

To address the invasiveness limitations while retaining the advantages of direct flux measurement, *in situ* sensor designs have been developed that enable air gap magnetic flux density monitoring without major modifications to machine structure. A notable approach integrates a stator slot wedge with fiber Bragg grating pressure sensing technology using magnetostrictive materials such as Terfenol-D [38]. The magnetostrictive material experiences dimensional changes in response to the magnetic field in the air gap, and these mechanical deformations are detected by the fiber Bragg grating optical sensor. This sensor configuration allows for relatively easy installation and retrofitting to existing machines without invasive actions affecting the machine's core electromagnetic elements. A fault index derived from the *in situ* magnetic sensor measurements enables quantitative assessment of the relative severity of permanent magnet demagnetization faults. The primary advantage of this approach lies in its ability to provide continuous online monitoring of air gap flux with minimal machine modification, though the requirement to retrofit sensor-equipped slot wedges still represents a degree of invasiveness compared to purely noncontact methods.

Search coil methods, with configurations employing one coil per armature tooth for detailed spatial resolution, enable direct online flux measurement during demagnetization monitoring [39]. In unsaturated conditions, the search coil captures the vector sum of PM flux and armature winding flux; linear time-invariant filtering followed by signal processing separates these components and eliminates inverter switching noise. Measurements from multiple search coils reveal magnetic flux drops due to demagnetization in both local and uniform fault patterns. An important advantage for demagnetization diagnosis is that on-load and transient operating modes have minimal impact on detection capability, and no advanced pattern recognition algorithms are needed for basic fault detection. However, the requirement to install auxiliary windings during machine fabrication or through subsequent retrofit limits applicability primarily to new designs or high-value machines.

External stray flux monitoring using fluxgate sensors provides a noninvasive alternative that avoids internal access requirements. Analysis of stray flux patterns for demagnetization fault diagnosis has demonstrated reliability independent of machine configuration, number and location of defective permanent magnets, and operating point [40]. Fault patterns in the stray flux spectrum can be analyzed across different torque and speed conditions. However, the stray flux magnitude is typically much smaller than the internal air gap flux, limiting detection sensitivity for low-severity faults, and stray flux patterns can be influenced by external magnetic fields and nearby ferromagnetic structures.

For air-cored permanent magnet axial flux generators, auxiliary winding-based flux monitoring provides a specialized direct measurement approach, as discussed in detail in the Back-EMF Analysis section. This technique exploits the normalized peak-to-peak voltage in the auxiliary winding as a direct demagnetization severity metric, though it is limited to partial demagnetization detection [23].

8.4.3.2 Indirect Flux Estimation Through Machine Models

As discussed in Chapter 4, model-based parameter estimation methods leverage terminal voltage and current measurements to estimate machine parameters without additional sensors. For demagnetization diagnosis, these methods target flux-related quantities including flux linkage in PMSMs, torque constant in BLDC motors, magnetic coefficients, and teeth magnetic flux in multiphase machines [33, 41–44]. These indirect methods can diagnose both local and uniform demagnetization faults. However, the general confounding factors described in Chapter 4, particularly temperature-dependent resistance variations (30–40% or more during typical operation) and saturation-dependent inductance changes, pose specific challenges for demagnetization detection: the small flux changes associated with incipient demagnetization often fall within the estimation error margins caused by these parameter uncertainties.

Magnetic equivalent circuit models combined with field reconstruction methods provide a more sophisticated approach to flux estimation and fault localization. These reduced-order models map the winding flux waveform onto the magnet magnetomotive force function as a function of angular position, enabling analysis of both fault occurrence and spatial fault location [45]. For permanent magnet synchronous machines with parallel stator windings, fault diagnosis strategies based on current spectrum data can be developed, and fault severity indices for permanent magnet demagnetization can be defined. In series winding configurations, demagnetization of permanent magnets causes back-EMF reduction that appears directly in the resistive load current. For parallel connection generators, local demagnetization manifests as new frequency components that are odd multiples of the main stator current frequency. These current components circulate within the machine windings but do not necessarily appear in the phase currents measured at the terminals, requiring internal current sensing for detection. The reduced-order magnetic equivalent circuit approach enables fault position determination under both symmetrical and asymmetrical conditions with fewer computational variables and faster processing compared to full finite element analysis. The reduction of the magnetomotive force function value at the fault location serves as a fault diagnosis index, indicating both fault occurrence and severity. Field reconstruction methods employed in magnetic field analysis enable estimation of flux linkage in each stator tooth and phase, providing spatially resolved diagnostic information [44].

Comprehensive analysis of permanent magnet demagnetization impact on stator variables using magnetic equivalent circuits has investigated the influence of winding configuration on fault signatures [11]. The flexibility of magnetic equivalent circuit models enables determination of how each rotor permanent magnet affects the stator windings, allowing definition of no-load voltage signals in the presence of faults and observation of fault-related frequency spectrum components. The computational efficiency of this approach shortens simulation time while maintaining sufficient accuracy, enabling rapid evaluation of multiple demagnetization fault scenarios with different numbers of defective magnets, spatial locations, orientations, and winding configurations. Tests and simulations conducted at different speeds and loads reveal that fault signal amplitudes exhibit relatively weak dependency on torque but strong dependency on rotational speed, a characteristic that must be carefully considered in variable speed control applications. Beyond amplitude information, the phase angles of fault-related harmonics can be calculated, helping to clarify slight differences between current spectrum and no-load voltage spectrum characteristics.

8.4.3.3 Advanced Flux Parameter Estimation Techniques

Advanced estimation techniques combine high-frequency signal injection with sophisticated filtering algorithms to improve flux and temperature parameter estimation accuracy beyond what basic model-based methods achieve. PM temperature estimation through high-frequency injection and expectation-maximization particle filtering is discussed in detail in the Filter-based Parameter Tracking subsection below.

For dual permanent magnet linear motors experiencing demagnetization faults, an alternative signal processing approach exploits three magnetic field signals measured in the air gap region as demagnetization fault indices [46]. Finite element analysis generates these magnetic signal waveforms under various fault conditions, providing training data for the diagnostic algorithm. Continuous complex wavelet transform processing extracts signal envelopes for feature identification and fault classification. The wavelet transform decomposes the time-varying magnetic signals into time-frequency representations that enhance fault feature visibility. Comparison studies have demonstrated that continuous complex wavelet transform methods outperform currently used techniques such as extreme value methods and Hilbert–Huang transform for demagnetization fault diagnosis in linear motor applications. The superior performance stems from the wavelet transform's ability to provide localized time-frequency resolution that clarifies transient fault signatures.

For uniform demagnetization detection in linear motor applications, adaptive MRAS flux-linkage observers combined with dual closed-loop sliding mode control have been developed, as discussed in the Filter-based Parameter Tracking subsection [47].

The fundamental trade-off for demagnetization flux monitoring mirrors the general accuracy vs. accessibility trade-off discussed in Chapter 4 for MFSA methods. A key demagnetization-specific consideration is that uniform demagnetization detection generally requires either direct measurement methods or model-based flux linkage tracking, as many indirect approaches lose sensitivity when the fault maintains magnetic symmetry. The comparison in Table 8.6 summarizes these capabilities for each technique.

8.4.4 Signal Injection Methods for Demagnetization Assessment

Parameter estimation methods based on signal injection address fundamental limitations of conventional diagnostic approaches, particularly their sensitivity to temperature variations and magnetic saturation effects. Signal injection techniques not only overcome the model sensitivity problem against temperature and saturation but also enable precise estimation of fault severity through controlled excitation and systematic response measurement. These methods can be applied to all inverter-fed machines, where signals with programmable amplitude and frequency are injected through the power electronics converter. By processing the obtained response signals, machine parameters are estimated or diagnostic indices are extracted for fault detection [48].

The fundamental principle of signal injection for demagnetization assessment involves applying controlled electrical or magnetic excitation to the machine and analyzing the resulting electromagnetic response. By injecting different waveform signals into faulty machines, the impact of permanent magnet demagnetization manifests in other machine signals in the form of meaningful patterns that can be exploited for diagnostic purposes. These patterns differ systematically between healthy and demagnetized conditions, providing reliable fault indicators. The controlled nature of the injected signals enables diagnostic measurements to be performed under standardized conditions, reducing variability introduced by load fluctuations and operating point changes that complicate other diagnostic methods.

Several distinct signal injection approaches have been developed for demagnetization diagnosis, each exploiting different electromagnetic phenomena. One practical approach analyzes the sum of positive and negative peaks of triangular current waveforms produced by pulse-width modulation switching [2]. The asymmetry in these current peaks provides diagnostic information about magnetic flux distribution changes caused by demagnetization. Alternative methods measure the differential inductance of the d-axis through application of controlled PWM voltage patterns [49], or exploit variations in permanent magnet resistance at high frequencies by injecting a high-frequency rotational vector combined with pulsating sinusoidal voltage [50]. Each technique offers distinct advantages in terms of implementation complexity, sensitivity to different demagnetization patterns, and robustness under varying operating conditions.

Among these signal injection methods, the differential inductance measurement approach has received particular attention due to its sensitivity to magnetic saturation changes that accompany demagnetization. Differential inductance is defined as the slope of each point on the flux linkage vs. current characteristic curve, mathematically expressed as the derivative $d\lambda/dI$ [50]. This parameter provides information about the incremental magnetic response of the machine to current variations, which is directly influenced by the magnetization state of the permanent magnets. The differential inductance is estimated by injecting a stator AC field in the d-axis direction and calculating the reactance from the fundamental component of the voltage and current response. Since variations of d-axis differential inductance in the saturated machine are particularly visible and sensitive to magnetic state changes, the measurement is typically conducted with the operating point positioned in the saturated region of the magnetization characteristic.

To achieve operation in the saturated region where differential inductance variations are most pronounced, the AC excitation signal is superimposed on a DC field applied in the d-axis direction. This DC bias field shifts the operating point into the saturated region of the B–H characteristic, where the relationship between flux and current becomes nonlinear, and the differential inductance exhibits strong sensitivity to permanent magnet flux variations. Demagnetization reduces the total flux in the magnetic circuit, effectively shifting the saturation characteristic and altering the differential inductance values measured at specific current levels. The pattern generated by differential inductance variations in both linear and saturated regions serves as a diagnostic index for fault detection.

The diagnostic procedure involves measuring differential inductance L_d as a function of d-axis current I_d for both healthy and potentially faulty conditions. In the presence of demagnetization or other rotor faults such as eccentricity, the L_d vs. I_d characteristic curve shifts relative to the healthy baseline. The difference between the fault condition curve and the healthy case curve, denoted as ΔL_d, exhibits characteristic peaks that are used to diagnose the specific fault type and severity [49]. Demagnetization produces a distinctive ΔL_d pattern that differs from signatures of other rotor faults, enabling discrimination between fault types. The magnitude of the differential inductance variation correlates with demagnetization severity, providing quantitative fault assessment capability.

Figure 8.13 illustrates the differential inductance characteristic for healthy, demagnetized, and eccentricity conditions. The L_d vs. I_d curves show that demagnetization shifts the saturation knee to lower current magnitudes, as the reduced PM flux lowers the total flux level at which the magnetic circuit saturates. Eccentricity produces a different shift pattern due to the nonuniform air gap geometry. The difference ΔL_d between each fault condition and the healthy baseline exhibits characteristic peaks at distinct current levels, enabling discrimination between demagnetization and eccentricity faults.

The differential inductance method offers several significant advantages for demagnetization fault diagnosis. High sensitivity to magnetic state changes enables detection of relatively low-severity demagnetization that might be missed by other diagnostic techniques. The method exhibits independence from operating conditions such as load torque and rotational speed, as the measurement is performed through controlled injection rather than relying on operational signals that vary with operating point. No additional hardware beyond the existing inverter and current sensing infrastructure is required, reducing implementation cost and complexity. The ability to recognize different rotor fault types through distinctive ΔL_d patterns provides diagnostic specificity

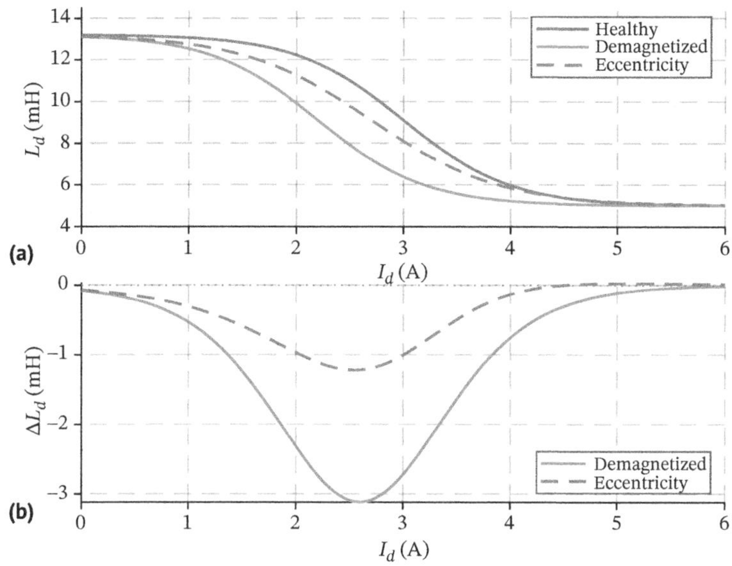

Figure 8.13 Differential inductance diagnostic characteristics for signal injection-based demagnetization assessment. (a) *D*-axis differential inductance L_d as a function of *d*-axis current I_d for healthy, demagnetized, and eccentricity conditions, showing the shift in saturation knee position. (b) Inductance difference ΔL_d relative to the healthy baseline, illustrating distinct peak patterns that enable fault type discrimination.

that distinguishes demagnetization from eccentricity, misalignment, and other mechanical faults. Critically, the method can diagnose both local and uniform demagnetization patterns, overcoming a fundamental limitation of many harmonic-based diagnostic approaches that are insensitive to symmetric fault conditions [49].

High-frequency injection methods extend signal injection concepts to enable simultaneous estimation of permanent magnet temperature and magnetic state. Since PM temperature directly affects demagnetization risk, temperature monitoring through high-frequency signal injection provides valuable complementary diagnostic information. The EM-PF and Kalman filter approach for PM temperature estimation through signal injection is discussed in detail under Filter-based Parameter Tracking in the Model-based Parameter Estimation section.

Field reconstruction methods, discussed in detail in the Multisignal Fusion section, can also be applied within a signal injection framework to optimize stator phase excitation for fault-tolerant control under demagnetization [51]. By combining field reconstruction with optimization algorithms, modified control strategies enable continued operation with maximum torque output and minimum ripple despite degraded magnetic conditions.

Despite the numerous advantages of signal injection methods for demagnetization assessment, certain limitations constrain their applicability in specific operational contexts. Continuous monitoring of permanent magnet quality is not possible with most signal injection techniques, as the measurement procedures typically require interruption of normal operation to apply the test excitation and acquire the diagnostic response. Consequently, signal injection methods are most suitable for systems with repeated stop and start cycles, where diagnostic measurements can be performed during startup sequences or scheduled maintenance intervals. For applications requiring continuous online monitoring during uninterrupted operation, alternative diagnostic approaches based on analysis of operational signals may be more appropriate, despite their potentially reduced sensitivity or increased complexity. The optimal diagnostic strategy often involves combining signal injection techniques for periodic detailed assessment with continuous monitoring methods for real-time fault detection, creating a comprehensive multimodal diagnostic framework. Table 8.7 compares the principal signal injection techniques for demagnetization assessment.

8.4.5 Model-based Parameter Estimation for Flux Linkage

Chapter 4 presents the general framework for model-based parameter estimation, including lumped-parameter approaches (EKF, RLS, Luenberger observers) and high-fidelity geometry-based methods (MEC, IWFT, hybrid FEM-analytical). This section examines how these estimation paradigms have been specifically adapted and extended for demagnetization severity quantification through flux linkage tracking, addressing the unique challenges of separating demagnetization-induced parameter changes from confounding thermal and saturation effects.

8.4.5.1 Observer-based Estimation Methods

For demagnetization detection, lumped-parameter models that separately represent PM and winding contributions to back-EMF can capture both local and uniform demagnetization phenomena [16]. When calibrated against healthy machine back-EMF waveforms through offline optimization, these models provide baselines against which fault-induced deviations are detected. The development of observer-based methods for demagnetization detection has focused particularly on eliminating the confounding effects of parameter uncertainties that can mask or mimic fault signatures. A notable approach employs a sliding mode disturbance observer specifically designed for interior permanent magnet motors, where the observer tracks the total disturbance caused by both flux linkage changes and stator parameter mismatches without directly constructing a flux linkage observer [52]. The methodology begins by formulating a motor model that explicitly includes disturbance terms representing demagnetization effects and parameter variations, then develops an SM observer to estimate this combined disturbance with stability guaranteed through Lyapunov analysis. The key innovation lies in a current-based extraction method that separates flux linkage disturbances from parameter-induced effects within the total observed disturbance, enabling computation of actual flux linkage and quantitative assessment of demagnetization severity.

Building upon the observer concept, robust fault-tolerant predictive current control algorithms have integrated hybrid observer architectures that combine the advantages of multiple estimation approaches [53]. These hybrid structures typically unite a sliding mode observer for robustness against uncertainties with a Luenberger observer for smooth state prediction, creating a complementary estimation framework. In the context of demagnetization diagnosis, the hybrid observer simultaneously estimates a compensating voltage to counteract parameter disturbances and predicts current values one step ahead to reduce control delay effects. The mathematical model underlying such observers explicitly represents parameter disturbances in inductance and demagnetization-induced flux variations, analyzing their impact on the current vector dynamics. By incorporating the observer-generated compensating voltage into the predictive current control law, these methods achieve the dual objectives of maintaining robust motor control performance while providing continuous parameter tracking for fault detection.

The integration of parameter estimation within the DQ reference frame offers particular advantages for permanent magnet synchronous machines operating under vector control [54]. In this approach, finite element analysis provides detailed characterization of the magnetic field distribution and consequent flux linkage and inductance variations in the DQ coordinate system under various demagnetization conditions. The FEM-derived parameter maps serve as lookup tables or analytical approximations within the control and diagnosis system, enabling real-time parameter

Table 8.7 Signal injection methods comparison for demagnetization assessment.

Method	Injected signal	Measured response	Detects uniform	Operating mode	Sensitivity	Hardware required	Advantages	Applications
PWM current peak analysis	Triangular PWM switching	Current peak asymmetry	Yes	Standstill, low speed	Medium-high	Standard inverter	Simple implementation; existing PWM; asymmetry indicator	Periodic testing; startup diagnostics
Differential inductance (d-axis)	AC + DC d-axis field	d-axis voltage and current for L_d	Yes	Standstill, rotating	High	Standard inverter	Detects uniform and local; fault discrimination; saturation sensitivity	Periodic assessment; automated test
High-frequency rotational vector	HF rotating voltage	HF current response	Yes	Any speed	High	Standard inverter with HF	PM resistance variation detection; speed-independent	Online with HF capability
Pulsating sinusoidal voltage	HF pulsating voltage	HF current harmonics	Yes	Standstill, rotating	Medium-high	Standard inverter	Magnetic state sensitivity; thermal assessment	Temperature and demagnetization monitoring
Field reconstruction + optimization	Optimized phase excitation	Torque, ripple, force	Yes	Under load	High	Standard inverter + optimization	Fault-tolerant control integration; performance optimization	Continued operation; compensation
EM-PF temperature estimation	HF signal injection	Temperature-dependent resistance	Yes (indirect)	Under load	High	Standard inverter	Simultaneous temperature and parameter estimation; thermal prediction	Thermal management; predictive maintenance

tracking through comparison of measured electrical quantities against model predictions. When combined with spectral analysis of stator currents and voltages using Fourier or wavelet transforms, this framework enables both detection of demagnetization onset and characterization of fault severity progression.

8.4.5.2 Filter-based Parameter Tracking

For demagnetization diagnosis, filtering techniques augment the machine state vector with flux linkage parameters, creating extended state-space representations where both dynamic states and slowly varying magnet parameters are estimated simultaneously. Spatial state models offer a sophisticated approach to permanent magnet temperature estimation, which serves as a proxy for thermal demagnetization assessment in applications where direct temperature measurement is impractical [24]. These models represent the spatial distribution of temperature through a multicriteria linear equation that captures temporal correlations, though the presence of unknown model parameters introduces nonlinearity into the estimation problem. The expectation-maximization particle filter addresses this challenge by simultaneously estimating both the temperature state and the model parameters, providing initial values that capture the system's nonlinear dynamics. Following this initialization phase, the state-space model can be linearized around the EM-PF estimates, enabling deployment of a Kalman filter for computationally efficient online temperature tracking. This two-stage approach balances the accuracy requirements of parameter initialization against the real-time constraints of continuous monitoring, with the EM-PF handling the complex offline estimation and the Kalman filter providing efficient online updates.

Model reference adaptive systems offer an alternative filtering architecture particularly suited to applications requiring consistent performance across varying operating conditions [47]. In the context of permanent magnet linear motors operating in high-precision motion control applications, dual closed-loop sliding mode control structures have been augmented with adaptive MRAS flux linkage observers to detect uniform demagnetization while maintaining speed performance. The adaptive observer employs an adaptive linear neuron (ADALINE) to adjust observer parameters in response to operating condition changes, ensuring that flux linkage estimation remains accurate despite variations in load, speed, or other operational factors. For detection of partial demagnetization under both stationary and nonstationary operating conditions, recursive least squares algorithms provide a computationally efficient alternative that extracts fault-related harmonic components from measured currents [36]. By tracking the ratio of these harmonic amplitudes relative to the fundamental component, RLS-based methods create a dimensionless fault index that remains relatively insensitive to operating point variations.

8.4.5.3 Advanced Modeling and Reconstruction Techniques

Beyond real-time observer and filter implementations, certain applications benefit from more detailed modeling approaches that trade computational complexity for enhanced spatial resolution or reduced dependence on extensive finite element simulations. Semi-analytical models based on permeance network representations offer a middle ground between purely analytical approaches and full finite element analysis [13]. These methods partition the machine geometry into a network of magnetic circuit elements, with permeances calculated from geometric and material properties following an initial FEM characterization of the flux paths. Once the permeance network is established, variations in magnet properties due to demagnetization can be evaluated with computation times substantially shorter than iterative FEM solution, while maintaining reasonable accuracy for many engineering purposes.

For applications requiring detailed spatial information about the magnetic field distribution, field reconstruction methods provide an efficient alternative to repeated three-dimensional finite element simulations [55]. The reconstruction approach begins with FEM computation of base magnetic flux density distributions produced by each independent source of magnetic field in the machine, typically evaluating a set of base current excitations. Once these fundamental field patterns are established, the flux density distribution for arbitrary operating conditions can be synthesized through linear superposition of the base patterns, weighted by the actual current values. This technique proves particularly valuable for axial flux permanent magnet machines, where three-dimensional field distributions are essential for accurate performance prediction, yet the computational burden of repeated 3D-FEM analysis for multiple operating points would be prohibitive. When demagnetization occurs, the altered field distribution can be detected by comparing reconstructed fields against measurements or by tracking changes in the superposition coefficients required to match observed behavior.

The general trade-offs among observer-based, filter-based, and advanced modeling approaches for parameter estimation are discussed in Chapter 4. For demagnetization diagnosis specifically, search coils positioned on specific machine components can improve estimation accuracy by providing localized flux measurements that reduce dependence on global machine parameters, though this reintroduces hardware modification [33, 41–44]. The selection among approaches for demagnetization applications depends particularly on whether the application requires detection of uniform demagnetization (favoring model-based flux linkage tracking) or spatial localization of local demagnetization (favoring geometry-based reconstruction methods).

8.4.6 Multisignal Fusion for Demagnetization Diagnosis

The general principles and architectures for multisignal fusion in fault diagnosis, including data-level, feature-level, and decision-level fusion strategies, are presented in Chapter 4. For demagnetization specifically, fusion techniques are particularly valuable because no single diagnostic modality reliably detects both local and uniform demagnetization across all operating conditions. Demagnetization-specific fusion implementations encompass three principal categories: neural network-based fusion of heterogeneous sensor data, observer-based integration of model predictions with measurements, and signal-level fusion combining electromagnetic, mechanical, and acoustic domains. Table 8.8 compares principal fusion architectures developed specifically for demagnetization diagnosis.

8.4.6.1 Neural Network-based Fusion Approaches

Addressing the limited labeled fault data challenge common to data-driven methods (discussed in Chapter 4), several semi-supervised and few-shot learning approaches have been developed specifically for demagnetization detection. One such approach employs a conditional random field-based graph attention network (CRF-GAT) specifically designed for fault detection from vibration or current signals acquired under varying load and speed conditions [56]. The methodology transforms raw time-domain signals into frequency-domain representations through FFT processing, then constructs a graph structure using the clustering adaptive neighbor (CAN) method to capture relationships among samples. The CRF-GAT architecture, a type of graph neural network, performs semi-supervised fault detection through graph learning, with optimization achieved through the variational expectation-maximization algorithm. This graph-based approach exhibits computational efficiency suitable for online implementation while achieving accuracy competitive with fully-supervised methods despite training on datasets containing flux density distributions from only three arbitrary lines in the air gap.

Table 8.8 Multisignal fusion approaches for demagnetization diagnosis.

Approach	Signal sources	ML method	Training data needs	Real-time capable	Key advantages	Best application
CRF-GAT (graph neural network)	Vibration or current FFT	Conditional random field – graph attention network	Semi-supervised (limited labeled)	Yes	Graph-based relationships; limited labels; VEM optimization	Variable load/speed with sparse fault examples
Acoustic + current fusion (BPNN)	Acoustic indices + current	Back propagation neural network	Supervised (moderate)	Yes	Noninvasive; parameter-insensitive; psychoacoustic indices	Electrical access limitations; quiet environments
Heterogeneous multisensor meta-learning	Vibration + current (RP-CSDP images)	Improved relation network	Few-shot learning	Yes	Heterogeneous sensors; rapid adaptation; unified 2D representation	Limited training data; varying conditions; multisensor
WSCN + SSDRB	Radial magnetic leakage flux	Wavelet scattering conv. + semi-supervised deep rule-based	Semi-supervised (small labeled)	Yes	Hierarchical features; deformation stability; small sample efficiency	Magnetic flux sensing; limited labeled data
Sliding mode observer (EID)	Voltage, current, speed	Equivalent input disturbance observer	Model-based (no ML training)	Yes	Physics-informed; fault-tolerant integration; GUUB stability; speed-independent	Control integration; model available; real-time needs
FEM + analytical hybrid	Current, voltage, FEM parameters	Fourier/wavelet + DQ model	Model-based (FEM pre-characterization)	Yes	High accuracy with real-time speed; physics-based; interpretable	High-accuracy with real-time constraints
Symbolic dynamics	Voltage + current time series	Symbolic sequence analysis	Model-based (discretization)	Yes	Computational robustness; qualitative features; discrete event	Efficiency priority; system identification focus
Multidomain field analysis	Flux, torque, vibration, acoustic	Field reconstruction + force analysis	Physics-based models	Partial	Comprehensive cross-domain correlation; root cause analysis	Detailed characterization; research applications

The fusion of acoustic and current signals presents another avenue for noninvasive demagnetization detection that avoids the parameter sensitivity challenges affecting many electrical signature-based methods. An approach utilizing seven objective and psychoacoustic indices has been developed to evaluate acoustic noise characteristics of healthy and demagnetized permanent magnet synchronous machines across various speed and load conditions [57]. The diagnostic framework employs a flux-based acoustic noise model in conjunction with a back propagation neural network (BPNN) trained on acoustic noise data to detect uniform demagnetization in surface-mounted permanent magnet machines. Laboratory experiments confirm that demagnetized motors emit abnormal and louder sounds compared to healthy units, with the seven psychoacoustic indicators providing quantitative metrics for evaluating the acoustic manifestations of demagnetization. The noninvasive nature of this acoustic-based approach offers particular advantages in applications where direct electrical access is limited, and its insensitivity to variations in electrical machine parameters enhances robustness across different operating conditions.

For applications involving heterogeneous multisensor systems, the challenge of computational complexity arising from diverse data types has been addressed through specialized fusion frameworks that transform disparate signal modalities into unified representations amenable to efficient processing. A few-shot cross-domain fault detection method based on meta-learning specifically targets the challenges of integrating vibration and current sensor data for demagnetization detection under limited training data availability [58]. The heterogeneous multisensor information fusion framework, termed low-pass pyramidal ratio-color symmetric dot pattern (RP-CSDP), reconstructs one-dimensional time series signals from both vibration and current sensors into two-dimensional color images, which are subsequently fused using the RP-CSDP algorithm. This transformation mitigates the exponential data increase typically caused by heterogeneous sensor information integration while creating a unified visual representation that captures both vibration and current characteristics. An improved relation network (RN) architecture provides the meta-learning component, enabling knowledge transfer across tasks and rapid adaptation to new operating conditions with minimal additional training. The meta-learning strategy proves particularly advantageous for fault detection under varying operating conditions, as the RN structure computes distances and matches degrees between samples through an embedding module for feature extraction and a correlation computation module for similarity assessment. Experimental validation with motors exhibiting local demagnetization of 10%, 30%, and 100% severity levels, as well as uniform demagnetization of 10%, demonstrates the framework's capability to accurately classify fault conditions from fused sensor data across different speeds and loads.

Advanced signal processing techniques that extract robust features from sensor data before fusion can further enhance diagnostic performance while maintaining reasonable computational requirements. A method based on wavelet scattering convolution networks (WSCN) combined with semi-supervised deep rule-based (SSDRB) classification addresses this objective by processing radial magnetic leakage flux signals derived from finite element analysis and magnetic equivalent circuit models [59]. To avoid complex signal processing operations, the symmetrical point pattern method converts one-dimensional fault signals into two-dimensional images suitable for convolutional network processing. The WSCN architecture extracts hierarchical features from these images through multiscale wavelet scattering transforms that provide stability to deformations and translations, while the SSDRB classifier performs fault detection using both labeled and unlabeled samples. Experimental validation with prototype motors confirms the superiority and efficiency of this approach using only small numbers of labeled samples under various operating conditions, demonstrating the value of combining advanced feature extraction with semi-supervised learning for practical deployment scenarios where extensive labeled fault data may be unavailable.

8.4.6.2 Observer-based Fusion Methods

Complementing the data-driven approaches, observer-based fusion methods integrate physical machine models with real-time measurements for physics-informed demagnetization diagnosis. The sliding mode observer with EID estimation, described in the Back-EMF Analysis section, exemplifies this paradigm by enabling simultaneous fault detection and fault-tolerant control through reconstruction of demagnetization-induced disturbances [26].

Hybrid FEM-analytical methods, discussed as a general paradigm in Chapter 4, have been specifically applied to demagnetization by computing magnetic field distributions, flux linkages, and DQ-frame inductances through finite element methods, then employing these parameters in analytical control models for real-time Fourier and wavelet analysis of stator currents and voltages [54, 60, 61]. These hybrid approaches achieve a compromise between the high accuracy of FEM and the computational speed needed for real-time demagnetization signature extraction.

8.4.6.3 Signal-level Fusion and Multidomain Analysis

Demagnetization manifests simultaneously across electromagnetic, mechanical, and acoustic domains, motivating signal-level fusion strategies that correlate signatures across these domains. Comparison of MCSA with vibration monitoring confirms complementary sensitivities for demagnetization: MCSA detects electrical asymmetry while vibration captures the mechanical consequences [62]. Precise interferometric measurements demonstrate that local magnet degradation causes asymmetrical radial forces leading to rotor path deviation, creating detectable vibration signatures. The presence of UMP and force harmonics in the system makes vibration and acoustic signal monitoring valuable diagnostic modalities, though these methods require permanent installation of vibration sensors on the machine. Modal analysis provides essential context for vibration-based demagnetization detection by characterizing resonance phenomena that amplify certain fault signatures. The analysis examines vibration spectra, acceleration, and displacement measurements to estimate resonance frequencies corresponding to peak response, with these modal frequencies indicating conditions where excitation forces approach natural frequencies of the stator reference system or where spatial harmonics of electromagnetic forces match stator side modes. Local demagnetization contributes to generation of low-mode vibrations indicative of unbalanced magnetic pull, with the amplitude pattern of acoustic noise power caused by demagnetization differing from signatures of other fault types in ways that enable fault discrimination [51, 63, 64].

Sophisticated signal processing techniques enable extraction of demagnetization signatures from torque ripple measurements combined with current analysis, providing diagnostic information even under nonstationary operating conditions where traditional spectral analysis methods struggle. A diagnostic approach based on permanent magnet machine torque ripple modeling that accounts for electromagnetic noise employs wavelet transformation to eliminate high-frequency electromagnetic disturbances from the torque converter, then applies Vold–Kalman filtering order tracking (VKF-OT) to extract torque ripple features reflecting magnet condition [65]. The order-tracked torque ripple features serve as inputs to a dynamic Bayesian network (DBN) trained for demagnetization detection and prediction during motor operation. This noninvasive online method offers the potential for embedding within motor controllers to provide continuous monitoring. Extension to partial demagnetization detection under nonstationary conditions has been achieved through tracking of demagnetization fault orders in stator current using the Vold–Kalman filter, with the fault order envelope amplitude serving as a robust fault indicator [66]. For enhanced time-frequency representation of fault signatures, the Choi–Williams distribution provides advanced processing capabilities despite the associated high computational load, enabling detailed characterization of fault evolution under transient operating conditions.

Symbolic dynamic methods represent an alternative approach to multisignal fusion that discretizes continuous sensor signals into symbolic sequences, enabling application of formal language theory and discrete event system analysis to fault diagnosis. A method based on symbolic dynamics for demagnetization monitoring converts voltage and current time series data into sequences of symbols through systematic discretization, achieving higher computational robustness and efficiency compared to continuous signal processing [41]. The symbolic representation provides a qualitative description of system features that facilitates system identification through inference of permanent magnet machine dynamic behavior. Demagnetization detection and severity estimation proceed by comparing the symbolic representation of observed system output behavior against the nominal expected behavior encoded in the symbolic model, with deviations providing quantitative fault estimates. Comparative evaluation against machine current signature analysis (MCSA) and particle filter-based fault diagnosis methods establishes the relative performance characteristics of the symbolic dynamic approach in terms of sensitivity, computational requirements, and robustness to operating condition variations.

Multidomain field analysis provides the most comprehensive fusion approach through coordinated electromagnetic, mechanical, and acoustic modeling that captures the full range of demagnetization effects across physical domains. An integrated analysis framework examining relationships among flux distribution, torque oscillations and ripple, vibration, and acoustic behavior of permanent magnet machines enables correlation of fault manifestations across domains [51]. Force analysis based on field reconstruction methods provides an efficient technique to examine electromagnetic field behavior relevant to electromechanical energy conversion and its acoustic consequences. Analysis reveals that local demagnetization elevates almost all force harmonics compared to healthy operation, directly linking electromagnetic field distortions to mechanical force generation and consequent acoustic emissions. However, the relationship between fault signatures and acoustic noise proves complex, as static eccentricity faults generate relatively high noise levels despite harmonic force patterns that differ from demagnetization, indicating that force harmonics alone do not deterministically predict noise generation. This recognition of the complex multidomain relationships governing fault manifestations underscores the value of fusion approaches that integrate multiple diagnostic modalities rather than relying on single-domain analysis.

The general selection principles among fusion architectures are discussed in Chapter 4. For demagnetization specifically, the choice is influenced by the need to detect both uniform and local demagnetization: neural network-based methods excel with sufficient training data, observer-based approaches leverage control system integration for real-time flux tracking, and multidomain analysis provides the strongest confirmation through cross-domain signature correlation. Hybrid architectures combining physics-based models with data-driven learning to represent the most promising direction for robust demagnetization monitoring across varying operating conditions.

8.4.7 Acoustic and Vibration Signatures

The general principles, sensor types, and signal processing methods for vibration and acoustic fault diagnosis are presented in Chapter 4. This section focuses on the specific vibration and acoustic manifestations of demagnetization faults, which arise from the asymmetric magnetic flux distribution creating unbalanced radial forces that translate into characteristic mechanical and acoustic signatures.

8.4.7.1 Vibration Monitoring and Analysis

Local demagnetization introduces asymmetrical radial forces that cause measurable rotor path deviation, as demonstrated through precision interferometer measurements [62]. The physical mechanism underlying this vibration generation stems from the UMP created by flux distribution asymmetry. When permanent magnet degradation affects specific rotor positions nonuniformly, the resulting spatial variation in magnetic flux density produces net radial forces that vary with rotor position, exciting mechanical vibrations at frequencies determined by the rotor speed and the spatial distribution of the demagnetized regions. Force harmonic analysis reveals that local demagnetization elevates nearly all force harmonics compared to healthy operation, directly linking electromagnetic field distortions to mechanical force generation [51]. However, the relationship between force harmonics and resultant vibration amplitude proves complex, as static eccentricity faults generate relatively high vibration levels despite harmonic force patterns that differ from demagnetization, indicating that force harmonics alone do not deterministically predict vibration severity across different fault types.

The electromagnetic behavior and consequent vibration characteristics depend significantly on the relative position of demagnetized permanent magnets within the rotor structure. Analysis of permanent magnet synchronous machines with oblique-fixed permanent magnet configurations requires to be specialized two-dimensional finite element methods that account for the three-dimensional geometry through equivalent geometrical transformations [10]. These modeling approaches enable prediction of position-dependent vibration characteristics while maintaining computational efficiency through implementation in conventional two-dimensional finite element software, offering advantages in simplicity, accuracy, and computational speed compared to full three-dimensional finite element analysis.

Modal analysis provides essential context for interpreting vibration measurements by identifying resonance phenomena that amplify specific vibration frequencies. The analysis examines vibration spectrum characteristics, acceleration profiles, and displacement measurements to estimate resonance frequencies of the stator support structure and frame. The frequencies corresponding to resonance peaks identify modal frequencies at which the mechanical system exhibits maximum response to exciting forces. Resonance conditions arise when the exciting frequency of electromagnetic forces approaches one of the natural frequencies of the stator reference system, or when the spatial harmonic order of the electromagnetic force matches stator-side vibration modes. Local demagnetization contributes to generation of low-mode vibrations that indicate the presence of unbalanced magnetic pull, providing a diagnostic indicator that distinguishes demagnetization from other fault types. The permanent mounting of vibration sensors on the machine enables continuous monitoring, though practical deployment typically reserves vibration analysis as a supplementary diagnostic method applied when acoustic emissions or severe vibration disorders become evident during operation.

8.4.7.2 Acoustic Emission Analysis

The acoustic signatures of demagnetization arise from the same electromagnetic force harmonics that generate vibration, as time-varying magnetic forces excite structural vibrations that radiate acoustic energy. Analysis of acoustic power patterns reveals that local demagnetization produces distinctive amplitude patterns that differ from eccentricity or bearing faults, enabling fault type differentiation [51, 63, 64].

Experimental observations confirm that permanent magnet machines operating with demagnetization faults emit abnormal and loud acoustic signatures compared to healthy machines [57]. To quantify these acoustic anomalies beyond simple amplitude measurements, seven objective and psychoacoustic indices provide comprehensive characterization of acoustic noise under various speed and load conditions. These indices capture both objective acoustic properties measured through standard sound level metrics and psychoacoustic characteristics that reflect human perception of sound quality, enabling assessment of whether demagnetization produces not only louder noise but also acoustically distinct signatures in terms of tonality, sharpness, roughness, and other perceptual dimensions. This multi-index approach proves particularly valuable for early fault detection where amplitude changes remain subtle but spectral or perceptual characteristics exhibit measurable deviations from healthy baseline behavior.

Advanced diagnostic architectures employ flux-based acoustic noise models combined with BPNN to enable online demagnetization detection from acoustic measurements [57]. The modeling approach establishes relationships between magnetic flux distribution characteristics and resultant acoustic emissions, providing a physics-informed foundation for the neural network training process. Detection proceeds by comparing measured acoustic signals against the predictions of the healthy machine model, with deviations in specific frequency bands or psychoacoustic indices indicating probable demagnetization. Laboratory validation demonstrates that the acoustic diagnostic method operates successfully as a noninvasive approach unaffected by variations in machine parameters, offering robustness advantages compared to electrical parameter-based methods that require compensation for resistance and inductance variations with temperature and saturation.

The practical limitations of acoustic monitoring discussed in Chapter 4 apply to demagnetization diagnosis as well, particularly regarding ambient noise in industrial settings. However, for demagnetization detection specifically, acoustic monitoring offers a valuable alternative when current signature analysis faces limitations such as uniform demagnetization invisibility or winding configuration harmonic cancellation effects, providing a complementary diagnostic pathway through an independent physical domain.

8.4.8 Experimental Validation and Case Studies

Experimental validation of demagnetization fault diagnosis techniques requires careful preparation of test specimens, precise measurement equipment, and systematic comparison between healthy and faulty machine behavior. Laboratory studies on permanent magnet synchronous motors provide critical verification of the theoretical models and simulation predictions discussed in previous sections, while also revealing practical challenges and considerations that arise in real-world fault detection scenarios.

8.4.8.1 Experimental Setup and Fault Implementation

A representative experimental validation study employed a 9-slot/8-pole surface-mounted permanent magnet motor with concentrated windings and all-teeth winding configuration. The motor specifications included rated power of 1.27 Nm at 3,000 rpm, phase resistance of 2,250 mΩ, self-inductance of 13.2 mH, and mutual inductance of negative 1.06 mH. The rotor diameter measured 28 mm with an air gap length of 0.5 mm and stack length of 60 mm. These specifications represent a typical small-scale permanent magnet motor suitable for laboratory investigation of demagnetization phenomena.

The experimental setup integrated several key components to enable comprehensive fault characterization, as illustrated in Figure 8.14. A high-end hysteresis dynamometer coupled mechanically with the motor shaft provided adjustable load application while measuring shaft torque and rotational speed with high accuracy. A three-phase motor drive supplied controlled excitation to the stator windings, implementing vector control strategy to regulate motor speed under varying load conditions. Data acquisition employed a 16-bit resolution acquisition card operating at 1 MS/s sampling rate, capturing voltage and current waveforms with sufficient bandwidth and precision

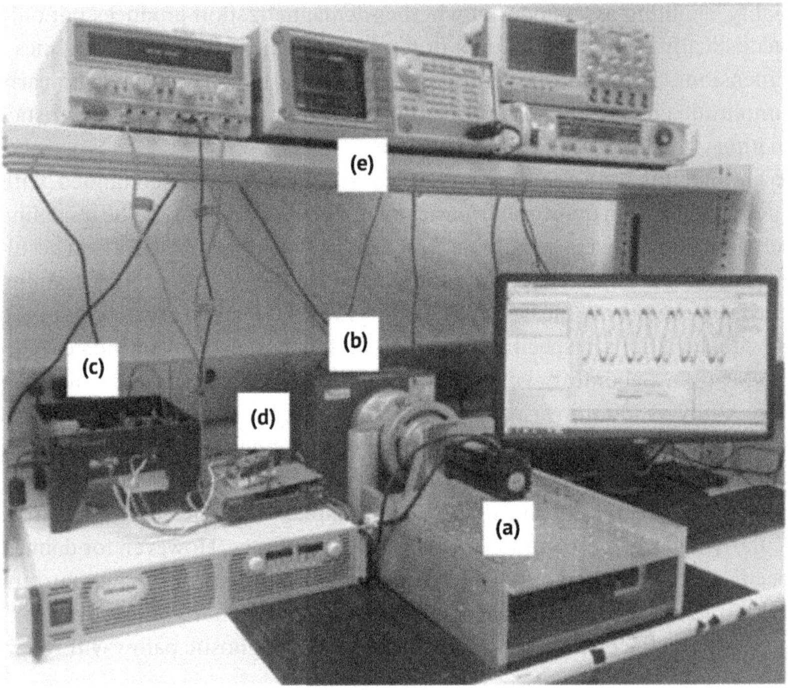

Figure 8.14 Experimental test setup for demagnetization fault validation showing (a) permanent magnet motor under test, (b) hysteresis dynamometer for load application and torque measurement, (c) three-phase motor drive with vector control implementation, (d) high-speed data acquisition system for voltage and current capture, and (e) spectrum analyzer for frequency domain analysis.

for subsequent frequency domain analysis. The measurement system recorded three-phase currents, line-to-line voltages, and rotor position signals synchronously, enabling reconstruction of electromagnetic quantities in both stationary and rotating reference frames.

Implementation of controlled demagnetization faults in laboratory conditions presents significant challenges compared to simulation environments where fault severity and location can be specified arbitrarily. The experimental study created localized demagnetization by carefully removing a portion of one rotor permanent magnet using a precision cutting tool, as shown in Figure 8.15. This mechanical removal approach produced a 25% reduction in magnet volume on a single pole, creating an asymmetric rotor condition that closely approximates the electromagnetic effects of partial irreversible demagnetization. While this mechanical fault implementation method differs from thermal or field-induced demagnetization mechanisms that occur in practice, it produces equivalent electromagnetic signatures because the fundamental phenomenon involves reduction of magnetic flux contribution from the affected pole, regardless of the physical mechanism causing that reduction.

The localized nature of the implemented fault deserves particular attention because it produces more complex electromagnetic signatures compared to uniform demagnetization affecting an entire pole. When only a portion of one magnet experiences flux reduction, the back-EMF waveform exhibits nonuniform amplitude reduction patterns that vary depending on which stator coils align with the defected magnet region at each rotor position. This spatial variation in flux reduction creates a richer harmonic spectrum compared to uniform faults, with the specific harmonic content depending on the fault span relative to the pole pitch and the winding distribution factor for each harmonic order.

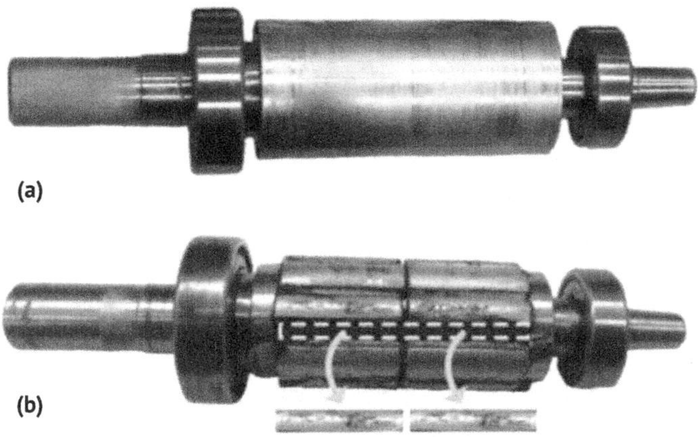

Figure 8.15 Rotor configurations for experimental demagnetization fault studies: (a) healthy rotor with complete permanent magnets enclosed in protective sleeve showing symmetric magnetic flux distribution, and (b) faulty rotor with one permanent magnet exhibiting 25% localized demagnetization created through precision material removal to simulate irreversible flux reduction.

8.4.8.2 Spectral Analysis of Back-EMF Signatures

Frequency domain analysis of measured back-EMF waveforms reveals the characteristic fractional harmonic signatures predicted by analytical models. For the 9-slot/8-pole motor configuration under study, demagnetization faults theoretically introduce sideband harmonics at frequencies given by $f_d = (1 \pm k/(P/2))f_e$ where k represents integer values, P denotes the pole pair count, and f_e represents the fundamental electrical frequency. This relationship predicts the appearance of harmonics at orders including 1/4, 2/4, 5/4, 7/4, and higher fractional values, in addition to the fundamental and existing integer harmonics present in healthy operation.

Experimental power spectra of phase back-EMF waveforms confirm the presence of all predicted fractional harmonic components when the motor operates with the 25% localized demagnetization fault. The spectrum measured from the faulty motor exhibits clearly visible peaks at fractional harmonic orders that remain absent or negligible in the healthy motor spectrum. Amplitude analysis shows that the dominant fault-related harmonics appear at orders 1/4 and 2/4, with progressively decreasing amplitudes for higher-order fractional components. The 5/4 and 7/4 harmonics remain visible above the noise floor but exhibit significantly lower magnitudes compared to the fundamental sideband components.

Comparison between localized and uniform demagnetization patterns reveals important differences in harmonic signature characteristics. When demagnetization affects only a portion of the pole pitch rather than the complete pole, the resulting harmonic spectrum exhibits higher amplitude fractional components, particularly for harmonics associated with three times the fundamental frequency. This phenomenon occurs because localized faults create sharper flux density transitions in the air gap compared to uniform faults, generating stronger high-frequency harmonic content. Additionally, localized demagnetization introduces harmonic components at specific orders that theoretical analysis predicts should vanish for uniform faults affecting complete poles. These spectral differences provide diagnostic information about fault spatial extent, potentially enabling discrimination between uniform thermal demagnetization and localized mechanical damage to magnets.

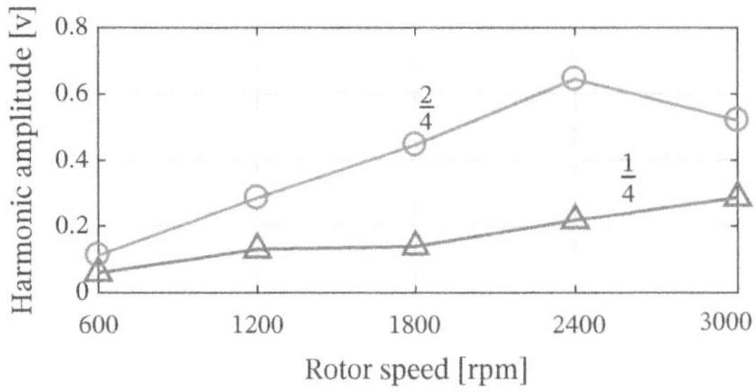

Figure 8.16 Variation of back-EMF sideband harmonic amplitudes across the operational speed range from 600 to 3,000 rpm for 25% localized demagnetization fault, demonstrating linear scaling relationship between harmonic magnitude and rotational speed for fractional orders 1/4 and 2/4, confirming that fault-induced harmonics originate from fundamental electromagnetic induction principles rather than nonlinear magnetic saturation or eddy current effects.

The amplitude of fractional harmonics in the back-EMF spectrum exhibits linear dependence on both fault severity and rotational speed, as illustrated in Figure 8.16. Experimental measurements at multiple speeds ranging from 600 to 3,000 rpm demonstrate that harmonic amplitudes scale proportionally with speed while maintaining constant ratios between different harmonic orders. The amplitude of the harmonics at the experimental speed range is approximately half of those predicted by the simulation model for 50% demagnetization severity, consistent with the 25% experimental fault severity. This proportional relationship confirms the linear scaling predicted by the lumped-parameter model in the preceding section. The rate of change in the amplitude of the harmonics at lower speeds is larger than that at higher speeds, a behavior also consistent with the model predictions. This linear scaling behavior confirms that the fractional harmonics arise from fundamental electromagnetic phenomena captured by the induced voltage law rather than nonlinear effects such as saturation or eddy currents.

An important observation from experimental back-EMF measurements concerns the presence of inherent asymmetries even in ostensibly healthy motors. Manufacturing tolerances, slight variations in magnet properties between poles, and minor eccentricities create small but measurable fractional harmonic components in the spectrum of motors without intentional faults. These manufacturing-related harmonics typically exhibit amplitudes one to two orders of magnitude below the fault-induced harmonics for 25% demagnetization severity, but they establish a baseline noise floor that limits the minimum detectable fault severity. Diagnostic algorithms must account for these inherent asymmetries through baseline measurements or statistical thresholds to avoid false alarm indications. The experimental power spectra presented in Figure 8.17 clearly demonstrate these fault signatures in both back-EMF and current measurements.

8.4.8.3 Current Signature Analysis and Validation

While back-EMF measurements provide direct assessment of magnetic flux variations, stator current analysis offers greater practical utility because current signals remain accessible during normal motor operation without requiring special test conditions. The fractional harmonic components present in back-EMF waveforms propagate into stator current spectra, but the star connection of stator windings filters certain harmonic orders that exhibit identical phase shifts across all three phases.

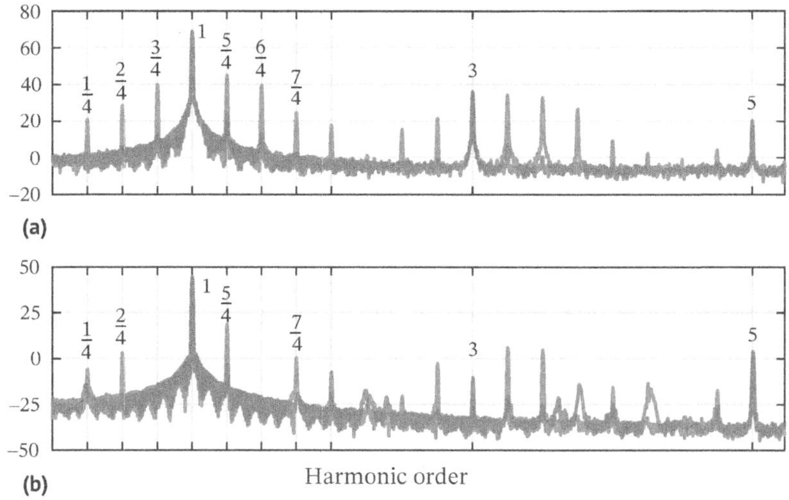

Figure 8.17 Experimental power spectra comparing healthy and faulty motor operation at nominal speed: (a) back-EMF frequency spectrum showing emergence of fractional harmonic components at orders 1/4, 2/4, 5/4, 7/4, and higher orders in the demagnetized condition, and (b) stator current frequency spectrum demonstrating selective appearance of fractional harmonics with suppression of 3/4 and 6/4 orders due to star-winding configuration while 1/4, 2/4, 5/4, and 7/4 harmonics remain clearly visible above the noise floor.

Experimental current spectra from the 9-slot/8-pole motor with 25% localized demagnetization reveal selective appearance of fractional harmonics compared to the complete set visible in back-EMF spectra. Specifically, the harmonics at orders 3/4 and 6/4 that appear prominently in back-EMF measurements vanish from the current spectrum because these components maintain zero-degree phase relationships in all three phases of the star-connected winding. This filtering behavior is precisely the same as predicted by the lumped-parameter model for the 9s/8p configuration in the frequency domain analysis section. The vectorial sum of three equal-amplitude, zero-phase-shift components equals zero, preventing these particular harmonics from producing circulating currents. In contrast, fractional harmonics at orders including 1/4, 2/4, 5/4, and 7/4 exhibit different phase relationships between phases and therefore appear clearly in measured current waveforms.

The selective filtering of certain harmonic orders by winding configuration has important implications for diagnostic algorithm design. Methods based exclusively on current signature analysis cannot detect all electromagnetic asymmetries visible through alternative measurement approaches such as search coil flux sensors or external magnetic field measurements. However, current-based detection offers significant practical advantages including noninvasive measurement through existing current sensors, immunity to mechanical vibration artifacts, and compatibility with motor drive control systems that inherently measure stator currents for feedback control purposes.

Comparison between simulation predictions and experimental current measurements demonstrates excellent agreement for the dominant fractional harmonic amplitudes. The measured magnitudes of 1/4, 2/4, 5/4, and 7/4 current harmonics match finite element simulation results within 5–10% error margins across the tested speed range. This close correspondence validates both the electromagnetic modeling approaches discussed in earlier sections and the analytical expressions derived for harmonic amplitude prediction. Minor discrepancies between measured and simulated harmonic content arise primarily from end-winding leakage effects not captured in two-dimensional finite element models and from measurement noise affecting the smallest harmonic components near the detection threshold.

Time-domain current waveforms exhibit characteristic amplitude modulation patterns that provide qualitative fault indicators complementary to frequency-domain harmonic analysis. The experimental measurements show that phase current peaks increase by approximately 8–12% compared to healthy operation when the demagnetized pole aligns with peak flux-linkage positions for each phase winding. This current increase compensates for reduced magnet flux contribution to maintain the electromagnetic torque demanded by the speed controller. The current amplitude variation repeats at the mechanical rotation frequency rather than the electrical frequency, creating a distinctive low-frequency envelope modulation visible in time-domain recordings. Automated fault detection algorithms can exploit this amplitude modulation as an additional diagnostic feature, particularly for variable-speed applications where frequency-domain analysis becomes complicated by time-varying fundamental frequency.

8.4.8.4 Validation of Model-based Diagnostic Approaches

The experimental validation study confirmed the effectiveness of lumped-parameter phase variable models for predicting electromagnetic behavior under demagnetization faults. Starting from measured back-EMF waveforms of the healthy motor, the model parameter identification process employed optimization algorithms to determine the trapezoidal waveform parameters including flat-top span, transition slope region span, and peak amplitude that best matched measured data. For the 9-slot/8-pole motor, the optimized parameters included transition region span of 0.74 radians, zero dead zone span, peak amplitude of 0.082 V at nominal speed, and Fourier series representation using 20 harmonic terms.

With these identified parameters, the phase variable model accurately reproduced both healthy and faulty motor behavior without requiring detailed geometric information about rotor and stator dimensions, magnet properties, or winding details beyond basic slot and pole counts. This capability demonstrates the practical value of such models for diagnostic applications where detailed machine design information may be proprietary or unavailable. The model predicted the appearance of specific fractional harmonic orders, their relative amplitudes, and their variation with fault severity and operating speed, with errors comparable to measurement uncertainty bounds.

Time-domain comparison between measured and modeled current waveforms shows that the lumped-parameter approach captures the essential dynamic behavior including current rise and fall rates, peak current magnitudes, and the low-frequency amplitude modulation induced by demagnetization. The model slightly underestimates current ripple amplitudes at high speeds due to neglect of saturation effects and inverter switching harmonics, but it reproduces the fundamental current components and fault-related fractional harmonics with sufficient accuracy for diagnostic purposes. Simulation time for the phase variable model proved approximately three orders of magnitude faster than equivalent finite element transient analysis, enabling real-time model-based fault detection implementations and rapid parameter sensitivity studies.

Observer-based diagnostic approaches that estimate rotor flux linkage from measured terminal quantities also received experimental validation. Implementations of extended Kalman filters and sliding mode observers successfully tracked the progressive amplitude reduction in estimated flux linkage as demagnetization severity increased. The estimated flux waveforms exhibited the expected asymmetry patterns with localized amplitude dips corresponding to the defected pole position. Comparison of estimated flux harmonic content against baseline reference values established during healthy operation enabled fault detection with sensitivity to approximately 10–15%

equivalent uniform demagnetization. This sensitivity threshold reflects the combined influences of measurement noise, model uncertainty, and inherent manufacturing asymmetries that create small flux variations even in healthy motors.

The experimental validation confirms that multiple diagnostic approaches, including frequency-domain current signature analysis, time-domain amplitude monitoring, and model-based flux estimation provide complementary information about demagnetization faults. No single method proves universally superior across all operating conditions and fault scenarios. Frequency-domain analysis excels for steady-state operation at constant speed and load where long measurement windows enable high-resolution spectral estimation. Time-domain amplitude monitoring offers faster response to transient fault development but requires careful threshold setting to balance sensitivity against false alarm rates. Model-based estimation methods provide fault localization information and operate effectively across varying speeds but depend critically on accurate parameter identification and computational resources for real-time observer implementation. Industrial diagnostic systems increasingly adopt multimethod fusion approaches that combine evidence from multiple diagnostic indicators to maximize detection reliability while minimizing false indications.

8.5 Summary

This chapter has presented a comprehensive treatment of permanent magnet demagnetization faults in brushless machines, encompassing failure mechanisms, modeling approaches, diagnostic techniques, and experimental validation. Demagnetization represents a particularly challenging fault mode because it degrades machine performance progressively while remaining difficult to detect through conventional monitoring approaches, particularly when degradation affects all poles uniformly. The economic significance of demagnetization faults stems from the high cost of permanent magnet materials and the potential for cascading failures that develop when initial flux reduction triggers electromagnetic-thermal feedback mechanisms. Understanding demagnetization phenomena requires integration of magnetic material science, electromagnetic field theory, thermal analysis, and fault diagnosis methodologies.

Permanent magnet demagnetization occurs when external factors force the operating point beyond the knee of the demagnetization curve into the nonreversible region, causing permanent reorientation of magnetic dipoles and reduced residual flux density. The primary failure modes include thermal stress-induced demagnetization resulting from temperature-dependent shifts in material coercivity, electrical fault-induced demagnetization generated by excessive armature currents producing strong demagnetizing fields, and manufacturing defect-related demagnetization arising from incomplete magnetization, improper adhesive bonding, material variations, mechanical damage, or corrosion. Demagnetization manifests in two distinct spatial patterns with fundamentally different diagnostic signatures. Uniform demagnetization affecting all poles equally maintains magnetic symmetry while reducing overall flux magnitude, making detection challenging as electrical signals remain balanced. Local demagnetization affecting specific poles or pole regions creates asymmetric flux distribution that generates harmonic components in voltages and currents and produces unbalanced magnetic pull forces on the rotor. Progressive demagnetization mechanisms create cascading failure through electromagnetic-thermal feedback where initial flux reduction requires higher stator current to maintain torque, which increases both demagnetizing magnetomotive force and resistive losses, leading to elevated temperature that further shifts the demagnetization curve and reduces thermal margins. The electromagnetic impact includes reduced air gap flux magnitude, introduction of spatial asymmetries in field distribution, decreased torque production capability, unbalanced magnetic pull forces, increasing bearing loading, distorted flux distribution introducing characteristic harmonic frequencies, and efficiency degradation through increased copper losses, altered core loss distribution, and power factor reduction.

Modeling and analysis of demagnetization faults employs multiple complementary approaches balancing accuracy against computational efficiency. Finite element methods provide the most comprehensive analysis with detailed magnetic field distribution accurately representing machine geometry, winding spatial distribution, slot effects, and nonlinear material behavior. Two principal finite element approaches exist, including fixed mesh modeling with sinusoidal field assumptions suitable for steady-state analysis and time-stepped modeling where the rotor physically rotates between time steps to capture transient behavior precisely. Studies examining permanent magnet breakage with different gap geometries reveal that demagnetization faults generate characteristic frequency components at integer multiples of mechanical frequency, with higher demagnetization levels significantly increasing even harmonics in stator current and torque ripple harmonics. Analytical modeling techniques provide computationally efficient alternatives, including magnetic equivalent circuits offering reduced-order modeling with shorter computation time, permeance network methods enabling rapid electromagnetic torque and induced voltage estimation, advanced analytical models based on vector potential formulation allowing detailed representation of local demagnetization patterns, and reluctance network models providing excellent compromise between accuracy and computation time while considering nonlinear characteristics and slot effects. Lumped-parameter models address limitations of long simulation times by incorporating winding and slot position effects directly in back-EMF calculations through Fourier series representations, with demagnetization effect modeled by multiplying an inverted pulse waveform with healthy coil back-EMF. This approach combines quick computation with sufficient accuracy, enabling implementation of different control techniques and facilitating fault-tolerant control studies. Frequency domain analysis reveals that demagnetization faults induce characteristic fractional harmonics following the relationship in stator current and back-EMF, with winding type significantly impacting demagnetization signatures. Harmonic component amplitudes increase linearly with both fault severity and rotor speed, with some harmonics neutralized in three-phase star-connected currents while others become more prominent depending on slot-to-pole ratio and winding configuration.

Fault diagnosis of demagnetization employs diverse methodologies addressing the challenge of monitoring magnetic flux degradation under varying operating conditions. Back-EMF analysis represents a physically intuitive diagnostic pathway directly reflecting permanent magnet flux reduction, with direct measurement during open-circuit operation providing accurate assessment but requiring specific test conditions, while model-based back-EMF estimation techniques overcome this limitation by reconstructing back-EMF from terminal measurements during normal operation. Motor current signature analysis leverages existing current sensors to detect demagnetization through characteristic fractional harmonic components, offering noninvasive online monitoring without additional hardware, though significant limitations include fundamental insensitivity to uniform demagnetization, harmonic signature overlap with other faults, and cancellation effects in symmetrical winding configurations. Flux monitoring techniques encompass direct measurement using Hall-effect sensors, search coils, and magnetostrictive sensors providing the highest accuracy but requiring invasive access, and indirect flux estimation through machine models enabling continuous online monitoring using existing measurements but with accuracy depending on temperature-sensitive and saturation-dependent parameters. Signal injection methods address fundamental parameter sensitivity limitations by applying controlled excitation and analyzing the electromagnetic response, with differential inductance measurement through d-axis signal injection exhibiting particular effectiveness by exploiting magnetic saturation changes accompanying demagnetization.

Model-based parameter estimation approaches provide physics-informed diagnostic capabilities integrating machine dynamics with real-time measurements, including hybrid observer architectures combining sliding mode observers with Luenberger observers, filter-based parameter tracking employing particle filters for temperature estimation, model reference adaptive systems with adaptive linear neurons maintaining accurate flux linkage estimation, and advanced field reconstruction methods offering efficient alternatives to iterative finite element analysis. Multi-signal fusion techniques leverage complementary information from diverse sensor modalities to overcome single-signal limitations, employing neural network-based fusion using graph attention networks for semi-supervised learning, acoustic and current signal fusion utilizing psychoacoustic indices, few-shot cross-domain meta-learning frameworks transforming heterogeneous sensor data into unified representations, and observer-based fusion methods integrating finite element analysis with analytical models. Vibration and acoustic monitoring provide complementary diagnostic modalities with local demagnetization generating asymmetrical radial forces creating measurable rotor path deviation, modal analysis identifying resonance phenomena where electromagnetic forces approach structural natural frequencies, and acoustic emission analysis offering noninvasive fault detection through microphone sensors with objective and psychoacoustic indices quantifying abnormal signatures.

Experimental validation employing controlled demagnetization in laboratory permanent magnet synchronous motors confirms theoretical predictions and validates diagnostic methodologies across multiple approaches. Frequency-domain analysis of measured back-EMF and current waveforms demonstrates characteristic fractional harmonic components with amplitudes exhibiting linear scaling with rotational speed and fault severity, validating analytical models predicting fundamental electromagnetic induction behavior. Lumped-parameter phase variable models accurately reproduce healthy and faulty motor behavior without detailed geometric information, achieving substantially faster computation than finite element transient analysis while maintaining diagnostic accuracy. Observer-based flux estimation methods successfully detect demagnetization with practical sensitivity thresholds reflecting measurement noise, model uncertainty, and manufacturing asymmetries. The experimental results confirm that no single diagnostic method proves universally superior across all operating conditions and fault scenarios, with frequency-domain analysis excelling for steady-state operation, time-domain monitoring offering faster transient response, and model-based estimation providing fault localization information. Industrial diagnostic systems increasingly adopt multi-method fusion approaches combining evidence from multiple indicators to maximize detection reliability while minimizing false indications, representing the most promising pathway toward robust demagnetization monitoring in practical applications where operating conditions vary widely, and diagnostic confidence requirements remain stringent.

References

1 Jacek F. Gieras and Mitchell Wing. *Permanent Magnet Motor Technology*. Marcel Dekker, New York, USA, 2002.

2 Jongman Hong, Doosoo Hyun, Sang Bin Lee, Ji-Yoon Yoo, and Kwang-Woon Lee. Automated monitoring of magnet quality for permanent-magnet synchronous motors at standstill. *IEEE Transactions on Industry Applications*, 46(4):1397–1405, July 2010. https://doi.org/10.1109/TIA.2010.2049972.

3 Ki-Chan Kim, Sang-Bae Lim, Dong-Hun Koo, and Ju Lee. The shape design of permanent magnet for permanent magnet synchronous motor considering partial demagnetization. *IEEE Transactions on Magnetics*, 42(10):3485–3487, October 2006. https://doi.org/10.1109/TMAG.2006.879077.
4 James D. McFarland and Thomas M. Jahns. Investigation of the rotor demagnetization characteristics of interior PM synchronous machines during fault conditions. *IEEE Transactions on Industry Applications*, 50(4):2768–2775, July 2014. https://doi.org/10.1109/TIA.2014.2309718.
5 Gunho Choi and Thomas M. Jahns. Interior permanent magnet synchronous machine rotor demagnetization characteristics under fault conditions. In *2013 IEEE Energy Conversion Congress and Exposition*, pages 2500–2507, Denver, CO, USA, 2013. IEEE. https://doi.org/10.1109/ECCE.2013.6647040.
6 Abdel A. El-Moneim, Andreas Gebert, Frank Schneider, Oliver Gutfleisch, and Ludwig Schultz. Grain growth effects on the corrosion behavior of nanocrystalline NdFeB magnets. *Corrosion Science*, 44(5):1097–1112, May 2002. https://doi.org/10.1016/S0010-938X(01)00123-8.
7 Babak Vaseghi, Noureddine Takorabet, and Farid Meibody-Tabar. Transient finite element analysis of induction machines with stator winding turn fault. *Progress In Electromagnetics Research*, 95:1–18, 2009. https://doi.org/10.2528/PIER09052004.
8 Ming Dai, Ali Keyhani, and Tom Sebastian. Fault analysis of a PM brushless DC motor using finite element method. *IEEE Transactions on Energy Conversion*, 20(1):1–6, March 2005. https://doi.org/10.1109/TEC.2004.841516.
9 Jawad Faiz, Bashir M. Ebrahimi, Bulent Akin, and Hamid A. Toliyat. Finite-element transient analysis of induction motors under mixed eccentricity fault. *IEEE Transactions on Magnetics*, 44(1):66–74, 2008. https://doi.org/10.1109/TMAG.2007.908479.
10 Juan-Carlos Urresty, Josep-Robert Riba, Luis Romeral, and Antonio Garcia. A simple 2-D finite-element geometry for analyzing surface-mounted synchronous machines with skewed rotor magnets. *IEEE Transactions on Magnetics*, 46(11):3948–3954, 2010. https://doi.org/10.1109/TMAG.2010.2062529.
11 Mohammad Zafarani, Tolga Goktas, and Bulent Akin. A comprehensive magnet defect fault analysis of permanent magnet synchronous motors. *IEEE Transactions on Industry Applications*, 52(2):1331–1339, March 2016. https://doi.org/10.1109/TIA.2015.2487440.
12 Jian Wang, Weiya Wang, Kais Atallah, and David Howe. Demagnetization assessment for three-phase tubular brushless permanent-magnet machines. *IEEE Transactions on Magnetics*, 44(9):2195–2203, September 2008. https://doi.org/10.1109/TMAG.2008.2001074.
13 Javed Farooq, Saïd Srairi, Abdelaziz Djerdir, and Abderrezak Miraoui. Use of permeance network method in the demagnetization phenomenon modeling in a permanent magnet motor. *IEEE Transactions on Magnetics*, 42(4):1295–1298, April 2006. https://doi.org/10.1109/TMAG.2006.870936.
14 Javed A. Farooq, Abdelaziz Djerdir, and Abderrezak Miraoui. Analytical modeling approach to detect magnet defects in permanent-magnet brushless motors. *IEEE Transactions on Magnetics*, 44(12):4599–4604, December 2008. https://doi.org/10.1109/TMAG.2008.2001751.
15 Tovo Raminosoa, Javed A. Farooq, Abdelaziz Djerdir, and Abderrezak Miraoui. Reluctance network modelling of surface permanent magnet motor considering iron nonlinearities. *Energy Conversion and Management*, 50(5):1356–1361, May 2009. https://doi.org/10.1016/j.enconman.2009.01.012.
16 Ehsan Mazaheri-Tehrani, Jawad Faiz, Mohsen Zafarani, and Bilal Akin. A fast phase variable *abc* model of brushless PM motors under demagnetization faults. *IEEE Transactions on Industrial Electronics*, 66(7):5070–5080, July 2019. https://doi.org/10.1109/TIE.2018.2868320.
17 Javier A. Rosero, Jordi Cusido, Angel Garcia, Juan A. Ortega, and Luis Romeral. Study on the permanent magnet demagnetization fault in permanent magnet synchronous machines. In *IECON 2006 - 32nd Annual Conference on IEEE Industrial Electronics*, pages 879–884. IEEE, 2006.

18 Yasser Gritli, Claudio Rossi, Domenico Casadei, Luca Zarri, and Fiorenzo Filippetti. Demagnetizations diagnosis for permanent magnet synchronous motors based on advanced wavelet analysis. In *2012 XXth International Conference on Electrical Machines*, pages 2397–2403. IEEE, 2012.

19 Domenico Casadei, Fiorenzo Filippetti, Claudio Rossi, and Andrea Stefani. Magnets faults characterization for permanent magnet synchronous motors. In *2009 IEEE International Symposium on Diagnostics for Electric Machines, Power Electronics and Drives*, pages 1–6. IEEE, 2009.

20 Hyung-Kyu Kim, Dong-Hyeok Kang, and Jin Hur. Fault detection of irreversible demagnetization based on space harmonics according to equivalent magnetizing distribution. *IEEE Transactions on Magnetics*, 51(11):1–4, November 2015. https://doi.org/10.1109/TMAG.2015.2439716.

21 Domenico Casadei, Michele Mengoni, Giovanni Serra, Angelo Tani, and Luca Zarri. Behavior of a five-phase surface-mounted permanent magnet motor under magnet demagnetization. In *8th IEEE Symposium on Diagnostics for Electrical Machines, Power Electronics and Drives*, pages 265–271. IEEE, 2011. https://doi.org/10.1109/DEMPED.2011.6063615

22 Karim Abbaszadeh, Seyedamin Saied, Siroos Hemmati, and Alberto Tenconi. Inverse transform method for magnet defect diagnosis in permanent magnet machines. *IET Electric Power Applications*, 8(3):98–107, March 2014. https://doi.org/10.1049/IET-EPA.2013.0176.

23 Giorgos A. Skarmoutsos, Konstantinos N. Gyftakis, and Markus Mueller. Detecting partial demagnetization in AFPM generators by monitoring speed and EMF induced in a supplemental winding. *IEEE Transactions on Industrial Informatics*, 18(5):3295–3305, May 2022. https://doi.org/10.1109/TII.2021.3053993.

24 Guodong Feng, Chunyan Lai, and Narayan C. Kar. Expectation-maximization particle-filter- and Kalman-filter-based permanent magnet temperature estimation for PMSM condition monitoring using high-frequency signal injection. *IEEE Transactions on Industrial Informatics*, 13(3):1261–1270, June 2017. https://doi.org/10.1109/TII.2016.2591509.

25 Jiaming M. Liu and Zi Q. Zhu. Improved sensorless control of permanent-magnet synchronous machine based on third-harmonic back EMF. *IEEE Transactions on Industry Applications*, 50(3):1861–1870, May 2014. https://doi.org/10.1109/TIA.2013.2284299.

26 Kaihui Zhao, Ruirui Zhou, Jinhua She, Changfan Zhang, Jing He, Gang Huang, and Xiangfei Li. Demagnetization-fault reconstruction and tolerant-control for PMSM using improved SMO-based equivalent-input-disturbance approach. *IEEE/ASME Transactions on Mechatronics*, 27(2):701–712, April 2022. https://doi.org/10.1109/TMECH.2021.3069787.

27 Luca Vancini, Michele Mengoni, Gabriele Rizzoli, Luca Zarri, and Angelo Tani. Local demagnetization detection in six-phase permanent magnet synchronous machines. *IEEE Transactions on Industrial Electronics*, 2023.

28 Jan De Bisschop, Peter Sergeant, Ahmed Hemeida, Hendrik Vansompel, and Luc Dupre. Analytical model for combined study of magnet demagnetization and eccentricity defects in axial flux permanent magnet synchronous machines. *IEEE Transactions on Magnetics*, 53(9):1–12, September 2017. https://doi.org/10.1109/TMAG.2017.2709267.

29 Julio-César Urresty, Jordi-Roger Riba, Miguel Delgado, and Luís Romeral. Detection of demagnetization faults in surface-mounted permanent magnet synchronous motors by means of the zero-sequence voltage component. *IEEE Transactions on Energy Conversion*, 27(1):42–51, March 2012. https://doi.org/10.1109/TEC.2011.2176127.

30 Gerald B. Kliman and J. Stein. Methods of motor current signature analysis. *Electric Machines & Power Systems*, 20(5):463–474, September 1992. https://doi.org/10.1080/07313569208909609.

31 William T. Thomson and Mark Fenger. Current signature analysis to detect induction motor faults. *IEEE Industry Applications Magazine*, 7(4):26–34, July 2001. https://doi.org/10.1109/2943.930988.

32 Wiehan le Roux, Ronald G. Harley, and Thomas G. Habetler. Detecting rotor faults in low power permanent magnet synchronous machines. *IEEE Transactions on Power Electronics*, 22(1):322–328, 2007. https://doi.org/10.1109/TPEL.2006.886620.

33 Satish Rajagopalan, Wiehan le Roux, Thomas G. Habetler, and Ronald G. Harley. Dynamic eccentricity and demagnetized rotor magnet detection in trapezoidal flux (brushless DC) motors operating under different load conditions. *IEEE Transactions on Power Electronics*, 22(5):2061–2069, September 2007. https://doi.org/10.1109/TPEL.2007.904183.

34 Taner Goktas, Mohsen Zafarani, and Bilal Akin. Discernment of broken magnet and static eccentricity faults in permanent magnet synchronous motors. *IEEE Transactions on Energy Conversion*, 31(2):578–587, June 2016. https://doi.org/10.1109/TEC.2015.2512602.

35 Haifang Yu, Zhiqiang Liu, and Shumei Cui. Research on fault analysis and diagnosis of PMSM in HEV. In *2011 International Conference on Electrical Machines and Systems*, pages 1–4. IEEE, 2011.

36 Ze Jiang, Xiaoyan Huang, and Wenping Cao. RLS-based algorithm for detecting partial demagnetization under both stationary and nonstationary conditions. *Energies*, 15(10):3509, May 2022. https://doi.org/10.3390/en15103509.

37 Matthaeus Brela, Markus Michalski, Hans-Joerg Gebhardt, and Joerg Franke. Hall measurement method for the detection of material defects in plastic-embedded permanent magnets of rotors. In *IECON 2013 - 39th Annual Conference of the IEEE Industrial Electronics Society*, pages 3928–3934, Vienna, Austria, 2013. IEEE. https://doi.org/10.1109/IECON.2013.6699763.

38 Anees Mohammed, Juan I. Melecio, and Sinisa Djurović. Electrical machine permanent magnets health monitoring and diagnosis using an air-gap magnetic sensor. *IEEE Sensors Journal*, 20(10): 5251–5259, May 2020. https://doi.org/10.1109/JSEN.2020.2969362.

39 Yao Da, Xiaodong Shi, and Mahesh Krishnamurthy. A new approach to fault diagnostics for permanent magnet synchronous machines using electromagnetic signature analysis. *IEEE Transactions on Power Electronics*, 28(8):4104–4112, August 2013. https://doi.org/10.1109/TPEL.2012.2227808.

40 Taner Goktas, Mohsen Zafarani, Kun Wang Lee, Bilal Akin, and Terry Sculley. Comprehensive analysis of magnet defect fault monitoring through leakage flux. *IEEE Transactions on Magnetics*, 53(4):1–10, April 2017. https://doi.org/10.1109/TMAG.2016.2617318.

41 Subhadeep Chakraborty, Eric Keller, Asok Ray, and Jeffrey Mayer. Detection and estimation of demagnetization faults in permanent magnet synchronous motors. *Electric Power Systems Research*, 96:225–236, March 2013. https://doi.org/10.1016/j.epsr.2012.11.005.

42 Jewon Lee, Yong-Ju Jeon, Doo-chul Choi, Seung H. Kim, and Sang W. Kim. Demagnetization fault diagnosis method for PMSM of electric vehicle. In *IECON 2013 - 39th Annual Conference of the IEEE Industrial Electronics Society*, pages 2709–2713, Vienna, Austria, 2013. IEEE. https://doi.org/10.1109/IECON.2013.6699559.

43 Jan De Bisschop, Ahmed Abouelyazied, Peter Sergeant, and Luc Dupré. Identification of demagnetization faults in axial flux permanent magnet synchronous machines using an inverse problem coupled with an analytical model. *IEEE Transactions on Magnetics*, 50(11):1–4, November 2014. https://doi.org/10.1109/TMAG.2014.2316851.

44 Amir Khoobroo and Babak Fahimi. A novel method for permanent magnet demagnetization fault detection and treatment in permanent magnet synchronous machines. In *2010 25th Annual IEEE Applied Power Electronics Conference and Exposition (APEC)*, pages 2231–2237, Palm Springs, CA, USA, 2010. IEEE. https://doi.org/10.1109/APEC.2010.5433492.

45 Cristian Ruschetti, Carlos Verucchi, Guillermo Bossio, Cristian De Angelo, and Guillermo García. Rotor demagnetization effects on permanent magnet synchronous machines. *Energy Conversion and Management*, 74:1–8, October 2013. https://doi.org/10.1016/j.enconman.2013.05.001.

46 Juncai Song, Jiwen Zhao, Xiaohu Zhang, Fei Dong, Jing Zhao, Liang Xu, and Zheng Yao. Accurate demagnetization faults detection of dual-sided permanent magnet linear motor using enveloping and time-domain energy analysis. *IEEE Transactions on Industrial Informatics*, 16(10):6334–6346, October 2020. https://doi.org/10.1109/TII.2019.2962730.

47 Jiwen Zhao, Lijun Wang, Liang Xu, Fei Dong, Juncai Song, and Xing Yang. Uniform demagnetization diagnosis for permanent-magnet synchronous linear motor using a sliding-mode velocity controller and an ALN-MRAS flux observer. *IEEE Transactions on Industrial Electronics*, 69(1): 890–899, January 2022. https://doi.org/10.1109/TIE.2021.3050360.

48 Byunghwan Kim, Kwanghwan Lee, Jinkyu Yang, Sang Bin Lee, Ernesto J. Wiedenbrug, and Manoj R. Shah. Automated detection of rotor faults for inverter-fed induction machines under standstill conditions. *IEEE Transactions on Industry Applications*, 47(1):55–64, January 2011. https://doi.org/10.1109/TIA.2010.2090931.

49 Jongman Hong, Sanguk Park, Doosoo Hyun, Tae june Kang, Sang Bin Lee, Christian Kral, and Anton Haumer. Detection and classification of rotor demagnetization and eccentricity faults for PM synchronous motors. *IEEE Transactions on Industry Applications*, 48(3):923–932, May 2012. https://doi.org/10.1109/TIA.2012.2191253.

50 David Diaz Reigosa, Daniel Fernandez, Zi-Qiang Zhu, and Fernando Briz. PMSM magnetization state estimation based on stator-reflected PM resistance using high-frequency signal injection. *IEEE Transactions on Industry Applications*, 51(5):3800–3810, September 2015. https://doi.org/10.1109/TIA.2015.2437975.

51 Dimitri Torregrossa, Amir Khoobroo, and Babak Fahimi. Prediction of acoustic noise and torque pulsation in PM synchronous machines with static eccentricity and partial demagnetization using field reconstruction method. *IEEE Transactions on Industrial Electronics*, 59(2):934–944, February 2012. https://doi.org/10.1109/TIE.2011.2151810.

52 Yaofei Han, Shaofeng Chen, Chao Gong, Xing Zhao, Fanggang Zhang, and Yunwei Li. Accurate SM disturbance observer-based demagnetization fault diagnosis with parameter mismatch impacts eliminated for IPM motors. *IEEE Transactions on Power Electronics*, 38(5):5706–5710, May 2023. https://doi.org/10.1109/TPEL.2023.3245052.

53 Changfan Zhang, Gongping Wu, Fei Rong, Jianghua Feng, Lin Jia, Jing He, and Shoudao Huang. Robust fault-tolerant predictive current control for permanent magnet synchronous motors considering demagnetization fault. *IEEE Transactions on Industrial Electronics*, 65(7):5324–5334, July 2018. https://doi.org/10.1109/TIE.2017.2774758.

54 Takeo Ishikawa, Yutaro Seki, and Nobuyuki Kurita. Analysis for fault detection of vector-controlled permanent magnet synchronous motor with permanent magnet defect. *IEEE Transactions on Magnetics*, 49(5):2331–2334, May 2013. https://doi.org/10.1109/TMAG.2013.2243135.

55 Ehsan Ajily, Mohammad Ardebili, and Karim Abbaszadeh. Magnet defect and rotor eccentricity modeling in axial-flux permanent-magnet machines via 3-D field reconstruction method. *IEEE Transactions on Energy Conversion*, 31(2):486–495, June 2016. https://doi.org/10.1109/TEC.2015.2506819.

56 Yao Tang, Xiaofei Zhang, Yujia Zhai, Guojun Qin, Dianyi Song, Shoudao Huang, and Zhuo Long. Rotating machine systems fault diagnosis using semisupervised conditional random field-based graph attention network. *IEEE Transactions on Instrumentation and Measurement*, 70:1–10, 2021. https://doi.org/10.1109/TIM.2021.3091212.

57 Min Zhu, Wensong Hu, and Narayan C. Kar. Acoustic noise-based uniform permanent-magnet demagnetization detection in SPMSM for high-performance PMSM drive. *IEEE Transactions on Transportation Electrification*, 4(1):303–313, March 2018. https://doi.org/10.1109/TTE.2017.2755549.

58 Xiaofei Zhang, Jingbo Tang, Yinpeng Qu, Guojun Qin, Lei Guo, Jinping Xie, and Zhuo Long. Few-shot fault diagnosis based on heterogeneous information fusion and meta learning. *IEEE Sensors Journal*, 23(18):21433–21442, September 2023. https://doi.org/10.1109/JSEN.2023.3299707.

59 Fengqin Huang, Xiaofei Zhang, Guojun Qin, Jinping Xie, Jian Peng, Shoudao Huang, Zhuo Long, and Yao Tang. Demagnetization fault diagnosis of permanent magnet synchronous motors using magnetic leakage signals. *IEEE Transactions on Industrial Informatics*, 19(4):6105–6116, April 2023. https://doi.org/10.1109/TII.2022.3165283.

60 Murat Yilmaz. Limitations/capabilities of electric machine technologies and modeling approaches for electric motor design and analysis in plug-in electric vehicle applications. *Renewable and Sustainable Energy Reviews*, 52:80–99, December 2015. https://doi.org/10.1016/j.rser.2015.07.033.

61 Ali Sarikhani and Osama A. Mohammed. Inter-turn fault detection in PM synchronous machines by physics-based back electromotive force estimation. *IEEE Transactions on Industrial Electronics*, 60(8):3472–3484, August 2013. https://doi.org/10.1109/TIE.2012.2222857.

62 Julio-Cesar Urresty, Reza Atashkhooei, Jordi-Roger Riba, Luis Romeral, and Santiago Royo. Shaft trajectory analysis in a partially demagnetized permanent-magnet synchronous motor. *IEEE Transactions on Industrial Electronics*, 60(8):3454–3461, August 2013. https://doi.org/10.1109/TIE.2012.2213565.

63 Zhi Yang, Xiaodong Shi, and Mahesh Krishnamurthy. Vibration monitoring of PM synchronous machine with partial demagnetization and inter-turn short circuit faults. In *2014 IEEE Transportation Electrification Conference and Expo (ITEC)*, pages 1–6. IEEE, 2014.

64 Ali Rezig, Mohammed Rachid Mekideche, and Abdesslem Djerdir. Impact of eccentricity and demagnetization faults on magnetic noise generation in brushless permanent magnet DC motors. *Journal of Electrical Engineering and Technology*, 6(3):356–363, May 2011. https://doi.org/10.5370/JEET.2011.6.3.356.

65 Min Zhu, Bin Yang, Wensong Hu, Guodong Feng, and Narayan C. Kar. Vold–Kalman filtering order tracking based rotor demagnetization detection in PMSM. *IEEE Transactions on Industry Applications*, 55(6):5768–5778, November 2019. https://doi.org/10.1109/TIA.2019.2932692.

66 Chao Wang, Miguel Delgado Prieto, Luis Romeral, Zhe Chen, Frede Blaabjerg, and Xiao Liu. Detection of partial demagnetization fault in PMSMs operating under nonstationary conditions. *IEEE Transactions on Magnetics*, 52(7):1–4, 2016.

9

Power Converter Faults in PM Machine Drives

9.1 Introduction

Most permanent magnet motors are driven by power converters, specifically variable speed drives, to achieve continuous and efficient operation across a wide range of speeds and torques. The power converter, as an integral component of the motor drive system, performs the critical function of converting DC power from the supply or energy storage system into variable frequency, variable amplitude three-phase AC power to control the motor. A typical motor drive system comprises a power electronic converter with semiconductor switches, gate driver circuits, DC-link capacitors, control circuits with sensing components, and protection systems (Figure 9.1).

Power converters and their constituent components, including semiconductors such as insulated gate bipolar transistors (IGBTs), metal-oxide-semiconductor field-effect transistors (MOSFETs), and diodes, gate drivers, circuit boards, sensors, and DC-link capacitors, are susceptible to various failure modes. Recent statistical analyses of power converter failures in industrial applications reveal that 38% of overall drive system failures are attributable to power electronic converters. Within the converter itself, 58% of failures stem from control circuit malfunctions. When examining power converter components specifically, semiconductor devices and their soldering, capacitors, and printed circuit boards account for 38%, 30%, and 26% of all converter failures, respectively [1]. These statistics underscore the critical role that semiconductor devices play in the reliability and availability of power converters and motor drive systems.

The failure of power converter components can have significant ramifications for permanent magnet motor operation. Unlike some motor-side faults that may develop gradually, converter faults can lead to immediate performance degradation, increased harmonic distortion in motor currents, elevated torque ripple, reduced efficiency, and in severe cases, complete system shutdown. Furthermore, converter faults can interact with and accelerate the degradation of motor components. For instance, harmonic currents resulting from faulty switching can increase motor winding temperatures and core losses, while asymmetric voltage supply due to switch failures can induce unbalanced magnetic forces similar to eccentricity conditions.

This chapter provides comprehensive coverage of power converter faults in permanent magnet machine drives. The failure modes and mechanisms section examines the physical degradation processes affecting power semiconductor devices, passive components, and auxiliary circuits. The modeling and analysis section develops mathematical representations of faulty converter operation and their electromagnetic interaction with the permanent magnet motor. The fault diagnosis section presents detection methodologies, signal processing techniques, and fault-tolerant control strategies for maintaining operation under degraded conditions. Understanding these

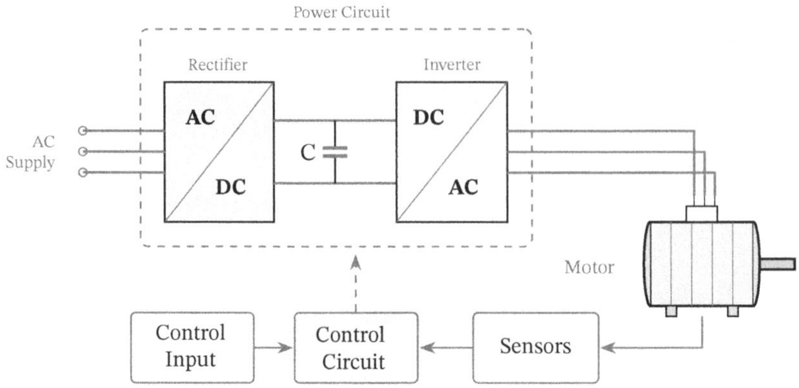

Figure 9.1 A typical motor drive system.

converter fault phenomena is essential for developing robust motor drive systems capable of reliable operation in demanding industrial applications.

9.2 Failure Modes and Mechanisms

9.2.1 Overview of Power Converter Reliability

Power electronic converters represent a critical reliability bottleneck in permanent magnet motor drive systems. The multilayer structure of power modules, operating under repetitive thermal cycling, electrical stress, and mechanical vibration, experiences progressive degradation that eventually manifests as functional failures. Understanding the physical mechanisms underlying these failures is essential for developing effective condition monitoring strategies and improving converter design for enhanced reliability.

9.2.2 IGBT and MOSFET Switch Failures

IGBTs are widely employed as semiconductor switching devices in electric drive applications above 1 kV and 1 kW, owing to their efficient switching performance combined with relatively low conduction losses [1]. IGBTs are available in various configurations and package types to suit different power levels and thermal management requirements. In motor drive applications, IGBTs are typically employed in modular form, where one or more semiconductor switches comprising IGBTs and freewheeling diodes are packaged together. The module packaging technology varies significantly depending on the required heat dissipation capabilities and power rating.

9.2.2.1 IGBT Module Structure and Thermomechanical Stress

The structure of a typical IGBT module comprises multiple layers with different thermal and mechanical properties, as illustrated in Figure 9.2. From bottom to top, the structure includes a heat-dissipating base (heatsink), a layer of thermal grease filling the gap between the heatsink and module, a copper baseplate, a ceramic substrate (typically Direct Copper Bonded or DCB), semiconductor chips, and aluminum bond wires connecting the chips to the copper traces on the substrate [1].

Table 9.1 summarizes the layer structure and characteristics of IGBT modules, highlighting the material properties that contribute to thermomechanical stress.

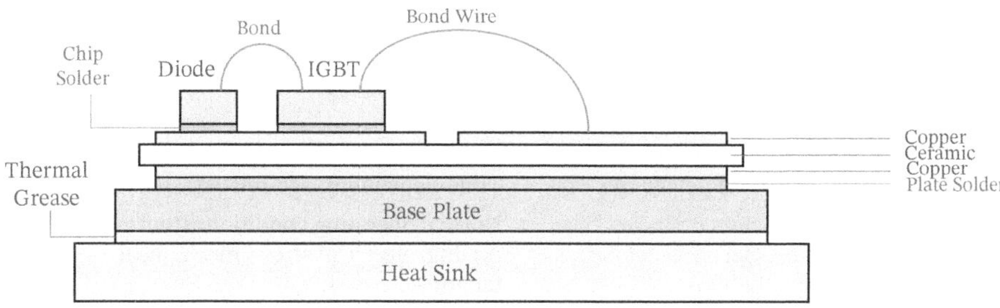

Figure 9.2 IGBT multi-chip module structure.

Table 9.1 IGBT module multilayer structure and characteristics.

Layer	Component/material	Primary function	Key characteristic
1	Heat-dissipating base (heatsink)	Heat removal	High thermal conductivity
2	Thermal grease	Thermal interface	Gap filling, thermal transfer
3	Copper baseplate	Mechanical support, heat spreading	High CTE
4	Ceramic substrate (DCB)	Electrical isolation, heat transfer	Low CTE, high dielectric strength
5	Semiconductor chips (IGBT/diode)	Power switching	Silicon, medium CTE
6	Aluminum bond wires	Electrical connection	Very high CTE

The multilayer structure of IGBT modules significantly impacts the thermomechanical stress experienced by module components. Mismatches in coefficient of thermal expansion (CTE) between adjacent layers, combined with temperature variations during operation, generate mechanical stress at material interfaces. The most significant CTE mismatch occurs between aluminum bond wires and silicon chips, presenting a critical reliability challenge due to their direct contact. Bond wires consequently face a higher risk of failure compared to other components in IGBT modules. However, the bond wire regions in contact with copper traces on the substrate experience smaller temperature swings and reduced mechanical stress.

Additional CTE mismatches exist between the ceramic substrate and the aluminum oxide-copper baseplate, as well as between silicon chips and the ceramic substrate. These mismatches subject the solder layers between chips and substrate, and between substrate and baseplate, to repetitive thermomechanical stress during thermal cycling. The magnitude of stress depends on the temperature swing amplitude, cycle frequency, medium junction temperature, and the mechanical properties of the materials involved.

9.2.2.2 Failure Classification: Random Failures and Wear-out Failures

For IGBT modules, various environmental and operational stresses can impact performance and reliability, including temperature variations, voltage transients, current overload, mechanical vibration and shock, humidity, and cosmic radiation. Failure mechanisms in power semiconductor devices are broadly classified into two categories: random failures and wear-out failures.

Random failures are characterized by sudden, catastrophic events that exceed the device strength threshold. These failures are not influenced by the device age or accumulated operational time,

Table 9.2 Comparison of random failures and wear-out failures in power semiconductor devices.

Characteristic	Random failures	Wear-out failures
Nature	Sudden, catastrophic	Gradual, progressive
Time dependence	Independent of device age	Time-dependent, age-related
Triggering factors	External factors (voltage transients, electrical overstress, particle radiation)	Normal operating conditions (thermal cycling, electrical stress, mechanical fatigue)
Manifestation	Immediate destruction (short-circuit or open-circuit)	Gradual parameter degradation before failure
Predictability	Unpredictable	Potentially predictable with monitoring
Examples	Voltage transients, electrical overstress, cosmic radiation	Bond wire lift-off, chip metallization reconstruction, solder fatigue, DCB cracking, gate oxide breakdown
Mitigation approach	Conservative design margins, enhanced protection systems	Condition monitoring, predictive maintenance
Monitoring opportunity	None (occurs too rapidly)	Yes (gradual degradation observable)

but rather are triggered by external factors such as voltage transients, electrical overstress, and particle radiation leading to electrical discharge. Random failures typically result in immediate device destruction with short-circuit or open-circuit manifestation.

Wear-out failures, in contrast, occur due to the accumulation of incremental physical damage beyond the device endurance limits under normal operating conditions. These failures are time-dependent and accelerated by thermal cycling, electrical stress, and mechanical fatigue. Examples of wear-out failures include bond wire heel cracking and lift-off, chip metallization reconstruction, solder layer fatigue and delamination, cracking of the DCB substrate, and gate oxide breakdown. Wear-out failures often exhibit a gradual degradation of device parameters before complete failure, providing an opportunity for condition monitoring and predictive maintenance. Table 9.2 summarizes the key distinctions between these two failure categories.

To address random failures, the average failure rate is typically considered based on the IGBT module type and stress levels induced by abnormal overload conditions such as overcurrent, overvoltage, over-temperature, and short-circuit. Research in this area continues, as the root causes of some random failure mechanisms are not yet fully understood. Current approaches to managing random failures include enhancing device performance through conservative design margins and reducing electrical and thermal stress through improved thermal management and protection systems [1].

9.2.2.3 Chip-related Failures

Chip-related failures are intrinsic to the semiconductor and can potentially destroy the device. They are driven by over-stress conditions and wear-out mechanisms, and may be accelerated by structural changes in the package that affect thermal impedance and junction temperature. The primary chip-related failure mechanisms include:

- **Time-dependent Dielectric Breakdown (TDDB):** Chronic defect accumulation in the gate oxide layer, driven by high temperature and strong electric fields, produces hot carriers that become trapped in the oxide. This progressively degrades gate controllability by shifting the threshold voltage $V_{GE(th)}$, transconductance g_m, and saturation current I_S, eventually resulting in complete gate oxide breakdown [1].

- **Electromigration:** High current densities displace metal atoms in silicon interconnects and metallization layers, forming voids that create open circuits or high-resistance paths. In IGBT chips, this primarily affects the aluminum metallization and can interact with bond wire degradation to accelerate failure.
- **Transient Electrical Stresses:** Electrostatic discharge (ESD), electrical overstress (EOS), and power supply transients subject the semiconductor junction to brief but intense electric fields. While modern IGBT designs incorporate protection structures, repeated exposure causes cumulative damage.
- **Over-temperature Effects:** Commercially available silicon power chips have a maximum junction temperature specification of $T_j = 175\,°\text{C}$. As junction temperature rises, failure rates increase exponentially, and thermal runaway can occur when increased leakage current generates additional heat in a positive feedback loop [1].
- **Cosmic Radiation:** High-energy particles create electron-hole pairs that can trigger spurious turn-on events or device damage, imposing limitations on maximum DC blocking voltage and requiring derating of up to 50% in high-voltage applications [1].

9.2.2.4 Package-related Failures

Package-related degradation often poses more significant challenges in motor drive applications than chip-related failures. These faults elevate operating temperatures, leading to degraded safe operating area (SOA) and compromised device performance. Specifically, switching transitions experience increased stress in cases of bond wire and die-attach degradation.

In standard power electronic modules, thermal limitations are equally critical as maximum electrical ratings. The reliability and operational lifetime of the system depend on the component temperature profile, as well as the chip and packaging technology employed. Bond wire fatigue and solder fatigue, both in the die-attach and DCB substrate solder layers, are identified as critical failure mechanisms due to thermomechanical stress in IGBT power modules.

The main reliability issues arise from the modular packaging multilayered structure, with each layer possessing its own coefficient of thermal expansion. Shear stress is generated within the assembled structure due to CTE mismatch between dissimilar materials combined with temperature cycling, leading to thermal deformation. Repetitive thermomechanical stress at material interfaces results in fatigue, wear-out, and eventual failures. While thermomechanical stresses are theoretically distributed throughout the entire power module package, reliability testing and field experience have identified three major regions prone to wear-out and fatigue failures:

- Chip front side connection (bond wire-to-silicon interface).
- Chip-to-substrate connection, also known as die-attachment (chip-solder joint).
- Substrate-to-baseplate connection (DCB substrate-solder joint).

Die-attachment exhibits comparatively lower CTE mismatch and is less susceptible to wear-out compared to bond wire lift-off and DCB solder fatigue, which represent the most commonly observed failure mechanisms in current packaging technology [1].

9.2.2.4.1 Bond Wire and Chip Metallization Damage

A significant reliability bottleneck in standard power modules is the wire bond interface. Failures in bond wires and chip metallization primarily occur due to temperature swings and the large CTE mismatch between aluminum bond wires and the silicon die (Figure 9.3).

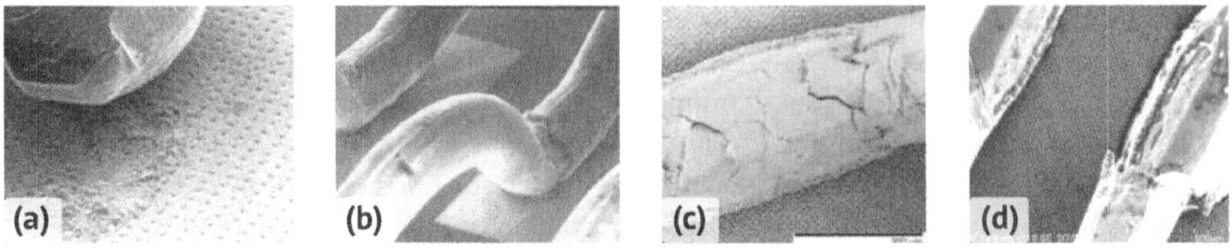

Figure 9.3 Bond wire damages: (a) lift-off, (b) crack, (c) metallurgical damage, (d) fracture. *Source*: [1]/with permission of IEEE.

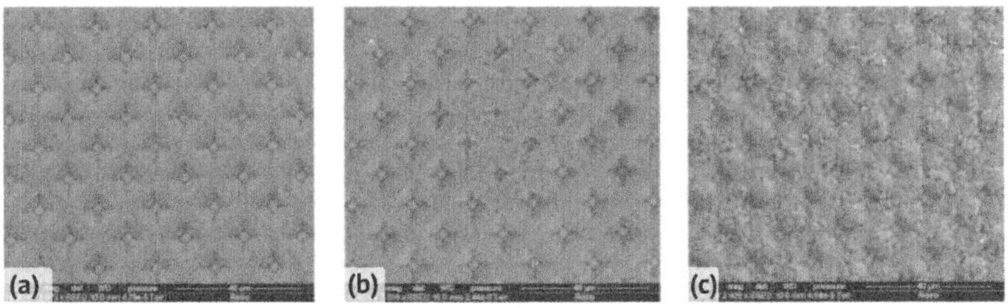

Figure 9.4 Metallization damage progression: (a) healthy condition, (b) initial damage, (c) progressive damage. *Source*: [1]/with permission of IEEE.

Bond wire lift-off, illustrated in Figure 9.3a, has been observed in both IGBTs and freewheeling diodes and is recognized as a major failure mode in wire-bonded power modules. The differential strain between aluminum and silicon leads to high thermomechanical stress at the interface during power cycling. Initial cracks typically form near the wire-to-chip interface and propagate along grain boundaries. Bond wire lift-off affects both emitter bond wires, which carry the main load current, and gate bond wires, which control the switching operation. Complete lift-off of emitter bond wires results in increased resistance and current redistribution to remaining wires, while gate bond wire failure causes loss of control and open-circuit fault behavior.

Another failure mode, bond wire heel cracking shown in Figure 9.3b, is less common in advanced IGBT power modules but can be observed after prolonged endurance testing or severe thermal cycling. Heel cracks initiate at the point where the bond wire transitions from horizontal to vertical orientation and can propagate until complete wire fracture occurs.

Damage to chip metallization and bond wires is closely associated with the medium junction temperature (T_m) and the junction temperature swing (ΔT_j). In devices with multiple bond wires, this failure mechanism primarily affects wires located near the chip center, where junction temperature reaches its maximum value. As the central emitter bond wire contact fails, the remaining bond wires must bear increased load with higher current density. As degradation progresses, more bond wire contacts fail, further increasing current density in remaining contacts. This progressive failure mechanism can lead to subsequent chip destruction, including localized overcurrent and chip burn-out failures.

Additionally, reconstruction of the aluminum metallization pad can occur on the chip top surface, resulting in extrusion of aluminum grains and increased surface roughness. Figure 9.4 illustrates different states of metal pads during chip metallization degradation. These images demonstrate the progression of aluminum metallization reconstruction during power cycling tests, which causes nonuniform current density distribution and increased metal sheet resistance.

Figure 9.5 Solder fatigue: (a) before rupture (b) after rupture. *Source*: [2]/with permission of Elsevier.

9.2.2.4.2 Solder Fatigue

Solder fatigue, a well-known factor affecting power module reliability, is considered a primary cause of failure in IGBT modules due to degradation of solder joints. As bonding techniques continue to improve, fatigue in both chip solder and substrate solder joints has emerged as a significant concern. The degradation of these solder layers leads to an increase in module thermal impedance, causing distortion of thermal flux and subsequent overheating of the chip itself [2].

The deterioration of solder layers can initiate from either the periphery or the center beneath the chip, influenced by the solder material composition and thermal operating conditions. Figure 9.5 shows solder fatigue progression, illustrating the condition before and after rupture.

The high thermomechanical stress experienced by chip-substrate solder and substrate-baseplate solder layers, caused by CTE mismatch and repetitive temperature swings, is not the sole factor contributing to damage. The size of solder joints, boundary constraints, material yield stress, and melting points also impact solder fatigue susceptibility.

Consequently, degradation in these solder layers manifests in the form of voids, cracks, and delamination. This leads to an increase in thermal resistance, accelerating degradation of wire bonding interfaces and directly affecting bond wire lifetime. Additionally, elevated junction temperature resulting from solder fatigue can induce hot spots and thermal runaway in affected areas of the module.

Beyond the chip-level solder failures, degradation of terminal connector-to-substrate solder has also been reported during temperature cycling. Crack propagation and deterioration contribute to increased thermal and electrical resistance. Cracks in the solder joint disrupt the current path, leading to increased on-state voltage drop across the solder joint.

Terminal connector-to-substrate joints must also withstand higher temperatures due to increasing power density trends. Nominal IGBT current ratings have tripled from the 1st to the 4th IGBT generation while maintaining similar module layouts, and this trend continues. As power electronics experience continuous growth in power density, power loss in terminal leads and terminal connector-to-substrate joints increases, resulting in higher temperatures and further impacting joint reliability.

Despite advancements in soldering techniques, such as void-free soldering processes and new baseplate materials, solder layer reliability has improved significantly. However, all solder joints including chip-to-substrate solder, substrate-to-baseplate solder, and terminal connector-to-substrate solder remain critical points during active and passive temperature cycling [1].

9.2.2.4.3 DCB Failures

In addition to bond wire and solder failures, high-temperature cycles have been observed to cause DCB ceramic cracks and DCB metallization delamination. DCB refers to a manufacturing process where copper foils are directly bonded to the top and bottom surfaces of a ceramic material at high temperatures. The CTE mismatch between copper metallization and ceramic material leads to mechanical stress during temperature variations. With new assembly and interconnection methods that either improve or bypass the solder layer, DCB substrates have emerged as potential weak points that can limit module lifetime.

Ceramic substrate cracks can occur due to excessive mechanical stress during thermal cycling or mechanical shock during handling and installation. These cracks compromise the electrical isolation between high-voltage conductors and the grounded baseplate, potentially leading to partial discharge or complete breakdown.

9.2.2.4.4 *Partial Discharge*

Partial discharge (PD) refers to the partial breakdown of insulation materials within the power module. Metallization edges and interfaces in silicone gel encapsulation at high voltages are the primary sources of PD. The shape of substrate metallization edges has a significant influence on PD inception voltage, and poor etching quality or solder residues can have adverse effects. The type of silicone gel and process parameters also contribute to the magnitude of PD.

Cracks in ceramic substrates can significantly lower PD resistance. If the applied voltage exceeds the breakdown voltage of air trapped within cracks or voids, a sudden flash-over discharge occurs across the substrate ceramic. Repeated partial discharge events cause progressive degradation of insulating materials and can eventually lead to complete insulation failure.

9.2.2.5 System-level Fault Manifestations

At the system level, failure modes in power semiconductor switches eventually manifest as open-circuit or short-circuit faults. Additionally, parameter drift, which refers to deviation of electrical characteristics of the power switch from its acceptable operating range, can be considered a third type of fault that precedes complete failure.

9.2.2.5.1 *Short-circuit Faults*

Short-circuit faults are typically catastrophic events, necessitating immediate system shutdown to prevent further damage caused by uncontrolled fault currents. Due to the significant consequences of short-circuit faults in semiconductor devices, they are usually detected by the protection system shortly after occurrence through monitoring of collector-emitter voltage saturation or gate driver desaturation detection.

Short-circuit faults can occur during different operating states of the power switch, with each state having its own root causes:

- **Turn-on State:** High gate voltages exceeding the gate-emitter voltage rating can cause gate oxide breakdown.
- **On-state:** Static latch-up due to excessive collector current.
- **Turn-off State:** Dynamic latch-up when gate voltage drops rapidly or due to high dv/dt.
- **Off-state:** Thermal runaway phenomenon due to excessive leakage current at elevated junction temperature.

High voltage spikes during turn-off, caused by the inherent parasitic inductance of the IGBT module and interconnections, can substantially increase the electric field and lead to breakdown of IGBT cells. Following each voltage spike, both collector-emitter and gate-emitter voltages increase, which can result in short-circuit fault during the subsequent turn-on transition.

Latch-up in IGBT operation refers to a condition where the collector current is not controlled by gate voltage. It can be categorized as static or dynamic. Static latch-up occurs as a result of high collector currents exceeding the device rating, while dynamic latch-up occurs when gate voltage drops rapidly or the off-state collector-emitter voltage is sufficiently high. Due to recent advances in IGBT technology, new generations of IGBTs are designed with immunity to latch-up, making it an uncommon fault mode in modern devices [1].

9.2.2.5.2 Open-circuit Faults

In contrast to short-circuit faults, open-circuit faults do not have immediate catastrophic consequences and may remain undetected for a period of time. However, they significantly impact motor performance and serve as the main cause for secondary and more severe faults in other components. Open-circuit faults in IGBT modules commonly occur due to breakage in connections of IGBT module components or gate driver malfunction.

The main reason for bond wire lift-off and rupture, leading to open-circuit faults, is a combination of vibration, CTE mismatch, and thermal cycling due to load variations and switching losses. Bond wire lift-off begins with development of a crack at the bonding interface and eventually results in separation of the bond wire from the surface of the silicon chip.

Moreover, failures in the gate driver circuit and its connection with the switch gate cause absence of switching signals on the power switch gates, resulting in open-circuit faults. Damage to PCB traces between the drive circuit and power switch, as well as power stage device failures, are common reasons for gate driver-related open-circuit faults.

9.2.3 Diode Failures and Reverse Recovery Issues

Freewheeling diodes are integral components of IGBT modules and experience similar thermo-mechanical stress conditions as IGBTs, including bond wire lift-off and solder fatigue. However, diodes have unique failure mechanisms related to their reverse recovery characteristics and the high dv/dt experienced during commutation that distinguish them from IGBT failures. Table 9.3 summarizes the various diode failure modes and their distinguishing characteristics.

9.2.3.1 Reverse Recovery and dv/dt-related Failures

Power diodes switch from forward-conducting to blocking by recovering stored minority carrier charge. With stress and aging under high-temperature operation, the effective minority carrier lifetime can change, altering reverse recovery characteristics. Degradation may manifest as either a longer reverse recovery time or a snappier recovery characterized by abrupt current cessation. Both scenarios cause larger voltage overshoots and increased device stress during switching transients [3].

A steep voltage rise (dv/dt) across a diode or transistor during switching can induce failure via parasitic mechanisms. In a MOSFET, high dv/dt across the intrinsic body diode can drive displacement current through the parasitic bipolar transistor structure, causing a destructive short circuit [4]. Snappy recovery combined with circuit stray inductance produces high voltage spikes that can force the diode into avalanche breakdown and, with repeated occurrences, cause localized heating and eventual failure [3].

9.2.3.2 Diode Short-circuit Failure Progression

The end result of catastrophic diode failure is typically a short circuit across the junction. A power diode subjected to severe overvoltage or avalanche enters thermal runaway, where high reverse current heats the junction until the silicon melts and forms a low-resistance conductive path, creating a recrystallized silicon filament that bridges the electrodes. Most power diodes fail in a shorted condition, and a failed freewheeling diode often results in a phase-leg short through the corresponding transistor, making fast fault detection critical for system protection [5].

Table 9.3 Diode failure modes and characteristics.

Failure mechanism	Physical cause	Manifestation	Consequences	Material-specific issues
Reverse recovery degradation	Temperature/aging effects on carrier lifetime	Longer or snappier recovery	Voltage overshoots, increased device stress	All silicon diodes
dv/dt-induced failures	High voltage rise rate during switching	Parasitic turn-on, displacement current	Destructive short circuit	MOSFETs, diodes with poor recovery
Reverse recovery overshoot	Abrupt current cutoff, stray inductance	Snappy recovery, high voltage spike	Avalanche breakdown, hotspots, melting	High di/dt sensitive devices
Short-circuit progression	Overvoltage, avalanche, thermal runaway	PN junction destruction, conductive path	Phase-leg short, follow-on currents	Most power diodes fail shorted
Body diode degradation (MOSFETs)	Repetitive stress, parasitic BJT activation	Increased $R_{DS(on)}$, leakage, lower breakdown	Latch-up, eventual failure	Silicon MOSFETs
SiC avalanche/surge	Surge current, voltage rating exceeded	Local overheating, junction damage	Thermal runaway, short or open	Limited surge capability in SiC Schottky
SiC cosmic ray	High-energy particle strikes at high voltage	Single-event burnout	Random instantaneous failure	SiC at 1200 V+ blocking voltage

9.2.3.3 Body Diode Degradation in MOSFETs

MOSFETs in PM drives often rely on their intrinsic body diodes for freewheeling currents. Repeated body-diode conduction at high current and temperature activates the parasitic bipolar transistor during reverse recovery, causing loss of gate control through latch-up and eventual failure. Aging studies show a gradual rise in on-state resistance $R_{DS(on)}$ followed by a sudden drop when the parasitic BJT triggers, indicating imminent failure. Repeated avalanche or current stress of the body diode can permanently reduce the drain–source breakdown voltage [6].

9.2.3.4 SiC Diode Considerations

Silicon Carbide Schottky barrier diodes have negligible reverse recovery due to the absence of minority carrier storage, representing a significant advantage over silicon diodes. However, SiC diodes have limited surge current capability and can suffer avalanche breakdown if driven beyond their voltage rating [7]. At high blocking voltages (1200 V and above), SiC devices can experience single-event burnout from cosmic ray particle strikes, though modern designs have reduced the Failure-In-Time rate by 100 times through improved device ruggedness [8].

9.2.4 Gate Driver Circuit Faults

Gate driver circuits interface between low-power control signals and high-power semiconductor switches, providing the necessary gate charge and discharge currents while maintaining electrical isolation between control and power stages. Failures in gate driver circuits can cause absence of switching signals on power switch gates, leading to motor performance degradation or catastrophic failure.

9.2.4.1 Bootstrap Capacitor and Driver Supply Failures

Many gate driver ICs use a bootstrap circuit to supply the high-side gate voltage. If the bootstrap capacitor fails by going open, losing capacitance, or developing high ESR, the high-side gate may not receive sufficient voltage, causing the device to turn on partially with excessive heating or to intermittently turn off under load [9]. A degraded bootstrap capacitor exhibits excessive voltage droop during high-side on-time, which can approach the undervoltage lockout threshold [10].

Similarly, sag or failure of the driver power supply can trigger undervoltage lockout, causing missing PWM pulses or asymmetrical drive. If the supply drops slowly, the gate drive may operate in a linear region, making the transistor operate in high conduction loss mode. Overvoltage transients on the supply can exceed the gate's maximum voltage and damage the gate oxide [11]. Complete supply loss means transistors receive no gate drive, often defaulting to off or, worse, to high impedance that allows noise-induced turn-on.

9.2.4.2 Isolator Degradation

Gate drivers use optocouplers or digital isolators for signal isolation. In optocouplers, LED luminous efficiency degrades over time, causing current transfer ratio reduction and slower switching that introduces timing delays [12]. Eventually, the optocoupler can fail to transmit signals entirely. The insulation barrier can also break down under high voltage stress, creating a low-resistance path between primary and secondary sides [13]. Modern digital isolators using capacitive or magnetic coupling can similarly suffer insulation breakdown if overstressed.

9.2.4.3 Gate Resistor Failures

Gate resistors shape the rise and fall times of gate voltage, controlling di/dt and dv/dt. An open-circuit gate resistor, typically caused by thermal fracture or solder issues, leaves the gate floating and can cause shoot-through via the Miller effect [14]. Increased gate resistance slows switching and can cause crossover conduction when the complementary device turns on before the first fully turns off [15]. Conversely, a shorted gate resistor causes excessively fast switching with voltage ringing and potential Miller-effect-induced false turn-on of the opposite device.

9.2.4.4 Shoot-through and Timing Degradation

Shoot-through, the unintended simultaneous conduction of both transistors in a bridge leg, effectively shorts the DC bus. While normally prevented by deadtime circuits, degraded gate driver timing can cause overlap. Propagation delay, increases in optocouplers, slowed rise and fall times from weakened driver stages, and component aging all erode the deadtime safety margin. A system safe at start-of-life may enter shoot-through at end-of-life if turn-off delay increases and encroaches on the fixed deadtime. Modern drives use desaturation or overcurrent sensing to detect shoot-through within microseconds, but even a single event can crack a chip or bond wire.

9.2.5 DC-link Capacitor Degradation

DC-link capacitors, often electrolytic or film capacitor banks, perform the essential function of filtering DC bus voltage ripple and providing instantaneous energy during transient load conditions. Statistical data indicates capacitors account for 30% of converter failures [1], making them a significant reliability concern.

9.2.5.1 Electrolytic Capacitor Lifetime and Degradation

Aluminum electrolytic capacitors have a liquid electrolyte that gradually evaporates or decomposes. Their lifetime follows an Arrhenius-based relationship where for roughly every 10 K decrease in core temperature, the expected life doubles [16]. The overall life can be estimated as $L_X = L_0 \times K_T \times K_R \times K_V$, where L_0 is the base life at rated conditions, $K_T = 2^{(T_0 - T_a)/10}$ accounts for actual temperature, K_R for actual ripple current relative to rated value, and K_V for operating voltage [16]. The AC ripple current flowing through the capacitor ESR causes internal heating that accelerates electrolyte loss, with low-frequency ripple near 100–120 Hz producing the worst-case heating [17].

9.2.5.2 ESR Increase and End-of-life

The hallmark of electrolytic capacitor aging is progressive ESR increase as the electrolyte dries and the dielectric degrades [17]. Higher ESR means more heat for the same ripple current, accelerating further aging in a positive feedback loop. Manufacturers typically define end-of-life as ESR reaching 2–3 times its initial value, or capacitance falling below 80% of nominal [17, 18]. In a motor drive, rising ESR manifests as increased DC bus ripple. Many modern drives perform online ESR monitoring to predict failure before catastrophic events such as capacitor venting or rupture [19].

9.2.6 Thermal Management System Failures

Thermal management systems in power converters maintain junction temperatures within acceptable limits through heatsinks, cooling fans or liquid cooling systems, thermal interface materials, and temperature sensors. Failures in these systems accelerate the thermomechanical degradation mechanisms described in the IGBT module section, as elevated junction temperatures exponentially increase failure rates and can trigger thermal runaway. Cooling system degradation, such as fan bearing wear, coolant leaks, or thermal interface material dry-out, raises the baseline operating temperature and reduces the thermal margin available during transient overloads. Loss of active cooling during high-load operation can cause junction temperatures to exceed safe limits within seconds, making thermal management integrity essential for converter reliability [1].

9.2.7 Control System and Sensor Faults

The control system of a PM drive relies on various sensors including current, voltage, temperature, and position sensors, along with a processor such as a DSP or microcontroller. Statistical data indicates that 58% of failures stem from control circuit malfunctions. Faults in these components can degrade performance or even damage the power stage if not handled properly. Sensor failures can lead to incorrect control decisions, potentially causing damage to both converter and motor.

9.2.7.1 Current Sensor Failure Modes

PM machine drives use current sensors such as Hall-effect sensors, shunts with amplifiers, or fluxgate sensors to measure phase currents for control and protection. Key failure modes include offset drift, gain error, saturation, EMI noise, and outright failure. Offset drift in one phase current feedback introduces a DC bias in the current controller, causing torque ripple at the fundamental electrical frequency. Many drives periodically auto-calibrate offsets at start-up or during PWM idle vectors. Gain errors from component tolerance shifts or temperature-dependent sensitivity changes cause misestimation of torque and flux, with asymmetric gain error between phases introducing torque ripple at twice the electrical frequency [20]. Current sensors can also saturate during fault conditions, potentially missing overcurrent events [21, 22], or fail outright with either no output or a stuck reading that can rapidly destabilize the current control loop [23]. Drives often employ plausibility checks such as verifying that the sum of phase currents equals zero.

9.2.7.2 Voltage Sensor Calibration Issues and Drift

DC bus voltage and phase voltages are sensed through resistor dividers, isolation amplifiers, or voltage transducers. Over time, resistor dividers may drift with humidity or temperature cycling, causing modulation saturation if the control overestimates bus voltage or false overvoltage alarms if it underestimates. Phase voltage sensors used for sensorless control observers are particularly sensitive, as small gain or offset errors can corrupt rotor angle estimation.

9.2.7.3 Temperature Sensor Fault Scenarios

Temperature sensors on heatsinks, motor windings, and semiconductor junctions provide critical thermal protection. A sensor that reads too low due to failure, such as a failed-open thermistor, prevents the system from curtailing power when needed, potentially leading to thermal runaway. A sensor that reads too high causes nuisance trips or unnecessary derating. Loss of thermal contact between sensor and monitored surface is particularly dangerous, as the sensor reads ambient temperature while the device overheats.

9.2.7.4 Position and Speed Sensor Faults

PM machines, especially synchronous motors, often use position or speed sensors like optical encoders, resolvers, or Hall effect position sensors for commutation. If these sensors fail or produce erroneous data, the consequences are immediate. The controller might apply currents out-of-phase with the rotor, leading to reduced torque, high current draw, vibration, and potentially demagnetization or overheating.

9.3 Modeling and Analysis

Understanding the behavior of power electronic converters under fault conditions requires comprehensive modeling that captures electrical, thermal, and electromagnetic interactions with the permanent magnet machine. This section presents mathematical models and simulation approaches for analyzing faulty converter operation, quantifying the impact on motor performance, examining harmonic content in distorted waveforms, developing thermal models for component temperature prediction, and utilizing finite element analysis to study the coupled converter-motor system behavior.

9.3.1 Converter Topology Modeling Under Fault Conditions

Modeling a two-level voltage source inverter under switch fault conditions requires modifying the ideal topology to reflect the failed component. For an open-circuit fault, the affected IGBT is effectively removed from the circuit as it cannot conduct current in either direction. In practice, this is equivalent to a switch that is permanently off, leaving only its antiparallel diode to conduct in one polarity [24]. Thus, if an upper device fails to open, that phase leg can no longer connect the DC positive rail to the phase. Current can only return through the lower diode when the phase voltage exceeds the DC-link voltage during freewheeling action. Conversely, a short-circuit fault in a switch is equivalent to a permanent connection across that device [24]. In this case, protection circuits will turn off the complementary transistor in the same leg to prevent a shoot-through of the DC bus [24]. The result is that the faulted phase node becomes clamped to one of the DC bus rails at all times.

Under healthy operating conditions, the phase voltage of a three-phase inverter is determined by the switching state of the upper and lower devices in each leg. For phase i, the pole voltage with respect to the DC-link midpoint can be expressed as:

$$V_{i0} = S_i V_{DC} \tag{9.1}$$

where S_i is the switching function taking values of 1 when the upper switch is on and 0 when the lower switch is on, and V_{DC} is the DC-link voltage. The line-to-neutral voltage is then:

$$V_{in} = V_{i0} - \frac{1}{3}\left(V_{a0} + V_{b0} + V_{c0}\right) \tag{9.2}$$

Under an open-circuit fault in the upper switch of phase a, the switching function becomes constrained. The phase can no longer be actively driven high, and the voltage depends on the diode conduction state:

$$V_{a0} = \begin{cases} 0 & \text{if lower switch is on} \\ V_{DC} & \text{if } D_1 \text{ conducts (current from phase node to } V_{DC}^{+}\text{)} \\ \text{floating} & \text{otherwise} \end{cases} \tag{9.3}$$

For a short-circuit fault where the upper switch is permanently on, the switching function becomes:

$$S_a = 1, \quad S_a^{lower} = 0 \quad \text{(forced off for protection)} \tag{9.4}$$

resulting in a clamped pole voltage:

$$V_{a0} = V_{DC} \tag{9.5}$$

These fault representations are illustrated in Figure 9.6, which shows a single inverter phase leg under three conditions: (a) healthy operation with normal current flow through the upper switch S_1, (b) an open-circuit fault where S_1 cannot conduct and current flows only through the lower switch or the freewheeling diode D_1, and (c) a short-circuit fault where S_1 is permanently on and the lower switch S_2 is forced off for protection, clamping the phase output to V_{DC}^+.

Mathematically, the inverter's state-space model must be adapted for the asymmetrical switching caused by the fault. Under healthy conditions, a three-phase two-level inverter has eight possible switching states with one redundant zero state that symmetrically synthesize the AC voltages. Under a single switch open fault, certain voltage output states become unreachable or occur only through the diode freewheeling path, making the system piecewise linear depending on whether the diode is conducting. This can be handled by a hybrid state-space model or a mixed logical dynamical representation [25, 26]. The system is no longer balanced, so common modeling approaches like dq-transformations yield a time-varying system or require splitting into positive and negative sequence components. Researchers often derive separate differential equations for the faulted phase with diode conduction logic and for the healthy phases [24]. The overall model captures the line-to-line voltage distortion. For instance, if an upper switch is open, the phase can only be pulled down to ground via the lower switch but never driven up to V_{DC} except through the diode when the phase back-EMF forces current into the DC-link. This produces a truncated output voltage on that phase and hence unbalanced line voltages.

When a switch fault occurs, the PWM strategy may need modification to maintain some level of performance. If a single switch is lost, one strategy is to abandon the usual sinusoidal modulation for that leg and instead drive the remaining healthy switch in the phase or the other two phases with a revised pattern to balance the line voltages. For example, after an open-circuit fault in one leg, the inverter can be reconfigured into a pseudo two-phase operation forming an effective four-switch inverter, and the pulse-width modulation is adjusted accordingly [27]. Such post-fault PWM schemes aim to minimize the DC-link voltage ripple and negative-sequence currents. Indeed, loss of one phase's active drive causes the DC bus to source power in pulses at lower frequency than in the healthy case, since the three-phase draw is no longer balanced. This leads to increased low-frequency ripple on the DC-link voltage and current [28]. The modeling must account for this by including the DC-link capacitor dynamics and the altered current waveform. Studies have shown that under open-phase conditions, the DC-link current contains a significant double-frequency component, twice the fundamental frequency, due to the power pulsation, which translates to a 2× line-frequency ripple on the DC voltage [28].

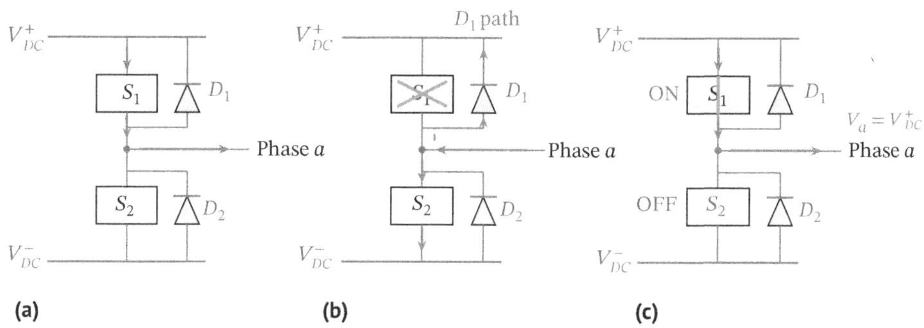

Figure 9.6 Single inverter phase leg under different operating conditions: (a) healthy operation with current through the upper switch, (b) open-circuit fault on S_1 with current limited to the lower switch and diode freewheeling path (dashed), (c) short-circuit fault on S_1 with phase clamped to V_{DC}^+ and S_2 forced off.

The DC-link voltage ripple increases significantly under fault conditions due to unbalanced power draw. The DC-link voltage can be modeled by considering the capacitor equation:

$$V_{DC}(t) = V_{DC,avg} + \frac{1}{C_{DC}} \int (i_{source} - i_{load})dt \tag{9.6}$$

Under open-phase conditions, the load current contains a prominent second-harmonic component at $2\omega_e$ (where ω_e is the electrical frequency), leading to a DC-link voltage ripple:

$$\Delta V_{DC} \approx \frac{I_{2\omega}}{2\omega_e C_{DC}} \tag{9.7}$$

where $I_{2\omega}$ is the amplitude of the double-frequency current component drawn from the DC-link.

Finally, the current trajectory deviations in the motor, such as in the dq plane, can be modeled by observing that the inverter can no longer maintain a balanced three-phase sinusoidal output. Under fault conditions, the stator currents can be decomposed into positive-sequence and negative-sequence components:

$$\vec{i}_s = \vec{i}_s^{+} + \vec{i}_s^{-} \tag{9.8}$$

where the negative-sequence component rotates opposite to the rotor and is absent in healthy operation. In the synchronous reference frame, the negative-sequence current appears as a second-harmonic oscillation:

$$i_{dq}^{-} = I^{-}e^{-j2\omega_e t} \tag{9.9}$$

The current vector will deviate from the ideal circular trajectory, instead showing oscillations or drops when the faulted phase is supposed to conduct. Figure 9.7 illustrates this deviation in the stationary $\alpha\beta$ reference frame: under healthy conditions in (a), the stator current space vector traces a perfect circle, whereas in (b) the open-switch fault distorts the trajectory into an irregular path due to the half-wave rectification in the faulted phase. The dashed reference circle in (b) highlights the extent of the deviation from the healthy locus. For instance, an open upper switch means the phase current cannot be driven positive and will drop to zero aside from diode freewheeling intervals. This results in an asymmetric current waveform, effectively half-wave rectification in

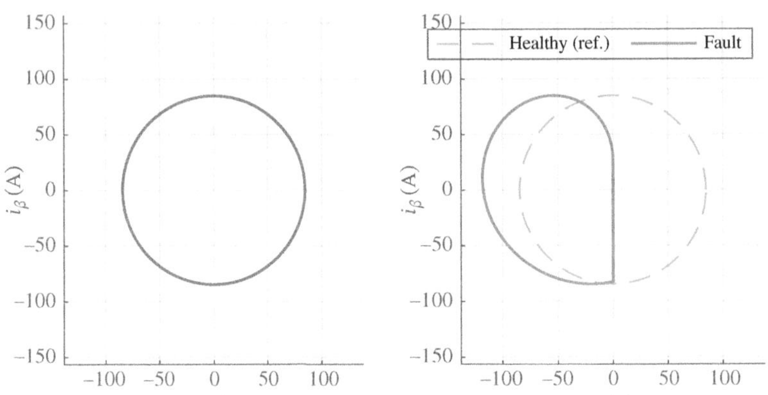

Figure 9.7 Stator current space vector trajectory in the stationary $\alpha\beta$ frame: (a) healthy operation showing a circular locus, (b) open upper switch fault producing a distorted trajectory. The dashed circle indicates the healthy reference.

the faulted phase, and a distortion of the rotating current vector. Modeling these trajectories may involve time-domain simulation of the phase currents using the modified switching functions or, analytically, by superimposing a negative-sequence current component to represent the imbalance.

The converter's topology under faults is represented by removing or bypassing the failed switch and updating the inverter's mathematical model, including circuit equations and switching logic, to capture the constrained voltage outputs, altered PWM scheme, increased DC-link pulsations, and unbalanced currents [24].

9.3.2 Switch Fault Impact on Motor Performance

When a power electronic switch fails in a PM drive inverter, the motor is subjected to unbalanced and distorted excitation, leading to multiple adverse effects on performance. The most immediate impact is a rise in electromagnetic torque pulsations. With one phase not contributing correctly or stuck at a DC level, the three-phase balanced torque production is lost. For example, open-circuiting one transistor effectively removes or halves one phase's current for part of each cycle, producing a large oscillatory torque component.

Under healthy operation, the electromagnetic torque in a permanent magnet synchronous motor can be expressed as:

$$T_{em} = \frac{3p}{2}\psi_m i_q \tag{9.10}$$

where p is the number of pole pairs, ψ_m is the permanent magnet flux linkage, and i_q is the quadrature-axis current. Under a switch fault, the presence of negative-sequence currents introduces a pulsating torque component. The total torque can be decomposed as:

$$T_{em,fault} = T_{avg} + T_{ripple}\cos(2\omega_e t + \phi) \tag{9.11}$$

where T_{avg} is the average torque (typically reduced from nominal) and T_{ripple} is the amplitude of the torque oscillation. The negative-sequence current i^- interacts with the positive-sequence flux to produce a torque pulsation at twice the electrical frequency:

$$T_{ripple} = \frac{3p}{2}\psi_m I^- \sin(\delta) \tag{9.12}$$

where δ is the phase angle between positive and negative sequence components. It has been observed that under an open-switch fault, torque ripple becomes inevitable due to the absence of one switch in the bridge arm, causing a loss of the balanced magnetomotive force [29]. Quantitatively, researchers have reported significant increases in torque ripple, on the order of 20–25% of the nominal torque, in fault conditions [29]. This increased torque ripple is mainly at twice the fundamental frequency for a single-phase open fault in a three-phase machine because the fault causes a second-harmonic oscillation in torque production as the remaining two phases carry oscillating load. Such torque pulsations, if severe, can induce speed oscillations and vibration in the drive.

A switch fault leads to severely distorted phase currents. The healthy phases may carry extra current to compensate for the missing phase contribution, and the faulted phase current may be zero or only flow in one direction. For an open switch, current might only circulate through the diode. This results in a nonsinusoidal, asymmetric current pattern with high harmonic content [26]. Figure 9.8 illustrates this effect for an upper-switch open-circuit fault in phase A: the healthy three-phase currents in (a) are balanced and sinusoidal, whereas in (b) the faulted phase current is clipped to its negative half-cycle only, since positive current can no longer be sourced through the open switch. The remaining phases carry additional current to compensate, resulting in visible

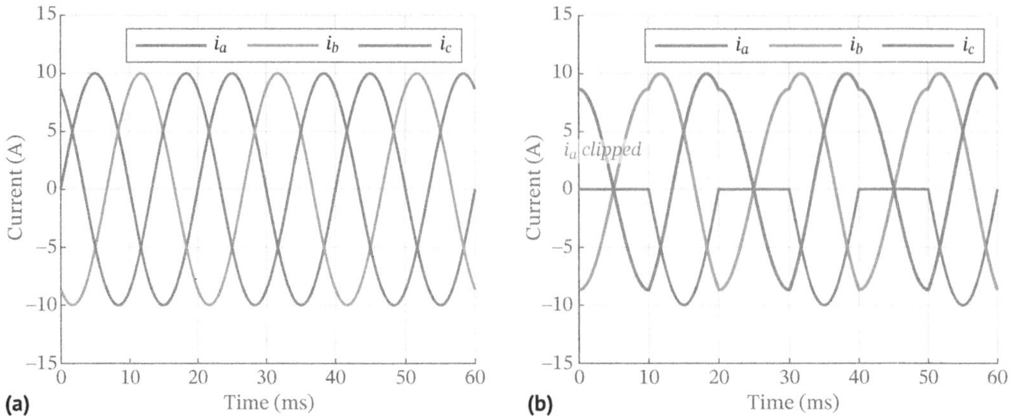

Figure 9.8 Three-phase stator currents: (a) healthy balanced operation, (b) upper-switch open-circuit fault in phase A showing half-wave rectification pattern and current redistribution to phases B and C.

amplitude asymmetry. In fact, once a fault occurs, many more frequency components appear in the three-phase currents compared to normal operation [26]. The line currents will contain not only the fundamental component but also increased even harmonics and possibly a DC component if one phase is partially conducting via a diode. The total harmonic distortion of the current rises substantially under fault conditions. One study noted that THD can increase by tens of percent when an inverter transistor is open-circuited [28]. This current imbalance also means the motor's magnetic field is no longer symmetric, which introduces a negative-sequence flux rotating opposite to the rotor. The negative-sequence currents create additional loss and pulsating torques. Due to these asymmetrical currents and fluxes, closed-loop current control becomes challenging and often must be retuned or supplemented with fault-tolerant control, since the standard dq regulators assume balanced conditions [28].

The motor's efficiency deteriorates during converter faults because of higher losses. The remaining healthy phases usually conduct more current to deliver the demanded torque, raising stator copper losses. The total copper loss can be expressed as:

$$P_{cu} = 3R_s I_{rms}^2 = R_s\left(I_a^2 + I_b^2 + I_c^2\right) \tag{9.13}$$

Under fault conditions with unbalanced currents, the RMS current in the remaining phases increases substantially, leading to:

$$P_{cu,fault} = R_s \sum_{i=a,b,c} \left[\left(I_i^+\right)^2 + \left(I_i^-\right)^2 + \left(I_i^{harm}\right)^2\right] \tag{9.14}$$

where the superscripts denote positive-sequence, negative-sequence, and higher harmonic components, respectively. Moreover, the distorted current waveforms induce higher harmonic flux in the iron, leading to increased core losses. The iron losses can be estimated using a modified Bertotti equation accounting for harmonics:

$$P_{Fe} = \sum_{h=1}^{N} \left[k_h f_h B_h^{\alpha} + k_c f_h^2 B_h^2 + k_e f_h^{1.5} B_h^{1.5}\right] \tag{9.15}$$

where h is the harmonic order, f_h and B_h are the frequency and flux density of the h-th harmonic, and k_h, k_c, k_e are the hysteresis, classical eddy current, and excess loss coefficients. Additional rotor eddy current losses arise in permanent magnets or rotor steel due to the negative-sequence field rotating at twice synchronous speed relative to the rotor:

$$P_{rotor,eddy} \propto (I^-)^2 f_{slip}^2 \tag{9.16}$$

where $f_{slip} = 2f_e$ for the negative-sequence component. The overall motor efficiency under fault becomes:

$$\eta_{fault} = \frac{P_{out}}{P_{out} + P_{cu,fault} + P_{Fe,fault} + P_{rotor,fault}} \tag{9.17}$$

A coupled field-circuit analysis of an induction machine under open-phase fault showed that both iron loss and rotor joule loss increased under fault operation compared to healthy conditions [28]. The overall losses in a PMSM drive under an open-switch fault are mainly caused by the asynchronous negative-sequence harmonic field and the increased RMS current in the remaining phases [28]. Consequently, the machine runs hotter for the same load. Experiments have confirmed a drop in efficiency and a need to derate the load. For instance, one study found that under a worst-case open-phase fault, the continuous load torque had to be reduced to approximately 60% of nominal to avoid overheating the motor [28]. Thus, without fault mitigation, both electrical and thermal efficiency of the motor are significantly compromised.

In a PMSM, the back-EMF of the machine can interact with the fault in unintended ways. For an open-circuit fault, when a phase is disconnected from the inverter's control, its terminals are essentially floating except for the diode path. The motor's back-EMF in that phase may drive a current through the diode into the DC-link during certain intervals, which is regenerative current. This can cause unexpected braking torques or additional bus voltage spikes. Essentially, the phase with an open switch behaves like it is connected to the DC bus only via a diode, so it will sink current when its EMF exceeds the bus voltage [24]. This uncontrolled current can add to drag on the rotor and also disturb the other phases' currents via the DC-link coupling. In the case of a shorted switch where a transistor is stuck on, the phase is effectively tied to a DC rail. The motor back-EMF then drives a large fault current until the other switch in that phase is turned off and the current commutates. During the brief fault interval, severe torque transients occur. The motor's EMFs can feed into the faulted leg, causing additional torque dips or spikes beyond what the controller commands.

The loss of one or more inverter switches also impairs the drive's ability to regulate speed smoothly. The increased torque ripple translates to speed oscillation, since the rotor will accelerate and decelerate with the torque pulsations if not perfectly stiff. Also, the standard control algorithms such as field-oriented control or direct torque control assume a balanced three-phase input. A fault breaks this assumption, potentially leading to integrator wind-up or oscillatory response in the speed and current controllers. Without intervention, the drive might experience hunting in speed or slower dynamic response under load changes. Fault-tolerant control strategies are therefore introduced to maintain operation. For example, one fault-tolerant approach reconfigures the inverter and adjusts the PWM and vector control so that the motor can still produce steady torque and maintain accurate speed even with one leg of the inverter lost [27]. This shows that in the absence of such measures, we would indeed see poor speed regulation and imbalanced phase currents under fault. With these measures, experiments have shown the system can continue to deliver steady torque and keep speed error minimal despite the fault [27]. But if no action is taken, the drive's speed controller might become unstable due to the unmodeled disturbance of the fault. A converter switch fault in a PM machine drive will increase torque ripple, inject abnormal harmonics into currents, reduce efficiency through higher losses, create potential back-EMF induced issues, and degrade the overall control of the drive. These impacts underscore the need for immediate fault detection and activation of fault-tolerant control to limit performance degradation and prevent equipment damage.

9.3.3 Harmonic Analysis of Faulty Converter Operation

Under fault conditions, the output voltage and current waveforms of the inverter contain distinctive harmonic signatures. In healthy three-phase operation with sinusoidal PWM, the line voltages consist of a fundamental frequency component and switching-frequency harmonics and their sidebands, but no even harmonics or DC, and the three phases are balanced so that many harmonics cancel out in the line-to-line voltages. A fault upsets this symmetry, introducing new frequency components. Different faults tend to produce specific harmonic patterns. With a single open-phase or open-switch fault in a three-phase inverter, the system effectively operates like a two-phase supply for the motor. This often yields a strong second-order harmonic in the three-phase instantaneous power and in the torque. In the currents and voltages, even harmonics that are normally absent can appear. For example, if phase A is open, the remaining two phases carry currents that are no longer 120° apart but 180° apart during certain intervals, generating a prominent second harmonic around 2× the fundamental frequency in the line currents. Similarly, a DC component at 0 Hz can appear in the faulted phase current because that phase may conduct only unipolar current through a diode. In contrast, a short-circuit fault with a stuck-on switch might impose a DC offset on the phase voltage if clamped to a DC rail, which can introduce a DC component and even harmonics in the output. Each type of fault thus has a fingerprint in the frequency domain. Researchers have noted that comparing the line current spectra of healthy versus open-fault conditions reveals clear new peaks at certain frequencies that were not present in healthy operation [26, 30]. These include increased even-order harmonics such as second and fourth, and negative-sequence fundamental, which in a stationary frame, looks like a backward rotating fundamental frequency component.

Faulty operation greatly increases the total harmonic distortion (THD) of both currents and voltages. The THD is defined as:

$$\mathrm{THD} = \frac{\sqrt{\sum_{h=2}^{\infty} I_h^2}}{I_1} \times 100\% \tag{9.18}$$

where I_h is the RMS value of the h-th harmonic component and I_1 is the fundamental component. An open-switch fault can increase current THD by roughly 30% compared to healthy operation, with even harmonics (second and fourth) dominating the distortion spectrum [31]. Figure 9.9 compares the harmonic content of phase A current under healthy and open-switch fault conditions. The fault introduces a DC component and prominent even harmonics at $2f_1$ and $4f_1$ that are entirely absent in the healthy spectrum, providing a clear diagnostic fingerprint. The appearance of non-triplen even harmonics is a clear indicator of asymmetry, as these cancel out in a balanced three-phase system. The increased THD correlates directly with extra heating in the motor from elevated iron and copper losses.

Faults also affect switching-frequency sidebands. Normally, a balanced PWM inverter produces harmonics at $f_{sb} = nf_{sw} \pm mf_1$. Under fault, new sidebands appear at $f_{sb,fault} = f_{sw} \pm 2f_1, f_{sw} \pm 4f_1$, etc., due to the altered periodicity of the three-phase waveform. Interharmonic components (noninteger multiples of fundamental) can also appear from amplitude or phase modulation induced by the fault.

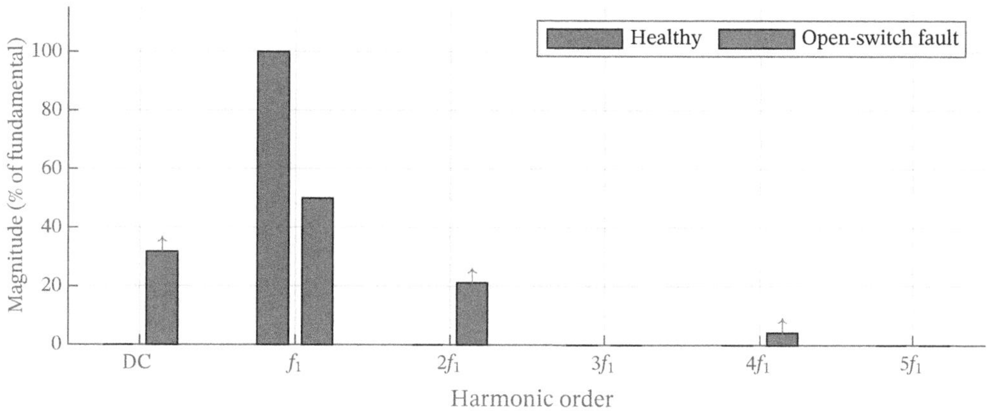

Figure 9.9 Harmonic spectrum of phase A current under healthy operation and open upper switch fault. Even harmonics ($2f_1$, $4f_1$) and a DC component appear exclusively under the fault condition.

Because switch faults occur suddenly, time-frequency analysis tools are valuable for studying harmonic evolution during and after the fault event. Short-time Fourier transform, wavelet transforms, and Hilbert–Huang Transform capture the transient burst of broadband content at fault onset followed by the emergence of new steady-state harmonics [26, 32]. Spectral kurtosis and wavelet packet energy have been used to detect and quantify fault-induced frequency components [26].

Table 9.4 provides a harmonic content comparison across different operating conditions, serving as a diagnostic fingerprint reference.

The spectral entropy, which quantifies the degree of disorder in the frequency spectrum, can be calculated as:

$$H = -\sum_{i=1}^{N} p_i \log_2(p_i) \tag{9.19}$$

where $p_i = P_i / \sum_{j=1}^{N} P_j$ is the normalized power spectral density at frequency bin i. An increase in spectral entropy, indicating more dispersed energy among frequencies, has been used as a signature of open-switch faults [26].

9.3.4 Thermal Modeling of Power Electronic Components

Accurate thermal models are essential for evaluating fault impact on power electronic components, as switch failures often lead to localized heating that can cause cascading damage. Semiconductor devices are characterized by a transient thermal impedanceZ_{th} curve describing the junction-to-case thermal response over time [33].

Two principal RC network approaches are used to model this thermal behavior. The Foster network, consisting of multiple parallel R-C branches fitted to the Z_{th} curve, expresses the transient thermal impedance as:

$$Z_{th}(t) = \sum_{i=1}^{n} R_{th,i}\left(1 - e^{-t/\tau_i}\right) \tag{9.20}$$

Table 9.4 Harmonic content comparison for different converter operating conditions.

Operating condition	Dominant harmonics	DC component	Even harmonics	Typical THD	Diagnostic signature
Healthy operation	Fundamental (f_1), switching harmonics ($nf_{sw} \pm mf_1$)	No	Absent (balanced system)	Low baseline	Balanced spectrum, triplen harmonics cancel
Single open-phase fault	Fundamental, strong second harmonic ($2f_1$), fourth harmonic	Yes (faulted phase)	Present (second, fourth)	Baseline + ~30%	Even harmonics, negative-sequence fundamental
Open-switch fault	Similar to open-phase, asymmetric current	Yes (unipolar diode conduction)	Present (second, fourth dominant)	Significantly elevated	Half-wave rectification pattern, beat frequencies
Short-circuit fault	Fundamental, DC-offset induced harmonics	Yes (phase clamped)	Present	Elevated	Phase voltage clamped to rail, severe imbalance
Post-fault reconfigured	Fundamental, reduced high-order harmonics	Minimized	Present (residual)	Moderately elevated	Modified sideband pattern: $f_{sw} \pm k \cdot 2f_1$

where $\tau_i = R_{th,i}C_{th,i}$ is the thermal time constant of the i-th branch. Foster models are useful for rapid junction temperature calculation but cannot easily be extended to include external thermal components such as heatsinks. The alternative Cauer model, structured as an RC ladder corresponding to physical device layers, provides physically meaningful temperature distributions and can be extended by adding thermal resistance elements for the complete heat flow path from junction through packaging to ambient (Figure 9.10) [33]. Table 9.5 compares these two approaches.

The steady-state junction temperature is determined by the total thermal resistance from junction to ambient:

$$T_j = T_{amb} + P_{loss} \cdot R_{th,ja} = T_{amb} + P_{loss} \cdot (R_{th,jc} + R_{th,ca}) \tag{9.21}$$

Thermal modeling is frequently coupled with electrical behavior to form electrothermal models. The instantaneous power loss feeds as a heat source into the thermal network, which updates junction temperature, which in turn affects device electrical characteristics such as IGBT saturation voltage $V_{CE}(T)$ or MOSFET on-resistance $R_{ds}(T)$. This two-way coupling is particularly important under fault conditions. During a short-circuit fault, junction temperature can rise tens of degrees in milliseconds, potentially leading to thermal runaway. During an open-circuit fault, remaining devices carry higher current, causing a thermal feedback loop where increased temperature raises losses further [25].

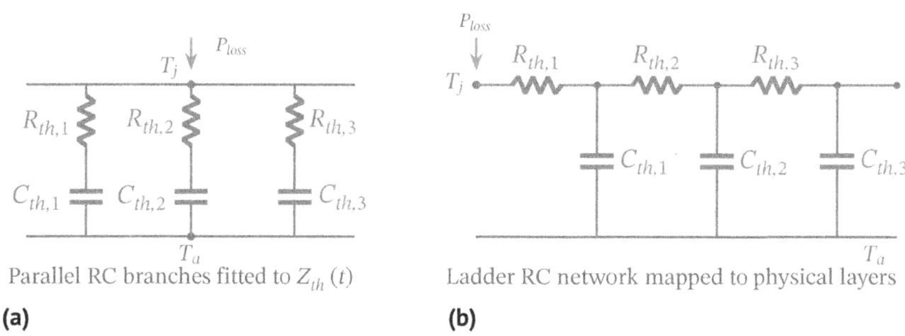

Figure 9.10 Comparison of thermal RC network structures: (a) Foster model using parallel RC branches fitted to transient impedance, (b) Cauer model using a ladder RC network mapped to physical layers.

Table 9.5 Comparison of Foster and Cauer thermal network models.

Characteristics	Foster network	Cauer network
Structure	Multiple parallel R-C branches	R-C ladder (series-parallel)
Mathematical form	$Z_{th}(t) = \sum_{i=1}^{n} R_{th,i}(1 - e^{-t/\tau_i})$	Heat diffusion through nodes: $C_{th,i}\dfrac{dT_i}{dt} = \dfrac{T_{i-1} - T_i}{R_{th,i-1}} - \dfrac{T_i - T_{i+1}}{R_{th,i}}$
Physical interpretation	No direct mapping to physical layers	Each node/element corresponds to a physical layer
Advantages	Quick calculations, simple fitting to Z_{th} curves	Physically meaningful, extensible to heatsink/ambient
Limitations	Cannot easily add external thermal components	More complex to parameterize
Typical applications	Rapid transient thermal analysis, datasheet-based modeling	System-level thermal design, coupled device-cooling simulation

Repeated thermal stress drives component degradation through fatigue mechanisms quantified by the Coffin-Manson relationship:

$$N_f = A \cdot (\Delta T_j)^{-n} \cdot e^{E_a/k_B T_m} \tag{9.22}$$

where N_f is cycles to failure, ΔT_j is the junction temperature swing, n is the Coffin–Manson exponent (typically 5–6 for bond wires and 2–3 for solder), and T_m is the mean junction temperature. Solder fatigue manifests as an increase in junction-to-case thermal resistance $R_{th,jc}$, with a 20% increase commonly used as a failure indicator [34]. By combining cycle-by-cycle thermal simulations with these fatigue models, the lifetime of power electronic components can be estimated. Fault conditions accelerate this process dramatically, as a single short-circuit event can impose extreme thermal shock that shortens remaining device life even if the device survives the event.

9.3.5 FEM Analysis of Converter-motor Interaction Under Faults

Finite element method simulations provide detailed analysis of the interplay between the power converter and the PM machine under fault conditions by coupling electrical, magnetic, and thermal physics. Under fault conditions, the interaction between inverter circuit behavior and motor electromagnetic response becomes critical, requiring co-simulation approaches that integrate circuit solvers for inverter switching with electromagnetic FE solvers for motor behavior [25]. Such coupled analysis captures phenomena that purely analytical models cannot adequately represent, including spatial field harmonics, localized saturation, and nonuniform heating. For instance, co-simulation can determine the maximum permissible fault duration before motor insulation is at risk, directly informing protection system design [25].

Under a faulty inverter supply, the air-gap magnetic field becomes distorted with new harmonics. The air-gap flux density can be decomposed into spatial and temporal harmonics:

$$B_g(r, \theta, t) = \sum_{m=1}^{\infty} \sum_{n=1}^{\infty} B_{mn} \cos(np\theta - m\omega_e t + \phi_{mn}) \tag{9.23}$$

where m is the temporal harmonic order, n is the spatial harmonic order, p is the pole pairs, and ω_e is the electrical frequency. Under fault conditions, additional backward-rotating components appear that induce eddy currents in the rotor at twice the synchronous frequency, contributing to elevated rotor losses. FE analysis of PMSM air-gap Maxwell stress under open-phase faults has confirmed the presence of additional harmonic components absent during healthy operation [35]. The Maxwell stress tensor from the FE solution:

$$\sigma_r = \frac{1}{2\mu_0}(B_r^2 - B_t^2), \quad \sigma_t = \frac{1}{\mu_0} B_r B_t \tag{9.24}$$

reveals unbalanced radial forces that introduce time-varying force patterns. Open-phase faults produce a prominent twice-line-frequency alternating radial force that can excite structural vibration modes and cause audible noise [35].

FEM-based loss computation under fault conditions shows that both stator and rotor losses increase substantially. The negative-sequence asynchronous field is a major contributor, as it induces currents in rotor regions that carry no current during healthy operation [28]. By feeding these computed losses into thermal models, hotspot localization identifies critical regions where insulation thermal limits may be exceeded, guiding how long a faulted drive can operate in degraded mode. FEM also enables verification of fault-tolerant control strategies, confirming that reconfigured inverter operation, such as four-switch mode, can restore near-constant torque output [27]. This coupled analysis confirms that converter faults propagate into the motor, causing higher losses, uneven heating, pulsating torques, and vibrational forces, making FEM an essential tool for designing robust PM machine drive systems.

9.4 Fault Diagnosis

Fault diagnosis for power converters in permanent magnet machine drives has become a critical research area due to the need for high reliability in applications ranging from electric vehicles and wind turbines to industrial drives and aerospace systems. Failures in the converter can lead to unexpected drive shutdowns, costly downtime, and safety hazards. Statistical studies in wind energy show that converter faults are a leading cause of wind turbine downtime [36], while in automotive electric vehicles, timely fault detection in the motor drive inverter is crucial for safety, as undetected faults can lead to malfunctions or accidents [37]. This has driven an upsurge in research on advanced diagnostic methods to detect and identify faults, as well as prognostic methods to predict failures and estimate remaining useful life.

Converter fault diagnosis faces several challenges. Converter faults can be fast-acting and may not produce easily distinguishable symptoms until failure occurs, requiring high-speed detection. The switching action of inverters introduces electrical noise and harmonic distortion in measured signals, complicating fault signature extraction [38]. Operating conditions in many applications are variable, such as wind turbine speed changes and electric vehicle driving cycles, which can mask or alter fault indicators. Any diagnostic system must be robust to such nonstationary conditions and noise. Moreover, developing accurate diagnostic models is difficult due to the wide variety of fault types and operating conditions. Different diagnostic approaches have been pursued to tackle these challenges, ranging from model-based methods to advanced artificial intelligence strategies.

9.4.1 Voltage and Current Signature Analysis for Switch Faults

Voltage and current signature analysis represents one of the most widely adopted approaches for detecting switch faults in power converters. Traditional motor current signature analysis (MCSA) has been extensively used to detect abnormalities such as missing current pulses or spectrum changes under fault conditions [39]. When a switch fails in open-circuit mode, the disrupted current flow in one phase creates characteristic patterns in the phase currents that can be detected through careful analysis. Short-circuit faults produce different signatures, often characterized by sudden current spikes and unbalanced phase voltages.

Advanced signal processing methods have significantly enhanced the capability to extract fault signatures from voltage and current measurements. Time-frequency transforms such as wavelet analysis can detect transient current spikes associated with switch failures, while ensemble empirical mode decomposition (EEMD) combined with permutation entropy has shown effectiveness in handling nonlinear and nonstationary signals typical of converter operation [38]. One notable EEMD and entropy-based scheme achieved approximately 98% accuracy for 22 different converter fault scenarios in a wind turbine application, demonstrating strong noise and speed robustness. These approaches extract distinctive features including specific harmonic amplitudes, statistical moments, and entropy measures that serve as reliable indicators of switch failures.

Clarke and Park transformations of the measured currents provide additional diagnostic information by revealing imbalances and asymmetries caused by switch faults. In healthy operation, the transformed quantities exhibit smooth, predictable patterns, while fault conditions introduce characteristic distortions and offsets. The DC-link current also carries valuable diagnostic information, as switch faults alter the ripple characteristics and average current draw. Analysis of DC-link current ripple frequency and amplitude can help identify specific fault types and distinguish between different failure modes.

Detection threshold determination remains a critical challenge in signature-based diagnostics. Thresholds must be set carefully to ensure reliable fault detection while minimizing false alarms. Adaptive thresholding approaches that account for varying operating conditions have shown promise, adjusting detection criteria based on load level, speed, and temperature. False alarm mitigation strategies often involve confirmation logic that requires fault signatures to persist for a minimum duration or to appear consistently across multiple diagnostic features before triggering an alarm.

9.4.2 DC-link Voltage Monitoring Techniques

DC-link voltage monitoring provides valuable diagnostic information for both switch faults and capacitor degradation. The DC-link voltage exhibits characteristic ripple patterns that change under fault conditions, making voltage ripple frequency analysis an effective diagnostic tool. In healthy converter operation, the ripple frequency is primarily determined by the switching frequency and load conditions, with predictable amplitude variations. Switch faults alter these ripple characteristics in distinctive ways, with open-circuit faults typically increasing ripple amplitude at specific frequencies and short-circuit faults causing transient voltage deviations.

Average DC-link voltage deviation detection serves as a straightforward but effective diagnostic indicator. Model-based approaches use state observers or Kalman filters to estimate the expected DC-link voltage based on operating conditions and compare it to measured values [36]. Significant deviations indicate potential faults, either in the switches or in the DC-link components themselves. Joint state-parameter estimation techniques can isolate faults in real time by simultaneously tracking both the system state and key parameters that change under fault conditions.

Capacitor health monitoring through DC-link voltage analysis has gained considerable attention, as DC-link capacitors are among the most failure-prone components in power converters. Capacitor degradation manifests as increased equivalent series resistance (ESR), which can be estimated from ripple measurements. As ESR increases due to aging or thermal stress, the voltage ripple amplitude increases for a given ripple current. Online ESR estimation algorithms track this parameter continuously, providing early warning of capacitor degradation before catastrophic failure occurs [40].

Peak-to-peak voltage monitoring and RMS voltage tracking complement average voltage deviation detection by capturing transient events and cyclic variations. These metrics are particularly sensitive to intermittent faults and degradation processes that manifest gradually over time. The correlation between voltage ripple and load conditions must be carefully considered, as normal load variations produce voltage changes that must be distinguished from fault-induced deviations. Advanced diagnostic systems incorporate load-normalized voltage monitoring that accounts for expected voltage variations under changing operating conditions.

9.4.3 Harmonic Distortion Analysis

Harmonic distortion analysis leverages the fact that power converter faults introduce characteristic changes in the harmonic content of voltage and current waveforms. Total harmonic distortion (THD) calculation provides a single metric that quantifies the overall distortion level, with threshold setting based on baseline measurements under healthy operation. However, THD alone may not uniquely identify specific fault types, as different faults can produce similar overall distortion levels while affecting different frequency components.

Individual harmonic magnitude tracking offers greater diagnostic specificity by monitoring the amplitude of specific harmonic components known to be sensitive to particular fault types. Switch open-circuit faults typically produce characteristic patterns in odd harmonics, while short-circuit faults may affect even harmonics differently. Systematic analysis of the harmonic spectrum can create fault signatures that enable discrimination between different failure modes. Frequency spectrum analysis using fast Fourier transform (FFT) or more advanced methods provides the foundation for extracting these harmonic features.

Harmonic ratio-based fault indicators normalize harmonic magnitudes relative to the fundamental component or to other harmonics, providing metrics that are less sensitive to load variations. These ratios change in characteristic ways under fault conditions, offering robust diagnostic features that maintain effectiveness across varying operating points. The sensitivity of harmonic indicators to operating conditions must be carefully characterized, as speed and load changes naturally affect harmonic content even in healthy operation.

Baseline harmonic levels under healthy operation serve as reference points for fault detection, requiring comprehensive characterization across the full operating envelope of the drive system. Statistical analysis of healthy operation data establishes the normal variation ranges for each harmonic component, enabling fault detection when measurements deviate significantly from these established baselines. Modern data-driven approaches can learn these baseline patterns automatically from operational data, reducing the need for manual characterization while improving robustness to system-specific variations [41].

9.4.4 Thermal Monitoring and Temperature-based Diagnosis

Thermal monitoring represents a critical diagnostic modality, as thermal stress is a primary cause of power semiconductor failure. Junction temperature estimation methods have evolved to include both direct measurement approaches using embedded sensors and indirect estimation techniques based on thermal models. Direct temperature measurement requires specialized power modules with integrated temperature sensors, which may not be available in all systems. Indirect methods estimate junction temperature from measurable quantities such as on-state voltage drop, which exhibits predictable temperature dependence, or from thermal models that compute temperature based on power dissipation and thermal resistance.

Case and heatsink temperature sensor placement strategies significantly influence diagnostic effectiveness. While these external measurements are easier to implement than junction temperature sensing, they suffer from thermal lag and may not accurately reflect rapid transient heating events at the semiconductor junction. Optimal sensor placement considers thermal resistance paths and time constants, positioning sensors to maximize correlation with junction temperature while maintaining practical feasibility. Multiple sensor locations can provide redundant measurements and improve estimation accuracy through sensor fusion techniques.

Thermal model-based observer design has emerged as a powerful approach for junction temperature estimation in systems without direct temperature sensing. These observers use measured electrical quantities and thermal models to compute real-time temperature estimates [40]. Digital twin implementations take this concept further by maintaining a virtual replica of the converter that simulates thermal behavior in parallel with the real system, using measured data to update and refine temperature predictions continuously. Such approaches have demonstrated the capability to estimate IGBT junction temperature and cumulative thermal stress in real time, providing online remaining life estimation for wind turbine converters [40].

Thermal signature analysis focuses on detecting sudden temperature rises that indicate fault conditions such as short-circuit faults or excessive switching losses. Temperature rate-of-change fault indicators prove particularly effective for detecting fast-developing faults, as normal operational temperature variations typically occur gradually while fault-induced heating can be rapid. Threshold-based detection on temperature derivatives provides early warning before temperatures reach dangerous levels. Thermal imaging using infrared cameras enables noncontact hotspot detection, particularly valuable for identifying localized failures in multidevice converter assemblies, though the requirement for optical access limits applicability in some installations.

9.4.5 Machine Learning Approaches for Converter Fault Detection

Machine learning has gained significant traction in converter fault diagnosis with the Industry 4.0 trend of large-scale data collection and analysis [41]. Data-driven methods can automatically learn fault patterns from operational data, reducing reliance on explicit analytical models and potentially discovering subtle fault indicators that might be overlooked in traditional approaches. These methods span a spectrum from classical machine learning algorithms to cutting-edge deep learning architectures.

Support vector machines (SVMs) have been successfully applied to classify inverter switch faults from current waveform features, achieving high accuracy under varying load conditions [42]. SVMs excel at finding optimal decision boundaries between fault classes even when those boundaries are complex and nonlinear. Random Forest and other ensemble classifiers leverage multiple decision trees to improve classification robustness and handle the inherent uncertainty in fault signatures. These methods prove particularly effective when combined with carefully engineered features extracted from voltage and current measurements.

Deep learning represents the cutting edge of machine learning-based diagnostics, offering the ability to automatically extract features directly from raw signals without requiring manual feature engineering. Convolutional neural networks (CNNs) can identify spatial patterns in current spectra and time-domain waveforms, while recurrent neural networks including long short-term memory (LSTM) and gated recurrent unit (GRU) architectures capture temporal dependencies crucial for tracking fault progression over time. Recent advanced architectures combine these approaches, with CNN-LSTM models achieving over 99.5% classification accuracy in detecting various AC/DC microgrid faults while maintaining over 92% accuracy even under heavy noise conditions [43].

Attention mechanisms represent a recent innovation that allows deep learning models to focus on the most informative portions of input signals, improving accuracy in noisy environments. These mechanisms have proven particularly effective in converter fault diagnosis where critical fault signatures may appear briefly or be embedded within complex background patterns [43]. Transfer learning and few-shot learning address the perennial challenge of limited fault data by enabling models trained on one system or fault type to adapt to others with minimal new training data. This approach has shown promise in wind farm applications, where models trained on one turbine were successfully adapted to others [44].

Despite their impressive performance, deep learning methods face challenges including their "black box" nature that can limit trust in safety-critical applications, substantial computational requirements that may exceed available embedded processing capabilities, and the need for large, diverse training datasets [45]. Current research addresses these limitations through explainable AI techniques that provide insight into model decision-making, lightweight model architectures optimized for real-time embedded execution, and data augmentation strategies including synthetic fault data generation to overcome training data scarcity.

9.4.6 Real-time Fault-tolerant Control Strategies

Fault-tolerant control strategies enable continued operation of PM machine drives after converter faults occur, maintaining functionality in degraded mode until repairs can be performed or safely shutting down when continued operation poses risks. These strategies are particularly critical in automotive applications where safety depends on maintaining some level of drive capability, and in remote installations like wind turbines where immediate repair may be impractical.

Phase loss compensation techniques reconfigure the control system to operate with reduced phase count when a complete phase leg fails. Four-switch three-phase operation represents a well-studied reduced topology that maintains three-phase AC output using only four functioning switches instead of the normal six, achieved through reconfigured modulation strategies and neutral point connections [37]. While this degraded mode reduces available torque and increases current harmonics, it enables continued operation that can be essential for safety-critical applications.

Control reconfiguration under single-switch failure involves adapting the modulation strategy to work around the failed device while maintaining as much normal functionality as possible. Model predictive control (MPC) has proven particularly amenable to fault accommodation, as fault detection can be integrated directly into the control algorithm's cost function. Studies have demonstrated successful detection and accommodation of all 21 possible open-switch faults in a PMSM inverter by monitoring the MPC cost and current residuals [37]. The control system responds to detected faults by modifying constraints and objectives to reflect the reduced capabilities of the faulted converter.

Hardware redundancy approaches including parallel switch configurations and backup phase legs provide the highest level of fault tolerance at the cost of increased component count and system complexity. Parallel switches can take over when one device fails, while redundant phase legs enable immediate switchover to healthy circuits. Software redundancy techniques complement these hardware approaches through analytical redundancy that uses mathematical models to reconstruct unmeasured quantities or verify sensor readings, enabling continued operation even when sensor faults accompany switch failures.

Graceful degradation strategies define operating modes with progressively reduced performance as fault severity increases, ensuring that the system safely transitions through capability levels rather than experiencing abrupt failure. Post-fault de-rating reduces current and torque demands to limit stress on remaining healthy components, while thermal management becomes particularly critical in fault-tolerant operation as current redistribution often increases losses and local heating. Fault isolation through protective measures prevents fault propagation, such as quickly disconnecting a short-circuited switch before the fault damages other components or the motor.

Table 9.6 summarizes fault-tolerant control strategies for power converter faults.

9.4.7 Condition Monitoring of Power Electronic Components

Condition monitoring extends beyond fault detection to track the gradual degradation of power electronic components, enabling predictive maintenance and preventing unexpected failures. This prognostic capability has become increasingly important as industries move toward condition-based rather than schedule-based maintenance, aiming to replace components just before failure rather than prematurely or after catastrophic breakdown.

Table 9.6 Fault-tolerant control strategies for power converter faults.

Strategy	Applicable fault types	Performance impact	Implementation complexity	Hardware requirements	Safety rating	Typical applications
Four-switch three-phase operation	Complete phase leg failure	Reduced torque, increased harmonics	Medium	Minimal (uses existing switches)	Good	Automotive, critical drives
Single-switch failure reconfiguration	Individual switch open-circuit	Moderate torque reduction	Medium–High	None (software only)	Medium	General industrial drives
Model predictive control (MPC) with fault accommodation	All 21 possible open-switch faults	Minimized through optimization	High	None (computation)	High	Advanced control systems
Parallel switch redundancy	Any switch failure	Minimal (seamless switchover)	Low (hardware)	Doubled switch count	Excellent	High-reliability systems
Backup phase legs	Phase leg failures	Minimal (immediate switchover)	Low (hardware)	Extra phase legs	Excellent	Aerospace, mission-critical
Analytical redundancy	Sensor + switch faults	Variable	High	None (computation)	Medium–High	Sensor-limited systems
Post-fault de-rating	All fault types	Reduced capability	Low	None	Good	General protection strategy

Online ESR monitoring for DC-link capacitors tracks one of the most reliable indicators of capacitor aging. As capacitors age due to electrolyte evaporation and dielectric degradation, ESR increases progressively. Monitoring this parameter enables early detection of capacitor health decline well before functional failure occurs. Junction temperature estimation for IGBTs and MOSFETs, as discussed in the thermal monitoring section, serves both diagnostic and prognostic purposes. Cumulative thermal stress calculated from temperature cycling drives physics-of-failure models that predict device lifetime based on well-characterized failure mechanisms including bond-wire fatigue and solder joint cracking [40].

Gate charge degradation tracking monitors changes in the gate characteristics of power semiconductors, as gate oxide degradation can occur over time and affect switching performance. On-state voltage drop monitoring provides a sensitive indicator of device condition, with $V_{CE,sat}$ for IGBTs and $R_{DS,on}$ for MOSFETs both increasing as devices degrade. These parameters can be measured during normal operation without interrupting drive function, enabling continuous health assessment. Switching loss increase detection identifies degradation that affects dynamic performance, as increased switching losses indicate changes in device characteristics that may precede complete failure.

Prognostics and remaining useful life (RUL) estimation represent the ultimate goal of condition monitoring, quantifying how long components can continue operating before failure. Physics-based prognostic models combine measured stress indicators with failure mechanism models to compute accumulated damage and project future health trajectories. Digital twin approaches integrate real-time operational data with these physics models to provide continuous RUL updates [40]. Data-driven prognostic methods use machine learning to learn degradation patterns from historical failure data, while hybrid approaches combine physics-based and data-driven elements to leverage the strengths of both paradigms [45].

Degradation indicators and trend analysis track multiple component health metrics over time to identify progressive deterioration before it reaches critical levels. Statistical process control methods applied to precursor parameters can detect subtle trends that indicate developing problems. The challenge lies in distinguishing true degradation trends from normal operational variations and environmental effects. Advanced analytics including Bayesian methods and particle filters enable probabilistic prognostics that quantify uncertainty in RUL estimates, providing decision-makers with confidence levels rather than single-point predictions [46].

9.5 Summary

This chapter examined power converter faults in PM machine drives, covering failure mechanisms, modeling approaches, and diagnostic methodologies. Statistical data shows that power converters account for 38% of drive system failures, with semiconductor devices, capacitors, and PCBs responsible for 38%, 30%, and 26% of converter-level failures, respectively.

The failure modes section distinguished between random failures (sudden catastrophic events from voltage transients, overstress, or cosmic radiation) and wear-out failures (progressive degradation from thermal cycling, CTE mismatch, and mechanical fatigue). Package-related mechanisms, particularly bond wire lift-off and solder fatigue, represent the most prevalent reliability challenges. DC-link capacitor degradation through electrolyte evaporation follows well-characterized Arrhenius relationships, while gate driver circuit faults can trigger catastrophic shoot-through events. Sensor failures in current, voltage, temperature, and position measurement channels can lead to incorrect control decisions that damage both converter and motor.

The modeling section developed mathematical representations of faulty converter operation, showing that open-circuit and short-circuit switch faults produce negative-sequence currents that introduce torque pulsations at twice the electrical frequency, with reported torque ripple increases of 20–25% and THD increases of approximately 30%. Thermal modeling using Foster and Cauer RC networks enables junction temperature prediction under fault conditions, while Coffin–Manson fatigue models quantify the accelerated component degradation from thermal cycling. FEM co-simulation captures the propagation of converter faults into the motor through distorted air-gap fields, elevated losses, and unbalanced radial forces.

The diagnosis section presented detection methods ranging from voltage and current signature analysis to machine learning approaches, with advanced techniques such as EEMD with permutation entropy achieving 98% accuracy across diverse fault scenarios. DC-link voltage monitoring, harmonic distortion analysis, and thermal monitoring provide complementary diagnostic information. Fault-tolerant control strategies, including four-switch operation and MPC, enable continued operation under degraded conditions. Condition monitoring and prognostics extend these capabilities toward remaining useful life estimation through physics-based models, digital twins, and data-driven approaches, supporting the transition from schedule-based to condition-based maintenance in PM machine drive systems.

References

1 Vanessa Smet, Francois Forest, Jean-Jacques Huselstein, Frédéric Richardeau, Zoubir Khatir, Stéphane Lefebvre, and Mounira Berkani. Ageing and failure modes of IGBT modules in high-temperature power cycling. *IEEE Transactions on Industrial Electronics*, 58(10):4931–4941, 2011. ISSN 0278-0046. https://doi.org/10.1109/tie.2011.2114313.

2 Lydia Feller, Samual Hartmann, and Daniel Schneider. Lifetime analysis of solder joints in high power IGBT modules for increasing the reliability for operation at 150 °C. *Microelectronics Reliability*, 48(8–9):1161–1166, 2008. ISSN 0026-2714. https://doi.org/10.1016/j.microrel.2008.07.019.

3 Muna T. Rahimo and Noel Y. A. Shammas. Freewheeling diode reverse-recovery failure modes in IGBT applications. *IEEE Transactions on Industry Applications*, 37(2):661–670, 2001. ISSN 0093-9994. https://doi.org/10.1109/28.913734.

4 Rohm Semiconductor. About dV/dt destruction - mechanisms of MOSFET destruction. https://techweb.rohm.com/product/transistors-diodes/transistors/10772/.

5 Cadence PCB Solutions. Common diode failure modes in circuits. https://resources.pcb.cadence.com/blog/common-diode-failure-modes-in-circuits.

6 NASA Technical Reports Server. MOSFET aging and reliability. Technical Report 20100037964, NASA, 2010. https://ntrs.nasa.gov/api/citations/20100037964/downloads/20100037964.pdf.

7 Ilyas Dchar, Cyril Buttay, and Hervé Morel. Avalanche robustness of SiC Schottky diode. *Microelectronics Reliability*, 64:494–501, 2016. https://doi.org/10.1016/j.microrel.2016.07.086.

8 Wolfspeed Inc. Designed to last even in the harshest environments. https://www.wolfspeed.com/knowledge-center/article/designed-to-last-even-in-the-harshest-environments/.

9 Infineon Community Forum. Bootstrap capacitor failure (dead short). https://community.infineon.com/t5/MOSFET-Si-SiC/Boot-strap-capacitor-failure-dead-short/td-p/658803.

10 Reddit r/AskElectronics. Why is the voltage across the bootstrap capacitor dropping so much? https://www.reddit.com/r/AskElectronics/comments/1gmiozg/.

11 Dynex Semiconductor. IGBT module failure mechanisms. Technical Report AN6442, Dynex Semiconductor. https://www.dynexsemi.com/Portals/0/PDF/DNX_AN6442.pdf.
12 HVM Technology. 8 signs your high-voltage opto-coupler needs replacing. https://www.hvmtech.com/post/8-signs-your-high-voltage-opto-coupler-needs-replacing.
13 Texas Instruments. Understanding failure modes in isolators. Technical Report SLYY081, Texas Instruments, Rev. B. https://www.ti.com/lit/pdf/slyy081.
14 Akane Electronics Industry. Resistor failure modes. https://www.akaneohm.com/english/column/resistor-failure-modes/.
15 Toshiba Electronic Devices & Storage Corporation. How does the series gate resistor affect the MOSFET? https://toshiba.semicon-storage.com/us/semiconductor/knowledge/faq/mosfet/.
16 Jianghai America. Electrolytic capacitor lifetime estimation. Technical report, Jianghai Capacitor Co. https://jianghai-america.com/uploads/technology/JIANGHAI_Elcap_Lifetime_-_Estimation_AAL.pdf.
17 Cadence PCB Solutions. The causes of electrolytic capacitor degradation. 2022. https://resources.pcb.cadence.com/blog/2022-the-causes-of-electrolytic-capacitor-degradation.
18 XP Power. Electrolytic capacitors determine the lifetime of a power supply. https://www.xppower.com/resources/blog/electrolytic-capacitor-lifetime-in-power-supplies.
19 Wikipedia. Capacitor plague. https://en.wikipedia.org/wiki/Capacitor_plague.
20 Analog Devices. Gain and offset temperature drift compensation for the AD7403/AD7405. Technical Report AN-1377, Analog Devices. https://www.analog.com/en/resources/app-notes/an-1377.html.
21 Socomec. Current sensors and transformers. Technical report, Socomec, April 2025. https://www.socomec.us/sites/default/files/2025-04/Current-sensors-and-transformers_TECHNICAL-NOTE_2025-04-16-16-47-37_English_PLURI_0.pdf
22 CX Ability. What are the common faults of current sensors? https://cxability.com/index.php?m=home&c=View&a=index&aid=125.
23 Chipsense. How to diagnose a fault in a hall effect current sensor and what are the common problems. https://chipsense.net/en/news/how-to-diagnose-a-fault-in-a-hall-effect-current-sensor-and-what-are-the-common-problems.
24 National Instruments. Three-phase inverter - VeriStand electric motor simulation toolkit. https://www.ni.com/docs/en-US/bundle/veristand-electric-motor-simulation-toolkit/page/three-phase-inverter.html.
25 Yue Shi, Jiabin Wang, and Bhaskar Mani. Electromagnetic-thermal coupled simulation under various fault conditions of a triple redundant 9-phase PMASynRM. *IEEE Transactions on Industry Applications*, 56(1):128–137, 2020. https://doi.org/10.1109/TIA.2019.2945295.
26 Shuangshuang Lin, Zhigang Liu, and Keting Hu. Traction inverter open switch fault diagnosis based on Choi–Williams distribution spectral kurtosis and wavelet-packet energy Shannon entropy. *Entropy*, 19(9):504, 2017. ISSN 1099-4300. https://doi.org/10.3390/e19090504.
27 Byoung-Kuk Hyon, Dongsu Kim, Kyeonghwan Lee, In Jae Jang, and Ki Jin Han. A low-cost fault-ride-through strategy for electric vehicle inverters using four-switch topology. *Electronics*, 13(11):2113, 2024. https://doi.org/10.3390/electronics13112113.
28 Huili Jia, Jiaqiang Yang, Rongfeng Deng, and Yan Wang. Loss investigation for multiphase induction machine under open-circuit fault using field–circuit coupling finite element method. *Energies*, 14(18):5686, 2021. ISSN 1996-1073. https://doi.org/10.3390/en14185686.
29 Lei Chen, Min Chen, Bodong Li, Xinnan Sun, and Feng Jiang. Optimized fault-tolerant control of dual three-phase PMSM under open-switch faults. *Energies*, 17(20):5198, 2024. ISSN 1996-1073. https://doi.org/10.3390/en17205198.

30 Bilal Djamal Eddine Cherif and Azeddine Bendiabdellah. Detection of two-level inverter open-circuit fault using a combined DWT-NN approach. *Journal of Control Science and Engineering*, 2018:1–11, 2018. ISSN 1687-5249. https://doi.org/10.1155/2018/1976836.

31 Muhammad Mahmood Ul Hassan, Syed Abdul Rahman Kashif, Muhammad Shahzad, Khurram Saleem Alimgeer, and Zeeshan Anjum. An analysis of voltage source inverter switches fault classification using machine learning techniques. arXiv preprint arXiv:2111.06566, 2021. https://arxiv.org/pdf/2111.06566.

32 Bilal Djamal Eddine Cherif, Azeddine Bendiabdellah, Mohamed-Said Bendiabdellah, and Abdelhak Kaddouri. Diagnosis of IGBT open-circuit fault by Hilbert-Huang transform. *Traitement du Signal*, 36(3):263–269, 2019. https://doi.org/10.18280/ts.360310.

33 Nexperia. Nexperia's Foster and Cauer thermal models accelerate thermal design. https://efficiencywins.nexperia.com/efficient-products/nexperia-s-foster-and-cauer-thermal-models-accelerate-thermal-design.

34 Dan Luo, Minyou Chen, Wei Lai, Hongjian Xia, Xueni Ding, and Zhenyu Deng. A study on the effect of bond wires lift-off on IGBT thermal resistance measurement. *Electronics*, 10(2):194, 2021. ISSN 2079-9292. https://doi.org/10.3390/electronics10020194.

35 Zaixin Song, Yu Wang, David Gerada, and Chris Gerada. Analysis of vibration in modular fault-tolerant PMSM under one-phase open-circuit fault. In *2018 XIII International Conference on Electrical Machines (ICEM)*, pages 548–554. IEEE, 2018. https://doi.org/10.1109/ICELMACH.2018.8506807.

36 Jinping Liang, Ke Zhang, Ahmed Al-Durra, S. M. Muyeen, and Daming Zhou. A state-of-the-art review on wind power converter fault diagnosis. *Energy Reports*, 8:5341–5369, 2022. ISSN 2352-4847. https://doi.org/10.1016/j.egyr.2022.03.178.

37 Mohammad Zamani Khaneghah, Mohamad Alzayed, and Hicham Chaoui. Fault detection and diagnosis of the electric motor drive and battery system of electric vehicles. *Machines*, 11(7):713, 2023. ISSN 2075-1702. https://doi.org/10.3390/machines11070713.

38 Jinping Liang and Ke Zhang. A robust fault diagnosis scheme for converter in wind turbine systems. *Electronics*, 12(7):1597, 2023. ISSN 2079-9292. https://doi.org/10.3390/electronics12071597.

39 Yong Chen, Siyuan Liang, Wanfu Li, Hong Liang, and Chengdong Wang. Faults and diagnosis methods of permanent magnet synchronous motors: A review. *Applied Sciences*, 9(10):2116, 2019. ISSN 2076-3417. https://doi.org/10.3390/app9102116.

40 Ole Christian Spro. Digital twins: Condition-monitoring of power electronic converters. SINTEF Energy Blog, September 2020. https://blog.sintef.com/energy/digital-twins-condition-monitoring-of-power-electronic-converters/.

41 David Gonzalez-Jimenez, Jon del Olmo, Javier Poza, Fernando Garramiola, and Patxi Madina. Data-driven fault diagnosis for electric drives: A review. *Sensors*, 21(12):4024, 2021. ISSN 1424-8220. https://doi.org/10.3390/s21124024.

42 Soumya Rani Mestha and Nagesh Prabhu. Support vector machine based fault detection in inverter-fed electric vehicle. *Energy Storage*, 6(1):e576, 2024. ISSN 2578-4862. https://doi.org/10.1002/est2.576.

43 Qiangsheng Bu, Pengpeng Lyu, Ruihai Sun, Jiangping Jing, Zhan Lyu, and Shixi Hou. Fault diagnosis method using CNN-Attention-LSTM for AC/DC microgrid. *Modelling*, 6(3):107, 2025. ISSN 2673-3951. https://doi.org/10.3390/modelling6030107.

44 Fan Wu, Gen Qiu, Lefei Zhang, Kai Chen, Li Wang, Jinxu Yu, and Jinqi Gao. Disturbance auto-encoder generation model: Few-shot learning method for IGBT open-circuit fault diagnosis in three-phase converters. *Frontiers in Energy Research*, 10:1077519, 2023. ISSN 2296-598X. https://doi.org/10.3389/fenrg.2022.1077519.

45 Ion-Stelian Gherghina, Nicu Bizon, Gabriel-Vasile Iana, and Bogdan-Valentin Vasilică. Recent advances in fault detection and analysis of synchronous motors: A review. *Machines*, 13(9):815, 2025. ISSN 2075-1702. https://doi.org/10.3390/machines13090815.

46 Chetan S. Kulkarni and Matteo Corbetta. Health management and prognostics for electric aircraft powertrain. In *AIAA Propulsion and Energy 2019 Forum*. American Institute of Aeronautics and Astronautics, 2019. https://doi.org/10.2514/6.2019-4474.

Index

d

W

X

Z

Wiley The manufacturer's authorized representative according to the EU General Product Safety Regulation is Wiley-VCH GmbH, Boschstr. 12, 69469 Weinheim, Germany, e-mail: Product_Safety@wiley.com.

Printed and bound by CPI Group (UK) Ltd, Croydon, CR0 4YY
19/09/2026
02219301-0004